Neutron Physics for Nuclear Reactors

Unpublished Writings by Enrico Fermi

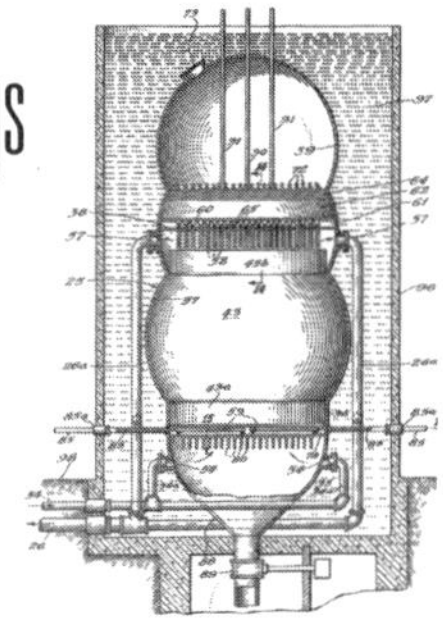

Neutron Physics for Nuclear Reactors

Unpublished Writings by Enrico Fermi

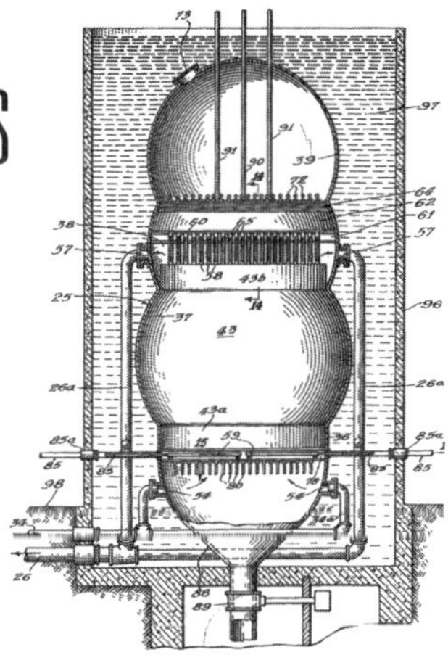

editors

S Esposito

O Pisanti

Università di Napoli "Federico II" & INFN Sezione di Napoli, Italy

World Scientific

NEW JERSEY • LONDON • SINGAPORE • BEIJING • SHANGHAI • HONG KONG • TAIPEI • CHENNAI

Published by

World Scientific Publishing Co. Pte. Ltd.

5 Toh Tuck Link, Singapore 596224

USA office: 27 Warren Street, Suite 401-402, Hackensack, NJ 07601

UK office: 57 Shelton Street, Covent Garden, London WC2H 9HE

British Library Cataloguing-in-Publication Data
A catalogue record for this book is available from the British Library.

NEUTRON PHYSICS FOR NUCLEAR REACTORS
Unpublished Writings by Enrico Fermi

ISBN-13 978-981-4291-22-4
ISBN-10 981-4291-22-6

Printed in Singapore by World Scientific Printers.

Contents

Introduction xi

Some Background to my Notes on Fermi's Neutron Physics Lectures (Los Alamos, 1945), by Anthony P. French xxxi

Neutron Physics 1

Neutron Physics – A Course of Lectures by E. Fermi 3

1. Sources of Neutrons 5
    - 1.1 Natural Sources 5
        - 1.1.1 Alpha Particle Sources 5
        - 1.1.2 Photo-Neutron Sources 7
    - 1.2 Artificial Sources 7

2. The Isotopic Chart: Nuclear Masses and Energies 9
    - 2.1 The Isotopic Chart 9
    - 2.2 Energy Balance of Reactions 10
    - 2.3 The Binding Energies of Nuclei 12
    - 2.4 The Packing Fraction Curve 12

3. The Scattering of Neutrons (Part 1) 17
    - 3.1 General Considerations 17
    - 3.2 Elementary Theoretical Treatment 19
        - 3.2.1 Elastic Scattering 20

3.2.2 Inelastic (n, n) Scattering 21
3.2.3 Inelastic (n, m) Scattering 21
3.2.4 Inelastic (n, γ) Scattering 22

4. Resonance: Models of the Nucleus 25

4.1 Resonance in Nuclear Reactions 25
4.2 Two Models of the Nucleus 27

5. The Scattering of Neutrons (Part 2) 31

5.1 The Solution of Schrödinger's Equation 31
5.2 The Scattering Cross Section 34
5.3 Neutron-Proton scattering 35
5.4 The Breit-Wigner Formula 36
5.5 The Effect of Chemical Binding on Scattering 38
5.6 Scattering Cross Section for Other Elements 39

6. Slow Neutrons as Waves 41

6.1 Introduction 41
6.2 Isotopic Effects 43
6.3 Penetration by Thermal Neutrons 44
6.4 The Production of Very Slow Neutrons 46
6.5 Reflection and Refraction 47

7. The Slowing Down of Neutrons 49

7.1 Introduction 49
7.2 The Energy Loss in One Collision 50
7.3 Many Collisions 52
7.4 The Spatial Distribution of Slowed Neutrons 54
7.5 Theory of the Spatial Distribution 58

8. The Age Equation 65

8.1 Derivation of the Age Equation 65
8.1.1 Diffusion 65
8.1.2 Energy Drift 66
8.2 The Problem of a Point Source 68

9. Thermal Neutron Distributions 71

9.1 The Basic Equation 71

9.2 Point Source of Slow Neutrons 72
9.3 Point Source of Fast Neutrons 72
9.4 Bounded Media 73
9.5 The Measurement of Diffusion Length 76
9.6 Diffusion in Graphite 77
9.7 Some Useful Quantities and Relationships 80

10. The Reflection of Neutrons 81

10.1 Introduction 81
10.2 Approximate Solution of the Escape Problem 81
10.3 Exact Solution by Integral Equation Method 82
10.4 The Albedo 85
10.5 Measurement of the Albedo 85

11. The Stability of Nuclei 87

11.1 The Binding Energy of a Nucleus 87
11.1.1 The Liquid Drop Model 87
11.1.2 Nuclear Composition 88
11.1.3 Coulomb Forces 89
11.1.4 The Odd-Even Effect 89
11.2 Determination of Coefficients 90
11.3 The Binding of Neutrons to Nuclei 91

12. Nuclear Fission 93

12.1 The Possibility of Fission 93
12.2 Limitations on the Occurrence of Fission 95
12.3 The Liquid Drop Model in Fission 97
12.4 The Particles of Fission 99
12.5 Cross Sections for Fission and Other Processes 102

13. The Possibility of a Chain Reaction 105

13.1 The Properties of Natural Uranium 105
13.1.1 The High Energy Region 106
13.1.2 The Thermal Region 107
13.2 Moderators 107
13.3 Homogeneous and Lumped Graphite Piles 108
13.4 The Possibility of a Homogeneous Pile 112

14. The Heterogeneous Pile 115

14.1 The Design of a Lumped Pile 115
14.2 The Determination of Pile Constants 116
14.2.1 The Magnitude of ϵ 116
14.2.2 $(1 - f_R)$ 116
14.2.3 Calculation of f_T 117
14.3 Reproduction Factor and Critical Size 119

15. The Time Dependence of a Pile 123

15.1 The Time-Dependent Equations 123
15.2 Evaluation of the Period 125

16. Practical Aspects of Pile Physics 131

16.1 The Determination of k 131
16.2 The Study of Pile Materials 132
16.3 Energy and Radiation Production 133
16.4 Shielding 134
16.5 Other Types of Piles 136

17. Fast Reactors 139

17.1 Elementary Considerations 139
17.2 The Integral Equation to the Neutron Distribution 141
17.3 The Critical Size for a Fast Reactor 144
17.4 Supercritical Reactors 151

Problems and Exercises 153

Nuclear Reactors 159

1. Process for the Production of Radioactive Substances
by E. Fermi, E. Amaldi, B. Pontecorvo, F. Rasetti and E. Segré 161

2. Test Exponential Pile
by E. Fermi 181

3. Method of Operating a Neutronic Reactor
by E. Fermi and L. Szilard 217

3.1 The Chain Fission Reaction 223

3.1.1 Neutron Loss by Absorption in Uranium 226
3.1.2 Neutron Loss by Absorption in the Moderator . . 228
3.1.3 Neutron Loss by Absorption by Impurities in the System 229
3.1.4 Exterior Neutron Loss in a Neutronic Reactor of Practical Size 233
3.2 Measurement of Neutron Losses 234
3.3 An Illustrative Neutronic Reactor having a Solid Moderator 244
3.4 Graphite in the Reactor 258
3.5 An Illustrative Neutronic Reactor having a Liquid Moderator 262
3.6 Neutronic Reactors with Other Moderators 269
3.7 Reduction of Losses Due to Resonance Capture 271
3.8 Reduction of Neutron Losses Due to Impurities in the Material 278
3.9 Effect of a Cooling System in a Neutronic Reactor 285
3.10 An Illustrative Gas-Cooled Neutronic Reactor 288
3.11 An Illustrative Liquid-Cooled Neutronic Reactor 298
3.12 Use of Different Lattices in the Same Neutronic Reactor . 303
3.13 Critical and Operating Sizes of Neutronic Reactors 307
3.14 Control of the Neutronic Reaction by Variation of Neutron Losses in or from the Neutronic Reactor 315
3.15 Uses of Neutronic Reactors 323

4. Chain Reacting System
by E. Fermi and M.C. Leverett 329

5. Neutronic Reactor I
by E. Fermi 379

6. Air Cooled Neutronic Reactor
by E. Fermi and L. Szilard 409

7. Testing Material in a Neutronic Reactor
by E. Fermi and H.L. Anderson 447

8. Neutron Velocity Selector
by E. Fermi 475

9. Neutronic Reactor II
by E. Fermi and L. Szilard 493

10. Neutronic Reactor III
by E. Fermi, W.H. Zinn and H.L. Anderson 513

11. Neutronic Reactor IV
by E. Fermi and W.H. Zinn 523

12. Method of Testing Thermal Neutron Fissionable Material for Purity
by E. Fermi and H.L. Anderson 535

13. Method of Sustaining a Neutronic Chain Reacting System
by E. Fermi and M.C. Leverett 549

14. Neutronic Reactor Shield
by E. Fermi and W.H. Zinn 583

Appendix A: Other Documents not Reported in this Volume 593

Appendix B: The Complete List of Fermi's Papers 621

Bibliography 653

Index 657

Introduction

Historical prelude

> It has been known for many years that vast amounts of energy are stored in the nuclei of many atomic species and that their release is not in contradiction with the principle of the conservation of the energy, nor with any other of the accepted basic laws of physics. In spite of this recognized fact, it was the general opinion among physicists until recently that a large scale release of the nuclear energy would not be possible without the discovery of some new phenomenon. [Fermi (1946b)]

The discovery of the fission of uranium and the possibility to produce a chain reaction

Such a new phenomenon, as mentioned by Fermi in one of his reviews of 1946, was that observed by Otto Hahn and Fritz Strassmann in the Fall of 1938 at the Kaiser Wilhelm Institute in Berlin, when bombarding the uranium nucleus with neutrons from a radium-beryllium source. The correct explanation of the Hahn and Strassmann experiments was soon given by Lise Meitner and Otto R. Frisch who interpreted the observed phenomenon as due to the splitting of uranium, from which two elements formed, each of approximately half of its original mass. The mass which "disappeared" was assumed to be converted into energy, according to Einstein's theory of relativity.

The news of the novel phenomenon reached the other side of the Atlantic Ocean just after Fermi and his family arrived in America, after receiving the Nobel Prize in Stockholm.

> Niels Bohr, who had come for a stay at Princeton, was on his

> way to attend a conference in Washington. [...] By the time he was ready to leave Princeton, Bohr had heard the results of Frisch's experiments. It was a most exciting development. [Anderson (1984)]

Willis Lamb was in Princeton at that time and, after heard from Bohr of this breaking news, he went to Columbia University and communicated it to Fermi [Segrè (1970)]. Quite independently, according to Anderson's recollections,

> on his way to Washington, Bohr thought it would be a good idea to drop by and see Fermi to tell him about the exciting new physics. He came to the Pupin Physics Laboratory looking for Fermi. [...] He didn't find Fermi; he found me instead. I was the only person around. He hadn't see me before but that didn't stop him. He grabbed me by the shoulder and said, "Young man, let me tell you about fission." [...] I had heard enough to catch the excitement. [...] When Bohr left I felt I had something to tell Fermi. [...] "Professor Fermi, I've come to tell you that I have just seen Professor Bohr. He was looking for you and he told me some very interesting things." Fermi interrupted me. A smile broke out and he said, "Let ME tell you about fission." Then I heard again, but this time much more graphically, how the energy would appear when the uranium was split and the pieces flew apart by Coulomb repulsion. [Anderson (1984)]

After the news spread out, many physicists (including Fermi and collaborators) confirmed the results by Hahn and Strassmann and proved true the interpretation and suggestions by Frisch, working rapidly for a better understanding of the phenomenon.

> In the spring of 1939 it was generally known that a fission that can be produced by the collision of a single neutron with a uranium atom was capable of producing more than one new neutron, probably something of the order of two or three. It was felt at that time by many physicists that a chain reaction based on the uranium fission was a possibility well worth investigating. [Fermi (1946b)]

The idea of a nuclear chain reaction able to liberate energy on a large scale came to Leo Szilard as early as in 1933-34, when it was believed that beryllium (instead of uranium) was unstable and that neutrons would split off when this element disintegrated. This proved soon incorrect, but the possibility to create a process that would emit more neutrons than were absorbed (or, in other words, with a multiplication or reproduction factor greater than one) came back into the picture when the fission of uranium was discovered. This was promptly recognized by Szilard who, according to Anderson [Anderson (1984)], "was very anxious to work with Fermi, or at least to have discussion with him" in order to achieve effectively a chain reaction.

Natural uranium and graphite

In 1939 a number of experiments were performed to put the problem of fission on a quantitative basis. The first important fact to be realized was that the cross section for neutron fission was higher for low energy neutrons, while the second one was that the key isotope of uranium involved in the fission induced by slow neutrons was the rare one of mass 235, instead of the most abundant ^{238}U. The problem was, however, complicated by the fact that, besides producing fission, slow neutrons can also give rise to the production of the radioactive isotope ^{239}U by simple capture. In particular the capture of neutrons with thermal energies (thermal neutrons) was proved to be due to a strong resonance absorption at somewhat higher energies [Anderson (1939a)]. Such a process competes with fission in taking up the neutrons which are needed to sustain a chain reaction, so that a major problem in making the chain reaction to be effective was to avoid losses due to this absorption.

In any case, the first basic point to be cleared up was the choice of the fissile material to be used and, in this respect, two alternatives were opened at the end of 1939. The first one was the separation of ^{235}U from the natural uranium, thus eliminating the absorption by the most abundant isotope ^{238}U. Obviously, for this method to work, the major difficulty for that time was to produce large quantities of the isotope needed. The alternative choice was, instead, to use directly natural uranium, with the evident drawback caused by the undesirable absorption of neutrons by the most abundant isotope, which may lower significatively the multiplication factor for the self-sustaining reaction to be achieved. The problems with both the alternative methods were serious, and Fermi chose to work out the one where more physical effects should be understood and kept under control, i.e. he decided to study the possibility of a chain reaction with natural uranium. It is quite interesting to observe that Fermi was very confident that such a way was the right one:

> "Herbert," he said, "if you stick with me we'll get the chain reaction first. The other guys will have to separate those isotopes

> first, but we'll make it work with ordinary uranium." [Anderson (1984)]

Such an attitude, as usual for him, came from the appropriate quantitative results he and his collaborators obtained from an extensive experimental work. Here, as already mentioned, the discriminating factor was the slowing down of the incident neutrons, which makes more effective the cross section for fission with respect to that for absorption.

The problem of the slowing down of neutrons and its effect on the development of neutron-induced nuclear reactions (and, in particular, the production of radioactive elements) had been the subject of intense and fruitful researches by Fermi and his group in Rome as early as 1934 [De Gregorio (2006)], and lead to several important papers, collected in [Fermi (1962)]. A patent for the practical applications of the results obtained was as well issued; the interesting subsequent anecdotes related to this patent have been narrated in [Segrè (1970)]. It was recognized that the most efficient way to slow down neutrons was to pass them through hydrogen, the lightest chemical elements present in water, paraffin, etc., so that the obvious conclusion for getting a reproduction factor high enough for a chain reaction was to disseminate uranium powder in water. However, measurements revealed [Anderson (1939b)] that thermal neutron absorption by hydrogen was too large for water to make it a usable medium for slowing down neutrons in a chain reaction, since that absorption (leading to formation of deuterium) would lower substantially the multiplication factor. Thus, other light elements should be taken into consideration.

> Out of Szilard's thinking came the idea of using graphite instead of water to slow down the neutron. [...] Fermi had also been thinking about graphite. [Anderson (1984)]

Measurements showed [Anderson (1940)] that the absorption of neutrons on graphite was small enough to make it the obvious choice for a material for slowing down the neutrons, so that Fermi set forth also the basic theoretical techniques for describing the behaviour of neutrons in such substances. It was also shown that, after the neutrons reached thermal energies, a second diffusion process began in which the neutrons continued to diffuse through the material until they either escaped or were absorbed. The advantages of graphite against water, as a moderator for neutrons, came out from experiments with a pile of graphite aimed at measuring the absorption of carbon [Anderson (1941b)]. In such a pile the neutrons were slowed down more slowly than in water, but once they reached thermal energies the neutrons would diffuse longer and reach greater distances from the source. As a consequence, a physical separation of the thermal neutrons from higher energy ones could be obtained, and this property was later used by Fermi in many different ways.

Experimental piles

At this point of the story, the next step was to design a chain reacting pile that would work, and, to this end, a number of experimental piles were built, early at Columbia University in New York and then at Chicago, to study directly the properties of uranium and graphite (or other moderators) in a pile.

The key ingredient was, of course, to work with sufficiently pure materials; these were obtained from different factories (with quite different degree of purity), and always were tested by Fermi and his collaborators. A chemical method, involving ether separation, was used to purify uranium [Anderson (1941a)] while the absorption of neutrons by graphite was especially measured.

Graphite bricks were stacked into the so-called "sigma pile" (denoted with the Greek letter "sigma"), designed to measure the absorption cross section. A neutron source was placed near the bottom of the pile and indium foils were exposed at various points on the vertical axis above the source; from the radioactivity induced in these foils the absorption cross section of graphite was deduced. To this regard, standard procedures were introduced [Anderson (1941c)] by which indium (and rhodium) foils could be calibrated in order that the measurement of their radioactivity could be used to give either the slow neutron density or the slowing down density in absolute units. The graphite column erected at Columbia was also used as a source of thermal neutrons in the measurement of the absorption cross section of boron. This element, in fact, had importance in absolute neutron measurement, because of its high neutron absorption cross section and its dependence on the inverse of the velocity of neutrons [Anderson (1942a)].

For uranium, apart from its purification, an important problem was that of resonance absorption, as mentioned before. The idea then came out of using uranium in lumps, just to reduce the resonance absorption. Also, Fermi measured the resonance absorption for uranium oxide compressed into spheres and, in particular, when these spheres were embedded in graphite [Fermi (1941)]. Evidently, he was already thinking about experiments to test a "complete" uranium-graphite reactor.

Meanwhile, the fission of uranium induced by fast (rather than slow) neutrons was as well investigated to some extent, not only for the possibility of obtaining a fast neutron chain reaction, but also for measuring the contribution of fast neutron-induced reactions to the slow neutron chain reaction [Anderson (1941d)].

Fermi and Szilard had the very important idea of placing the uranium oxide in a lattice in the graphite, instead of spreading it out uniformly [Wattenberg (1984)]. Here the problem was "to ascertain whether a given lattice of uranium oxide lumps embedded in graphite could give a divergent chain reaction if its dimensions were made sufficiently large" [Anderson (1942b)], by exercising the greatest care in keeping under control possible losses of neutrons.

In order to test with a smaller structure whether a larger one would work, Fermi invented the "exponential experiment". Uranium was placed among the graphite bricks in a cubic lattice array, with a radium-beryllium neutron source near the bottom and indium foils exposed at various distances from it on the vertical axis. The arrangement is, thus, similar to that of the sigma pile, but the exponential pile was much larger than the sigma pile. The exponential decrease in the neutron density along the axis is greater or less than that expected due to leakage according to whether the reproduction factor is less or greater than one.

Such exponential piles were developed at Columbia in Summer-Fall of 1941 [Fermi (1946b)]; they produced results indicating that even an infinite amount of material would not lead to a self-sustaining structure, this being due mainly to the impurities in the graphite. The situation changed when, during the following Spring (1942), some new graphite was available. The last two experiments performed at Columbia, before the move to Chicago, gave encouraging results [Fermi (1942c,d)], and definitively demonstrated an understanding of the physical effects being involved.

Achieving the first nuclear chain reaction

The National Academy of Science Committee, whose chairman was Arthur H. Compton of the University of Chicago, was charged to review the uranium projects of the United States and to judge their military importance. At the end of 1941 the Committee decided that the work made by the Fermi group using natural uranium was important and, one day before the Pearl Harbor attack on December 7, 1941, the Metallurgical Laboratory was established with Compton as its scientific head in Chicago. For people working on a chain reaction using natural uranium, Chicago became the only game going and, finally, Fermi and his group at Columbia definitively moved to Chicago in April 1942.

> Under Compton leadership a large number of people came too. Among them there was Szilard who worked hard getting the graphite free from neutron absorbing impurities, and Norman Hilberry, who did a marvellous job procuring what was needed. Soon large quantities of graphite began to appear for us to test. Equally strenuous efforts were expended getting uranium in forms sufficiently pure. First we worked with uranium oxide. Then various people worked to produce uranium metal. Outstanding among those was Frank Spedding from Iowa State University. [...] Spedding's uranium was an important component of the first chain reaction. [Anderson (1984)]

A number of engineers then came into the project to produce an appropriate and feasible design of a chain reacting system, so that a first practical problem was to "translate" the known physical achievements into a form suitable to them who had little knowledge in a field completely new. To this end, Fermi invented the notion of "danger coefficient" [Fermi (1942a)] for identifying the impurities which were dangerous for the realization of the chain reaction, due to their high neutron absorption cross section. The effect of such impurities was, then, taken into account directly on the evaluation of the multiplication factor through those danger coefficients. For example, it was determined the effect of gases in the interstices of the graphite, mainly concerning with the appreciable amount of nitrogen impurity in the porous graphite, or even the effect of the undesirable impurity of water in graphite or uranium [Wattenberg (1984)].

Another problem studied was the stability of the pile against temperature changes, since the heat production in the reactor would have altered the reactivity of the pile [Fermi (1942b)] [Christy (1942)].

The study of the uranium-graphite reactor was not the sole work carried out at the Metallurgical Laboratory; other possible systems were as well considered and some measurements made. This is the case, for example, of the so-called "water boiler", that is a reactor system made of a central uranium core enriched with ^{235}U and water around it serving as a moderator [Breit (1942)]. Also, the multiplication factor of a uranium oxide system with a beryllium metal as neutron moderator was measured [Wattenberg (1984)].

Turning back to the study of the main uranium-graphite reactor, the first important result was obtained in August 1942, when very pure uranium oxide was delivered to the Laboratory, making the reproduction factor K greater than one for the first time [Fermi (1942e)]. The 4% excess available ($K = 1.04$) effectively opened the road to the building of the first self-sustaining pile, the Chicago Pile No. 1 (CP-1).

The major engineering problem with it was the choice of an adeguate cooling system with sufficiently low neutron absorption, since the "official" motivation for the project was to produce plutonium, another fissile material (other than ^{235}U) to be used also for military purposes. Indeed, "a large effort was underway for planning the pilot and production reactors, on the assumption that CP-1 would succeed" [Wattenberg (1984)]. Alternative choices [Fermi (1942g)] were proposed to cool the system by gas (preferably helium), water, or even liquid bismuth, this ingenious proposal by Szilard being later set aside because of the lack of engineering experience with this material. The Chicago group definitively worked on the design for a helium cooled plant submitted by the engineers T.V. Moore and M. Leverett.

> So it happened that on 15th of November [1942] we started to build the pile in the West Stands [of the Stagg Field, in Chicago.] [...] Fermi wanted to build the pile with a shape as close to spherical as possible. This would minimize the surface/volume ratio and make the best use of the material which would became available. [...] A major change in design came when we had news that Spedding would be sending some of his high purity uranium metal. The best place for this was as close to the center as possible. As a result, the shape of the pile was changed as we went along. The spherical shape we started with got squashed somewhat as we went along because the purity of the material we were getting was better than we had anticipated. [Anderson (1984)]

The delivery of the Spedding's metal avoided the use of another ingenious trick proposed by Anderson, i.e. to build the pile inside an envelope made of ballon cloth to remove the air (and replace it with Carbon dioxide), in order to minimize the absorption of neutrons by the nitrogen in the air within the pile, with a gain of about 1% in the reproduction factor [Fermi (1942f); Anderson (1984)].

To initiate the chain reaction, it was not necessary (as in experimental piles) to introduce in the pile a separate neutron source since, as already experimentally measured, the uranium also had a non-vanishing probability for spontaneous fission, so that it emits a few neutrons of its own. However, when the pile was building, to keep it from becoming too reactive once it began to approach the critical size, some neutron absorber was needed to control the reactivity of the chain reaction. Control rods were, then, inserted within the pile, made simply of strips of cadmium, since such element was known to be a strong neutron absorber. The pile was controlled and prevented from burning itself to complete destruction just by these cadmium rods, which absorb neutrons and stop the bombardment process of uranium. Further safety arrangements were as well conceived and set up by Fermi for the first reactor (see [Anderson (1984)], for example), whose construction resulted to be completed about a week earlier than the director of the Metallurgical Laboratory had officially anticipated. In the afternoon of December 2, 1942, in fact, the Chicago Pile No. 1 finally got critical and a chain reaction successfully started for the first time.

> We had built the pile, and Fermi had established that we could get a self-sustaining nuclear reaction that we could control in a very predictable manner. [Wattenberg (1984)]

Further studies on nuclear piles during the war years

The further development in the studies on nuclear pile, during the three years 1943-45 was of course focused on the main objective of producing weapons, so that it is natural to expect very few detailed information on these classifies topics. Indeed, none of these appeared in the *Collected Papers* by Fermi [Fermi (1962)], and our source of information is only composed of eyewitnesses (see, for example, [Segrè (1970)]). However, quite fortunately, some reports exist that testify on part of Fermi's activity during these years, not strictly and directly related to military applications, though those reports had been classified for some time (see [Fermi (1962)]).

First of all, the pile was used as a suitable device for checking directly the purity of the uranium and for studying a number of features of the uranium-graphite lattice, unaccessible before [Fermi (1943a)] [Fermi (1943b)]. However, after about three months of operation, the original CP-1 pile was explored sufficiently to learn how to rebuild it with many improvements. A second pile, CP-2, was effectively built at the Argonne site, near Chicago, in March of 1943, and several studies started to be done. These were mainly aimed at designing an efficient pilot plant for producing plutonium or for isotope separation. Such plants were actually erected (at the end of 1943 and later on) at Oak Ridge, Tennessee (known as "Site

X") and at Hanford, Washington (known as "Site W"). An example is the designing and test of a radiation shield for the production piles to be built at Hanford, mentioned in [Fermi (1943d)].

The pile was also used as a tool to measure neutron absorption cross sections by several elements. Samples of these elements were put in the pile, and the compensating changes in control rod position were determined [Anderson (1947)]. This method also became a routine tool for checking for neutron absorbing impurities in the materials used in reactors.

Some explicit "physics works" was, furthermore, carried out when the so-called "thermal column" was devised by Fermi and incorporated in experimental piles [Fermi (1943e)]. A graphite column was, in fact, set up on the top of a pile, where thermal neutrons could be found with substantial intensity and essentially free from those of higher energy. This lead to the discovery of a novel phenomenon, that is the diffraction of thermal neutrons by graphite lattice [Anderson (1946)], which opened the road to investigate the wave properties of neutrons [Fermi (1944c)]. The increased neutron intensity available from a pile also allowed to obtain truly monochromatic beams of neutrons for different experiments (such as, for example, the measurement of the boron cross section at a well definite neutron velocity). This was made possible by a thermal neutron velocity selector designed by Fermi at the Los Alamos Laboratory (known as "Site Y") [Fermi (1947)] and then built at Argonne.

The fission spectrum of uranium was also measured accurately by exploiting the slow neutrons provided by a pile, which were then absorbed by a layer of uranium. Other physical properties of ^{235}U and ^{239}Pu were as well determined [Fermi (1944b)], and these measurements, performed at Los Alamos in 1944 with the active collaboration of Chicago's people, revealed somewhat unexpected properties of plutonium. In the same period some interesting work was also done on the theoretically possible phenomenon of "breeding" [Fermi (1944a)], namely of producing more fissionable material in a reactor than was consumed, clearly depending on the effective number of neutrons available in the chain reaction.

The increased production of heavy water in 1943 made possible to take seriously into account a proposal by H.C. Urey of April 1942 to use heavy water as neutron moderator. This lead to the construction of an experimental reactor, known as P-9 and later becoming CP-3 pile, which would have much more power than CP-2, thus extending the experimental possibilities [Fermi (1943f)].

Finally, other effects were studied during 1944, ranging from the dissociation pressure of water due to fission [Anderson (1944)] to the measurement of the amount of nitrogen in the first production pile at Hanford. An unexpected problem with the Hanford pile was also studied, and independently solved by Fermi and J. Wheeler, on the xenon poisoning, which caused the full stop of the chain reaction [Fermi (1944d)].

Further works by Fermi until the end of the Second World War concerned mainly the realization of the atomic bomb at the Los Alamos Laboratory, so that the corresponding written reports were strictly classified and not available for the *Collected Papers.* A relevant exception are the lecture notes [Fermi (1946a)] for a course that Fermi gave at Los Alamos just after the end of the war, in the fall of 1945. Here he summarized the results achieved on neutron physics, with particular reference to nuclear piles. These lectures are an example of the didactic ability of Fermi rather than a source of information about his research work.

The set of lectures on neutron physics

Once the pile program of the Metallurgical Project in Chicago was sufficiently advanced not to need a continuous attention by Fermi, he definitively moved to Los Alamos (in September 1944) to join the Manhattan Project. Here Fermi began to give isolated lectures on many different subjects [Anderson (1984); Segrè (1970)], related to that project, for the benefit of the people who worked at Los Alamos, many of them being just students or graduated guys. Then, after the end of the war, in the Fall of 1945 he taught a regular course on neutron physics to about thirty students: this was the first time that such a complete course was given, ranging over more than ten years of important discoveries, and also the first occasion for the scientists who contributed in those achievements to pause and reason a bit more on the results obtained.

We know about the content of this course from the notes taken down in class by one of the attending students, I. Halpern, who assembled them into a (classified) typescript on February 5, 1946. A first part of the Fermi lectures at Los Alamos, containing neutron physics without reference to chain reactions, was declassified on September 5, 1946, while the remaining part has been declassified only in 1962. Both parts have been later published in the *Collected Papers* by Fermi [Fermi (1946a)]. Leaving aside the pregnant didactic style by Fermi, the main relevance of such notes is, as we have already mentioned, that they present for the first time a complete and accurate treatment of neutron physics from its beginning, including a detailed study of the physics of the atomic piles. In this respect it is not surprising that especially the second part of the notes, dealing just with chain reactions and pile physics, was considered as "confidential" material by governmental offices.

However, one of us (S.E.) has recently recovered a *different* version of the Fermi lectures at Los Alamos, formerly belonged to James Chadwick and now deposited at the Churchill Archive Centre in Cambridge (U.K.). Two relevant (for the main topic of the present volume) folders exist in this archive. The first one (CHAD I 17/3) contains a letter from R.T. Batson of the Atomic Energy Commission (A.E.C.), a copy of the paper *Elementary Theory of the Pile* by Fermi[1] and a copy of only the *first part* of the Halpern notes of the Fermi lectures. The second folder (CHAD I 4/1) contains a version of the *complete* set of lectures made by A.P. French, dated June 23, 1947.

It is apparently not strange that the material of the first folder belonged to Chadwick, since he was the respected (also by Americans) leader of the British Mission in the United states. The biggest part of the British contingent was, in fact, at Los Alamos, and Chadwick himself was present at the world's first nuclear test at Alamogordo on July 16, 1945. Several scientists of the British Mission were very young and, among the others, it was Anthony P. French who graduated in Physics at the Cambridge University just in 1942. In the same year he joined the atomic bomb project ("Tube Alloys") at the Cavendish Laboratory, and was later sent to Los Alamos in October 1944 as a member of the British Mission. Here he worked with E. Bretscher, O.R. Frisch, J. Hughes, D.G. Marshall, P.B. Moon, M.J. Poole, J. Rotblat, E.W. Titterton, and J.L. Tuck in the field of experimental nuclear physics [Szasz (1992)][2], and returned to the United Kingdom in 1946, working for two years at the just newly formed Atomic Energy Research Establishment (A.E.R.E.). The second folder of the Chadwick papers mentioned above contains just the notes of Fermi course on neutron physics taken by French on his own, when he was at Los Alamos, and later (1947) re-organized into a final version when he came back to England.

In the first part of the present volume we report, for the first time, just the complete set of the French notes.

[1]This paper is reproduced in [Fermi (1962)]; in particular see page 538 of Volume II.

[2]The remaining part of the British Mission was composed by B. Davison, K. Fuchs, D.J. Littler, W.G. Marley, R.E. Peierls, W.G. Penney, G. Placzek, H. Sheard, and T.H.R. Skyrmes.

From these notes it comes out that our previous knowledge of the Fermi course was incomplete and, to some extent (limited to the Halpern notes), misleading. We have indeed performed a careful analysis of the mentioned documents, and the main results are summarized below.

First of all, our study has shown that the French notes do *not* depend on the Halpern ones, but French probably saw them (the organization of the introduction is similar). The topics covered are exactly the same, although to a certain (minor) extent the material is organized in a little different manner. Almost all the topics discussed were expounded by Fermi; according to French, when Fermi was absent, R.F. Christy and E. Segrè treated the scattering of neutrons and the albedo in the reflection of neutrons, respectively.

The text of the notes is different in the French and Halpern versions; in few cases, however, similar or even identical words or sentences are present in both versions, likely denoting quotes from an original wording by Fermi. In general, the French notes are much more detailed and accurate (as may be roughly deduced even looking at the table of contents), with a great number of shorter or larger peculiar additions[3] (explanations, calculations, data or other, and 5 more exercises) not present in the Halpern notes. It is quite interesting that the greater detail already present in the French notes increases even more in quality (especially figures and data) in the last part, directly related to chain reactions and their applications, and, moreover, explicit references to bomb applications are made (see below). By limiting ourselves to significative scientific remarks or discussions, the French version of the Fermi lecture notes contains about 100 additions, 18 of them being quite relevant while the remaining part accounts for minor remarks, calculation details or figures. Instead the peculiar additions present in the Halpern version but not in the French one are only about 30 (and 3 more exercises), only one of them being relevant. Also, the French paper contains the six questions which were set as a final examination at the end of the lecture course.

In order not to distract reader's attention, we have preferred not to indicate all these differences throughout the volume, but we report in the

[3]The case is completely different, for example, from that of the revision of the (first part of the) Halpern notes made by J.G. Beckerley in 1951 (document AECD 2664 of the Atomic Energy Commission). Here the author *re-wrote* the Fermi lectures by including several additions from *other* sources, "where clarity demanded more information and where the addition of recent data made the text more complete." Contrarily to the present case (as it is evident from the text of the notes), Beckerley "was not privileged to attend the course" by Fermi.

following the most relevant additions (the references correspond to what can be found in the present volume):

- the entire section 2.3, *The Binding Energies of Nuclei*, where the definition of the binding energy and an example for calculating it in a specific case is reported;

- the introduction of Sect. 5.1: "In this section we consider the solution by wave mechanics of a simple problem in nuclear scattering. The nucleus is considered as a centre of force, the force being of short range, so that it ceases to exist beyond a certain distance r_0 from the origin. The actual shape of the nuclear potential then approximates to a square well, as shown in Fig. 5.1. The potential U is negative and constant over most of the nucleus. This corresponds to the fact, as far as we know them, of the interaction between a neutron and a nucleus. The depth of the nuclear potential well is equal to the binding energy, that is about 8 MeV.";

- some details in Sect. 6.2 about the Bragg scattering of slow neutrons by an element with different isotopic composition, ending with the following remark: "The total scattered intensity is thus given by $I_{\rm sc} =$ const. $\left(\sigma_1 + \sigma_2 \pm 2\sqrt{\sigma_1 \sigma_2}\right)$, and may be seen to consist of coherent and incoherent contributions, the latter not being subject to interference";

- a long discussion in Sect. 7.4 (see pages 55-57), with detailed calculations about the spatial distribution of slowed neutrons, aimed at calculating the source strength in neutrons per second both for a thermal detector and for a resonance detector (final explicit expressions are reported);

- calculation details on pages 63-64 about the neutron scattering in a medium (with the determination of the mean free path), ending with a prediction for the neutron-proton scattering cross section (in water) of $\sigma \simeq 20$ barns which "agrees very closely with the accepted value";

- discussion in Sect. 9.6 (see pages 77-79) on calculation details aimed at solving the so-called (Fermi) age equation for the diffusion of neutrons from a Ra-Be source in a column of graphite of square section (with length of side a) and infinite length; the effective length of a side of the column, $a = a_{\rm geometrical} + 2 \cdot (0.67\,\lambda)$ (λ being the mean free path), and the range of the neutrons, $r_0 = \sqrt{4\tau}$ (τ being the age parameter), are introduced; the numerical values of r_0 (instead of only τ as in the Halpern notes) for three (instead of two) typical neutron energies are given; the addition in the French notes ends with the peculiar observation that "we have the somewhat paradoxical result that the system can be made infinite for fast neutrons being slowed down, but not for the same neutrons when they have become thermal";

- after some calculations in Sect. 10.5 (see page 86), this ends with the observation that "a thermal neutron in these media [paraffin and water] makes about 100 collisions before being captured. The distance it travels, measured along the path, is about 80 cm on the average, and the time it takes to do this, which is its lifetime as a thermal neutron, is something less than a millisecond";

- introductory remarks in Sect. 11.1 on the binding energy of a nucleus, with the theoretical expression for the measured mass of an atom in terms of A, Z and the said binding energy;

- several important additions on pages 89-92 related to the stability of nuclei (according to the even- and odd-ness of Z, A or both) and the accurate determination of the expression of the binding energy of nuclei in terms of Z and A, with several numerical data (and a graph);

- the inclusion in Sect. 12.5 of three graphs (Figs. 12.10-12.11) for the cross section of (n, γ) and $(n, \text{fission})$ processes on uranium as function of the incident neutron energy;

- relevant additions on pages 112-113 about homogeneous and lumped graphite piles: explicit calculations of the neutron absorption volume by uranium spheres of 3 cm radius (and of other quantities) lead to the conclusion that "no homogeneous pile of this type will work, and we must therefore devote our attention (if we are considering only the U-graphite combination) to heterogeneous piles";

- small introduction of Sect. 14.1 (with key comments) to the design of a lumped pile, with figures (Fig. 14.1) of lattice structures with spherical lumps or rods of uranium;

- the two relevant figures 14.4 (and related discussion) about efficient cooling systems (by blast of air or water flowing) for piles;

- introductory remarks to Sect. 14.3 dealing with the reproduction factor and critical size of a pile, with a graph (Fig. 14.4) of the actual neutron density in a finite pile as a function of the distance from the center of the pile; explicit expressions and related comments on pages 121-122 on the reproduction factor K as a function of geometrical and other parameters of the pile;

- "We have discussed the mechanism of thermal neutron chain reactions. The question now arises how to produce a nuclear explosion" (see page 139); introductory remarks in Sect. 17.1 about fast reactors starting from the calculated expression for the growth of neutron density in a reactor;

- on pages 141-142, definition, calculations and related discussion on the transport cross section and transport mean free path for neutrons in a fast reactor with a core of ^{235}U and a tamper (neutron reflector);

- discussion in Sect. 17.3 of equilibrium conditions (with explicit expressions) for a fast reactor and mathematical expressions for some quantities describing neutron losses (see pages 144-146);

- the end of Chap. 17 on fast reactors (and, then, of the lecture notes) is: "in this way one can calculate the e-folding time for a fast reactor, and its value thus found will be valid until mechanical effects set in – these having to be known before the efficiency etc. of the bomb can be estimated".

The only relevant addition in the Halpern version is, instead the description of a fast neutron detector, based on the scattering of a neutron flux by a paraffin layer (see page 471 of Volume II of [Fermi (1962)]).

Fermi's patents

A less known part of the history of the development of the Manhattan Project, with its military and non-military applications, is the controversial issue of patenting the novel scientific and technological discoveries [Wellerstein (2008)]. The utilization of patents was, indeed, an *ad hoc* attempt at legal control of the nuclear energy by governmental administrators, and patents themselves played an important role in the thinking of project administrators concerned with meaningful postwar control of the atomic energy.

As a matter of fact, when the Los Alamos laboratory was formed, it included a patent office as part of its administration, starting its operation in July 1943; captain R.C. Smith was appointed as advisor on patent matters. That office collected records of the researchers completed work, with a view to covering them by patents (Smith commanded that the personnel keep workbooks with records of recent discoveries) but, of course, the patent applications on "sensitive" inventions were placed under a "secrecy order", and the patents could not be issued until the applications were declassified. Smiths office worked on five hundred patent cases altogether; completed cases were then filed with the U.S. Patent Office.

Fermi himself became an "inventor", and his "inventions" were described in several patents on nuclear reactor designs, processes, methods and instrumentation. All but two among these patents (accounting for a total of 15 papers) were filed between May 1944 and January 1946, while the last one was a re-issue (with small corrections) of the fundamental patent on nuclear reactors filed by Fermi and Szilard on December 1944.

An intriguing, though particular, story is deserved for the first patent about the process for the production of radioactive substances, originally issued in Italy as early as in 1934 [Segrè (1970)]. Indeed, when in October 1934 Fermi and his group in Rome discovered the well-known effect of enhancing artificial radioactivity with slow neutron bombardment [Amaldi (1935)], the Director of the Institute of Physics, Orso Mario Corbino, recognized the potential practical importance of such effect and urged Fermi to file for a patent. This was effectively done, and on October 26, 1934 an Italian patent was granted (which was later extended to other countries, including U.S., Great Britain, Belgium, France, Germany and Canada) concerning a method of producing radioactive substances by neutron collisions and, in particular, covering the increase of efficiency obtainable by slowing the neutrons with multiple elastic collisions. After the discovery of fission, when nuclear energy development started in earnest, this patent became obviously fundamental to all industrial and military applications, and hence of considerable value. However, during the war it was neither possible nor desirable for Fermi and collaborators to raise questions about compensation for that patent. The question was pointed out only at the end of the war, but then the affair became even more complicated since, although the Atomic Energy Act sought to restore legality in the use of atomic patents by allowing their purchase, governmental managers were resistant to settling existing claims and eventually they used the new legislation to argue that no compensation should be paid. In addition, since Bruno Pontecorvo was among the inventors of this patent, his flight to Soviet Union in October 1950 further complicated matters. At the end, after much legal wrangling, the controversy was solved, and the government of the United States paid some compensation for the patent's rights [Turchetti (2006)].

The whole set of Fermi's patents were not included in the *Collected Papers* published in the 1960s, probably due to the fact that many of them were kept confidential at that time: all but two of them, indeed, deal with the technical and operative construction of nuclear reactors. This fact, unfortunately, has prevented us to know about the precise contributions of Fermi on pile physics and engineering, although the activity by himself on this subject was early well recognized from the accounts given by the living testimonies (see, for example, [Segrè (1970)]) and partially documented by several papers appeared in the Fermi's *Collected Papers*.

Nevertheless, from the direct reading of the text of the patents, a number of important scientific and technical points comes out, putting some new bright light on the Fermi's activity about the construction and functioning of nuclear reactors. In practice, what Fermi *effectively* did in this applicative field is here technically documented, and very clearly emerges from these papers. It is quite impressive the fact that, just from the accurate reading of the patents, anyone who has at his own disposal the necessary materials could effectively build a working reactor, with a number of possible alternatives.

Different topics are, instead, covered in the first and the eighth patents of the set, though they are strictly related to pile physics research. The first case refers to the Italian patent, dealing with the production of radioactive substances, already discussed above. Instead, in the second case, the corresponding patent presents a detailed description of the construction and operation of a velocity selector for neutrons [Fermi (1947)].

For any of these patents, reported in the present volume, a concise description of their content has been added as an introduction. We then refer the reader to these introductive texts for further information about the specific content of the given patents.

This volume

In the present book we reproduce, for the first time, the French notes of the Fermi's lectures on neutron physics and, in the second part, the complete set of the 15[4] patents authored (or co-authored) by Fermi. All these documents do not appear in the Fermi's *Collected Papers* published in the 1960s [Fermi (1962)].

For the French notes, we have used the only available document (to the best of our knowledge) reporting them, i.e. the copy belonged to James Chadwick and now kept at the Churchill Archive Centre in Cambridge (U.K.). In particular, such a copy (dated June 23, 1947) is conserved in the folder CHAD I 4/1 of the Papers of Sir James Chadwick at the archive mentioned.

Instead, for the text of the Fermi's patents we have employed the material conserved at the United States Patent and Trademark Office. For several of these patents, however, this Office is not the only available source of documents, which may also be found in the Patent Offices of Italy (patent no. IT324458, the original one), Great Britain (patent no. GB465045),

[4]Note that two patents belonging to this set are almost identical; see below.

Belgium (patent no. BE411973A), France (patent no. FR796795A), Germany (patent no. DE681540C) and Canada (patents no.s CA407558A, CA407559A, CA552312A, CA552693A, CA619065A, CA620923A). In general, these alternative versions do not contain more information[5] with respect to the U.S. patents, this motivating our choice.

The major effort we have made to carefully check and type all equations and tables was motivated by our desire to facilitate the reading of Fermi's unpublished papers as much as possible. However, we have tried to keep as much as possible the original editing, this holding, in particular, for sectioning and equations (especially in Part 2, however, for a better understanding we have introduced some punctuation marks). Additions or modifications are properly pointed out throughout the text; in particular, explicative non-original footnotes are marked with the symbol @. The figures present in the first part (lecture notes) have been redrawn electronically for a better display, without the use of photographic or scanning devices but are otherwise true in form to the original drawing. Figures appearing in the patents, instead, have been elaborated directly on the originals, in order not to lose important though tiny details. In both cases, the numbering of figures is different from that present in the original documents: here we have numbered them according to the sections where they are present, thus keeping traces of their relative position. Tables have been reproduced, again, as close as possible to the originals but, for editorial reasons, especially for Part 2 their position is not always the original one, the same being true for figures (in the original patents, all the figures are placed at the beginning of the patent itself). In few cases, whenever required by pagination, very long tables have been broken into two or more parts. For units, we have adopted the common modern notation instead of that present in the original documents (for example, "g" rather than "gr" for gram, "s" rather than "sec" for second, etc.).

Finally, for any of the patents reported, we have added (in italics) a brief introduction about their content, in order to present shortly the key points of what described in the given patent.

The 3rd and 15th patents (that is U.S. patents no. 2708656 and 2798847) refer to the same work, the text being practically the same. The second version was probably prepared by the same authors (Fermi and Szilard) in order to correct several misprints in the previous version. For these reasons, we have reported only the "definitive" second version replacing the

[5]With slight exceptions for Canada patents CA407558A and CA407559A with respect to the first U.S. patent 2206634, where two more claims and two tables appear.

text of the 3rd patent.

Appropriate bibliography, not present in the original documents, have been added at the end of the volume, before the index. Far from being exhaustive, it provides only some references about what touched upon in this introduction or in the specialized introductions preceding each patent.

An introductory note by Prof. A.P. French follows this Preface.

Acknowledgements

We are indebted to Prof. A.P. French for a very helpful exchange of information, his willingness to write the following historical note and, more in general, for his kind encouragement. We are as well very thankful to Dr. A. De Gregorio and Prof. E. Recami for valuable discussions, and to Prof. G. Miele for his kind support and encouragement.

The active and valuable cooperation of the staff of the Churchill Archive Centre, Cambridge (U.K.), the staff of the U.S. Patent and Trademark Office and of the Information Resource Center of the U.S. Embassy in Rome is here gratefully acknowledged.

S. Esposito
O. Pisanti

Naples, March 2010

Some Background to my Notes on Fermi's Neutron Physics Lectures (Los Alamos, 1945)

Introduction

In October, 1939 I entered Cambridge University as a freshman. World War II had recently begun, but the admission of students was not yet affected. I began attending lectures in physics at the Cavendish Laboratory as part of a program of studies called the Natural Sciences Tripos. My subjects besides Physics were Mathematics, Chemistry and Mineralogy. These four subjects (or some other combination of introductory subjects in science) were typically attended during the first two undergraduate years, after which the aim during the third year (Part II of the Tripos) was to concentrate on one subject, in my case physics, and complete a B.A. degree.

As the war proceeded, large numbers of students were recruited into the Armed Forces and into special projects, postponing the completion of their university studies until a later time. My own fate – and that of many others – was somewhat different. I completed my degree program but went into war work later. In my case I was directed to report to my physics tutor Egon Bretscher at the Cavendish Laboratory two weeks after receiving the B.A. He had given an exciting course of lectures about nuclear physics, but of course I had known nothing about the atomic bomb project and his involvement in it.

Tube Alloys

As the world learned later, the discovery of nuclear fission in 1938 was followed by intensive research in a number of laboratories on both sides of the conflict. In Britain the leading figures were Otto Frisch (co-discoverer of fission) and Rudolf Peierls (a senior theoretical physicist at the University of

Birmingham). Together they produced, in 1940, a seminal memorandum on the possibility of making a nuclear fission bomb. Using the available nuclear data, they made a quantitative estimate that a few kg of U^{235} should be sufficient. To separate this from the much more abundant isotope U^{238} would be a monstrous task. But the possibilities were judged to be sufficiently good to justify setting up a British atomic bomb project. It was given the code name "Tube Alloys" and was set up with headquarters in London. The staff for the project included most of the nuclear scientists working at universities in Britain and Canada, plus some major industrial participants. The chief academic partner was the Cavendish Laboratory, where James Chadwick discovered the neutron, and where Cockcroft and Walton had achieved the first artificial nuclear reactions in 1932.

The project had two main groups. One of them (in which my former tutor Egon Bretscher was a leader) was concerned with atomic bomb design, by producing fission in U^{235} by fast neutrons with bombarding energies of 1 MeV or so. The other group (slow neutron physics) was concerned with designing a nuclear reactor using a "pile" of U^{238} to generate power and neutrons in a large mass of ordinary uranium. Fermi led the team that succeeded in making the world's first chain-reacting pile in December 1942 at the University of Chicago.

The Manhattan Project

Much of the preceding section is familiar history, but it gained special significance later in the war. Britain had undoubtedly taken the lead initially, but the United States, with its size and immense resources, was ready and able to join the effort. It should be remembered, however, that the USA did not actually join the war until after the attack on Pearl Harbor in December, 1941. There had been intense consultation between Britain and the USA well before this, but the USA was unready or unable to become an actual combatant. Once the USA had joined the struggle, a formal partnership became possible. It was not all plain sailing, however. Specific national interests were involved. Valuable meetings between British and American scientists took place, but the going became rough at times. A crucial meeting between Roosevelt and Churchill took place in Quebec in 1943. It was recognized that it was impossible, in the conditions of enemy bombing, food rationing and a general shortage of supplies, to continue with the development of the atomic bomb in Britain. A new project – the Manhattan Project – was launched in the USA. This became an immense venture, including all the industrial-scale plants needed for separation of U^{235} and plutonium. But even the basic research in nuclear physics was terminated in Britain. The work in slow-neutron physics toward the design of nuclear reactors was transferred to Canada, and the basic research on fast-neutron fission (for the design of the atomic bomb itself) went to a newly created laboratory at Los Alamos, New Mexico.

Once the Manhattan Project had been established, it outstripped anything that could have been achieved in Britain. It became the most powerful (and expensive!) organization for research and development in atomic science that had ever been created. Only in the circumstances of the world war could it have been done. It benefited from the participation of some of the best scientists in the world who had left Europe as refugees from Nazi and Fascist oppression. One of these was Enrico Fermi, who fled from Italy with his family after receiving the Nobel Prize in physics in 1938.

The British Mission

A special British mission to Los Alamos was created under the leadership of James (later Sir James) Chadwick. This mission contained a little over 20 members, who were attached to specific parts of the atomic bomb project, including explosives. The most prestigious member of the British Mission was Niels Bohr, but there were half a dozen junior assistants such as myself. Thus it was that I found myself traveling to Los Alamos in October, 1944 to join Bretscher, who had preceded me by a few months. (To the best of my knowledge I am the only surviving member of that group.) I found that Bretscher himself was no longer doing research related to the bomb; he had joined forces with Edward Teller, who since the beginning of the Manhattan Project had pushed for working on the theory of thermonuclear reactions of the lightest nuclei, with the aim of creating a bomb called "the super". But Teller and his associates were all theorists, so it was appropriate for Bretscher to form a very small experimental research group, which he did, using two of us (M.J. Poole and A.P. French, junior staff from the British Mission) together with two American assistants. We built a miniature accelerator to provide up to 100 kV or so to study nuclear collisions of hydrogen isotopes, suitable at energies for thermonuclear reactions. It was reminiscent of the apparatus that Cockcroft and Walton had constructed and used in their pioneering researches in 1932.

The Special Engineering Detachment

An important component of the scientific work force at Los Alamos was the Special Engineering Detachment (S.E.D.) made up of young American university students with scientific and technical skills aspiring to attend graduate school after the war. Their total number, I believe, was of the order of 1000. They were attached to research groups in all branches of the project. The two assistants in Bretscher's group were S.E.D.s. Many of S.E.D.s became successful scientists and engineers in their own right after the war. One such assistant (Val Fitch) won a Nobel Prize in Physics later in his career.

"Los Alamos University"

The dropping of the two atomic bombs on Japan marked the end of the short-term goals of the Manhattan Project, but there were many ongoing projects worth continuing. Although many of the senior Los Alamos research staff were eager to return to their home institutions, there were many others glad to continue their research projects at Los Alamos. There was another significant factor. The large numbers of S.E.D.s were subject to military discipline and were not necessarily free to leave Los Alamos at will. It was natural to take advantage of this situation by setting up lecture courses in which the junior people could receive education from older experts. This opportunity was informally called "Los Alamos University". Students could sign up for two courses and have the possibility of receiving academic credit from a chosen college for doing so. I believe that about a dozen different courses were offered, taught by people some of whom were world experts.

The existence of this scheme was too good to miss, and I sought permission to enroll although I was not specifically a target of the program. The courses I chose were Electromagnetism, taught by Hans Bethe, and Neutron Physics, taught by Enrico Fermi. Both men were outstanding as physicists, but there was no doubt concerning my own preference as to subject matter.

Concluding Remarks

I left Los Alamos at the end of August, 1945. Before doing so I mailed a package of my personal Los Alamos books and records addressed to myself at the Cavendish Laboratory. However, I did not return directly to Cambridge until 1948. Like several other people who had been at Los Alamos, I went to the newly established Atomic Energy Research Establishment (A.E.R.E.) at Harwell, England, where Egon Bretscher became head of the Physics research division and remained until his retirement. I do not think that I wrote up a final copy of my notes on Fermi's lectures until after my return to the U.K. I may have sent a set of the notes to the London office of Tube Alloys, but I have no recollection of sending a copy of them directly to Chadwick (although I may have done so). It would have been natural for Chadwick to receive the notes from the Tube Alloys headquarters, as he had served as head of the British mission, and it is no surprise that the notes were found in Chadwick's papers many years later by Dr. Esposito. My own copy of the notes slept in my personal files until I was reminded of its existence when Dr. Esposito first wrote to me in 2008.

Anthony P. French

PART 1
Neutron Physics

Neutron Physics

A Course of Lectures
by
E. Fermi

Introduction

These notes are based on a course of about 30 lectures given at Los Alamos towards the end of 1945. The great majority of the lectures were given by Dr. E. Fermi; a few lectures on the subjects of scattering and the albedo were presented by Drs. R.F. Christy and E. Segré respectively.

For the most part, this version follows the original lectures quite closely. To secure a logical presentation the original order has, however, been slightly changed in one or two places. The text has also been amplified at a few points, where the treatment as given in the lectures appeared incomplete.

The division into chapters has been made, so far as possible, to permit each chapter to be read as a self-contained whole, although the complete course of lectures is, strictly speaking, a continuous development.

The exercises and problems given during the course have been collected together at the end of this volume.

13 June 1947 A.P. French.

Chapter 1

Sources of Neutrons

1.1 Natural Sources

Bombardments of certain elements with natural radiations may result in the production of neutrons. The reaction $Li^7(\alpha, n)\,B^{10}$ is a possible source of this kind. A certain minimum α-particle energy is required to initiate the reaction. The masses of the nuclei involved are set out below:

$_3Li^7$	7.01804	$_5B^{10}$	10.01605
$_2He^4$	<u>4.00388</u>	$_0n^1$	<u>1.00893</u>
	<u>11.02192</u>		<u>11.02498</u>

The total mass of the product nuclei exceeds that of the primary nuclei by 0.00306 mass units, which is equivalent to 2.85 MeV. This is the kinetic energy that must be supplied to the system in the centre of gravity frame in order that the reaction shall take place. The measured energy of the α-particle in the laboratory frame has to be greater than this by a factor $\frac{7+4}{7}$. that is 4.5 McV approximatcly. Thc rcaction thus has a threshold at this energy, and the yield rises sharply with increasing α-particle energy, as indicated in Fig. 1.1. Other reactions of this type are:

$$Be^9(\alpha, n)\,O^{12} + 5.5 \quad MeV;$$
$$Be^9(\alpha, n)\,3He^4\ .$$

The natural neutron sources may be divided into two groups, according to whether the disintegrations are produced by α-particle or γ-rays.

1.1.1 *Alpha Particle Sources*

1. Ra-Be. As usually prepared, this consists of a mixture of Be powder and a radium salt. A typical composition would be 5g Ne and 1g Ra. This is a

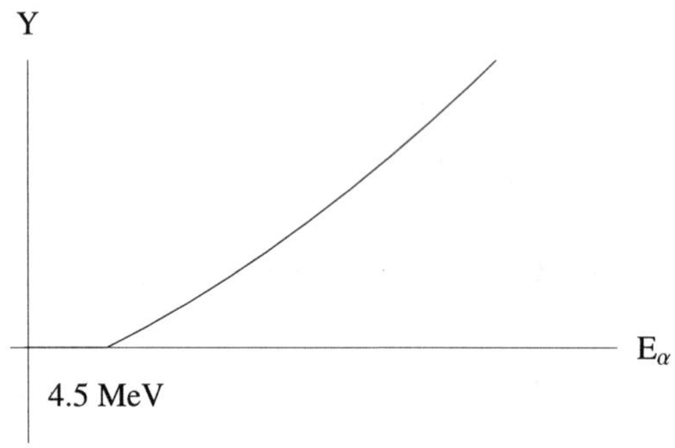

Fig. 1.1

good source, but it must be remembered that it also emits γ-rays.

2. Rn-Be. This takes the form of a vial of Rn gas surrounded by walls of Be. Again there is a certain amount of γ-radiation; there is the additional drawback that the source decays with the half life of radon itself - only 4 days. Both Ra-Be and Rn-Be sources give about $10 \div 15 \times 10^6$ neutrons/curie/s; the energy spectrum is continuous, with a peak at 1 or 2 MeV and an upper limit of about 13 MeV.

In Fig. 1.2 is shown the radioactive decay series[1] beginning with Ra, with the various α-particle energies marked. From this diagram it will be possible to infer which α-particle groups will be effective in any neutron-producing reaction under consideration, once the mass excess or deficiency for the process is known.

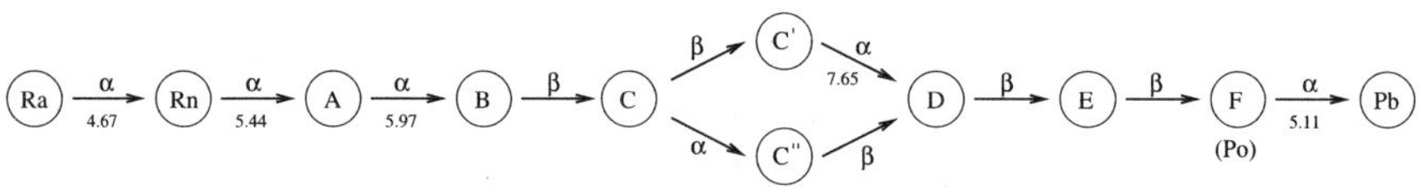

Fig. 1.2

3. Po-Be. There is very little γ-radiation associated with this source, but the neutron intensity is rather low - 2.8×10^6 neutrons/curie/s.

[1]@ In this figure, the old notation for radium decay products (A,B,C, etc.) is used; Rn (Radium emanation) corresponds to Rn^{222}, Ra A to Po^{218}, Ra B to Pb^{214}, Ra C to Bi^{214}, Ra C$'$ to Po^{214}, Ra C$''$ to Tl^{210}, Ra D to Pb^{210}, Ra E to Bi^{210} and Ra F to Po^{210}, respectively.

4. Other sources.

$B^{11}(\alpha, n)\ N^{14}$, 1.1×10^6 n/curie/s;
$F^{19}(\alpha, n)\ Na^{22}$, 10^5 n/curie/s;
Po-BF_3 is a good mock fission source.

1.1.2 *Photo-Neutron Sources*

1. The reaction $Be^9(\gamma, n)\ Be^8$ has a threshold at 1.61 MeV. Two of the radium γ-rays are above this energy, giving neutrons of 0.12 MeV and 0.51 MeV respectively.

2. $D^2(\gamma, n)\ H^1$ has a threshold at 2.17 MeV. Only one of the Ra γ-rays is above this energy, so the source is monochromatic.

3. The γ-rays from MsTh[2] give neutrons of 0.220 MeV from D_2O, and two groups of 0.82 and 0.16 MeV, from Be.

4. Using γ-ray sources prepared in the Argonne pile, the following neutron sources have been made:

γ-ray emitter	Target	Neutron energy
Sb	Be	0.030 MeV
Ga	D_2O	0.12
Na	D_2O	0.24
La	Be	0.62
Na	Bc	0.82

1.2 Artificial Sources

Probably the best known artificial source is the D-D reaction:

$$D + D \begin{cases} \nearrow & He^3 + n + 3.25\,\text{MeV}, \\ \searrow & H^3 + H^1 + 3.98\,\text{MeV}. \end{cases}$$

The yields of the alternative branches of the reaction are approximately equal. The neutron yield per incident deuteron from a thick D_2O (heavy

[2]@ Mesothorium denotes generically the first two decay products of thorium, that is Ra^{228} (mesothorium I) and Ac^{228} (mesothorium II).

ice) target is given below for various deuteron energies:

E_D(keV)	0	50	100	200	300	500	1000	2000
Yield $\times 10^7$	0	0.2	0.75	3.0	7.0	22	120	700

The reaction $Li^7(p,n)\ Be^7$ is useful in that it can be used to produce mono-energetic neutrons over a large range of energies (about 50 keV to several MeV). It is an endo-ergic[3] reaction, with a Q value of -1.62 MeV, and hence a threshold of $\frac{8}{7} \cdot 1.62 = 1.86$ MeV.

Another commonly used reaction, usually employed as a source for a cyclotron, is the process $Be^9(d,n)\ B^{10}$. The thick target yields at 1 MeV and 8 MeV incident deuteron energy are about 10^8 and 10^{10} neutrons/s/μA, respectively.

There are many other light particle reactions with good yields, but most of them give rise to complex neutron spectra.

[3]@ Here and in the following, the author uses this compact form for endo-energetic.

Chapter 2

The Isotopic Chart: Nuclear Masses and Energies

2.1 The Isotopic Chart

This consists of an array of squares, each representing an isotope. The ordinate is N and the abscissas Z. A given mass number $A = N + Z$ is thus shown as a line at 135° to the Z axis. (See Fig. 2.1). The line, or rather the band, representing the known isotopes, has an initial slope of 45° and then curves upward as shown.

The chart provides one of the most convenient means of presenting collected data on nuclear transitions. Consider, for example, the section of the chart shown in Fig. 2.2, for the isotopes near to Al^{27}. This is the only stable isotope of Al. We see from the figure that the (n, α) reaction would be

$$\mathrm{Al}^{27}(n, \alpha)\,\mathrm{Na}^{24}.$$

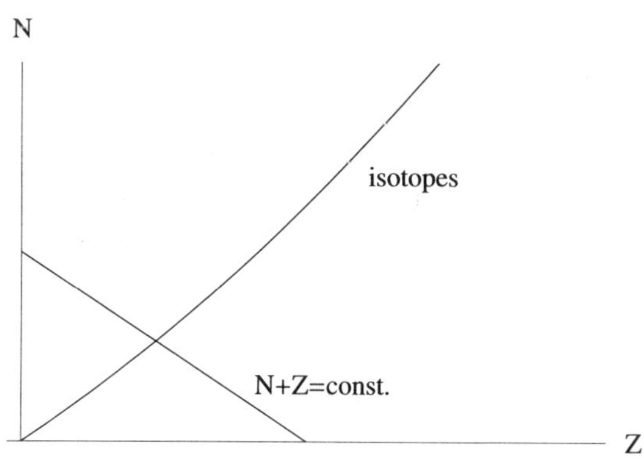

Fig. 2.1

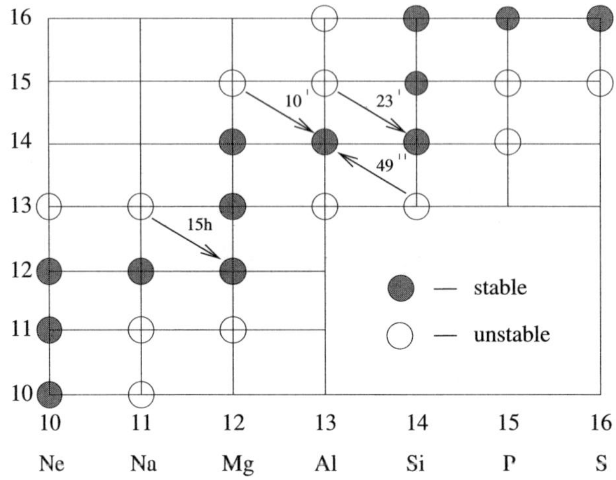

Fig. 2.2

We see that Na^{24} is unstable, being a β-particle emitter with a half period of 15 hrs.

The (n, p) reaction shows how the neutron can be exchanged for a proton and an electron. We have here

$$Al^{27}(n,p)\,Mg^{27}.$$

Mg^{27} is a β-emitter, reverting to Al^{27}.

The (n, γ) reaction is

$$Al^{27}(n,\gamma)\,Al^{28}.$$

Al^{28} is a β-emitter, with a half period of 23 min., and goes to Si^{28}, which is stable.

Any nucleus with an excess of protons will decay by positron emission or by K electron capture. In the latter, an electron is taken usually from the innermost level of planetary electrons and a proton is converted to a neutron inside the nucleus.

2.2 Energy Balance of Reactions

In α-particle decay, we find that the α-particle has a definite energy, because there is a definite jump from one energy level to another. We should expect to find the same result for β-decay, but this is not so. Figs. 2.3a and 2.3b

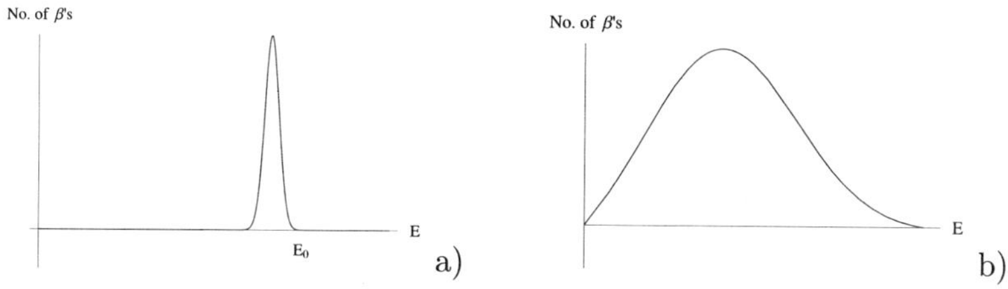

Fig. 2.3

indicate the expected and observed spectra. From a knowledge of the two energy levels involved in various cases of β-decay, it has been established that the upper end point of the continuous distribution corresponds to the total energy available. To preserve the conservation of energy, angular momentum, magnetic moment, Pauli invented the neutrino. We now write a β-transformation in the following way:

$$\mathrm{Mg}^{27} \to \mathrm{Al}^{27} + \beta^- + \nu$$

(to take a specific example). The properties ascribed to the neutrino are: a very small or zero rest mass, zero charge, and half-integral spin. Its kinetic energy is given by $(E_0 - E_\beta)$

To determine whether any given process is exo-[1] or endo-ergic, one may sum the masses involved. Thus in the process

$$\mathrm{Al}^{27} + n \to \mathrm{Mg}^{27} + p + Q,$$

one can evaluate Q by this means. There is, however, a more convenient method, based on the fact that the β-decay of Mg^{27} is known:

$$\mathrm{Mg}^{27} \to \mathrm{Al}^{27} + \beta^- + E$$

($E = 1.8$ MeV). Since by addition of these equations we obtain

$$n = p + \beta^- + E + Q = H + E + Q,$$

only the values of E and the neutron-hydrogen mass difference need be known. This latter is 0.0008 mass units, or 0.75 MeV. Thus we have

$$Q = 0.75 - 1.8 = -1.05\,\mathrm{MeV}.$$

Most (n, p) reactions are endo-ergic, like this present example.

All (n, γ) reactions, on the other hand, are strongly exo-ergic, because one can write these in the form

$$A + n \to A + 1 + Q,$$

giving $Q \sim 8$ MeV always.

[1]@ Here and in the following, the author uses this compact form for exo-energetic.

2.3 The Binding Energies of Nuclei

In this section we consider the energies of nuclei as manifested in their masses. We should perhaps begin with the usual reminder - that calculations are performed with the masses of neutral atoms rather than with the masses of the nuclei themselves. Thus in the example at the beginning of Chapter 1 it is in fact the atomic masses that have been used. It may easily be verified that this procedure is perfectly valid, so long as one is consistent, because in any reaction charge as well as total energy must be conserved. There is one notable exception to this rule, namely the process of positron emission, where two electron masses (= 1.02 MeV) must be created from the available energy.

It is well known that the mass of a stable nucleus is less than the sum of the masses of its constituents. Thus, in forming He^4 from two protons and two neutrons, we have

$$\begin{aligned} 2\mathrm{H} &= 2.01626 \\ 2n &= \underline{2.01786} \\ 2\mathrm{H} + 2n &= 4.03412; \qquad\qquad \mathrm{He}^4 = 4.00386, \end{aligned}$$

whence the mass difference is 0.030 mass units, giving an energy of formation of He^4 equal to about 28 MeV. It is stable against disintegration into two protons and two neutrons by just this amount, or, as it is usually put, the binding energy of the nucleus is 28 MeV, which is 7 MeV per nucleon. This represents a sort of mean value for stable nuclei.

2.4 The Packing Fraction Curve

The packing fraction is a rather arbitrary but convenient means of expressing the binding energy. If M is the mass of an atom, and A its mass number, then

$$f = \frac{M - A}{A} = \frac{M - (Z + N)}{(Z + N)} .$$

Thus, taking the case again of He^4, we have $A = 4$, $M = 4.0039$, giving $f = 0.002$. For H the value of f is almost ten times greater. The variation of f over the Periodic Table is roughly represented in Fig. 2.4. Over a considerable section in the middle of the table the packing fraction is nearly constant and equal to about -0.0005. The Coulomb repulsion between protons are the cause of the rise of the curve for high values of A. The energy of

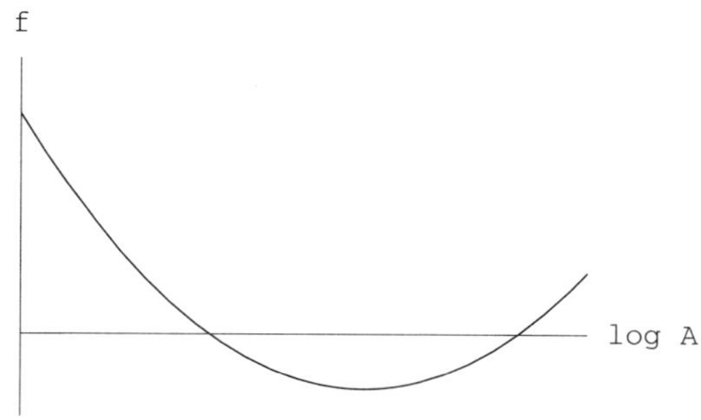

Fig. 2.4

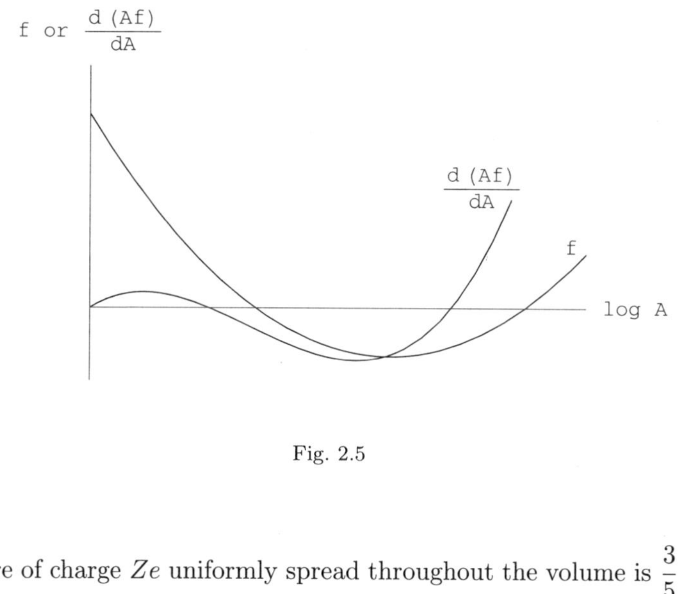

Fig. 2.5

a sphere of charge Ze uniformly spread throughout the volume is $\frac{3}{5}\frac{(Ze)^2}{R}$, and thus raises sharply with increase of Z. The curve is known accurately for the lighter elements, but not too well above about $A = 50$.

Suppose we put

$$f = \frac{M - A}{A} = f(A)\,.$$

Thus $M = Af(A) + A$. We can make a nucleus $(A + 1)$ by adding a nucleon, which on the average has a mass midway between n and H, i.e.

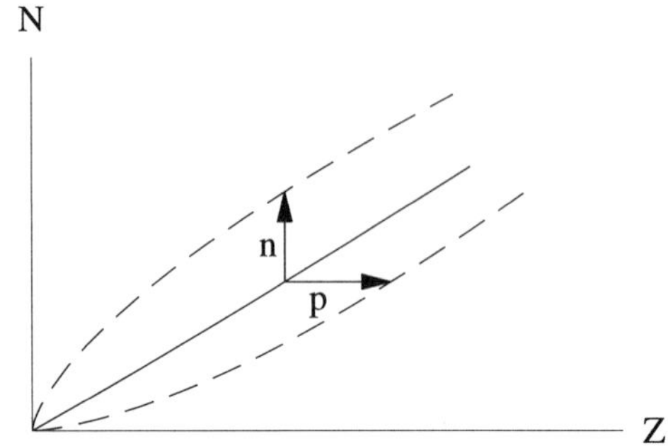

Fig. 2.6

1.0085 approximately. The total mass we start with is therefore

$$Af(A) + A + 1.0085.$$

But by the formula, the mass of the nucleus $(A+1)$ is

$$(A+1) + (A+1)f(A+1).$$

The binding energy of the added nucleon is thus

$$\Delta E = 0.0085 - [(A+1)f(A+1) - Af(A)] = 0.0085 - \frac{\mathrm{d}}{\mathrm{d}A} Af(A),$$

using the definition of a derivation and assuming A to be large. Fig. 2.5 indicates the variation of f and the derivative of Af with A. We would expect the derivative to be initially negative, and later positive, so that the binding energy is at first greater and subsequently less than 0.0085 mass units ($\sim$ 8 MeV). We would expect the binding energy to be about 8 MeV for all (n, γ) reactions, but there are the exceptions to this rule, viz. ${}_2\mathrm{He}^4$ and ${}_0n^1$. If the binding forces were 20% larger than we at present think, a di-neutron might be observed.

One could in theory add neutrons or protons indefinitely to a stable isotope, but those with too many neutrons would emit negative β-particles, and then with too many protons would emit positrons. In Fig. 2.6 the broken lines indicate the limits of addition of neutron (above) and protons (below). In general, fission fragments are close to the upper edge of this area . Of the stable nuclei, two with especially low binding energy are D and Be^9:[2]

$$\begin{array}{ll} \mathrm{D} \to n + p & -2.2\,\mathrm{MeV} \\ \mathrm{Be}^9 \xrightarrow{\gamma} n + \mathrm{Be}^8 & -1.6\,\mathrm{MeV} \end{array}$$

[2]@ More appropriately, the following reactions should be written as $\mathrm{D} + \gamma \to n + p$ and $\mathrm{Be}^9 + \gamma \to n + \mathrm{Be}^8$, where the photons have the (positive) energy indicated.

It should be generally truc that the cross section for any process is greater for nuclei which are overstuffed with neutrons.

Chapter 3

The Scattering of Neutrons (Part 1)

3.1 General Considerations

The term "scattering" embraces all types of collision processes, whether or not the outgoing particles are the same as the incoming. Thus for the scattering of neutrons by nuclei we have the following possibilities:

(n, n),	$(n, 2n)$,	
(n, γ),	$(n, np) \equiv (n, d)$	(not observed),
(n, p),	$(n, \text{fission})$,	
(n, α).		

The (n, n) process can take either of two forms. If the neutron energy is the same both before and after the collision, the scattering is said to be elastic. If the neutron energy alters, the scattering is inelastic. All other collisions (i.e. where something other than a neutron leaves the nucleus) are inelastic.

For slow neutrons it is certain that the collision is not just the impact of two small hard spheres. This may be readily be seen by comparing the cross sections calculated in terms of this model with those actually observed. The radius of an average nucleus is round about 10^{-12} cm, so that the classical scattering cross section would be $\sim 10^{-24}$ cm^2. But cases are known of cross sections many orders of magnitude greater than this (e.g. Xe^{135}, with $\sigma \sim 10^{-18}$). Such cross sections are only observed, however, for collisions of slow neutrons; their explanation lies in the fact that, according to wave mechanics, a neutron exerts its influence over a radius approximately equal to its de Broglie wavelength. We have:

$$\lambda = \frac{h}{mv}; \qquad \sigma \leq \lambda^2,$$

$$\frac{1}{2} m v^2 = E \times 1.6 \cdot 10^{-12},$$

where E is in eV.

$$m^2v^2 = E\,m \times 3.2 \cdot 10^{-12};$$

$$\lambda = \frac{h}{mv} = \frac{6.6 \cdot 10^{-27}}{\sqrt{E\,m \times 3.2 \cdot 10^{-12}}} = \frac{6.6 \cdot 10^{-27}}{\sqrt{E}\,\sqrt{3.2 \cdot 10^{-12} \cdot 1.57 \cdot 10^{-24}}}$$

$$= \frac{2.86 \cdot 10^{-9}}{\sqrt{E}}\ \text{cm}.$$

The practical limits of E are about 10^{-3} eV to a few MeV. The following table shows how σ may assume very large values at sufficiently low neutron energies:

E(eV)	λ	$\lambda^2 \geq \sigma$
0.025	$2 \cdot 10^{-8}$	$4 \cdot 10^{-16}$
1	$3 \cdot 10^{-9}$	$9 \cdot 10^{-18}$
$2 \cdot 10^{6}$	$2 \cdot 10^{-12}$	$4 \cdot 10^{-24}$

In order to determine a total scattering cross section experimentally, one measures the fraction of neutrons removed from a direct beam by a known amount of scatterer. In order to make the geometrical conditions well defined, the source and the detector of neutrons are placed a considerable distance apart compared to their size, and the scatterer is placed about midway between them. The total cross section thus determined is the sum of a number of partial cross sections covering all the possibilities enumerated at the beginning of this chapter. For the detection of neutrons in such experiments one normally uses a method of observing the recoil protons ejected from hydrogenous matter, if the neutrons are fast, or observes the products of a suitable exo-ergic reaction for slow neutrons. The most convenient reactions in this latter case are:

1) $B^{10} + n \to Li^7 + \alpha + 2.5$ MeV

(the Q of the reaction is less than the full energy release - 2.9 MeV - because in over 90% of the reactions the Li^7 is left in an excited state);

2) $Li^6 + n \to H^3 + \alpha + 4.5$ MeV.

Reaction no. 1 has a high cross section; at thermal energies it is 705 barns (1 barn $= 10^{-24}$ cm^2) per atom of the normal isotopic mixture of B.

3.2 Elementary Theoretical Treatment

The scattering process may be considered as a transition from an initial state, comprising an incident neutron and an initial nucleus, to a final state, consisting of a final nucleus and an emergent particle (neutron, heavy particle or quantum[1]). The initial energy state is single, being completely specified at the moment of the collision, but the final state will in general possess a number of energy levels, or perhaps a continuum. There is a certain probability of transition to any one of these final states. The probability is given by the product of the matrix element for the transition by the density of the energy levels. Denoting the matrix element by H, the number of levels by N, and the energy of the level by E, we have

$$\text{Transition probability for unit time} \;=\; H \times \frac{\mathrm{d}N}{\mathrm{d}E}\,.$$

We shall take it that the quantity H is not, in first approximation, a function of E, and shall confine ourselves to the evaluation of $\mathrm{d}N/\mathrm{d}E$.

The Uncertainty Principle asserts that there cannot be more than one state per volume h^3 of phase space. Than the number of states per volume Ω of configuration space is given by

$$\mathrm{d}N = \frac{1}{h^3}\,\Omega\,\mathrm{d}p_x\mathrm{d}p_y\mathrm{d}p_z\,,$$

where the p_i are the components of the total momentum p. Considering the magnitude of p only, regardless of direction, we have

$$\mathrm{d}N = \frac{1}{h^3}\,\Omega\,4\pi p^2\mathrm{d}p\,.$$

If the emergent particle is heavy, of mass M,

$$E = \frac{1}{2}\,Mv^2 = \frac{p^2}{2M}; \qquad \mathrm{d}E = \frac{p\mathrm{d}p}{M}\,.$$

Hence

$$\mathrm{d}N = \frac{1}{h^3}\,\Omega\,4\pi p^2\,\frac{M\mathrm{d}E}{p}\,,$$

and

$$\boxed{\frac{\mathrm{d}N}{\mathrm{d}E} = \text{const.} \times p\,.}$$

If the emergent particle is a quantum, of frequency ν,

$$E = h\nu; \qquad p = \frac{h\nu}{c} = \frac{E}{c}\,.$$

[1]@ That is, a photon.

Thus

$$dN = \frac{1}{h^3}\,\Omega\,4\pi\,\frac{E^2}{c^2}\,\frac{dE}{c},$$

and

$$\boxed{\frac{dN}{dE} = \text{const.} \times \nu^2\,.}$$

The level density for both heavy particles and quanta may be written in the form

$$\boxed{\frac{dN}{dE} = \text{const.} \times \frac{p^2}{v}\,.}$$

The number of transitions per unit time, or the product $h \times \frac{dN}{dE}$, is proportional to the product of incident neutron velocity with the cross section for the process concerned. This product (σv) has the dimension of volumes/s; it may be regarded as the volume of neutron swallowed up per second by the scattering nucleus. We thus have finally, as the general equation to a scattering process,

$$\boxed{\sigma_{(n,\text{particle})} = f(E) \times \frac{1}{v} \cdot \frac{p_{\text{out}}^2}{v_{\text{out}}}\,,}$$

where the form of $f(E)$ $(= H)$ is unspecified. With the aid of this crude formula we can obtain some useful information about the different scattering processes.

3.2.1 *Elastic Scattering*

If the scattering is elastic, v_{in} is equal to v_{out}, so that the cross section formula simplifies to

$$\sigma = f(E) \times M^2\,.$$

It may be noted that M is the reduced mass of the system. Elastic scattering will be the only process so long as the energy of the incoming neutron is less than the energy of the first excited state above the ground state of the system.

3.2.2 *Inelastic (n, n) Scattering*

If the energy of the excited state is W, and if the energy of the incident neutron exceeds this value, then we have

$$\frac{1}{2} M v_{\text{out}}^2 = \frac{1}{2} M v_{\text{in}}^2 - W\,.$$

Putting $W = \frac{1}{2} M v_0^2$, this may be written

$$v_{\text{out}} = \left(v_{\text{in}}^2 - v_0^2\right)^{1/2}\,,$$

showing the existence of a threshold at $v = v_0$. The trend of σ with v is shown in Fig. 3.1.

3.2.3 *Inelastic (n, m) Scattering*

This heading covers the emission of heavy charged particle. If M is the reduced mass of the system, m_0 is the neutron mass, Q is the energy release of the process, then

$$\frac{1}{2} M v_{\text{out}}^2 = \frac{1}{2} m_0 v^2 + Q,$$

whence

$$v_{\text{out}}^2 = \frac{m_0 v^2}{M} + \frac{2Q}{M}; \qquad p_{\text{out}}^2 = M^2 \left(\frac{m_0 v^2}{M} + \frac{2Q}{M}\right),$$

and

$$\sigma = f(E) \times \frac{1}{v} \cdot M^2 \left(\frac{m_0}{M} v^2 + \frac{2}{M} Q\right)^{1/2}.$$

If v is small, and Q is small and positive, it may readily be seen that σ is proportional to $\frac{1}{v}$. If Q is negative, we may put

$$-Q = P = \frac{1}{2} m_0 v_0^2\,,$$

and, just as for inelastic (n, n) scattering, there exists a threshold at a neutron velocity v_0. Above the threshold, we might expect the cross section to vary in the same way as $\sigma_{(n,n)}$. The fact that m is charged leads, however, to the presence of a Gamow factor in $f(E)$, and the curvature of the σ vs. v curve is thereby reversed. This is indicated in Fig. 3.1.

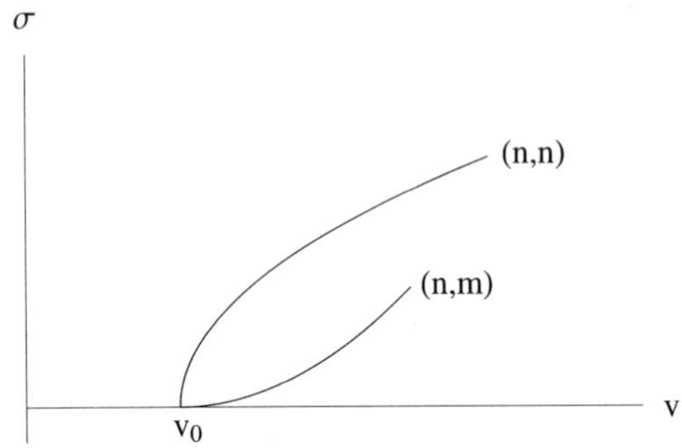

Fig. 3.1

3.2.4 *Inelastic* (n, γ) *Scattering*

The formula for σ in this case is written

$$\sigma = f(E) \times \frac{1}{v} \frac{h^2\nu^2}{c^3} .$$

We know (cfr. p. 11) that this process is exo-ergic, so that the incident neutron is likely to be captured into an excited state. Fig. 3.2 suggests the sort of energy level diagram involved. If W_1, W_2, etc. are the binding energies for the neutron for various energy states in the nucleus, and if the incident neutron velocity is v, then the possible quantum energies are

$$h\nu_i = \frac{1}{2} Mv^2 + W_i .$$

If v is small, $h\nu_i \sim W_i =$ const., so that we could expect σ to follow a $\frac{1}{v}$ law at low neutron energies.

One important point should be noted in connection with low energy collisions, namely that it is the relative velocity of neutron and nucleus which determines the cross section, not the neutron velocity alone. The effect of this may be seen from a simple example. Let us suppose that we have a process whose cross section is given by

$$\sigma = Af(v_{\text{rel}}) ,$$

where v_{rel} denotes the relative velocity of the neutron and the nucleus that it strikes. Moreover, let there be N nuclei per cm^3 of the scattering medium, and let the fraction of them which have the velocity v_{rel} relative to the incident neutron beam be $g(v_{\text{rel}})$. Then the total probability per second (and per neutron) of the occurrence of the process is

$$A\,N \sum f(v_{\text{rel}}) \cdot g(v_{\text{rel}}) \cdot v_{\text{rel}} .$$

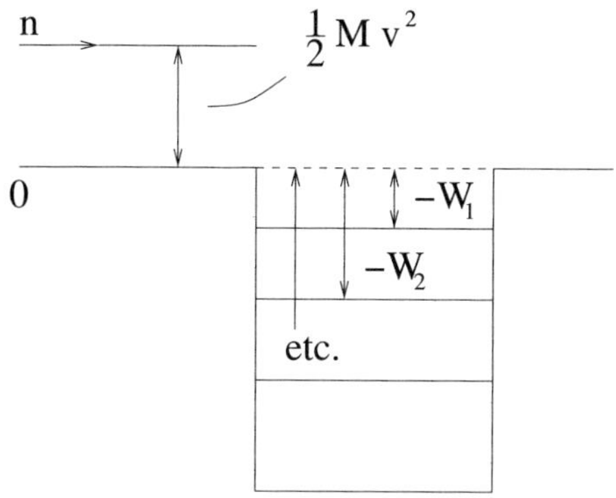

Fig. 3.2

If the nuclei are at rest and the neutron velocity is v_0, then the capture probability is

$$A\,N\,f(v_0)\cdot v_0\,.$$

These two expressions are identical only if

$$f(v)=\frac{1}{v}\,,$$

in which case they both reduce to $A\,N$. Thus the thermal agitations of the nuclei in a scattering medium may have a pronounced effect on the magnitude of a scattering process.

Chapter 4

Resonance: Models of the Nucleus

4.1 Resonance in Nuclear Reactions

We have seen that the (n, γ) reaction is strongly exo-ergic (Chapter 2). And if we except the elementary arguments of the last chapter, we have (cfr. p. 22) that the cross section as a function of energy should follow the $1/v$ law, at least approximately. This, however, only happens at the light end of the Periodic Table. Elsewhere the cross section exhibits sharp and strong resonance peaks, as indicated in Fig. 4.1. This fact throws an important light on the mechanism of the reaction. There are two possible ways of looking at any (n, particle) process. One may suppose that the neutron gives its energy to one particle in the struck nucleus, or else that it gives it to the nucleus as a whole, the incident neutron itself becoming a part of a compound system. On this latter view the re-emission of a neutron competes with alternative processes, and the relative frequency of these alternative depends on the probabilities with which they may severally occur. The fact of resonance has helped to confirm the conclusion, now universally accepted, that a compound nucleus is formed. The argument runs like this.

The shape of a resonance curve, as observed for an (n, γ) process, like that of an atomic spectral line, is like a bell, and can be characterized by its height and by its width at half the peak value. The width is marked as ΔE in Fig. 4.1, and for most γ-rays is about 1 eV. Now by the Uncertainty Principle, this degree of definition in energy is associated with a corresponding definition in time, namely

$$\Delta E \cdot \Delta t \sim h\,.$$

For $\Delta E = 1\text{eV} = 1.6 \cdot 10^{-12}$ ergs, one finds $\Delta t \sim 5 \cdot 10^{-15}$ s. By analogy with atomic spectra, again, we may say that this represents the time it

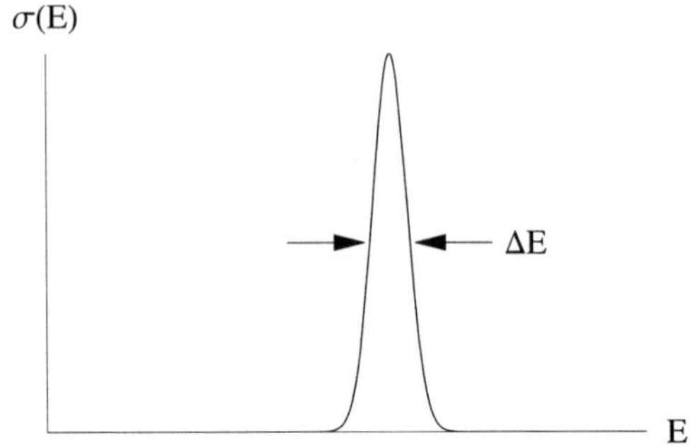

Fig. 4.1

takes for the γ-ray to be emitted, or in other words the lifetime of the nucleus from which it comes. Now this is a very long time indeed compared to the time a neutron spends in traversing a distance equal to a nuclear diameter. Even for a thermal neutron this time is only about $5 \cdot 10^{-18}$ s; for a neutron of 1 MeV it is down to $\sim 10^{-21}$ s. Thus it is pretty certain that a compound state is formed, and that an (n, γ) process in fact takes place in two stages:

$$A + n \rightarrow (A + n)^*$$
$$(A + n)^* \rightarrow (A + n) + \gamma \,,$$

the (*) indicating an excited state.

There is a further question; why should the cross section for the formation of a compound nucleus exhibit the numerous sharp peaks? So far we have merely adduced the existence of the compound state, but we can go a little further, on the basis of the evidence. On the quantum-mechanical view, a nucleus can exist in a number of discrete states, to each of which corresponds a certain energy. Thus the nucleus cannot accept energy in an arbitrary amount, and it is reasonable to suppose that an (n, γ) resonance occurs when the neutron energy is such that the total energy of the system coincides with the energy of one level in the compound nucleus. Now, since the capture of a neutron into a nucleus brings about an excitation of ~ 8 MeV, the observed spacing between the resonance peaks represents the level spacing at this energy above the ground state of the nucleus. A study of numerous medium weight nuclei has shown that, as the incident neutron energy is raised from zero, the first capture resonance is reached on the average at about 10 eV, so we may say that this is a fair value for

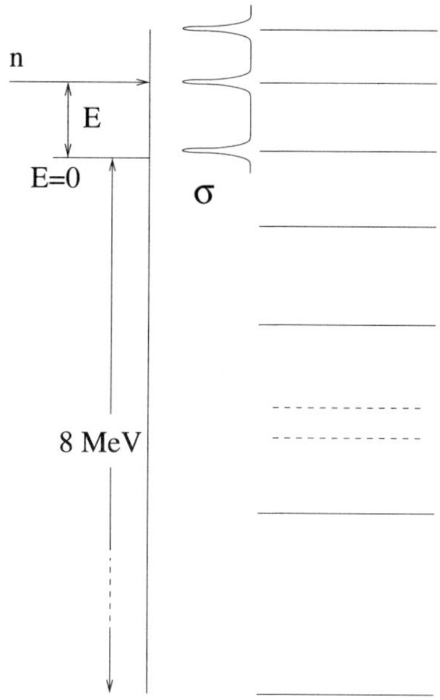

Fig. 4.2

the level spacing for such nuclei. Experiments dealing with the same nuclei near their ground states indicate that the spacing is very much greater for lower excitation - about $10^5 \div 10^6$ eV. The general rule is that, the higher the excitation, the closer are the levels. Fig. 4.2 is an attempt to suggest this scheme and to indicate also the trend of the cross section curve.

4.2 Two Models of the Nucleus

In making a nuclear model we must account for the two observed facts that have been discussed above, viz.

a) relatively stable nuclei are formed in nuclear collisions;

b) there is a rapid decrease of level spacing as the excitation energy increases.

The first fact can be accounted for on any nuclear model. The nucleus

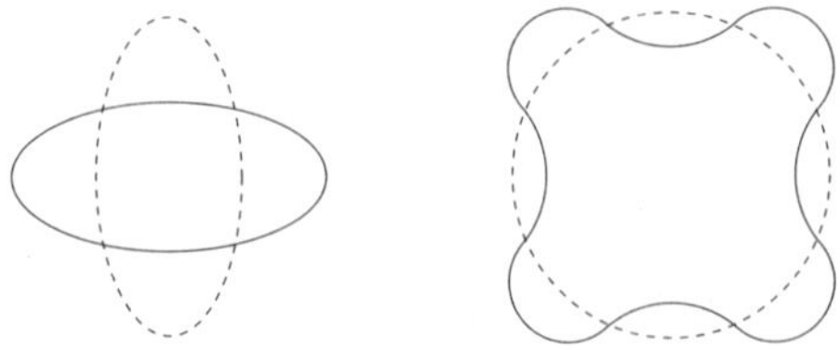

Fig. 4.3

is reckoned to be a mass of closely packed particles bound by short range forces. Interactions of particles with their immediate neighbours are strong. For the extra-nuclear electrons, a collision with an external particle is essentially a two-body problem, but in a nuclear collision the energy is shared amongst all particles in the nucleus. Thus the chance of a single nuclear particle acquiring sufficient energy to escape from the nucleus is small, and the event, being improbable, may not occur for some time after the collision - hence the long lifetime of some compound nuclei.

The problem of level spacing can be dealt with in two models. In the first, one considers the nucleus as a mechanical system of A members (A = mass number), each with 3 degrees of freedom. There will then be $3A$ characteristic frequencies ν_i, associated with energies $h\nu_i$, for the system as a whole. Thus in general the total energy may be written

$$E = a_1 h\nu_i + a_2 h\nu_2 + \cdots + a_{3A} h\nu_{3A} \, .$$

The a_i are integers. If $\nu_1 \neq \nu_2 \neq \nu_3$ etc., one can see that as E increases the number of levels per unit energy also increases. This is quite generally true, although the picture is crude; a system of many degrees of freedom will have levels closer together as the total energy increases.

The picture is refined by the use of Bohr's liquid drop model. Nuclei and drops are alike in being held together by forces of range small compared to the overall dimensions, and in having volumes proportional to their masses. A drops has, of course, numerous modes of oscillation, two of them being sketched in Fig. 4.3. The energy level density has the form $\rho = \mathrm{e}^{-f(E)}$, but $f(E)$ cannot be really defined.

The second nuclear model is based on considering the nucleus as a potential well, with a neutron + proton gas in it. The gas has a temperature T. At $T = 0$ the nucleus is in the ground state, although of course the Pauli exclusion principle demands that the individual particles are moving fast.

As energy is fed in, T increases, but the energy goes as T^2 rather than T:

$$U = \frac{1}{4}\pi^2 \left(\frac{A}{\zeta}\right)\tau^2 ,$$

where U = energy, A = no. of particles, ζ = ground state energy ($\sim$ 19 MeV), $\tau = kT$. If we put $U = a\tau^2$, then the entropy S is given by

$$S = \frac{\partial U}{\partial \tau} = 2a\tau = 2\sqrt{a\,U} = \frac{\pi}{\zeta^{1/2}}\sqrt{A\,U} .$$

One usually writes $S = \log P +$ const., where P is the probability of a given state. Statistically, P is the number of states per unit energy interval. Thus the density of levels is given by

$$\rho = C \cdot \mathrm{e}^{\frac{\pi}{\zeta^{1/2}}\sqrt{A\,U}} .$$

(Note that the expression $\frac{\pi}{\zeta^{1/2}}\sqrt{AU}$ corresponds to $-f(E)$ of the liquid drop model.) Applying the formula to a medium weight nucleus with 8 MeV excitation ($A \sim 100$, $\zeta \sim 19$ MeV), we have, for the number of states per unit energy interval,

$$\rho \sim C \cdot \mathrm{e}^{20} \sim C \cdot 10^8 .$$

Since for zero excitation ($U = 0$) the formula gives a level density of C, it follows that the level spacing at 8 MeV is, according to this model, 10^8 closer than at the ground level. This is much too high a ratio.

Both these pictures of the nucleus indicate only the trends, therefore, and any approach gives some kind of logarithmic expression for level density. For a more detailed account see Bethe's article (Rev. Mod. Phys. 9, April 1937) §53.

Chapter 5

The Scattering of Neutrons (Part 2)

5.1 The Solution of Schrödinger's Equation

In this section we consider the solution by wave mechanics of a simple problem in nuclear scattering. The nucleus is considered as a centre of force, the force being of short range, so that it ceases to exist beyond a certain distance r_0 from the origin. The actual shape of the nuclear potential then approximates to a square well, as shown in Fig. 5.1. The potential U is negative and constant over most of the nucleus. This corresponds to the facts, as far as we know them, of the interaction between a neutron and a nucleus. The depth of the nuclear potential well is equal to the binding energy, that is about 8 MeV.

If the kinetic energy of the neutron is denoted by E, Schrödinger's equation is

$$\Delta\psi + \frac{8\pi^2 m}{h^2}(E-U)\,\psi = 0,$$

$$\Delta\psi + \frac{2m}{\hbar^2}(E-U)\,\psi = 0.$$

We shall suppose that $\psi = \psi(r)$ only, so that we concern ourselves only with S-wave interaction - for which the angular momentum of the neutron relative to the nucleus is zero. Then we have

$$\psi'' + \frac{2}{r}\,\psi' + \frac{2m}{\hbar^2}(E-U)\,\psi = 0\,,$$

which can be written

$$\frac{\mathrm{d}^2}{\mathrm{d}r^2}(r\psi) + \frac{2m}{\hbar^2}(E-U)(r\psi) = 0\,.$$

Putting $r\psi = u$, $\dfrac{2m}{\hbar^2} = k$, the equation simplifies to

$$u'' + k(E-U)u = 0\,.$$

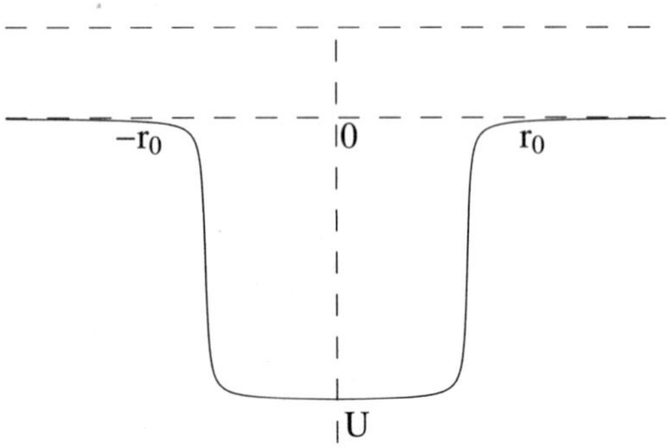

Fig. 5.1

If ψ is to be finite at the origin, $u(0) = 0$.

Some essential features of the solution may be seen at once, by considering the three possibilities for $(E - U)$:

a) $E - U > 0$. The equation may be re-written

$$u'' = -k(E - U)u\,.$$

We see that the curvature of u is everywhere away from the r-axis, so that the solution must approximate to a sinusoidal wave.

b) $E - U < 0$. We may in this case write

$$u'' = +k\,|(E - U)|\,u\,.$$

The curvature is everywhere away from the r-axis.

c) $E - U = 0$. The solution is a straight line of unspecified slope. From these general trends, and from the condition that the wave function and its derivative must everywhere be continuous, we can construct the shape of the solution for a given value of E.

1) $\underline{E = 0}$

The result is shown in Fig. 5.2. From $r = 0$ to $r = r_0$ we get waves, the wavelength depending on the depth of the well at any point. Outside r_0 we have $u = \text{const.} \times r$, i.e. $\psi = \text{const.}$, so that the neutron is no more likely to be found at a given point outside the nucleus than at any other.

2) $\underline{E > 0}$

In this case the solution is in the form of a wave for all r. Outside r_0, $(E-U)$ assumes a smaller value than it has inside the nucleus, which means that the curvature of the wave is reduced. In fact, for $r > r_0$, u is a pure sine curve whose wavelength is equal to the de Broglie wavelength of the neutron of energy E. This is depicted in Fig. 5.3.

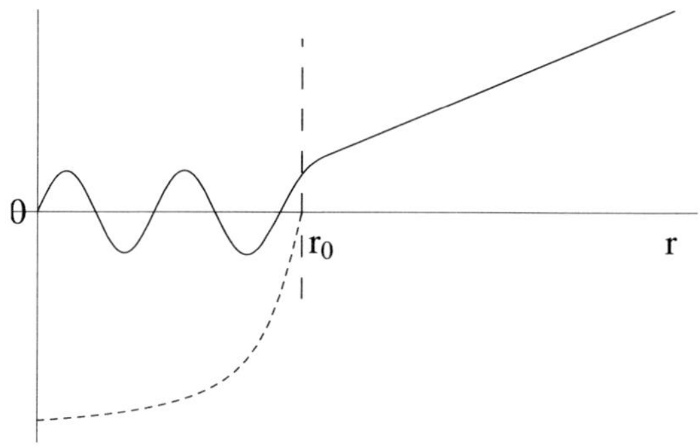

Fig. 5.2

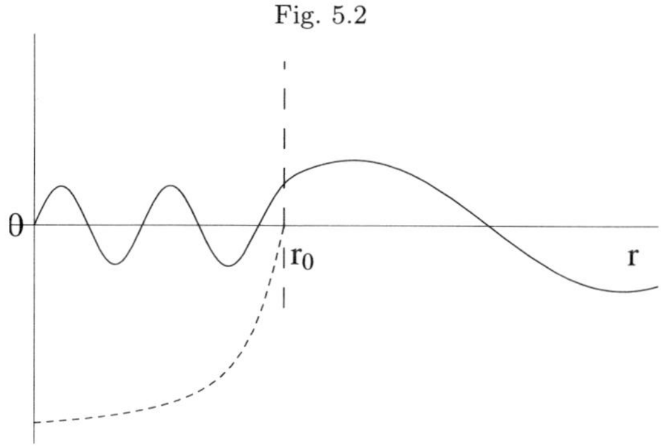

Fig. 5.3

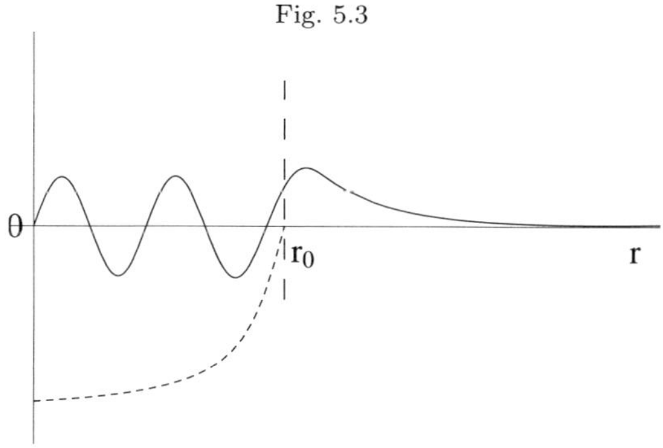

Fig. 5.4

3) $\underline{E < 0}$

The solution of this problem gives us the stationary energy states for a neutron bound to the nucleus. For $r > r_0$, we have

$$u'' = \frac{2m}{\hbar^2}(-E)\,u\,,$$

with the solutions

$$u = \mathrm{e}^{\pm\left[\frac{2m}{\hbar^2}(-E)\right]^{\frac{1}{2}} r}.$$

The positive exponent is inadmissible, so we get $u \to 0$ for $r \to \infty$. This exponential curve joins smoothly to the wave form of the curve inside r_0 (Fig. 5.4); the condition that it should do so defines the possible values of E.

5.2 The Scattering Cross Section

We can only have a scattering cross section if a neutron is approaching a nucleus, i.e. if $E > 0$. We are therefore interested in case 2 of the last section. We have seen that the form of the wave function outside the nucleus is sinusoidal with the de Broglie wavelength, and that this wave must join up smoothly with the waves inside the nucleus. This limitation demands that in general the wave outside r_0 should have a certain definite phase shift respect to 0 (see Fig. 5.5) so that it is written

$$u = A\,\sin\left(\sqrt{\frac{2mE}{\hbar^2}}\,r + \beta\right).$$

Now the scattering cross section is directly related to β, by the equation

$$\sigma_{\mathrm{sc}} = \frac{4\pi^2\hbar^2}{p^2}\,\sin^2\beta + \text{other terms.}$$

This result follows from a full treatment of the problem of wave mechanics - see, e.g., Mott & Massey, "The Theory of Atomic Collisions".[1] In the equation, p refers to the momentum of the neutron outside the nucleus. The first term of the equation represents the contribution to the scattering cross section of the S-waves alone. So long as we limit ourselves to consider slowish neutrons, this is the only term that counts. For a neutron with

[1]@ The complete reference is: N.F. Mott and H.S.W. Massey, The Theory of Atomic Collisions (Clarendon Press, Oxford, 1933).

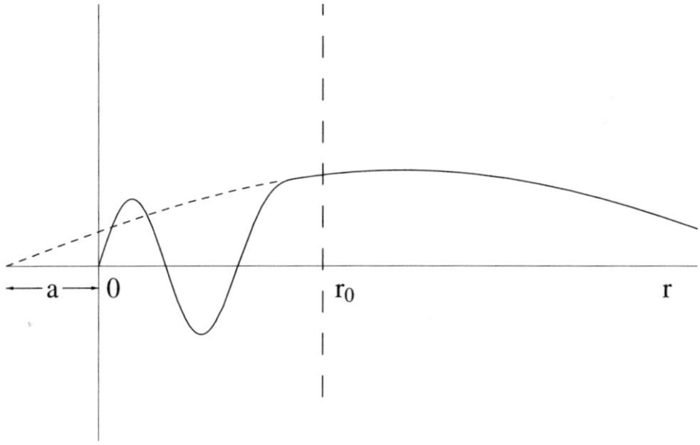

Fig. 5.5

quantum number ℓ to characterize its angular momentum, the collision parameter b is given by

$$b = \frac{h}{mv}\,\ell\,.$$

If $b \gg r_0$, scattering of this order in unimportant. It can be verified that the simple S-wave scattering formula should be good enough for neutron energies up to 10^5 eV at least.

If the neutron energy is very low, the cross section can be very simply expressed. If the intersection of the extra-nuclear wave with the r-axis (Fig. 5.5) is at a distance "a" from O, and if the wavelength is λ, then

$$\sin\beta \sim \beta = \frac{2\pi a}{\lambda} = \frac{2\pi a\, p}{h} = \frac{a\, p}{\hbar}\,.$$

The cross section formula then reduces to

$$\sigma_{\rm sc} = 4\pi a^2\,.$$

5.3 Neutron-Proton scattering

If bound states of the system neutron+nucleus exist, the formula

$$\sigma_{\rm sc} = \frac{4\pi^2\hbar^2}{p^2}\,\sin^2\beta$$

reduces to

$$\sigma_{\rm sc} = \frac{4\pi^2\hbar^2}{M(\epsilon + E/2)}\,,$$

where E is the kinetic energy of the incident neutron and ϵ is its binding energy in the compound nucleus. This holds equally for virtual states, but ϵ is then negative and one must replace it by $|\epsilon|$ in the formula.

In the case of (n, p) scattering, the compound nucleus is of course the deuteron, and this can exist in either of two states - triplet, with neutron and proton spins parallel, or singlet, with spins anti-parallel. The triplet has a higher value of ϵ than the singlet, and is the normal state of the deuteron. In fact the binding energy for the singles state is so small that it is hard to say whether it is positive or negative, We can, however, write the total scattering cross section for either case in the form

$$\sigma_{\rm sc} = \frac{4\pi^2\hbar^2}{m}\left[\frac{3}{4}\cdot\frac{1}{\epsilon_3 + E/2} + \frac{1}{4}\cdot\frac{1}{|\epsilon_1| + E/2}\right].$$

The states are here weighted according to the number of possible spin orientations ($= 2s + 1$).

One can find whether the singlet deuteron state is real or virtual by scattering neutrons from ortho- and para-hydrogen. In the latter, the spins of the two protons are anti-parallel, so that both the singlet and the triplet state are formed when a neutron interacts with the molecule. Now there exists a phase difference of nearly 180° between the scattered waves from real and virtual states so that, if the singlet state of the deuteron were virtual, destructive interference should occur between the two outgoing neutron waves from the collision of a neutron with p-H_2. This would mean a small scattering cross section. For the collision with o-H_2, on the other hand, the two atoms necessarily scatter in phase, since one gets two singlet states or two triplet states according to the relative orientations of the neutron and proton spins. The cross section for scattering from orthohydrogen has in fact been observed to be greater than from para-hydrogen, and the singlet state is therefore believed to be virtual, with a binding energy of about -100 keV.

5.4 The Breit-Wigner Formula

We have seen (Chapter 3) how the cross sections of nuclei for various processes exhibit resonances. The probability of any such process as a function of the energy of the incident particle is given by a formula similar to the dispersion formula for the scattering of light. When the resonance levels of the compound nucleus are far apart, the cross section in the neighbourhood of any one of them for an (a, b) process is given by

$$\sigma_{(a,b)} = \pi \lambda_a^2 \frac{\Gamma_a \Gamma_b}{(E_a - E_r)^2 + \Gamma^2/4}.$$

This is the famous formula that was first derived by Breit and Wigner. λ_a and E_a are the wavelength and energy respectively of the incident particle. E_r is the resonance energy and Γ is the width of the resonance peak at half the maximum height. The above expression should really be multiplied by a factor depending on the spins of the initial particles and the compound nucleus, but we shall assume that these are incorporated in the Γs. Γ_a and Γ_b are the partial widths of the resonance peaks associated with the emission of 'a' and 'b' particles. Their exact form is complicated, but since the probability of emission of a particle b with momentum p is proportional to $\frac{p_b^2(\text{out})}{v(\text{out})}$, so is Γ_b. The relative probability that the outgoing particle is a is given by Γ_a/Γ, where $\sum_{i=a,b,\ldots} \Gamma_i = \Gamma$. Also, $\Gamma_i \times \tau_i \sim \hbar$, where τ_i is the average lifetime of the compound nucleus against the emission of a given type i of particle. Thus the probability that no particle at all has been emitted up to time t is $\prod_i \mathrm{e}^{-t/\tau_i}$, which can alternatively be written $\mathrm{e}^{-t\sum_i 1/\tau_i}$. We see from this that the average life of the nucleus against all disintegrations is $\tau = \left[\sum_i 1/\tau_i\right]^{-1}$. Note that for τ, just as for the partial lifetimes, we may write $\Gamma \times \tau \sim \hbar$.

The Breit-Wigner formula, as given above for one level, can be applied to any nuclear process in which the resonance levels are fairly widely spaced. Consider, for example, an (n, γ) process - the resonance capture in Indium. We know $E_r = 1.44$ eV, $\sigma(n, \gamma)_{\text{res}} = 26000$ barns, $\Gamma = 0.05$ eV. What is the number of neutrons for each γ-ray emitted, given that neutron and γ-ray emission are the only processes involved, and that there are more γ-rays than neutrons? The experimental data tell us

$$\text{a)}\ \Gamma = 0.05 = \Gamma_\gamma + \Gamma_n\,,$$
$$\text{b)}\ \Gamma_\gamma > \Gamma_n\,.$$

Moreover, at resonance we have

$$\sigma_{\text{res}} = 4\pi \lambda_{n,\text{res}}^2 \frac{\Gamma_n \Gamma_\gamma}{\Gamma^2}.$$

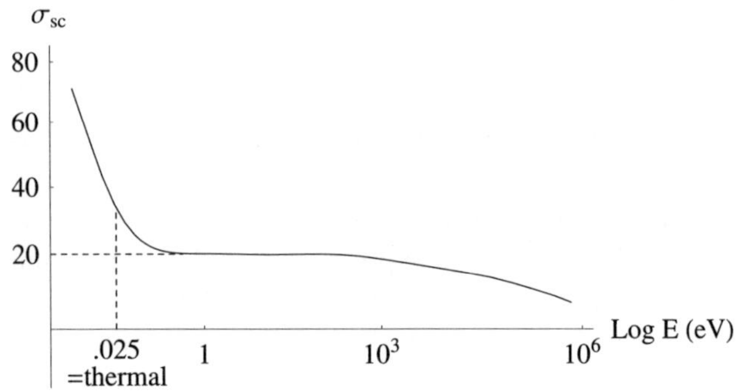

Fig. 5.6

Inserting the numerical values ($\lambda\!\!\!^- = 3.8 \cdot 10^{-10}$ cm) one finds $\Gamma_\gamma \Gamma_n = 0.36 \cdot 10^{-4}$. Combining this result with relation (a), we have[2]

$$(\Gamma_\gamma - \Gamma_n)^2 = 25 \cdot 10^{-4} - 1.4 \cdot 10^{-4} = 23.6 \cdot 10^{-4},$$
$$(\Gamma_\gamma - \Gamma_n) = 0.0485.$$

Thus $\Gamma_n \sim 0.00080$ eV, $\Gamma_\gamma = 0.049$ eV, $\Gamma_n/\Gamma_\gamma \sim 0.015$.

For the sake of interest, we append data on two other elements which have strong capture resonances and are therefore good detectors for slow neutrons:

For Au, $E_r = 4.8$ eV, $\sigma_{\text{res}} \sim 26000$ barns, $\Gamma = 0.07$ eV.
This gives $\Gamma_n = 0.0035$, $\Gamma_\gamma \sim 0.07$ eV.

For Ag, $E_r = 5.5$ eV, $\sigma_{\text{res}} = 7200$ barns, $\Gamma = 0.19$ eV.

5.5 The Effect of Chemical Binding on Scattering

From the Breit-Wigner formula, the elastic scattering cross section for low energy neutrons becomes $\sim \pi \lambda\!\!\!^-_n{}^2 \dfrac{\Gamma_n^2}{E_r^2}$. Now $\lambda\!\!\!^-_n$ goes as $1/v_n$, and Γ_n goes as $\dfrac{p_{\text{out}}^2}{v_{\text{out}}}$ or as v_n (since $v_{\text{out}} = v_n$). Thus σ should be independent of energy, if

[2]@ That is: $(\Gamma_\gamma - \Gamma_n)^2 = (\Gamma_\gamma + \Gamma_n)^2 - 4\Gamma_\gamma \Gamma_n$.

E_n is well below the first resonance energy. For paraffin the cross section does level out at about 20 barns, but at very low velocity there is a sudden increase (See Fig. 5.6). This is explained as follows. Since $\Gamma_n \propto m^2$, σ must go as m^2. In the application of the Breit-Wigner formula, m is the reduced mass and the velocities are relative velocities. Thus, for neutrons in hydrogen, $m = 1/2$ [3] unless the protons are bound in the scatterer. At very low velocities, however, the proton is effectively bound i.e. its mass becomes infinity instead of 1, and the reduced mass of the scattering system becomes 1. The cross section varying as the square of the mass, an increase by a factor 4 is to be expected at the neutron energy from above to below the binding energy of protons in paraffin. If measurements are made with chilled paraffin (say cooled to liquid air temperature) so as to reduce thermal agitation and hence the relative velocity of neutrons and protons, the cross section does in fact approach the value of 80 barns.

5.6 Scattering Cross Section for Other Elements

The scattering law for hydrogen has a particularly simple form. For the scattering of neutrons by more complex nuclei one might expect resonances and so forth to upset any smooth variation of the cross section with energy. This expectation is fulfilled in practice, as Figs. 5.7, 5.8 and 5.9 certify.

[3]@ That is, the reduced mass is half the nucleon mass.

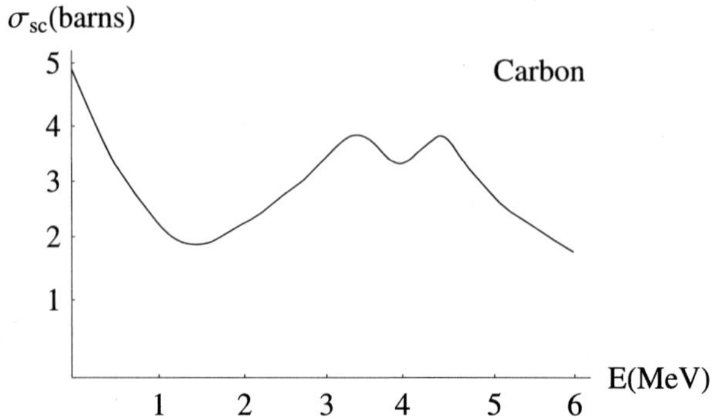

Fig. 5.7

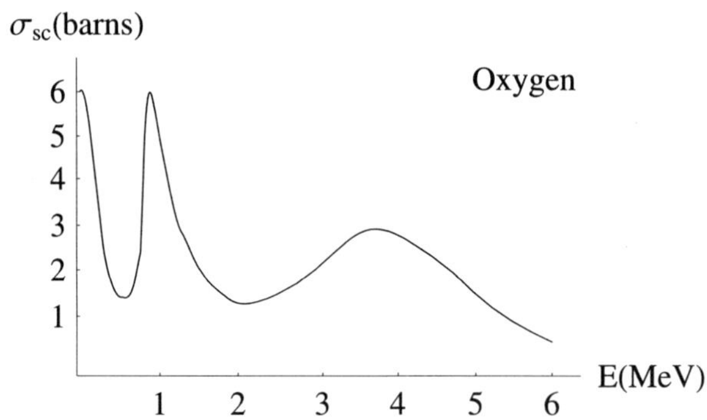

Fig. 5.8

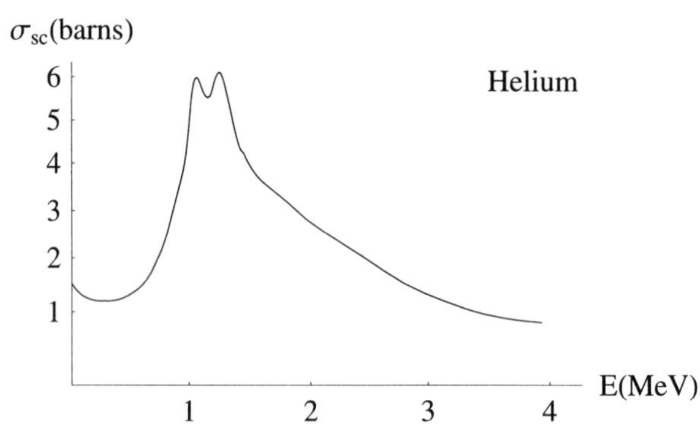

Fig. 5.9

Chapter 6

Slow Neutrons as Waves

6.1 Introduction

When neutrons are slow, their wavelengths are comparable with interatomic spacings in matter, for example:[1]

$$n\,\lambda_{1\,\mathrm{eV}} = 0.286 \cdot 10^{-8}\,\mathrm{cm},$$
$$n\,\lambda_{1/40\,\mathrm{eV}} = 1.82 \cdot 10^{-8}\,\mathrm{cm}.$$

In this circumstance, interference effects are to be expected and are in fact observed. Thus the neutron scattering produced by a given sample of graphite might exhibit the sort of variation with neutron energy that is shown in Fig. 6.1. Since we are here dealing with a light element there is no question of this being a resonance effect.

If one takes a crystal of a substance and allows a slow neutron beam to fall upon it, then it is possible (with, for example, a BF_3 chamber) to confirm that the Bragg relation is obeyed (see Fig. 6.2):

$$n\,\lambda = 2a\sin\theta\,.$$

For a first order reflection, one has

$$\lambda = 2a\sin\theta = \frac{h}{mv}\,; \qquad v = \frac{h}{2ma}\cdot\frac{1}{\sin\theta}\,.$$

Observation at a given θ selects a narrow band of the Maxwell distribution (assuming the neutrons to be thermalized), as indicated in Fig. 6.3. As one simultaneously increases the angle θ of the crystal and the angle 2θ of the counter, so one detects slower and slower neutrons. The correctness of this view can be checked by inserting a boron $1/v$ absorber between crystal and counter and observing the attenuation of the reflected neutron beam as a function of θ.

[1]@ Here n is the order number of the neutron reflection; see below.

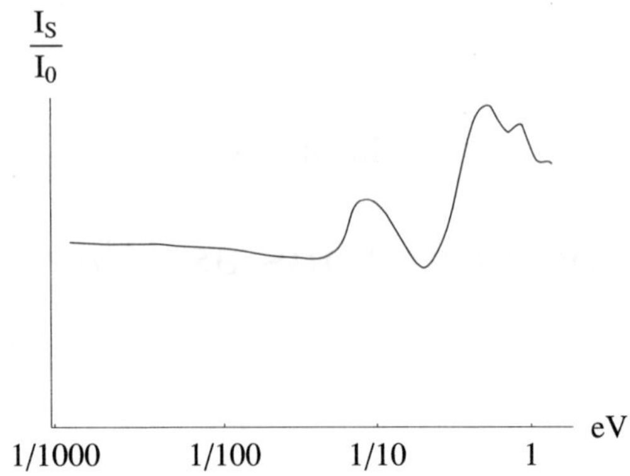

Fig. 6.1

The existence of this Bragg scattering makes it difficult to determine any purely nuclear cross section for thermal neutrons. The percentage loss of a thermal neutron beam as it traverses a given mass of scatterer depends on the crystal structure. This micro-crystalline forms of substances may exhibit a larger cross section per atom than large crystals of those same substances. This is because each micro-crystal, being randomly oriented, may remove a different band from the Maxwell distribution of the beam, with the result that neutrons of all velocities are attenuated, whereas a single crystal may entirely remove those few velocity groups for which the Bragg condition is satisfied, and leave the rest of the spectrum untouched. (We are supposing the measurement made with good geometry).

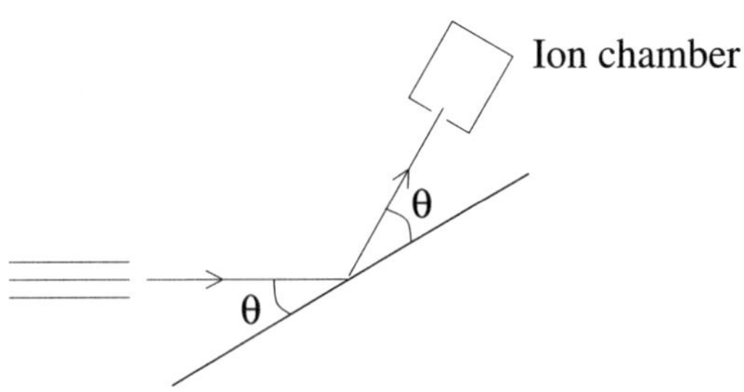

Fig. 6.2

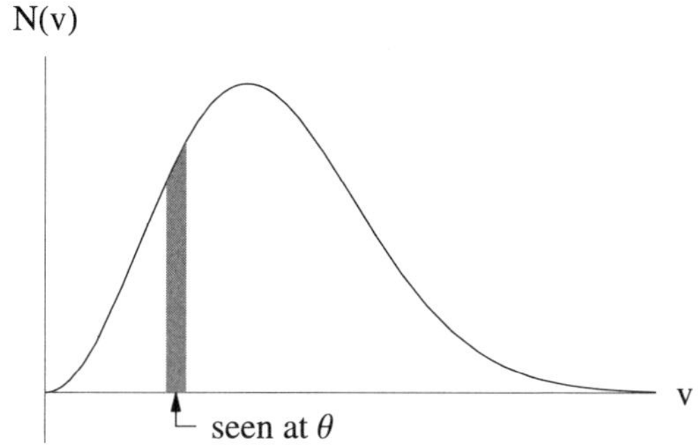

Fig. 6.3

6.2 Isotopic Effects

Although the Bragg scattering of neutrons can be compared with that of X-rays up to a point, the two are not precisely similar. This is because X-ray scattering is purely electronic, and is sensitive only to the nuclear charge of the atoms. With neutrons, one must expect to find variations according to the particular isotope of a given element that is considered as scatterer.

Suppose, for example, an element with two equally abundant isotopes, whose cross sections are σ_1 and σ_2. The amplitude of the scattered wave is proportional to the square root of the cross section, and the waves from two isotopes may be in or out of phase. This can be seen from the results of the last chapter (pp. 34-35) where we have

$$\sigma = 4\pi a^2,$$

whence

$$a = \left(\frac{\sigma}{4\pi}\right)^{1/2},$$

and may be a positive or a negative distance. For the scattered wave we may write

$$A = A_1 \sin kx \pm A_2 \sin kx\,,$$

where the quantities A are amplitudes, with $A_1 = \text{const.} \times |\sqrt{\sigma_1}|$, $A_2 = \text{const.} \times |\sqrt{\sigma_2}|$. The total scattered intensity is thus given by

$$I_{\text{sc}} = \text{const.}\ (\sigma_1 + \sigma_2 \pm 2\sqrt{\sigma_1\sigma_2})\,,$$

and may be seen to consist of coherent and incoherent contributions, the latter not being subject to interference.

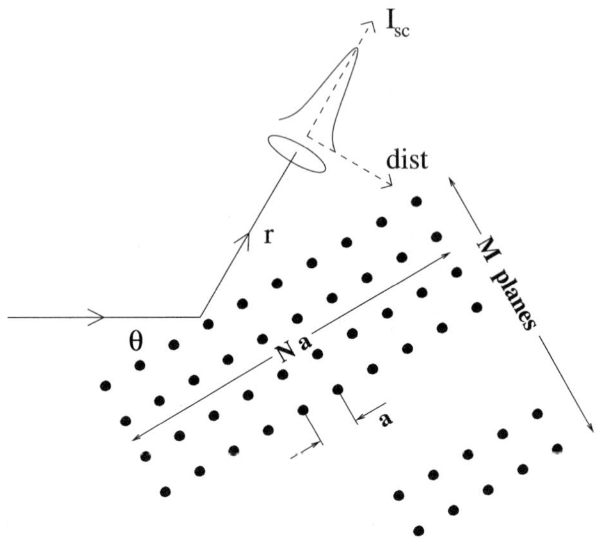

Fig. 6.4

6.3 Penetration by Thermal Neutrons

Suppose we have a neutron beam falling upon a crystal, and consider what happens to those neutrons for which the Bragg condition is satisfied. To fix our ideas, let us take it that scattering is confined to a limited region of the crystal, containing M planes of $N \times N$ atoms (cfr. Fig. 6.4). The number of atoms in the selected section is $M\,N^2$. Let I be the primary beam intensity, and let σ be the scattering cross section. We shall ensure that all the nuclei are identical and that the scattering is isotropic.

For a single atom, the scattered intensity is $I\,\sigma$. Thus the scattered intensity at a distance r from the crystal (Intensity = neutrons/cm^2/s) is given by

$$\mathrm{d}I_{\mathrm{sc}} = \frac{I\,\sigma}{4\pi r^2}\,.$$

The scattered amplitude is proportional to the square root of this:

$$\mathrm{d}A_{\mathrm{sc}} \propto \frac{\sqrt{I}\,\sqrt{\sigma}}{\sqrt{4\pi}r} \sim \frac{\sqrt{I}\,\sqrt{\sigma}}{r}\,.$$

At the Bragg angle, the amplitudes add, so that for the total scattered amplitude we have:

$$A_{\mathrm{sc}} \propto N^2\,M \cdot \frac{\sqrt{I}\,\sqrt{\sigma}}{r}\,.$$

The intensity at a distance r from the crystal in the direction of the reflected maximum is therefore

$$I_{\mathrm{sc}} = \frac{I\,\sigma}{r^2} \cdot N^4\,M^2\,.$$

Now the theory of the diffraction grating gives the well-known result that the angular width of the first order peak at wavelength λ, if the grating has N lines of spacing a, is $\sim \lambda/Na$, or approximately $1/N$ for any reasonable value of θ in Fig. 6.4. Thus the area of the patch at r through which the high intensity coherently scattered beam passes is $\sim \left(\frac{r}{N}\right)^2$. Combining this result with the above expression for I_{sc}, we find that the number of neutrons removed per second by Bragg scattering is given by

$$\boxed{S = \frac{r^2}{N^2} \cdot \frac{I\,\sigma}{r^2} \cdot N^4 M^2 = I\,\sigma\,M^2 N^2\,.}$$

Now, if specular reflection took place, the number of neutrons scattered per second would be

$$S_{\mathrm{r}} = I\,N^2 a^2,$$

and S cannot possibly exceed this value. This fact establishes an upper limit M_0 for M, corresponding to the fact that no neutrons fulfilling the Bragg condition can penetrate deeper into the crystal than a distance $M_0 a$. We have

$$I\,\sigma\,N^2 M_0^2 = I\,N^2 a^2,$$

giving

$$\boxed{M_0 = \frac{a}{\sqrt{\sigma}}\,.}$$

Our calculation is of course very crude, for we have assumed an attenuation of the beam as it penetrates the crystal (until it becomes suddenly zero!), whereas in fact the scattered amplitude must become steadily less for successive planes of atoms. Nevertheless our value for M_0 must approximate to the half-value thickness for the removal by Bragg scattering of neutrons from a beam. Inserting numerical values for a and σ, one finds that the depth for extinction of a beam is $\sim 10^{-5}$ to 10^{-4} cm.

Thus if we have a thermal neutron beam, with Maxwellian distribution, striking a large perfect crystal, all the components suffering Bragg reflections will be removed in the first fraction of a mm. To account for the total attenuation of the beam, one must take into account the following additional factors:

i) The incoherent scattering from isotopes will remove part of all the other components.

ii) If the nuclei of the crystal have spins, we again get incoherent scattering because of the random distributions of the spins.

iii) Thermal agitations will further smudge out the scattering.

iv) The crystal is bound to have a certain amount of absorption.

As we have mentioned before (p. 42), a micro-crystalline aggregate has the chance of removing a Bragg reflection band from all parts of the Maxwell distributions. The finer the grain of the micro-crystalline structure, the relatively more important is the Bragg contribution at the first few planes of each micro-crystal.

6.4 The Production of Very Slow Neutrons

Let us write the Bragg relation in the form

$$\lambda = \frac{2a \sin\theta}{n},$$

n being the order of the reflection. Since we have $\sin\theta \leq 1$, $n \geq 1$, it follows that we can only have Bragg reflection at all so long as

$$\lambda \leq 2a\,.$$

This suggests a means of obtaining very slow neutrons, for it means that they will pass through a micro-crystalline aggregate with less absorption than the slightly faster ones. A suitable material for producing a separation by this means is graphite. This contains 99% of C^{12}, which has no spin. Moreover, for graphite, $2a = 6.69 \cdot 10^{-8}$ cm. The energy corresponding to $\lambda = 2a$ is (cfr. formula on p. 18)

$$E = \left(\frac{0.286}{6.69}\right)^2 = 0.0018\,\text{eV},$$

more than ten times less than thermal (0.025 eV). Passing a thermal neutron beam (300°K) through a block of graphite, say a foot long, does filter it very effectively and gives an emergent beam of about 1/15 kT energy. The mean free path in graphite of the incident beam is $\sim$ 2.6 cm, and of the emergent beam $\sim$ 20 cm.

6.5 Reflection and Refraction

X-rays have a refractive index $n \sim 1 - 0.00001$ for material media, and total reflection is possible at very small glancing angles. The general process of refraction is a combination of the scattering by separate atomic centres, leading to a change of phase, and hence to the existence of an index of refraction.

Neutrons similarly undergo a change of phase, but because the scattering is nuclear instead of atomic, and the cross sections involved are much smaller than for X-rays, the value of n differ even less from unity. We have also the possibility of reversed phases, so that we may write

$$(n - 1)_{\text{neutron}} \sim \pm 10^{-5} \div 10^{-6}.$$

Thus some substances will give total internal reflection and some will not.

The fact of neutron refraction means that one could, in theory, focus neutron beams with suitable lenses. They would, however, need to have enormously high curvature (bi-convex or bi-concave) to give reasonable focal lengths.

Chapter 7

The Slowing Down of Neutrons

7.1 Introduction

In general, when a neutron undergoes an elastic collision, it suffers a loss of energy, and this loss can be calculated with the aid of simple mechanics. The energy after collision is a certain fraction of the initial energy, the magnitude of the fraction depending on the change of direction suffered by the neutron. If one knows the angular distribution of scattering, one can calculate the average fractional energy loss per collision.

The slowing down of neutrons in hydrogenous media is of special importance. This is because neutron-proton collisions provide the most efficient means of bringing neutrons down from high to thermal energies, for two reasons:

1) The $n - p$ scattering cross section is large.

2) The equality of neutron and proton masses leads to the highest possible value for the average energy loss per collision; if the neutron energy is E, it becomes on the average $\frac{1}{2}E$. Thus, if we have n successive "average" collisions, the neutron energy is reduced by a factor 2^n. It is easy to see from this that the process of slowing down is quite rapid. Suppose, for example, that we wish to slow a neutron from 1 MeV down to thermal energy (1/40 eV). This represents a factor $4 \cdot 10^7$. Equating this to 2^n, one finds $n \sim 25$. As will be seen later, the true average value of n, taking into account the facts that the fractional energy loss is not always $\frac{1}{2}$, but may be anything between 0 and 1, and that the significant quantity is $\log E$ rather than E, is somewhat less than this. The actual time taken for a 1 MeV neutron to become thermal through collisions in a dense hydrogenous medium (e.g. paraffin wax) is about 1 millisecond.

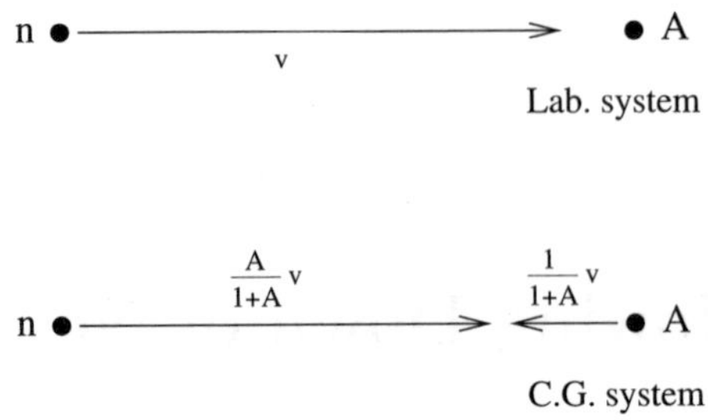

Fig. 7.1

7.2 The Energy Loss in One Collision

Let us consider the general elastic collision between a neutron of mass 1 and a nucleus of mass A, assuming isotropic scattering in the centre of gravity system and supposing A to be initially at rest. Let the initial neutron velocity be v. The first step is to move into the centre of gravity co-ordinate system (see Fig. 7.1) in which the velocity of the neutron is $\dfrac{A}{1+A}v$ and that of A is $\dfrac{1}{1+A}v$. The effect of the collision is to change the directions of the velocities without affecting their magnitudes. This is shown in Fig. 7.2, where the neutron velocity makes an angle θ with the original direction. Now, in order to find the magnitude and direction of the

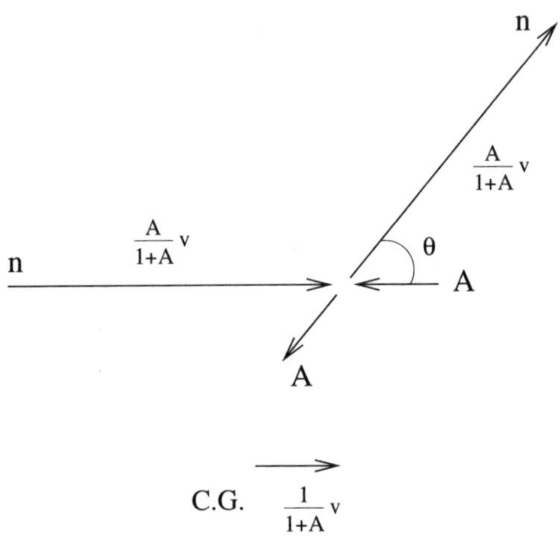

Fig. 7.2

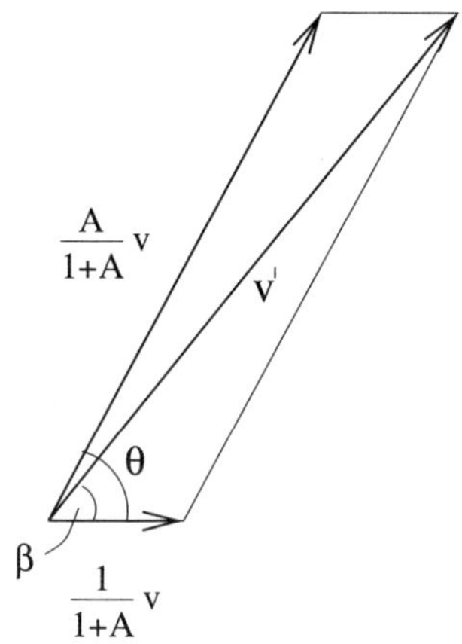

Fig. 7.3

neutron velocity as observed in the laboratory system after the collision, we must add to the neutron the velocity of the centre of gravity, which we subtracted from it in the first place. This is illustrated in Fig. 7.3, where the final neutron velocity is shown as (v', β).

We have:

$$v' = \frac{v}{1+A}\left[1 + A^2 + 2A\cos\theta\right]^2$$

$$E' = E \cdot \frac{1 + 2A\cos\theta + A^2}{(1+A)^2}$$

$$\cos\beta = \frac{1 + A\cos\theta}{(1 + 2A\cos\theta + A^2)^2}$$

$$E'_{\text{max}} = E$$

$$E'_{\text{min}} = E \cdot \left(\frac{A-1}{A+1}\right)^2 \sim \begin{cases} E \cdot \left(1 - \frac{4}{A}\right) & \text{for } A \text{ large,} \\ 0 & \text{for } A \approx 1. \end{cases}$$

For a given A, the probability of a given E' is uniform between E'_{min} and E'_{max}, as long as all θ are equally probable. This probability distribution is shown in Fig. 7.4; the proof of the result follows.

The element of solid angle contained between the angle of scattering θ and $\theta + \mathrm{d}\theta$ is $2\pi \sin\theta \, \mathrm{d}\theta$. The total solid angle being 4π, the probability for scattering between θ and $\theta + \mathrm{d}\theta$ is

$$\mathrm{d}p = \frac{1}{2} \sin\theta \, \mathrm{d}\theta \, .$$

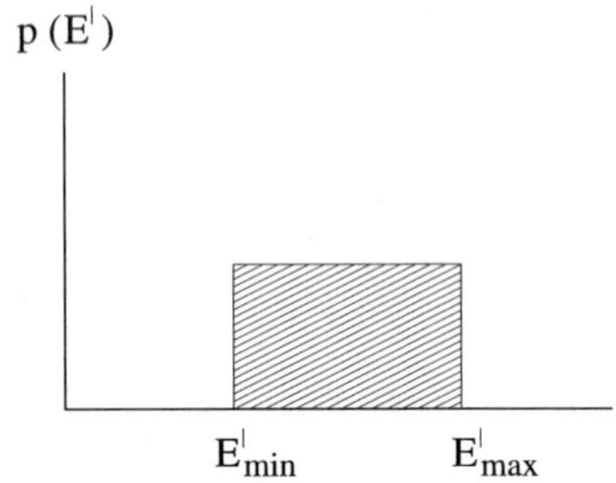

Fig. 7.4

Now we have also

$$dE' = \frac{-2AE}{(1+A)^2} \sin\theta \, d\theta \, .$$

It follows that[1]

$$\boxed{dp = -\frac{(A+1)^2}{4AE} \cdot dE' \, ,}$$

which proves the result. Note also that

$$E'_{\rm max} - E'_{\rm min} = E \cdot \frac{(A+1)^2 - (A-1)^2}{(A+1)^2} = \frac{4AE}{(A+1)^2}$$

or

$$\int_{E'_{\rm min}}^{E'_{\rm max}} dp = 1 \, ,$$

as required.

7.3 Many Collisions

We shall here derive the true value for the average energy loss in a number of collisions.

Put

$$\epsilon = \log_e E, \qquad d\epsilon = \frac{1}{E} \, dE \, .$$

[1]@ Note that the following minus sign is wrong; this mistake does not appear in the Halpern notes.

Put also

$$\xi = \overline{\log_e \frac{E}{E'}}\,,$$

where E and E' are the neutron energies before and after any single collision. Then if the neutron energy is reduced from E_1 to E_2 in n collisions, we have the identity:

$$\Delta\epsilon = \log E_1 - \log E_2 = n\,\xi,$$

i.e.

$$E_2 = E_1\,\mathrm{e}^{-n\xi}\,.$$

Assuming isotropic scattering as before, we have:

$$\begin{aligned}
\xi &= \int_{E'_{\min}}^{E'_{\max}} \log \frac{E}{E'} \cdot \mathrm{d}p(E') \\
&= \int_{\left(\frac{A-1}{A+1}\right)^2 E}^{E} \log \frac{E}{E'} \cdot \frac{(A+1)^2}{4A} \cdot \frac{\mathrm{d}E'}{E} \\
&= \frac{(A+1)^2}{4AE} \Big[E' \log E - E' \log E' + E' \Big]_{\left(\frac{A-1}{A+1}\right)^2 E}^{E}\,,
\end{aligned}$$

whence

$$\boxed{\xi = 1 - \frac{1}{2} A \left(1 - \frac{1}{A}\right)^2 \log_e \left(\frac{A+1}{A-1}\right).}$$

Inserting numerical values, one finds $\xi = 1$ for hydrogen ($A = 1$), and $\xi = 0.158$ for graphite ($A = 12$). For the slowing of neutrons from E_1 to E_2 in hydrogen one therefore has

$$E_2 = E_1\,\mathrm{e}^{-n}\,,$$

and putting $E_1 = 1$ MeV, $E_2 = 1/40$ eV, one finds $n = 17.5$. The corresponding result for graphite is 110 collisions.

Once a neutron has been reduced to thermal energies in a medium, it will just wander about until it is captured, moving short distances in random directions. Diffusions of neutrons from a fast neutron source embedded in a slowing down medium only occurs essentially during the first few mean free paths, while the slowing down is proceeding. This will be considered in more detail in the following sections.

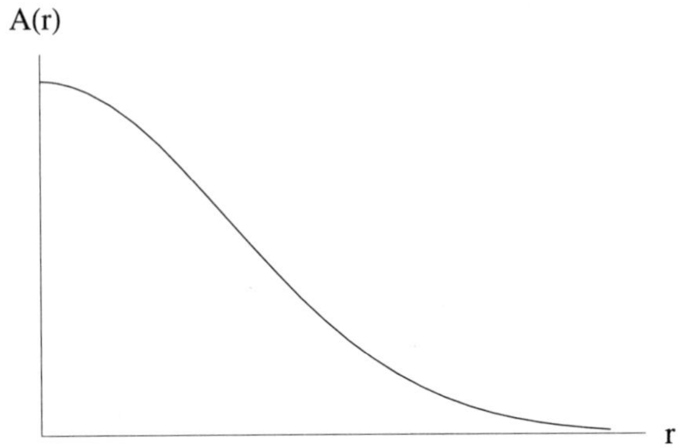

Fig. 7.5

7.4 The Spatial Distribution of Slowed Neutrons

Suppose that we place a Ra-Be neutron source, which is effectively a point, in a large tank of water. Then if we measure the activation of a neutron detector, held for a fixed time in the tank, we find a strong variation of activity with distance from the source, as indicated in Fig. 7.5. It is possible to arrange that the detector shall respond to one neutron energy only; in this case the activity curve represents the radial variation of density of neutrons of a certain definite energy. By using different detectors one can then observe a series of density distributions for various energies. The higher the energy to which the detector responds, the more rapidly does the activity fall off with increasing r.

As an example of a detector, we may consider In, which is much used. The cross section of In as a function of energy is shown in Fig. 7.6a. It has a sharp resonance at 1.44 eV, and follows the $1/v$ law at lower energies. It will thus respond markedly to neutrons of two energies (thermal and resonance). If, however, one sandwiches an indium foil between two thin layers of Cd (1/2 mm is enough), it will respond only to the resonance neutrons. This is because Cd absorbs very strongly indeed all neutrons below about 1 eV (Fig. 7.6b). Thus, by measuring the activity of an In foil with and without Cd backing, one can find the radial variations of 1.44 eV and thermal neutron densities separately. Many other detectors can be used in the same way, e.g. Rh, which responds at 1 eV. For studying higher energy neutrons, one may use iodine detectors, which respond to a rather

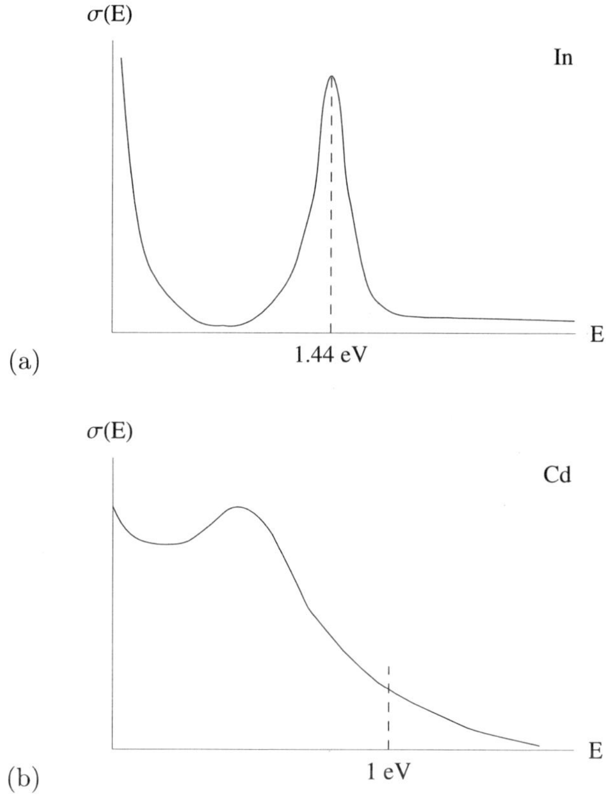

Fig. 7.6

broad band of neutron energies, $\sim 60 \div 70$ eV.

Fig. 7.7 indicates the density distributions for three different neutron energies. It will be seen that they are essentially similar but have different half-widths.

The density $\rho(r)$ of neutrons of a given energy at r is of course proportional to the detector activity $A(r)$. The number of neutrons between r and $r + \mathrm{d}r$ is therefore proportional to $4\pi r^2 \mathrm{d}r \cdot A(r)$. Assuming no neutron capture in the medium above the detector energy, every neutron leaving the source passes through this energy. Consequently, the area under the curve showing $r^2 A(r)$ as a function of r is proportional to the strength of the neutron source. This curve (Fig. 7.8) has a maximum at a certain value of r. There is another quantity of interest to be derived from the curve,

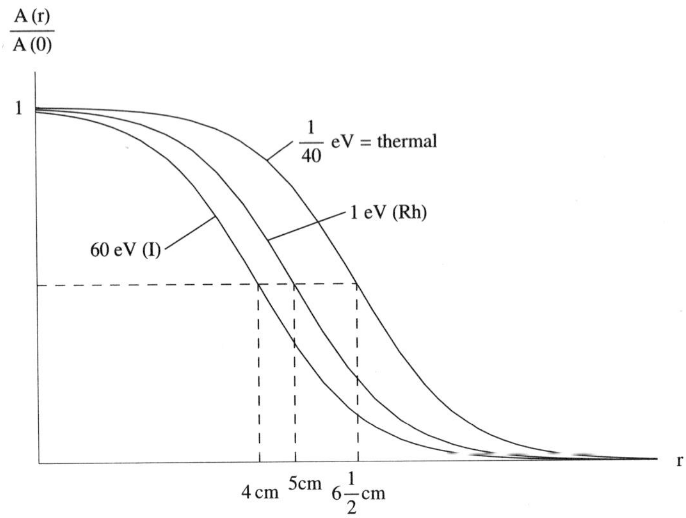

Fig. 7.7

and that is the mean square distance of neutrons from the source, which can be related to the neutron mean free path. We can write

$$\overline{r^2} = \frac{\displaystyle\int_0^\infty r^4 A(r)\,\mathrm{d}r}{\displaystyle\int_0^\infty r^2 A(r)\,\mathrm{d}r}\,.$$

For the sake of interest, we will derive the formula relating the strength of a neutron source to the initial activity (immediately after the end of an irradiation) of a detector that has been exposed to saturation, i.e. for many half-lives, in the neutron flux. Let us denote this activity by $A(r,0)$. It can be derived from the observed quantity, the β-particle counting rate $A(r,t)$ if one extrapolates the decay of the latter back to $t=0$ and then corrects for the overall efficiency of the counting system.

If n_0 is the number of nuclei in the detector, $\sigma(v)$ their capture cross section at the neutron velocity v, then

$$A(r) = n_0 \int \rho(r,v)\, v\, \sigma(v)\,\mathrm{d}v\,.$$

For a $1/v$ detector (e.g. In in the thermal region), this simplifies to

$$A(r) = n_0\, \rho(r)\; (\sigma_0 v_0)\,,$$

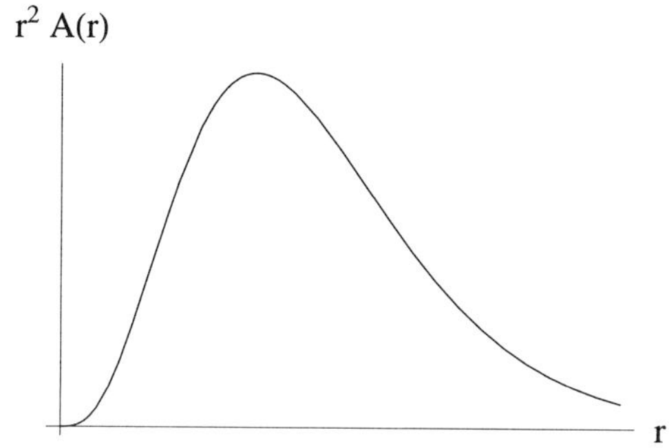

Fig. 7.8

where $\rho(r)$ means the density of all neutrons at r, irrespective of velocity. For a resonance detector, if we assume the resonance to be narrow, in the region $v \pm \mathrm{d}v$,

$$A(r) = n_0\, \rho(r, v)\, \sigma(v)\, \Delta v\,.$$

Integrating over all space,

$$4\pi \int_0^\infty A(r)\, r^2 \mathrm{d}r = n_0\, v\, \sigma(v)\, N(v)\, \Delta v\,,$$

where $N(v)$ is the total number of neutrons of velocity v present at any moment. But it can be shown (Bethe, Rev. Mod. Phys., Apr. 1937, Sect. 59) that

$$N(v)\, \Delta v \equiv N(E)\, \Delta E = Q\, \tau(E)\, \frac{\Delta E}{E}\,,$$

where Q is the source strength in neutrons per second, and $\tau(E)$ is the mean time for which a neutron has the energy E. We thus find, for the resonance case,

$$4\pi \int_0^\infty A(r)\, r^2 \mathrm{d}r = n_0\, v\, \tau(E)\, Q\, \overline{\sigma(E)}\, \frac{\Delta E}{E}\,,$$

where ΔE is the energy width of the resonance, and $\overline{\sigma(E)}$ is the average value of σ taken over the resonance.

We thus arrive at the following expressions for Q:

Thermal detector($1/v$ law) : $$Q = \frac{4\pi}{n_0\, (\sigma_0 v_0)\, \tau} \int_0^\infty A(r)\, r^2 \mathrm{d}r,$$

Resonance detector : $$Q = \frac{4\pi}{n_0\, v\, \tau(E)\, \overline{\sigma(E)}\, \dfrac{\Delta E}{E}} \int_0^\infty A(r)\, r^2 \mathrm{d}r$$

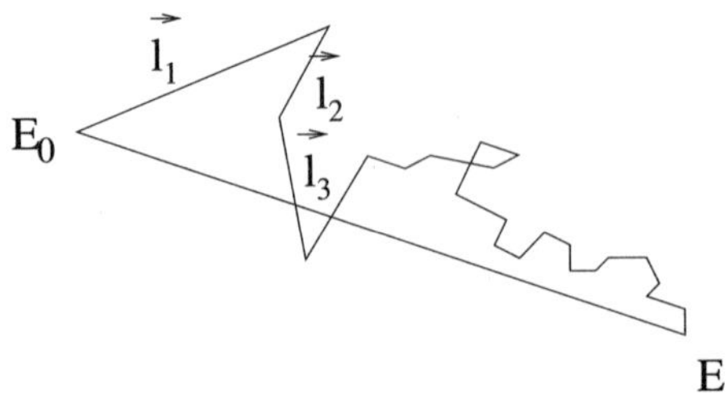

Fig. 7.9

($\sigma_0 v_0$ is the constant product of σ and v at any velocity, τ is the mean life of a thermal neutron).

7.5 Theory of the Spatial Distribution

Let us suppose that the scattering of neutrons by the nuclei of a medium is isotropic in the C.G. system. This is expressed by the statement

$$\overline{\cos\theta} = 0\,.$$

In this case the scattering is certainly not isotropic in the laboratory system. Denoting, as before, the laboratory angle by β, we have

$$\overline{\cos\beta} = \int_0^\pi \frac{1 + A\cos\theta}{\left(1 + 2A\cos\theta + A^2\right)^2} \cdot \frac{1}{2}\sin\theta\,\mathrm{d}\theta = \frac{2}{3A}\,.$$

Suppose now that we have an infinite extent of the medium, and that a point source of neutrons of energy E_0 is placed in it. Let us try to find the mean value of r^2 for those neutrons which have been slowed down from E_0 to some value E. The displacement r of the neutron from the starting point will have been reached in general after a number of collisions, so that the vector $\vec{r}$ is compounded of a number of vectors $\vec{\ell}_1$, $\vec{\ell}_2$, etc., as shown in Fig. 7.9. We then have

$$\vec{r} = \vec{\ell}_1 + \vec{\ell}_2 + \ldots + \vec{\ell}_n$$

$$\begin{aligned} r^2 &= \sum_i \ell_i^2 + \sum_{i,j} \vec{\ell}_i \cdot \vec{\ell}_j \\ &= \sum_i \ell_i^2 + \sum_{i,j} \ell_i\,\ell_j\,\cos(i\,j)\,, \end{aligned}$$

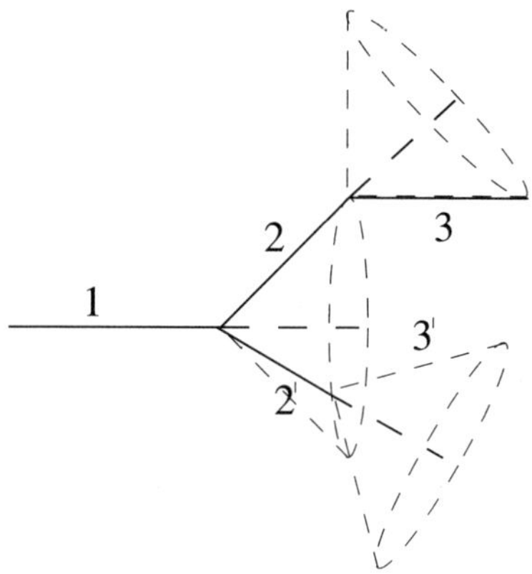

Fig. 7.10

where $(i\,j)$ denotes the angle between the vectors $\vec{\ell}_i$ and $\vec{\ell}_j$. We shall perform this summation by easy steps.

(1) Let us suppose first of all that the angle between any two adjacent vectors is fixed in size but can take any direction. Then, relative to one of the vectors as axis, the other can trace out a cone. If to this second vector we attach a third in the same way, it is clear that it may take up a whole variety of orientations relative to the first. We try to depict this in Fig. 7.10. We shall evaluate the average value of the cosine of the angle between two non-adjacent vectors, first when there are two intermediate vectors, and from that result to the general case.

Suppose, then, that we have four vectors, as shown in Fig. 7.11, and we wish to find the value of $\overline{\cos(3\,4)}$. Consider the spherical triangle formed by taking equal lengths along the vectors 1, 3, 4.

By a well-known theorem in spherical trigonometry,

$$\cos(1\,4) = \cos(1\,3)\,\cos(3\,4) + \sin(1\,3)\,\sin(3\,4)\,\cos(1\,3\,4)\,,$$

where $(1\,3\,4)$ is the angle between the planes $(1\,3\,0)$ and $(4\,3\,0)$. Now, since the vector 4 is free to rotate about the vector 3, the angle $(1\,3\,4)$ can take any value from 0 to 2π. Hence $\overline{\cos(1\,3\,4)} = 0$. It follows that

$$\overline{\cos(1\,4)} = \overline{\cos(1\,3)}\,\cos(3\,4)\,.$$

Considering now the triangle formed by 1, 2 and 3, we have in the same way

$$\cos(1\,3) = \cos(1\,2)\,\cos(2\,3) + \sin(1\,2)\,\sin(2\,3)\,\cos(1\,2\,3)\,.$$

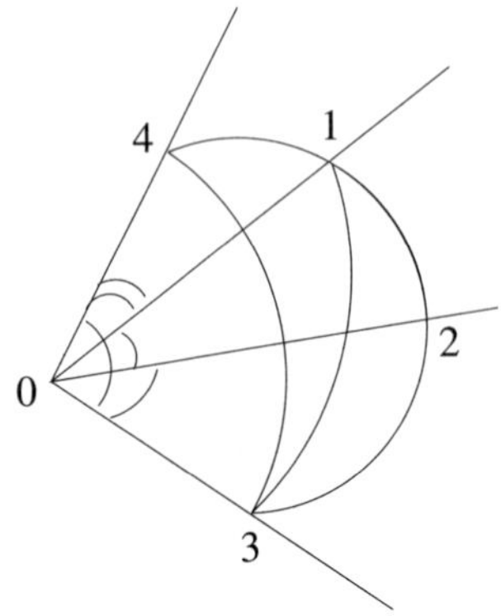

Fig. 7.11

The angle (1 2 3) is just like (1 3 4), so that $\overline{\cos(1\,2\,3)} = 0$. Thus

$$\overline{\cos(1\,3)} = \cos(1\,2)\,\cos(2\,3)\,,$$

so that

$$\overline{\cos(1\,4)} = \cos(1\,2)\,\cos(2\,3)\,\cos(3\,4)\,.$$

It is clear that we can proceed to write down a general expression, viz.

$$\overline{\cos(i,k)} = \cos(i,i+1)\,\cos(i+1,i+2)\,\cdots\,\cos(k-1,k)\,.$$

Using this result, we can write down a sort of first average of r^2, where we take into account this swivelling of the ℓ_i about each other, but ignore all other factors. Then in this case we have

$$\begin{aligned}\overline{r^2}_{(1)} &= \sum_i \ell_i^2 + \sum_{i,j} \ell_i \ell_j \cos(i,i+1)\,\cdots\,\cos(j-1,j)\\ &= \ell_1^2 + \ell_2^2 + \ldots + 2\,\{\ell_1\ell_2\cos(1\,2) + \ell_2\ell_3\cos(2\,3) + \ldots\\ &\quad + \ell_1\ell_3\cos(1\,2)\,\cos(2\,3) + \ldots\\ &\quad + \ell_1\ell_4\cos(1\,2)\,\cos(2\,3)\,\cos(3\,4) + \ldots\\ &\quad + \ldots\}\,.\end{aligned}$$

(2) We will now consider the effect of averaging over the ℓ_i. Let the mean free path of a neutron at some energy be λ_i. Then the probability of a path of length between ℓ_i and $\ell_i + \mathrm{d}\ell_i$ is given by

$$\mathrm{d}p = \mathrm{e}^{-\ell_i/\lambda_i}\,\frac{\mathrm{d}\ell_i}{\lambda_i}\,.$$

Using this result, we can readily evaluate $\overline{\ell_i}$ and $\overline{\ell_i^2}$:

$$\overline{\ell_i} = \int_0^\infty \ell_i \, \mathrm{e}^{-\ell_i/\lambda_i} \frac{\mathrm{d}\ell_i}{\lambda_i} = \lambda_i$$

$$\overline{\ell_i^2} = \int_0^\infty \ell_i^2 \, \mathrm{e}^{-\ell_i/\lambda_i} \frac{\mathrm{d}\ell_i}{\lambda_i} = 2\lambda_i^2 \, .$$

We will continue to assume that the angles between the adjacent vectors ℓ_i are fixed. This means that the neutron energy on any part, sat ℓ_j, of its path is completely determined by the number j and the actual values of the angular deflections it has suffered. If this is so, then it is legitimate to write

$$\overline{\ell_i \ell_j} = \overline{\ell_i} \cdot \overline{\ell_j} \, .$$

Our second stage in obtaining $\overline{r^2}$ therefore gives us

$$\begin{aligned} \overline{r^2}_{(2)} &= \sum_i 2\lambda_i^2 + \sum_{i,j} \lambda_i \lambda_j \cos(i, i+1) \cdots \cos(j-1, j) \\ &= 2\left(\lambda_1^2 + \lambda_2^2 + \ldots\right) + 2\left\{\lambda_1\lambda_2 \cos(1\,2) + \lambda_2\lambda_3 \cos(2\,3) + \ldots \right. \\ &\quad + \lambda_1\lambda_3 \cos(1\,2)\, \cos(2\,3) + \ldots \\ &\quad + \lambda_1\lambda_4 \cos(1\,2)\, \cos(2\,3)\, \cos(3\,4) + \ldots \\ &\quad \left. + \ldots \right\} . \end{aligned}$$

(3) We shall next introduce an approximation, by assuming A to be large, so that it takes several collisions to slow down the neutron appreciably. This assumption gives the same result as an equivalent assumption, namely that the scattering cross section, and hence the mean free path, should change little with neutron energy. Under either of this assumptions, we may write

$$\lambda_1\lambda_j \sim \lambda_1^2, \qquad \lambda_2\lambda_j \sim \lambda_2^2, \qquad \text{and so on.}$$

We can then re-write our expression for $\overline{r^2}_{(2)}$ in the following way:

$$\begin{aligned} \overline{r^2}_{(2)} &\sim 2\left(\lambda_1^2 + \lambda_2^2 + \ldots\right) + 2\left\{\lambda_1^2 \left[\cos(1\,2) + \cos(1\,2)\, \cos(2\,3) + \ldots\right] \right. \\ &\quad + \lambda_2^2 \left[\cos(2\,3) + \cos(2\,3)\, \cos(3\,4) + \ldots\right] \\ &\quad + \lambda_i^2 \left[\ldots + \cos(i, i+1) \cdots \cos(k-1, k) + \ldots\right] \\ &\quad \left. + \ldots \right\} . \end{aligned}$$

We now perform our third average, over the values of $\cos(i, i+1)$. Putting $\overline{\cos(i, i+1)} = c \; (= 2/3A)$, we have

$$\begin{aligned} &\lambda_i^2 \left[\cos(i, i+1) + \cos(i, i+1)\, \cos(i+1, i+2) \right. \\ &\left. + \cos(i, i+1)\, \cos(i+1, i+2)\, \cos(i+2, i+3) + \ldots\right] \\ &= \lambda_i^2 \left(c + c^2 + c^3 + \ldots\right) = \lambda_i^2 \cdot \frac{c}{1-c} \, . \end{aligned}$$

We thus obtain the third and final average:

$$\overline{r^2}_{(3)} = 2\sum_i \lambda_i^2 + 2\cdot\frac{c}{1-c}\sum_i \lambda_i^2 = \frac{2}{1-c}\sum_i \lambda_i^2 .$$

It remains now only to evaluate the summation over the λ_i^2.

(4) Let us put, as before,

$$\epsilon = \log E .$$

Then in the logarithmic energy interval $d\epsilon$ the number of collisions (cfr. p. 53) is given by $\frac{d\epsilon}{\xi}$. Using this fact, we can replace the summation over the λ_i^2 by an integral:

$$\sum_i \lambda_i^2 \equiv \int_{\log E}^{\log E_0} \lambda^2(\epsilon)\,\frac{d\epsilon}{\xi} .$$

We then arrive at the final form of $\overline{r^2}$:

$$\boxed{\overline{r^2} = \frac{2}{\xi\left(1-\frac{2}{3A}\right)}\int_{\log E}^{\log E_0} \lambda^2(\epsilon)\,d\epsilon = \frac{2}{\xi\left(1-\frac{2}{3A}\right)}\int_E^{E_0} \lambda^2(E')\,\frac{dE'}{E'} .}$$

For slowing down in hydrogen an exact formula can be written:

$$\begin{aligned}\overline{r^2} &= 2\lambda^2(0) + 2\lambda^2(a) + 2\int_0^a \lambda^2(x)\,dx + 2\lambda(0)\int_0^a \lambda(x)\,e^{-x/2}\,dx \\ &\quad +2\lambda(0)\lambda(a)e^{-a/2} + 2\lambda(a)\int_0^a \lambda(x)e^{-(a-x)/2}\,dx \\ &\quad +2\int_0^a \lambda(u)\,du\int_0^{a-u} \lambda(u+x)\,e^{-x/2}\,dx ,\end{aligned}$$

where

$$x = \log\frac{E_0}{E}, \qquad a = \log\frac{E_0}{E_1} ,$$

E_1 being the final energy considered and E any intermediate energy. ($\lambda(0)$ = m.f.p.[2] at $a = 0$, i.e. $E_1 = E_0$.)

From thermal energy up to about 1 keV the $n - p$ scattering cross section is nearly constant. This means that the mean free path is also nearly constant over this range of energies, as shown in Fig. 7.12. If, then,

[2]@ That is, the mean free path.

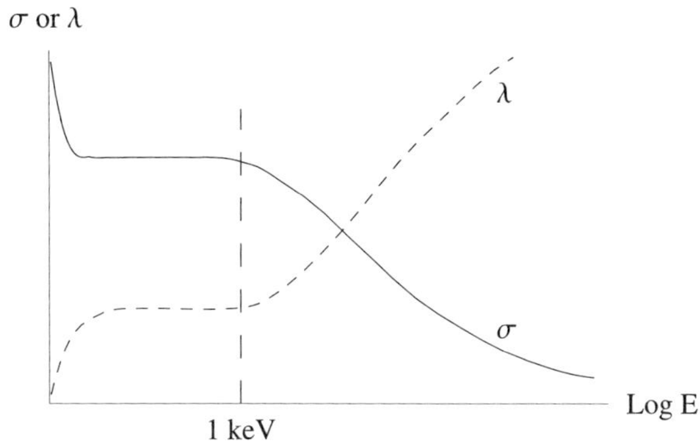

Fig. 7.12

we consider the slowing of neutrons of energy E_0, first to an energy E_1 and then to a lower energy E_2, where both E_1 and E_2 lie in the region of constant mean free path, we can approximate the above complicated expression for $\overline{r^2}$ by a sum of two terms, of which the first is a function of the upper limit E_0 only, whilst the second depends on E_1 or E_2 as the case may be. Precise treatment gives

$$\overline{r^2}(1) \sim f(E_0) - 6\lambda^2 \log E_1$$
$$\overline{r^2}(2) \sim f(E_0) - 6\lambda^2 \log E_2 .$$

Thus, by subtraction,

$$\Delta\overline{r^2} = 6\lambda^2 \log \frac{E_1}{E_2} ,$$

where λ is the constant low energy mean free path. This result can actually be reached by starting from our equation for $\overline{r^2}_{(2)}$ on p. 61. Suppose that the energy E_1 is reached after m collisions, and E_2 after n collisions $(n > m)$. Then by differencing the summations for n and m we obtain

$$\Delta\overline{r^2} = 2\sum_{i=m}^{n} \lambda_i^2 + \sum_{i=m,\, j=m+1}^{i=n-1,\, j=n} \lambda_i\lambda_j \cos(i, i+1)\cdots\cos(j-1, j) .$$

But in this expression we can now put $\lambda_i = \lambda_j = \lambda$, so that we obtain

$$\begin{aligned}\Delta\overline{r^2} &= 2\lambda^2(n-m) + 2\lambda^2 \sum_{i,j} c^{j-i} \\ &\sim 2\lambda^2(n-m) + 2\lambda^2(n-m)\frac{c}{1-c} \\ &= 2\lambda^2(n-m)\frac{1}{1-c} .\end{aligned}$$

But $(n-m)$ is the number of collisions required on the average to go from E_1 to E_2, and is therefore given by

$$(n-m) = \frac{1}{\xi}\log\frac{E_1}{E_2}\,.$$

For neutron-proton collisions, $\xi = 1$ and $c = 2/3$, so that

$$\Delta\overline{r^2} = 2\lambda^2\log\frac{E_1}{E_2}\frac{1}{1-2/3} = 6\lambda^2\log\frac{E_1}{E_2}$$

as before.

This change of mean square distance with energy for low energy can be linked up immediately with experimental results on neutron densities as described in the previous section. For example, using a Ra-Be source in water the following results have been obtained:

Rh detector	(1 eV)	$\overline{r^2} = 276.6\ \text{cm}^2.$
I detector	($\sim$ 65 eV)	$\overline{r^2} = 262.2\ \text{cm}^2.$
		$\Delta\overline{r^2} = 14.4\ \text{cm}^2\,.$

$$6\lambda^2\log_e 65 = 14.4$$

$$\lambda^2 = \frac{14.4}{6}\cdot\frac{1}{4.28} = 0.57$$

$$\lambda = 0.76\ \text{cm}.$$

From this result we can obtain the $n-p$ scattering cross section, for if n is the no. of scattering nuclei per cm^3,

$$\sigma = \frac{1}{n\,\lambda}\,.$$

For water, $n = 0.67\cdot 10^{23}$ (we are ignoring the contribution of O nuclei to the scattering, so that n refers to protons only).

Hence $\sigma = \dfrac{10^{-23}}{0.67\cdot 0.76}\ \text{cm}^2 = \underline{20\ \text{barns}}$, very nearly. This agrees very closely with the accepted value.

Chapter 8

The Age Equation

8.1 Derivation of the Age Equation

In this chapter we carry on a little further our study of the space distribution of neutrons emitted from a source which is placed in a moderating (= slowing down) medium. We shall first of all introduce a quantity n, which is defined as the density of neutrons of a given energy ϵ at a point ($\epsilon = \log E$, as usual). Then the number of neutrons per cm^3 at a point (x, y, z) with energies between ϵ and $\epsilon + \mathrm{d}\epsilon$ is $n(x, y, z, \epsilon)\,\mathrm{d}\epsilon$. We shall consider the dynamic equilibrium of n when the neutron distribution has become steady after the introduction of the source into the medium.

For this purpose we will single out a small volume element $(\mathrm{d}x, \mathrm{d}y, \mathrm{d}z)$ at the point (x, y, z). We have to consider two mechanisms by which n may change. The first of these is the bodily transfer of neutrons of energy ϵ into or out of our elementary volume; this is the process of diffusion. The second is the occurrence of collisions within the volume by which faster neutrons are brought into the energy band $\mathrm{d}\epsilon$ at ϵ, and neutrons already in the band fall out of it.

8.1.1 *Diffusion*

If λ is the mean free path of a neutron, and if v is its velocity, the diffusion coefficient D in the case of isotropic scattering is given by

$$D = \frac{\lambda v}{3}\,.$$

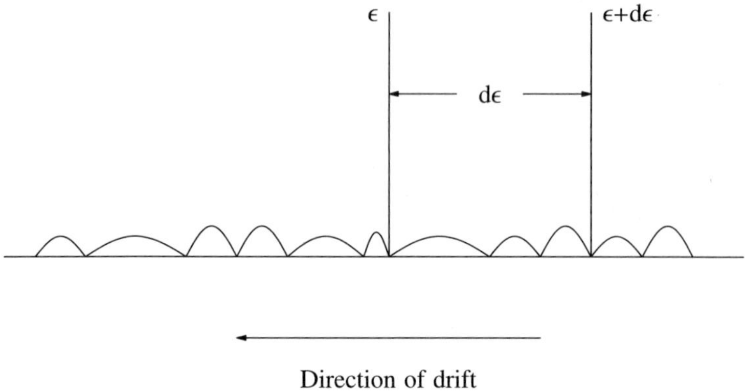

Fig. 8.1

In the case of anysotropic scattering with which we are concerned, this is replaced by

$$D = \frac{\lambda v}{3\left(1 - \overline{\cos\beta}\right)} = \frac{\lambda v}{3\left(1 - \dfrac{2}{3A}\right)}.$$

For the flow of diffusion along the x-axis alone, we have that the rate of change of the number of neutrons within the volume, in the energy range ϵ to $\epsilon + \mathrm{d}\epsilon$ is given by

$$\mathrm{d}y\,\mathrm{d}z\,\mathrm{d}\epsilon \cdot D\left[\frac{\partial n(x+\mathrm{d}x)}{\partial x} - \frac{\partial n(x)}{\partial x}\right] = \mathrm{d}x\,\mathrm{d}y\,\mathrm{d}z\,\mathrm{d}\epsilon \cdot D\,\frac{\partial^2 n}{\partial x^2}.$$

We have similar contributions from the other two pairs of faces. Since the product $\mathrm{d}x\,\mathrm{d}y\,\mathrm{d}z$ is the volume of the element, the rate of change of the number of neutrons per cm^3 with energies between ϵ and $\epsilon + \mathrm{d}\epsilon$ is simply

$$D\,\Delta n\,\mathrm{d}\epsilon\,.$$

8.1.2 *Energy Drift*

In Fig. 8.1 we have tried to depict the way in which a neutron hops down the ϵ-axis. The average length of one hop is ξ; the actual values are scattered about this. The number of collisions made per second by a neutron of velocity v and mean free path λ is v/λ. Thus the average change of ϵ per second is equal to $\xi \dfrac{v}{\lambda}$, and this is therefore the rate of drift along the axis of ϵ. The number of neutrons leaving the region $\mathrm{d}\epsilon$ per second is therefore $\xi\, n(\epsilon)\, \dfrac{v}{\lambda}$, where all quantities except ξ are functions of ϵ. A corresponding influx of neutrons occurs at the upper boundary $(\epsilon + \mathrm{d}\epsilon)$, so that the net number of neutrons entering $\mathrm{d}\epsilon$ per second and per cm^3 is given by

$$\xi \frac{\partial}{\partial \epsilon}\left[\frac{v}{\lambda}\, n(\epsilon)\right] \mathrm{d}\epsilon\,.$$

In the stationary state, therefore,

$$D\,\Delta n + \xi \frac{\partial}{\partial \epsilon}\left[\frac{v}{\lambda}\, n(\epsilon)\right] = 0\,,$$

or, substituting for D,

$$\boxed{\frac{\lambda v}{3\left(1 - \dfrac{2}{3A}\right)}\,\Delta n + \frac{\partial}{\partial \epsilon}\left[\xi\, \frac{v}{\lambda}\, n(\epsilon)\right] = 0\,.}$$

At this point we will introduce two more quantities. First, let us put

$$q = \xi\, \frac{nv}{\lambda}\,.$$

q is the mean number of neutrons per cm^3 per second that are slowed down at the energy ϵ. It is known as the Slowing Down Density. Substituting in our neutron balance equation, we find

$$\frac{\lambda v}{3\left(1 - \dfrac{2}{3A}\right)}\,\Delta\left(\frac{\lambda}{\xi v}\, q\right) + \frac{\partial q}{\partial \epsilon} = 0\,.$$

But ξ, λ and v are not functions of positions, so this can be rewritten

$$\frac{\lambda^2(\epsilon)}{3\xi\left(1 - \dfrac{2}{3A}\right)}\,\Delta q + \frac{\partial q}{\partial \epsilon} = 0\,.$$

Next we define a quantity τ, such that

$$\tau = \frac{1}{3\xi\left(1 - \dfrac{2}{3A}\right)} \int_{\epsilon}^{\epsilon_0} \lambda^2(\eta)\mathrm{d}\eta\,.$$

ϵ_0 corresponds to the initial fast neutron energy. Differentiating at the lower limit of integration,

$$\frac{\partial q}{\partial \epsilon} = \frac{\partial q}{\partial \tau} \cdot \frac{-1}{3\xi\left(1 - \dfrac{2}{3A}\right)} \cdot \lambda^2(\epsilon)\,.$$

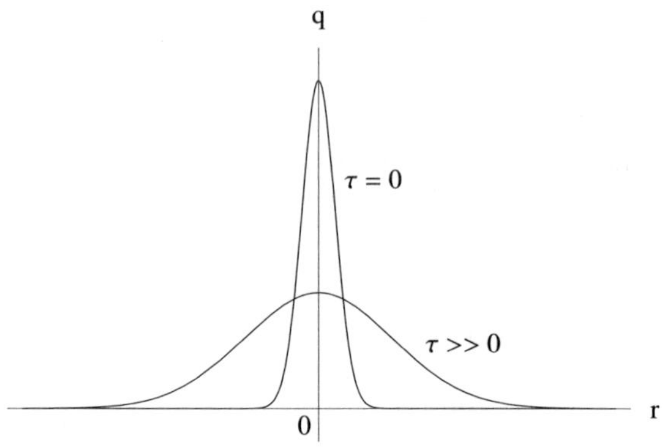

Fig. 8.2

Substituting in our equation, we arrive at the very simple form

$$-\Delta q + \frac{\partial q}{\partial \tau} = 0\,.$$

This is the famous "Age Equation" of Fermi. It is analogous to the fundamental equation of heat flow, with τ taking the place of time, and q taking the place of temperature. It is for this reason that τ is called the "age" of the neutrons, although in fact it has the dimension of an area. τ does increase as the time from birth of the neutrons at the source increases, but really it is a measure of the mean square displacement from the source of neutrons that have fallen to a given energy.

8.2 The Problem of a Point Source

Suppose now that we have a point source of neutrons of energy E_0 (or ϵ_0) in an infinite medium. We wish to find the spatial distribution of neutrons of energy E. Now at the age $\tau = 0$, all neutron are to be found at the origin with energy E_0. The distribution for E_0 therefore takes the form of a δ-function.

For any other value of τ, the distribution is smeared out, as shown in Fig. 8.2. We can take over our solution from the more familiar study of heat flow, and find

$$q(r,\tau) = \frac{Q}{(4\pi)^{3/2}\tau^{3/2}} \cdot \mathrm{e}^{\frac{-r^2}{4\tau}}\,.$$

Q is the strength of the neutron source measured in neutrons per second. We can confirm the correctness of this solution, first by sy substituting it in the Age Equation, and second by integrating q over all space, for any value of τ, and seeing that the answer is Q, as it must be if neutron capture during slowing down does not occur. The proofs of these identities are elementary and need not to be set out here. It may be noted that, for a given value of τ, the distribution of q in space follows a Gaussian error curve.

If we content ourselves with a study of n, the true neutron density for neutrons of a given energy, instead of q, the slowing down density, we can get a solution of sorts without recourse to the Age Equation. For suppose we want $n(x)$ at a certain τ. A particular τ means a certain number of collisions since birth, and therefore a particular number of path-lengths travelled. Now if the number of path-lengths is large, we can make use of a result in the theory of numbers, which says: If there is a large number of magnitudes M, such that

$$\sum M = 0\,, \qquad\qquad |M| \gg 0\,,$$

then the distribution of the end-points formed by adding up the magnitudes in a random way is a Gaussian. This random addition represents precisely the way in which a neutron track is built up, so we can at once put

$$n(x) = \text{const.} \times \mathrm{e}^{-x^2/k}\,,$$

where k defines the width of the distribution. It then follows that

$$n(x, y, z) = \text{const.} \times \mathrm{e}^{-r^2/k}\,.$$

To arrive at exact values of the constants demands, of course, a detailed study of the slowing down process.

A value of τ of particular interest is that corresponding to thermal energy. Inserting it into the Age Equation we obtain the space distribution of thermal neutron source density produced by a fast neutron source in a moderating medium. We shall have cause to use this fact in the next chapter, where we investigate the distribution and movement of the neutrons after they have become thermal.

Chapter 9

Thermal Neutron Distributions

9.1 The Basic Equation

In one important respect the problem of thermal neutron distributions is simpler than that for neutrons being slowed down - only one energy is involved, and it is the lowest energy that neutrons can have before being captured. Thus there is no longer any sliding along the energy axis, and the problem becomes one of spatial distribution only. There are three ways in which the neutron density at a point may change, viz.

a) Diffusion
b) Absorption (capture)
c) Entry into the thermal region of neutrons from higher energy groups.

It is thus easy to see that, in the stationary state, we have

$$D\Delta n - \frac{n}{\theta} + q_T - \frac{\partial n}{\partial t} = 0\,,$$

where D is the diffusion coefficient, θ is the mean life of a thermal neutron in the medium, and q_T is the value of q at thermal energy. This, as explained in the last chapter, must be obtained by solution of the Age Equation at the appropriate value of τ. Once the neutrons are thermal, the scattering is isotropic, so that D is equal to $\frac{\lambda v}{3}$. The equation to the stationary state thus becomes

$$\Delta n - \frac{3}{\lambda v \theta} \cdot n + \frac{3 q_T}{\lambda v} = 0\,.$$

In this equation, let us put $\frac{\lambda v \theta}{3} = \ell^2$, $v\theta = \Lambda$. ℓ is known as the Diffusion Length, and Λ is absorption mean free path (it evidently is equal to the mean lifetime, measured along its path, that is travelled by a neutron between becoming thermal and being captured). Then we arrive at the following, where ℓ^2 can be replaced by $\frac{\lambda\Lambda}{3}$ if required:

$$\boxed{\Delta n - \frac{n}{\ell^2} + \frac{3q_T}{\lambda v} = 0\,.}$$

9.2 Point Source of Slow Neutrons

In this case the value of q_T is zero everywhere except at the source, which we shall take as the origin of the co-ordinates. Remembering that $\Delta = \frac{\partial^2}{\partial r^2} + \frac{2}{r}\frac{\partial}{\partial r}$, it may readily be verified that the solution is

$$n = A \cdot \frac{e^{-r/\ell}}{r}\,.$$

Near to the source, $n = \frac{A}{r}$. From this fact we can deduce A, for the outward flux of neutrons across the surface of a sphere drawn around the source at a small distance from it is equal to the strength of the source[1]:

$$D \operatorname{grad} \frac{A}{r} \cdot 4\pi r^2 = Q\,,$$

whence

$$A = \frac{3Q}{4\pi\lambda v}\,.$$

9.3 Point Source of Fast Neutrons

We shall suppose that we have a point source of fast neutrons at S (Fig. 9.1), and that we wish to find the thermal neutron density at the point A. The answer has to be arrived at by integrating the contributions from all the elementary volumes such as that at P in the figure, which act as point sources of thermal neutrons.

From the Age Equation, we have

$$q_T = \frac{Q}{(4\pi\tau)^{3/2}} \cdot e^{-\rho^2/4\tau}\,,$$

[1]@ The following formula in the original text contains, incorrectly, the quantity grad $\frac{1}{r}$ instead of grad $\frac{A}{r}$.

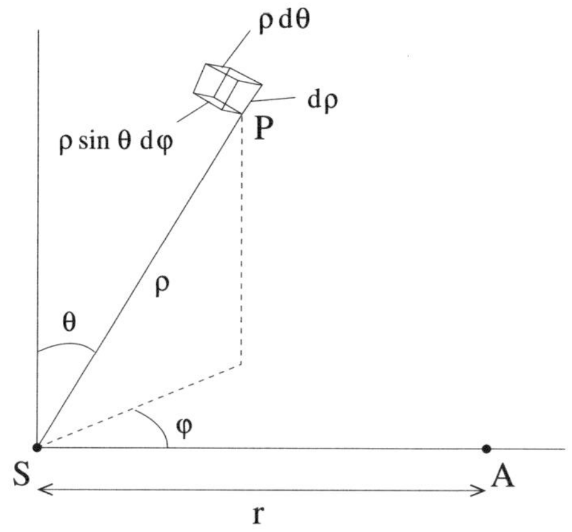

Fig. 9.1

τ being taken to correspond to thermal neutrons. Now, as we have just seen from the previous section, the thermal neutron density at A due to a source S at P is given by

$$n = \frac{3S}{4\pi\lambda v} \cdot \frac{\mathrm{e}^{-|\boldsymbol{\rho} - \mathbf{r}|/\ell}}{|\boldsymbol{\rho} - \mathbf{r}|},$$

where

$$|\boldsymbol{\rho} - \mathbf{r}| = \left(\rho^2 + r^2 - 2\rho r \sin\theta \cos\phi\right)^{1/2},$$

and

$$S = n\,\mathrm{d}v = n\rho^2 \mathrm{d}\rho \,\sin\theta\,\mathrm{d}\theta\,\mathrm{d}\phi\,.$$

Thus the total slow (thermal) neutron density at A is given by:

$$N = \int_{\phi=0}^{2\pi} \int_{\rho=0}^{\infty} \int_{\theta=0}^{\pi} \frac{Q\,\mathrm{e}^{-\rho^2/4\tau}}{(4\pi\tau)^{3/2}} \frac{3}{4\pi\lambda v} \cdot \frac{\mathrm{e}^{-|\boldsymbol{\rho} - \mathbf{r}|/\ell}}{|\boldsymbol{\rho} - \mathbf{r}|} \rho^2 \mathrm{d}\rho \,\sin\theta\,\mathrm{d}\theta\,\mathrm{d}\phi\,.$$

9.4 Bounded Media

If a moderating medium is finite in extent, we have to consider what happens to the neutron density at the boundaries. Let us assume that the curvature of the medium is everywhere convex, so that a neutron, after escaping from the medium, cannot get back into it. Then to a zero approximation we can put n or $q = 0$ at the boundary. A better approximation

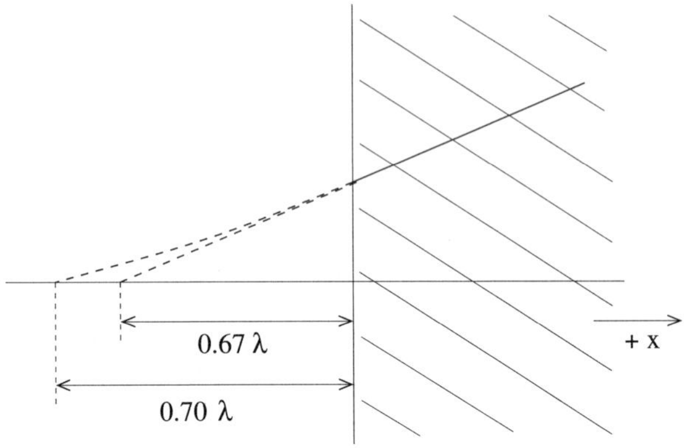

Fig. 9.2

is that the extrapolation of n as a function of distance gives $n = 0$ at a distance 0.667λ outside the boundary (Fig. 9.2); a still better value for the extrapolation is 0.70λ.

We shall derive here the first approximation to the true result. It is based on assuming that the neutron density as a function of the distance measured into the medium from the boundary is given by

$$n(x) = p(\alpha + x)\,,$$

so that n becomes zero at the distance $-\alpha$. We shall find α by calculating the flux of emergent neutrons in two ways and equating the results:

1) The outward flux ϕ of neutrons at the boundary is given by

$$\phi = D\,\frac{\mathrm{d}n}{\mathrm{d}x} = D\,p\,,$$

i.e.

$$\boxed{\phi = \frac{\lambda v}{3}\,p\,.}$$

2) Every neutron getting out of the medium must have suffered a last collision at a certain value x in the medium (See Fig. 9.3). If the neutron emerges in a direction making an angle θ with the normal to the boundary, the probability of its escaping rather than making a further collision is $\mathrm{e}^{-x/\lambda\cos\theta}$. The fraction of 4π contained between the directions θ and $\theta + \mathrm{d}\theta$ is $\dfrac{1}{2}\sin\theta\,\mathrm{d}\theta$. The number of collisions per second in a distance $\mathrm{d}x$ at x is

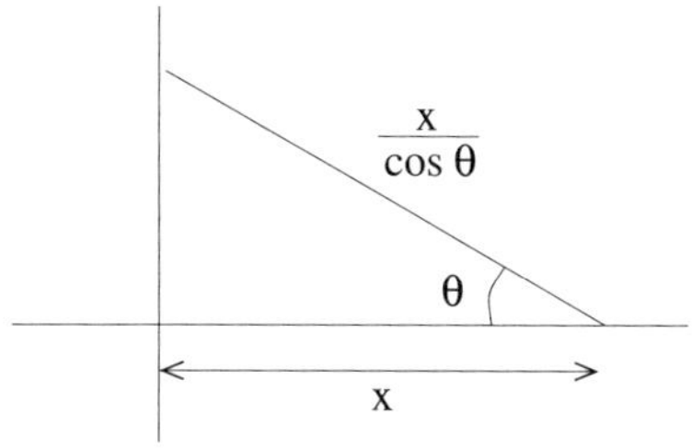

Fig. 9.3

equal to $\dfrac{n(x)\cdot v}{\lambda}$. Multiplicating these factors together, and integrating, we obtain

$$\begin{aligned}\phi &= \int_{\theta=0}^{\pi/2}\int_{x=0}^{\infty}\frac{n(x)\cdot v}{\lambda}\cdot \mathrm{e}^{-x/\lambda\cos\theta}\cdot\frac{1}{2}\sin\theta\,\mathrm{d}\theta\,\mathrm{d}x\\ &= -\frac{v}{2}\frac{p}{\lambda}\int_0^{\pi/2}\int_0^{\infty}(\alpha+x)\mathrm{e}^{-x/\lambda\cos\theta}\mathrm{d}x\,\mathrm{d}(-\cos\theta)\\ &= +\frac{v}{2}\frac{p}{\lambda}\int_0^{1}\int_0^{\infty}(\alpha+x)\mathrm{e}^{-x/\mu\lambda}\mathrm{d}x\,\mathrm{d}\mu\\ &= \frac{v}{2}\frac{p}{\lambda}\int_0^{1}(\alpha\lambda\mu+\lambda^2\mu^2)\mathrm{d}\mu\,,\end{aligned}$$

where $\mu=\cos\theta$, i.e.

$$\boxed{\phi=\frac{v}{2}\frac{p}{\lambda}\left(\frac{\alpha\lambda}{2}+\frac{\lambda^2}{3}\right).}$$

Thus, equating the expressions for ϕ,

$$\frac{\lambda}{3}=\frac{1}{2\lambda}\left(\frac{\alpha\lambda}{2}+\frac{\lambda^2}{3}\right),$$

whence

$$\alpha=\frac{2}{3}\lambda=0.667\,\lambda\,.$$

It must be borne in mind that n does not in fact tend to zero in the way suggested by the extrapolation formula. (n would become negative for $|-x|>\alpha$, if the formula were true outside the boundary!). There must be continuity of neutron flux, however, so that the product of neutron drift velocity and neutron density must be the same for two points close together on opposite sides of the boundary. If the outside medium is a gas (air) and the inside medium is a solid moderator, there will be a large drop of neutron density across the boundary.

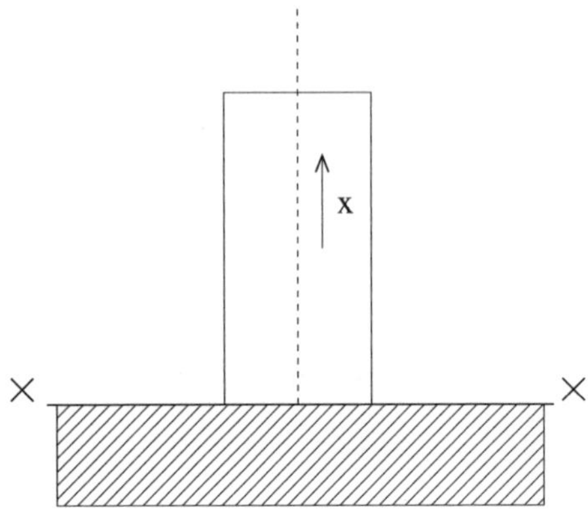

Fig. 9.4

9.5 The Measurement of Diffusion Length

Suppose we have a large plane source of slow neutrons - e.g. a pile - and put a column of the material being investigated on top, as shown in Fig. 9.4. If we can assume that the transverse dimensions of the column are large compared with ℓ , we have merely to consider variations of neutron density in the vertical direction, so that we have to solve the equation

$$\frac{\mathrm{d}^2 n}{\mathrm{d}x^2} - \frac{1}{\ell^2}\, n + q_T = 0,$$

where $q_T = 0$ except at the boundary $\times\times$. The solution is a sum of two exponentials, of which we discard the one which increases with increasing x, and thus obtain

$$n(x) = A\,\mathrm{e}^{-x/\ell}\,.$$

One has to make measurements of the neutron density up the column with and without a Cd sheet at $\times\times$. By difference one then obtains the effect due to thermal neutrons alone. The method is feasible for hydrogenous moderator, for which the transverse dimensions can readily be made large compared with ℓ (e.g. for water, $\ell \sim 2.8$ cm at 300°K); but for other materials, such as graphite ($\ell = 50$ cm) it cannot be used.

The value of ℓ has been found to vary with temperature; e.g. for water it obeys the relation

$$\ell = 2.64 + 0.0061\, t\,,$$

where t is in °C. The temperature of the neutrons in a hydrogenous medium is higher than that of the medium itself, because they are captured before complete slowing down; the temperature variation of ℓ refers to the temperature of the water, in the formula just quoted.

The importance of measuring ℓ is that, taken together with the mean free path λ, it yields the value of the absorption mean free path and hence the capture cross section:

$$\ell^2 = \frac{\lambda\Lambda}{3}\,, \qquad \Lambda = \frac{1}{\sigma_a n_a}\,,$$

n_a being the number of absorbing atoms per cm^3.

9.6 Diffusion in Graphite

Because it is impossible, or at any rate very difficult, to make experiments with a piece of graphite whose dimensions are large compared with ℓ, the finite bounded problem must be solved. Moreover, since it is of interest to do the measurements with Ra-Be sources, we shall investigate whether it is practicable to do so, rather than starting with slow neutrons from a pile. Fig. 9.5 shows our choice of axis, for a column of square section (length of side $= a$).

The first thing to do is to solve the Age Equation for this particular geometry. The solution has been obtained by Fermi, and is as follows, the fast neutron source being at the point $(0, u, v)$:

$$q(x,y,z,\tau) = \frac{4}{a^2}\cdot\frac{Q}{2\sqrt{\pi\tau}}\cdot \mathrm{e}^{-x^2/4\tau}\cdot\sum_{r,s=1}^{\infty} \mathrm{e}^{-\frac{\pi^2\tau}{a^2}(r^2+s^2)} \times \sin\frac{\pi r y}{a}\,\sin\frac{\pi s z}{a}\,\sin\frac{\pi r u}{a}\,\sin\frac{\pi s v}{a}\,.$$

The method of arriving at this result is to observe the boundary conditions, viz. $q = 0$ at $y = 0$ and a, $z = 0$ and a, for all x. This prompts one to try a solution of the type

$$q = \sum_{r,s} q(r,s)\,\sin\frac{\pi r y}{a}\,\sin\frac{\pi s z}{a}\,.$$

"a" is here the effective length of a side of the column, i.e.

$$a = a_{\text{geometrical}} + 2(0.67\,\lambda)\,.$$

If one further puts

$$q_{r,s} = p_{r,s}(x,\tau)\,\mathrm{e}^{-\frac{\pi^2\tau(r^2+s^2)}{a^2}}\,,$$

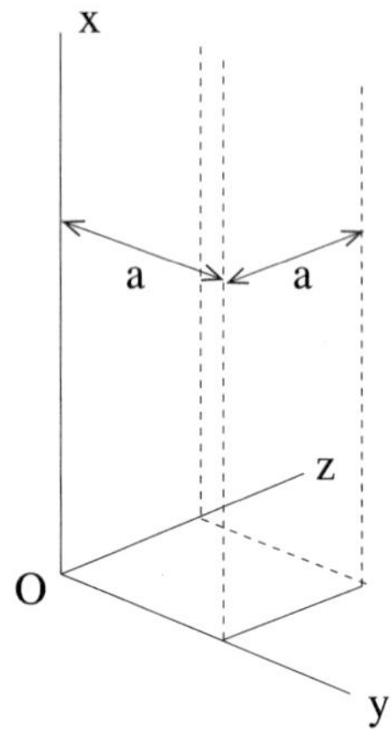

Fig. 9.5

then one finds

$$\frac{\partial^2 p}{\partial x^2} = \frac{\partial p}{\partial \tau},$$

and hence

$$p_{r,s} = \frac{1}{\sqrt{\tau}} \cdot e^{-x^2/4\tau}.$$

The form of the exponential term in the Fourier series for q, containing $(r^2 + s^2)$ in the exponent, makes for rapid convergence, so that one can in practice ignore all terms except that for which $r = s = 1$. If, also, one assumes the source to be placed at the point $(0, \frac{1}{2}a, \frac{1}{2}a)$, the solution takes the simpler form

$$\boxed{q(x, y, z, \tau) = \frac{4}{a^2} \cdot \frac{Q}{2\sqrt{\pi\tau}} e^{-x^2/4\tau} e^{2\pi^2\tau/a^2} \sin\frac{\pi y}{a} \sin\frac{\pi z}{a}.}$$

The interesting factor in this expression is evidently $e^{-x^2/4\tau}$, for it will tell us what sort of distance the neutrons have to travel, on the average, from the Ra-Be source before becoming thermal. Let us put $4\tau = r_0^2$. r_0 is called the range of the neutrons, for a given energy drop. Now the space distribution of neutrons from a Ra-Be source can be analyzed into 3 Gaussians for a given neutron energy, so that for each energy there are three values of τ and three values of r_0. These are given in the following table for three energies (thermal, In resonance, and I resonance)[2]:

[2]@ The values of τ in the following table are measured in cm^2 while, for r_0, cm are used as units.

Age and Range of Ra-Be Neutrons

Fraction	τ(I)	r_0(I)	τ(In)	r_0(In)	τ(Th)	r_0(Th)
0.152	54	14.7	130	20.8	183	27.0
0.693	268	32.7	340	36.8	386	59.7
0.157	736	54.1	815	57.0	867	58.9

It is thus clear that for $x > 50$ cm, there are not many neutrons left above thermal energy, so that it is nearly true to put $q_T = 0$. The upshot of all this is that the dimensions of the system are (or can easily be made) large compared with r_0, when at the same time they are not large compared with ℓ. In other words, we have the somewhat paradoxical result that the system can be made infinite for fast neutrons being slowed down, but not for the same neutrons when they have become thermal.

We can now proceed to solve the equation for a thermal neutron distribution with $q_T = 0$, viz.

$$\Delta n - \frac{1}{\ell^2}\, n = 0\,.$$

As for the Age Equation, we will try a solution

$$n = \sum_{r,s=1}^{\infty} n_{r,s}(x)\, \sin\frac{\pi r y}{a}\, \sin\frac{\pi s z}{a}\,.$$

By substituting in the equation we find

$$\frac{\mathrm{d}^2}{\mathrm{d}x^2}\, n_{r,s} - \left\{\frac{\pi^2}{a^2}\left(r^2 + s^2\right) + \frac{1}{\ell^2}\right\} n_{r,s} = 0\,.$$

Put

$$\frac{1}{b_{r,s}^2} = \frac{\pi^2}{a^2}\left(r^2 + s^2\right) + \frac{1}{\ell^2}\,.$$

Then the solution can be written

$$n_{r,s} = \text{const.}\ \mathrm{e}^{\pm x/b_{r,s}}\,,$$

and we discard the positive argument. In addition, we can again neglect all terms except $r = s = 1$. The final form of the solution is therefore

$$\boxed{n(x,y,z) = \mathrm{e}^{-x\sqrt{\frac{2\pi^2}{a^2}+\frac{1}{\ell^2}}}\, \sin\frac{\pi y}{a}\, \sin\frac{\pi z}{a}\,.}$$

9.7 Some Useful Quantities and Relationships

In the following table are given some data relating to commonly used moderating material[3]:

Material	Density	Atoms/10^{24}cm^3	ℓ (cm)	ℓ^2/Λ (cm)
H_2O	1.00	0.0334	2.85	0.142
D_2O	1.10	0.0331	100	0.80
Be	1.8	0.1235	31	0.87
C	1.62	0.0871	50.2	0.903

The only two fundamental quantities are the scattering and the absorption cross section. Using the same symbols as previously, we have the following relationships between the various quantities used to describe the behaviour of thermal neutrons:

$$v\theta = \Lambda = \frac{1}{\sigma_{\rm a} n_{\rm a}}$$

$$\ell^2 = \frac{\lambda\Lambda}{3}$$

$$\lambda = \frac{1}{\sigma_{\rm sc} n_{\rm a}}$$

$$N = \frac{\sigma_{\rm sc}}{\sigma_{\rm a}} = \frac{\Lambda}{\lambda} (= \text{etc.}) = \text{Average number of collisions before capture.}$$

For a general survey of methods for measuring ℓ and λ, see the article by Amaldi and Fermi in the Physical Review (Vol. 49, p.899, 1936).

[3]@ In the following table, the values of the density are measured in g/cm^3. Note that in the original manuscript the incorrect value 0.871 is reported instead of the correct one 0.0871 given here.

Chapter 10

The Reflection of Neutrons

10.1 Introduction

If we have a boundary between two semi-infinite media, and a neutron crosses the boundary from one medium into the other, the probability that it will ultimately return to the first medium is called the albedo, or whiteness, of the reflecting medium. (The term is taken by analogy from the language of optics). The albedo is a function of the angle of incidence of the neutron on the boundary, and one has to define the particular conditions in which it is measured.

Before we consider the albedo itself, we will deal with the simpler problem of finding the probability that a neutron already in the medium, at a given distance from the boundary, shall escape. This is treated in the next two sections.

10.2 Approximate Solution of the Escape Problem

First of all, we shall make the problem one-dimensional, calling the density of neutrons $\rho(x)$ at the point x. We shall suppose that a point source of neutrons, of strength Q, is situated at $x = a$. The neutrons are considered to be thermal throughout. Then the diffusion equation (p. 71) is written

$$\frac{\mathrm{d}^2\rho}{\mathrm{d}x^2} - \frac{1}{\ell^2}\,\rho + \frac{3}{\lambda v}\,q = 0\,,$$

where $q = 0$ except in the narrow region $a - \epsilon \le x \le a + \epsilon$, and $\int_{a-\epsilon}^{a+\epsilon} q(x)\,\mathrm{d}x = Q$. We must also impose the boundary condition $\rho = 0$ at $x = 0$ and ∞.

It can readily be verified that the solution of the diffusion equation outside the source region is given by

$$\rho = A\left(e^{x/\ell} - e^{-x/\ell}\right) \qquad \text{for } x < a,$$

$$\rho = B\,e^{-x/\ell} \qquad \text{for } x > a.$$

To find A and B, we integrate the diffusion equation over the source region

$$\int_{a-\epsilon}^{a+\epsilon} \left(\frac{d^2\rho}{dx^2} - \frac{1}{\ell^2}\rho\right) dx = -\frac{3Q}{\lambda v}.$$

If ρ is not to tend to ∞, the second term of the integral must be zero. Thus we have

$$\frac{d\rho}{dx}(a+\epsilon) - \frac{d\rho}{dx}(a-\epsilon) = -\frac{3Q}{\lambda v}.$$

We know also that ρ itself is continuous, so that, as ϵ tends to zero,

$$\rho(a+\epsilon) = \rho(a-\epsilon).$$

From these two conditions one finds

$$A = \frac{3\ell}{2\lambda v}\,Q\,e^{-a/\ell}, \qquad B = \frac{3\ell}{2\lambda v}\left(e^{a/\ell} - e^{-a/\ell}\right).$$

The rate of escape of neutrons from the medium is given by the value of $D\,\frac{d\rho}{dx}$ at $x = 0$, D being the diffusion coefficient $\frac{\lambda v}{3}$.

Now $\left(\frac{d\rho}{dx}\right)_0 = \frac{2A}{\ell}$, so that the outward flux is given by

$$\frac{2A}{\ell}\cdot\frac{\lambda v}{3} = Q\,e^{-a/\ell}.$$

Since the rate of production of neutrons is Q, the probability $p(a)$ of escape from the depth a is simply

$$\boxed{p(a) = e^{-a/\ell}.}$$

10.3 Exact Solution by Integral Equation Method

The result obtained above has been arrived at through the solution of a differential equation. This is not, however, the only method - nor is it even correct in all cases. We may digress a moment to discuss this point.

To find the neutron density in a particular volume V at a time t, one could investigate the density of neutrons moving towards V at earlier times t' such that they would be in V at time t. By adding up contributions of

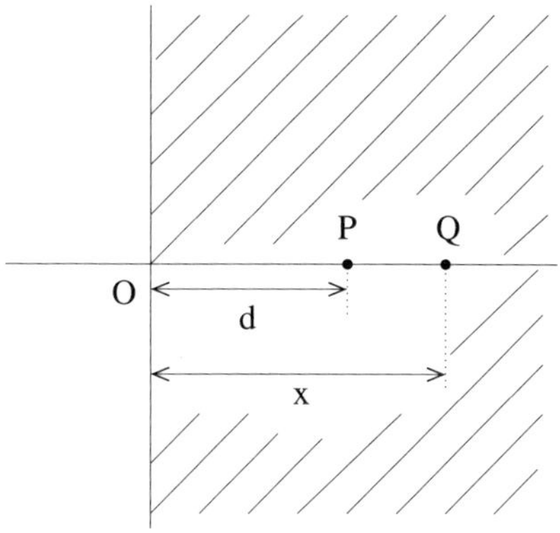

Fig. 10.1

this sort over all space, we arrive at an integral expression for the density in V. It is necessary to ask ourselves what are the relative merits of the differential and integral equation methods.

In setting up the diffusion differential equation, one assumes tacitly that n, $\frac{dn}{dx}$, etc. vary slowly with respect to the mean free path of the diffusing particles. One also assumes that n is large enough to give a meaning to $\frac{dn}{dx}$. One could not, for example, expect a significant result in solving by diffusion theory the problem of finding the neutron density $n(r, \theta, \phi)$ in a sphere of radius $\lambda/2$. But there are no such restrictions on integral equation methods. One needs no assumption about the nature of $\frac{dn}{dx}$ and so on. Thus integral equation methods are more general; but they are also more difficult. Where practicable it is best to get an approximate solution by differential equations and then to use the integral equation method as a refinement. In this present problem we shall assume that the form of the solution is that deduced in the previous section, but that the various coefficients may have to be adjusted. In other words, we shall suppose that the probability of escape from a depth x is given by

$$p(x) = A\,e^{-\alpha x}\,,$$

where A is not necessarily unity and α is not necessarily equal to $1/\ell$.

Consider a neutron at the point P (Fig. 10.1), distance d from the boundary. It may go to the left or the right from P. The chance that it will move in one direction in particular is $1/2$. If it goes to the left, it may escape before making a second collision (we assume that we begin looking at the neutron just as it makes a collision at P) or it may collide. The probability of escape without a further collision is therefore

$$\frac{1}{2}\,e^{-d/\lambda}\,.$$

However the neutron may also escape after a further collision. Suppose that next collision occurs at a point Q, distance x from the boundary (Q may of course be either right or left of P), and suppose that the probability of the collision occurring at Q is $\pi(x)$. Then the total probability of escape for our neutron, after it has suffered its first collision at P, is given by

$$p(d) = \frac{1}{2}\,e^{-d/\lambda} + \frac{1}{2}\sum_{\text{all } x} \pi(x)\,p(x)\,.$$

This assumes isotropic scattering, and ignores the possibility of capture. If there are on the average N collisions before capture, this becomes

$$p(d) = \frac{1}{2}\,e^{-d/\lambda} + \frac{1}{2}\sum_{\text{all } x} \pi(x)\,p(x)\,\frac{N-1}{N}\,.$$

$\pi(x)$ is the usual probability of a collision in the distance $d-x$ (or $x-d$ as the case may be). Thus

$$\begin{aligned} p(d) = \frac{1}{2}\,e^{-d/\lambda} &+ \frac{1}{2}\int_0^d e^{-(d-x)/\lambda}\,\frac{dx}{\lambda}\,p(x)\,\frac{N-1}{N} \\ &+ \frac{1}{2}\int_d^\infty e^{-(x-d)/\lambda}\,\frac{dx}{\lambda}\,p(x)\,\frac{N-1}{N}\,. \end{aligned}$$

Combining the second and third term,

$$p(d) = \frac{1}{2}\,e^{-d/\lambda} + \frac{1}{2}\int_0^\infty e^{-|x-d|/\lambda}\,\frac{dx}{\lambda}\,p(x)\,\frac{N-1}{N}\,.$$

Inserting into this equation our tentative solutions,

$$\begin{aligned} p(d) &= A\,e^{-\alpha d}\,, \\ p(x) &= A\,e^{-\alpha x}\,, \end{aligned}$$

we find

$$A = \frac{\sqrt{N}}{\sqrt{N}+1}\,, \qquad \alpha = \frac{1}{\lambda\sqrt{N}} = \frac{1}{\sqrt{3}\,\ell}\,.$$

Thus, by the integral equation method,

$$\boxed{p(x) = \frac{\sqrt{N}}{\sqrt{N}+1}\,e^{-x/\sqrt{3}\,\ell}\,.}$$

10.4 The Albedo

The albedo, which is usually denoted by the symbol β, follows readily from the above result. Suppose we have a beam of slow neutrons moving from left to right (Fig. 10.1) striking the plane $x = 0$. Then the probability that the first collision suffered by a neutron will be in a distance $\mathrm{d}x$ at a length x in the medium is $\mathrm{e}^{-x/\lambda}\,\dfrac{\mathrm{d}x}{\lambda}$. The probability of not being absorbed is $\frac{N-1}{N}$, and the probability of escape from the medium is $p(x)$.

Hence,

$$\beta = \int_0^\infty \mathrm{e}^{-x/\lambda}\,\frac{\mathrm{d}x}{\lambda}\,p(x)\,\frac{N-1}{N} = \frac{\sqrt{N}-1}{\sqrt{N}+1}\,.$$

The whole of the above discussion is one-dimensional. If the problem is treated in three dimensions, the albedo for a neutron incident at an angle θ to the normal is given by

$$\beta(\theta) = \frac{\sqrt{N}-1}{\sqrt{N}+\sqrt{3}\,\cos\theta}\,.$$

This does not agree with our first solution if we put $\theta = 0$, because the neutrons are no longer restricted to path along the x-axis; the consequence is that neutrons take a longer path on the average to get to the surface of the medium, and their chance of being captured instead of escaping is increased. Thus β becomes smaller, as can be seen from the two formulae.

Measurement of β is evidently a way of finding N, which is equal to the ratio of the scattering and absorption cross section.

10.5 Measurement of the Albedo

The measurement of β is fairly straightforward in water or paraffin wax, where the diffusion region is small - effectively only a few centimeters in extent. All one has to do is to find the activity of a thermal detector placed in the medium, first with a Cd plate on one side of it, and then without any Cd. In the first case, thermal neutrons can only get to the foil from the side opposite to the Cd. Suppose that n_0 thermal neutrons per second cross the detector with this arrangement, and let the activity produced be A_1. Then it is legitimate to write

$$A_1 = k\,n_0\,,$$

since the neutrons, after crossing the detector, run into the Cd and are completely absorbed; there is no chance that they will get back into the foil.

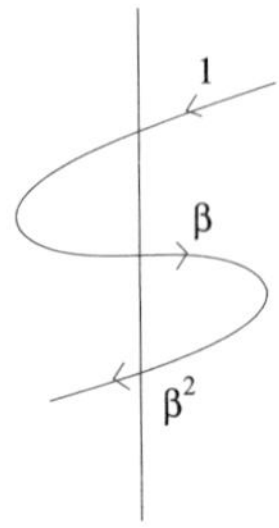

Fig. 10.2

Now let us consider what happens in the second experiment. The number of neutrons crossing the detector foil at least once is $2n_0$, since neutrons can now get at the foil from either side. There is also the effect due to the albedo, that neutrons may cross the detector several times. Consider a neutron crossing the detector for the first time, from right to left (Fig. 10.2). Then the probability that it will re-cross the foil from left to right is β, since we can regard the foil, if it is big enough, as dividing the medium into two semi-infinite media, each with albedo β. The probability that it will make a second crossing from right to left is β^2, and one evidently must continue this series to infinity. Thus the activity in this case is given by

$$\begin{aligned} A_2 &= 2k\, n_0 \left(1 + \beta + \beta^2 + \dots \text{ ad inf.}\right) \\ &= \frac{2k\, n_0}{1-\beta}\,. \end{aligned}$$

The ratio of the activities is thus equal to $\dfrac{2}{1-\beta}$. This argument is only valid if the detector does not absorb neutrons appreciably, i.e. $n\sigma\delta \ll 1$, where n is the number of atoms per cm^3, σ is their capture cross section, and δ is the thickness of the detector. If appreciable absorption takes place, β must be replaced by $\beta(1-\xi)$, where ξ is the fraction of neutrons removed from an incident group in one traversal of the foil. The value of the ratio A_2/A_1 has been found equal to about 11 for paraffin and water. This gives $\beta = 0.82$, and hence $N \sim 100$, so that a thermal neutron in these media makes about 100 collisions before being captured. The distance it travels, measured along the path, is about 80 cm on the average, and the time it takes to do this, which is its lifetime as a thermal neutron, is something less than a millisecond.

[*In connection with this chapter, see Fermi's paper "Sul moto dei neutroni (lenti) nelle sostanze idrogenate", Ric. Sci.* **7**, *13 (1936)*]

Chapter 11

The Stability of Nuclei

11.1 The Binding Energy of a Nucleus

Any nucleus is characterized by its mass number A and its atomic number Z. Its mass M can be expressed as a function of these two quantities. According to the convention in these matters, M is taken to include the masses of the electrons surrounding the nucleus, together with their binding energy; this latter is a small but appreciable contribution for heavy elements.

The mass of a neutron on the usual scale ($O^{16} = 16.00000000$) is 1.00893, and the mass of a proton+electron is 1.00812. The expression for M must therefore be of the form

$$M = 1.00893(A - Z) + 1.00812Z + (\text{binding energy in mass units}).$$

The relation between the various units of energy is as follows:

$$1 \text{ mass unit } = 931\,\text{MeV } = 1.493 \cdot 10^{-3}\,\text{erg}.$$

We shall consider the various terms which contribute to the binding energy.

11.1.1 *The Liquid Drop Model*

Experiments tell us that a nucleus has a certain fairly definite radius, beyond which the specifically nuclear forces have no effects and one sees only the Coulomb potential due to the positive nuclear charge. For the very heavy nuclei, the radius R is found from the Gamow theory for α-particle decay, which relates the energy and the decay constant to Z and R. For light nuclei, scattering experiments show where the field begins to vary more rapidly than the Coulomb field, and this again defines A. For A greater than about 10, one can write

$$R \sim 1.48 \cdot 10^{-13} A^{1/3}\,\text{cm}.$$

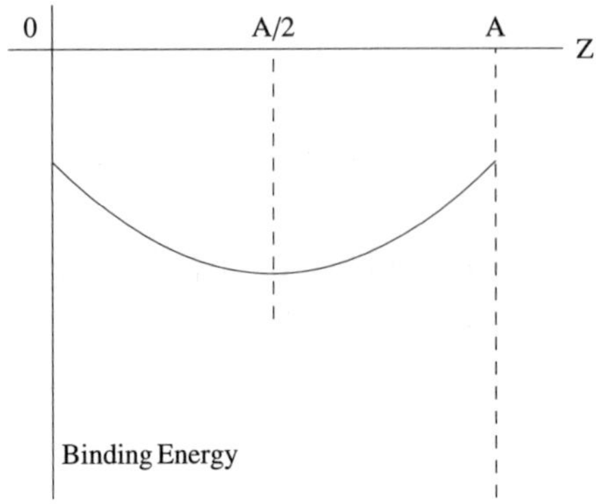

Fig. 11.1

This means that the nuclear density is approximately constant, and suggest that the properties of a small nucleus and of a portion of a large nucleus are similar. The result leads one to consider the analogous case of a liquid, and to ascribe to a nucleus, as a result of the comparison, a binding energy which is proportional to the total mass - and which is negative, to ensure stability. We then write, as our first approximation,

$$\text{Binding energy} \;=-a_1\,A\,.$$

If we pursue the liquid drop analogy, we must also suppose that there is a surface contribution to the total energy, since the forces between particles cease to balance near the surface, and there is a surface tension. The force on a particle becomes less at the surface, and reduces the binding energy of the nucleus as a whole. Introducing this capillary term,

$$\text{Binding energy} \;=-a_1\,A+a_2\,A^{2/3}.$$

11.1.2 *Nuclear Composition*

We know that in stable nuclei the numbers of protons and neutrons are about equal. The electrostatic forces between protons make $(A-Z)$ increases more rapidly than Z, but the electrostatic energy is a second order effect, going as Z^2, and we shall ignore it for the moment. Suppose then (neglecting electrostatic forces) that (p,p), (n,n) and (n,p) forces are equal, and let us plot the binding energy curve for a given A as a function of Z (Fig. 11.1). The binding energy is everywhere negative (meaning stability), and is bound to be symmetrical, with a minimum at the stable point

$Z \simeq \frac{1}{2}A$. The quantity $\left(\frac{A}{2} - Z\right)$ gives a measure of the deviation from the minimum, and the curve is to a first approximation a parabola. The energy per nucleon can be written as

$$+a_3\left(\frac{A}{2} - Z\right)^2,$$

and the contribution to the total nuclear energy is therefore

$$\boxed{+a_3 \cdot \frac{\left(\frac{A}{2} - Z\right)^2}{A}.}$$

11.1.3 *Coulomb Forces*

The electrostatic forces between protons lead to a positive contribution to the binding energy and hence to an increase of M. If we consider the nuclear charge Ze to spread uniformly through a sphere of radius R, the electrostatic potential energy of the nucleus is $+\frac{3}{5}\frac{Z^2e^2}{R}$. Using the expression for R given on p. 87, this gives

$$\boxed{+0.000627 \cdot \frac{Z^2}{A^{1/3}} \text{ mass units.}}$$

11.1.4 *The Odd-Even Effect*

Nuclei are most stable when $(A - Z)$ and Z are both even, and least stable when they are both odd. The intermediate combinations, of odd N and even Z or even N and odd Z, are about equally stable, and about half-way between the extreme. The abundances of stable nuclei of three four types testify to these facts. Thus, the 272 stable nuclei are distributed as follows:

Z even, N even,	160,
Z even, N odd,	56,
Z odd, N even,	52,
Z odd, N odd,	4.

The four odd-odd nuclei are the lightest possible nuclei of this type, viz. ${}_1D^2$, ${}_3Li^6$, ${}_5B^{10}$ and ${}_7N^{14}$.

These stability rules are probably connected with the spin properties of the nucleons, for according to the Pauli exclusion principle only two protons (or two neutrons) with their spins oppositely directed can exist in the same energy level; a third particle must be placed in the next higher level. In terms of this simple picture the observed trend of stabilities can readily be explained.

A term δ must be added to the binding energy to take account of this effect. If we choose our zero so that the correction vanishes for the odd-even and even-odd types, we have a contribution $+\epsilon$ for odd-odd nuclei, and $-\epsilon$ for even-even.

Collecting together all the component parts of the nuclear mass, we have finally

$$M(A,Z) = 1.00893\,(A-Z) + 1.00812\,Z - a_1\,A + a_2 A^{2/3} + a_3 \cdot \frac{\left(\frac{A}{2}-Z\right)^2}{A} + 0.000627 \cdot \frac{Z^2}{A^{1/3}} - \frac{1}{2}\,\epsilon\left[(-1)^Z + (-1)^{A-Z}\right].$$

11.2 Determination of Coefficients

In the above formula we have four undetermined quantities, a_1, a_2, a_3, and ϵ. These are all determined by fitting the formula to the known stable isotopes. First of all one differentiates M with respect to Z, keeping A fixed:

$$\frac{\partial M}{\partial Z} = -1.00893 + 1.00812 - a_3 \cdot \frac{2\left(\frac{A}{2}-Z\right)}{A} + 0.001254 \cdot \frac{Z}{A^{1/3}}.$$

Setting this expression equal to zero will give the value of Z corresponding to the minimum M i.e. to the most stable nucleus of mass A. (We have ignored the effect of δ in performing the differentiation). The result is

$$Z_A = \frac{0.00081 + a_3}{\frac{2a_3}{A} + 0.001254 \cdot \frac{1}{A^{1/3}}}.$$

This contains only one of the parameters, and can therefore be easily fitted to the data. One finds

$$\boxed{a_3 \simeq 0.083.}$$

Using this value, Z_A can now be put in the simpler form

$$Z_A = \frac{A}{1.981 + 0.015\,A^{2/3}}.$$

The coefficients a_1 and a_2 can now be obtained from a further study of nuclear masses. Their values are found to be

$$\boxed{a_1 = 0.01504;} \qquad \boxed{a_2 = 0.014.}$$

Substituting this in the mass formula gives

$$M(A,Z) = 0.99389\,A - 0.00081\,Z + 0.014A^{2/3} + 0.083 \cdot \frac{\left(\frac{A}{2} - Z\right)^2}{A}$$
$$+0.000627 \cdot \frac{Z^2}{A^{1/3}} + \delta,$$

where[1] $\delta = -\frac{1}{2}\,\epsilon\left[(-1)^Z + (-1)^{A-Z}\right]$. From a study of the odd-even alternation, one finds

$$\boxed{\epsilon \sim \frac{0.036}{A^{3/4}}.}$$

The quantities which are often used in considering nuclear stability can be easily found from the formulae. They are:

$$\text{Mass Defect} = M(A,Z) - A,$$

$$\text{Packing fraction} = \frac{M(A,Z) - A}{A}.$$

For the most recent packing fraction curve, see the article by Dempster, in Phys. Rev. **53**, p. 870 (1938).

11.3 The Binding of Neutrons to Nuclei

We shall now apply the formulae to an important problem - the energy of attachment of a neutron to various isotopes of uranium. Let us consider what happens if we add a neutron to the nucleus U^{235}. The total initial mass is given by

$$M_1 = M(235, 92) + 1.00893$$
$$= 235.11240 + 1.00893 = 236.12133.$$

The final mass is

$$M_2 = M(236, 92) = 236.11401.$$

[1] @ In the original manuscript, the overall minus sign in the expression for δ is missing.

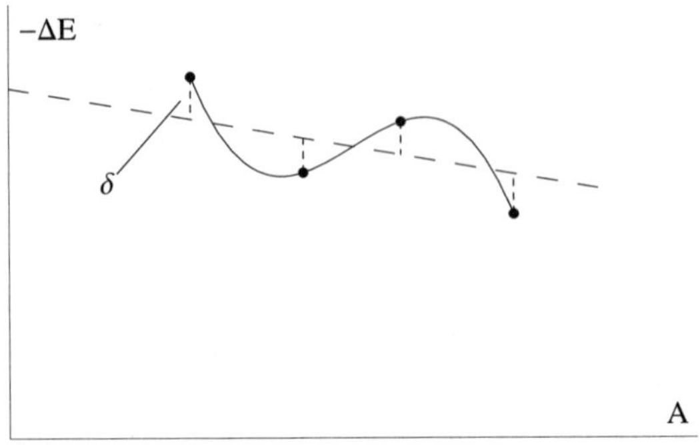

Fig. 11.2

From this one finds

$$\Delta M = -0.00732 \text{ mass units,}$$
$$\Delta E = -6.81 \text{ MeV.}$$

This means that the nucleus U^{236} is stable against disintegration into U^{235} and a neutron.[2] The binding energy of a neutron to the other isotopes can be calculated in a similar way. The results are tabulated on the following table and Fig. 11.2 shows this variation of binding energy with A graphically, with the characteristic alternation due to the odd-even effect superimposed on the general trend.

A	$-\Delta E$ (MeV)	$A - Z$	
235	6.81	143	odd
236	5.51	144	even
237	6.56	145	odd
238	5.31	146	even

[2]@ Since ΔE is negative.

Chapter 12

Nuclear Fission

12.1 The Possibility of Fission

Very heavy elements have more energy per particle than elements in the middle of the Periodic Table. This can be seen from the packing fraction curve, which we have already discussed in Chapter 2, and which we reproduce here (Fig. 12.1) for the sake of convenience. The packing fraction is usually expressed in units of 10^{-4} . On this scale, H has a p.f. of 80, and U^{235} has a p.f. of about 6. Elements in the middle of the periodic table have a p.f. of round about 7. The splitting of a heavy nucleus therefore liberates much energy.

Let us consider the case when a nucleus of mass A splits into two parts

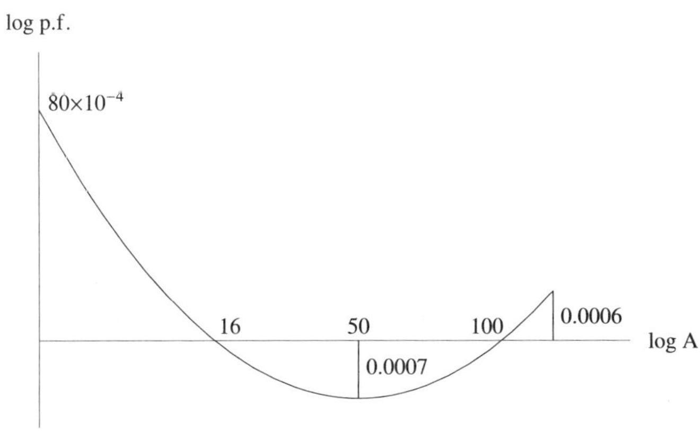

Fig. 12.1

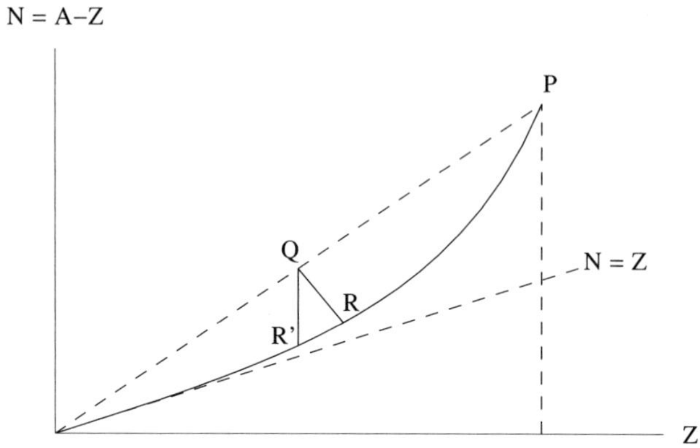

Fig. 12.2

each of mass $\frac{1}{2}A$. If E is the energy liberated,

$$\begin{aligned}E &= M(A) - 2M\left(\frac{1}{2}A\right)\\ &= [M(A) - A] - 2\left[M\left(\frac{1}{2}A\right) - \frac{1}{2}A\right]\\ &= A\left[\frac{M(A)-A}{A} - \frac{M\left(\frac{1}{2}A\right) - \frac{1}{2}A}{\frac{1}{2}A}\right]\\ &= A\left[f(A) - f\left(\frac{1}{2}A\right)\right],\end{aligned}$$

where $f(A)$ is the packing fraction for any nucleus of mass A. We can calculate this energy release from the formula for $M(A, Z)$ given in the last chapter. This does not, however, represent the total amount of energy given out in fission, because the nuclei of mass $\frac{1}{2}A$ are themselves unstable. This can be seen from Fig. 12.2. If we start at P in the figure, with a nucleus of charge Z and mass A, the division we have assumed will bring us to Q, with two nuclei of charge $\frac{1}{2}Z$ and mass $\frac{1}{2}A$. But these nuclei have an excess of neutrons with respect to the line of stable nuclei on the isotopic chart; they must therefore move by β-disintegration to the point R or by neutron emission to R'.

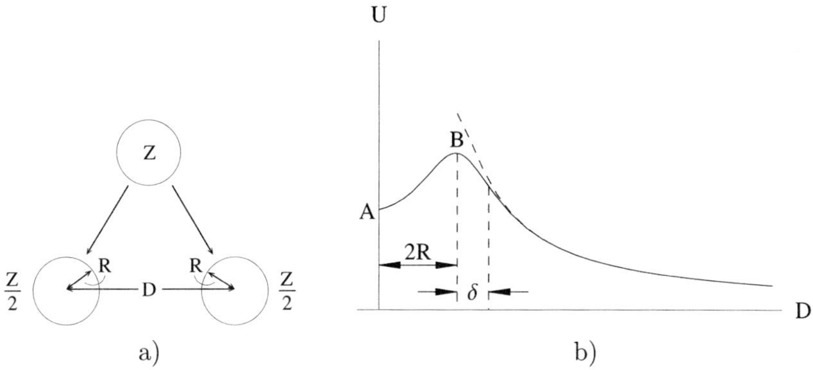

Fig. 12.3

To make the problem more specific, we will consider what happens to a nucleus of mass number 240, when it splits into two equal parts. From the formula on p. 91 we find that the most stable nucleus of mass 240 has a charge of 93.74. The splitting therefore gives rise to two nuclei each with mass 120 and charge 46.87. But if we again apply our formula, this time to the nucleus of mass 120, we find that the best value of Z is 51.15. Thus we see that something like four β-transformations are needed to bring about a stable condition in either of the half nuclei. This means eight β-transformations (or their equivalent in β-transformations and neutron emission combined) with a total energy release of about 15 MeV. This emission of débris will in general occur after the separation of the two fragments, so that the energy associated with it will not manifest itself in the kinetic energy of the initial fission process.

12.2 Limitations on the Occurrence of Fission

From inspection of the packing fraction curve, it may be seen that the energy release upon splitting a nucleus into two equal parts should be positive quite a long way down the periodic table – to about $A = 100$, in fact. Yet fission is observed to occur only for the very heaviest nuclei. Why should this be so? The answer is to be found from a consideration of the mutual potential energy of the fragments. Let the potential be 0 when the fragments are very far apart. Then at some separation D (Fig. 12.3a) the energy U is given by $\dfrac{Z^2e^2}{4D}$, Z being the charge of the initial nucleus. This

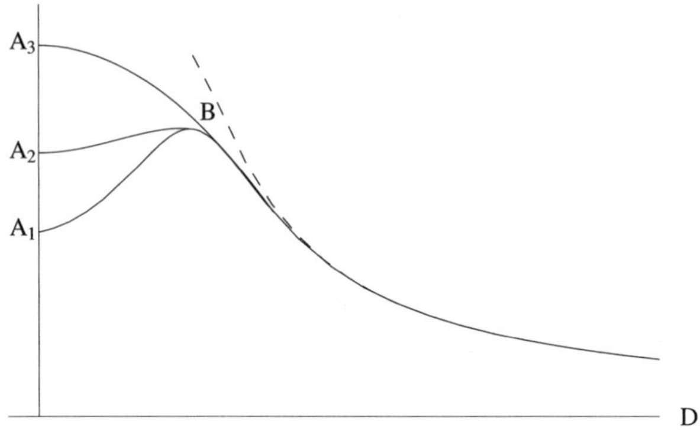

Fig. 12.4

is true as long as D is appreciably greater than the sum of the radii of the fragments, i.e. beyond the limit of nuclear interaction as distinct from the Coulomb field. For $D = 2R + \delta$, the potential is modified by an attractive component, and $D = 0$ corresponds to the initial nucleus. Thus the potential as a function of D appears as in Fig. 12.3b. The ordinate OA represents the energy which would appear as kinetic energy of the fragments if fission occurred. But before the fragments can roll away down the Coulomb slope to acquire this energy they must surmount the onset of the potential hill at B in the figure, and if the initial nucleus is sufficiently stable to be found in nature, B is higher than A. Now since the fragments are regarded as being at rest at A, and since their potential energy increased from A to B, their kinetic energy at B must be negative. This is the old impasse of classical physics, according to which a potential barrier of this kind is impenetrable. Wave mechanics has resolved the problem, and according to it the wave function has an exponentially decreasing amplitude over the region of negative kinetic energy. This means that a particle inside a nucleus has a small but finite chance of appearing on the other side of the barrier.

Now we may distinguish three possible forms of the potential function; they are shown in Fig. 12.4. In the first, A is considerably lower than B. The probability of penetration of the barrier is so small that it can be disregarded for all practical purposes. The second possibility corresponds to a spontaneously fissionable nucleus. A_2 is at about the same height as B, but the effect of the potential hump is to render fission rather unlikely. Finally, the third possibility is a nucleus formed by the excitation of a

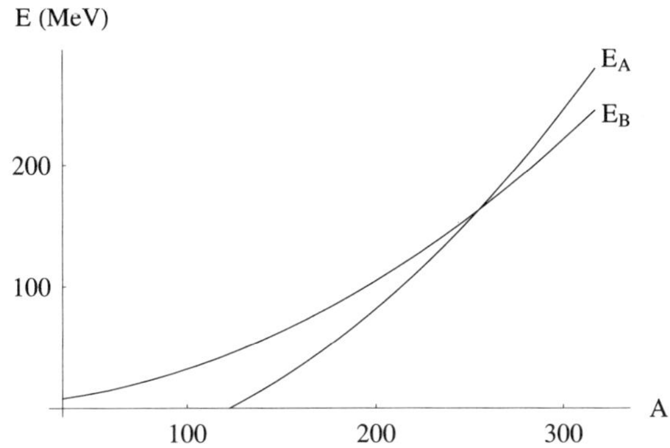

Fig. 12.5

nucleus of the second type (e.g. by capture of a neutron). Such a nucleus could not exist in nature, for it has no potential barrier to prevent its component particles from separating.

The chance of penetration of the barrier depends very sensitively on its height ($E_B - E_A$ in the figure), and it is this which confines fission to the heaviest nuclei. Fig. 12.5 shows E_B and E_A as functions of A. The curves cross at about $A = 250$. E_A itself becomes negative below $A \sim 85$, so that fission could not occur below this point even if the potential barrier were entirely absent.

A full account of the problem is to be found in the famous paper of Bohr and Wheeler, in the Physical Review, **56**, p. 426 (1939). Here the basis of the discussion is the liquid drop and its stability against deformations. We shall consider this picture in the next section.

12.3 The Liquid Drop Model in Fission

This leads to a more exact idea of the energy-separation curve. The drop, supposed initially spherical, can suffer deformations of various kinds. The simplest of there is an ellipsoid, with no change of volume. The energy changes are then due only to

(i) An increase of surface area, and hence of surface energy,
(ii) A decrease of electrostatic energy.

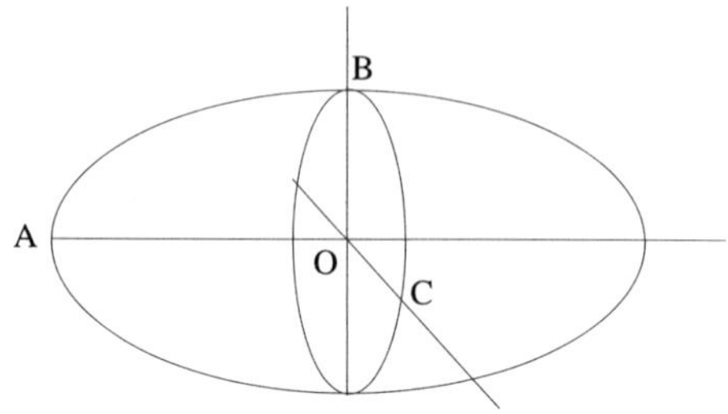

Fig. 12.6

Now the surface energy goes as $A^{2/3}$, whereas the electrostatic energy goes as $\frac{Z^2}{A^{1/3}} \sim A^{5/3}$. The electrostatic energy thus becomes more important for heavy nuclei, and this at once suggests the possibility of fission.

Suppose we have an ellipsoid of revolution, as shown in Fig. 12.6. $OA = a$, $OB = OC = b$. We imagine the charge Ze to be uniformly distributed through it. Then the volume energy of the charge is given by

$$W_v = \frac{3}{10} \frac{Z^2 e^2}{\sqrt{a^2 - b^2}} \log \frac{a + \sqrt{a^2 - b^2}}{a - \sqrt{a^2 - b^2}} .$$

The surface energy is given by

$$\begin{aligned} W_s &= \text{const.} \times \text{surface area} \\ &= \text{const.} \left[2\pi b^2 + 2\pi \frac{a^2 b}{\sqrt{a^2 - b^2}} \cos^{-1}\left(\frac{b}{a}\right)\right] . \end{aligned}$$

Now consider a sphere of radius R. We stretch it in one dimension to $a = R(1 + \epsilon)$. The volume stays constant, so that we have

$$\frac{4\pi}{3} a\, b^2 = \frac{4\pi}{3} R^3 ,$$

$$b = \frac{R}{\sqrt{1+\epsilon}} .$$

Substituting for a and b in terms of R in the above expression, we find

$$\Delta W_v = -\frac{3}{5} \frac{Z^2 e^2}{R} \left(\frac{1}{5}\epsilon^2\right) ,$$

$$\Delta W_s = \text{const.}\ 4\pi R^2 \left(\frac{2}{5}\epsilon^2\right) .$$

Using the appropriate numerical coefficients for these volume and surface energies from the formula for $M(A,Z)$ (p. 91) we find that the excess of energy of the ellipsoid over the sphere is:

$$\Delta W = \epsilon^2 \left[\frac{2}{5}\, 0.14\, A^{2/3} - \frac{1}{5}\, 0.000627\, \frac{Z^2}{A^{1/3}} \right] .$$

The energy change is positive if $\frac{Z^2}{A} < 44.7$, and this is the condition of stability since it means that the energy has to be put into the system to deform it. For U, $\frac{Z^2}{A} = 36$, and for all lighter elements it is lower. The theory is therefore not correct in detail. However, as we have pointed out in the previous section, instability will in fact set in before the condition of stability against deformation is violated, because the potential barrier is rather transparent and one would expect appreciable spontaneous fission. Even for U^{238} there are ~ 20 fissions/gm/hr, and for heavier nuclei it quickly becomes still more frequent - e.g. Pu^{238} has 10^7 spontaneous fissions/gm/hr.

Neutron-induced fissions are of more interest than the spontaneous effect. The capture of a neutron adds 5-6 MeV to the nuclear energy, as we have seen (p. 91), and this may give rise to fission before a γ-ray can be emitted. Since the binding energy of neutrons to nuclei with an odd number of neutrons is larger than to those with an even number, we expect fission for a thermal neutron striking a nucleus to happen more easily if the nucleus is of the former type. This is found to be so; U^{235}undergoes fission by thermal neutrons, whereas U^{238} does not. From such facts, and from the measurement of photo-fission thresholds, one can estimate that for U the height of the fission barrier is ~ 5 MeV.

The relative probabilities of fission and other processes are not simply a function of the energy. One must of course <u>have</u> the energy, but it must also be concentrated in the proper modes of motion for the fission. This may take time, and competing processes may occur. Since the number of modes - and therefore of useless non-fission modes - increases with the excitation energy, it may be that this is the reason why photo-fission has not been observed with 100 MeV γ-ray on lighter nuclei.

12.4 The Particles of Fission

Besides fission fragments, one can observe neutrons, β-particles, γ-rays and sometimes very fast α-particles following or accompanying fission. The

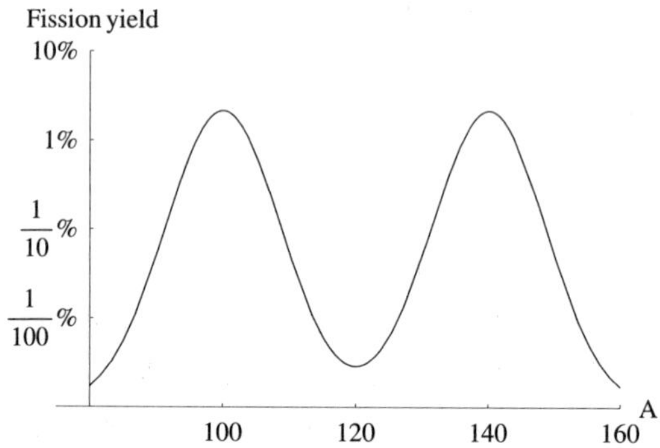

Fig. 12.7

fragments themselves exhibit two peaks in the mass distribution, as shown in Fig. 12.7. Exact halves are extremely rare, and this fact has not yet been explained theoretically.

The neutrons are classed as prompt or delayed. 'Prompt' implies that they are emitted in a time less than 10^{-15} seconds after the fission (shorter, that is to say, than we can detect). One can account for the prompt neutrons in terms of the excess energy possessed by the fission fragments. The parts of the original nuclear drop have much vibrational energy, and this may well be enough to evaporate a neutron, especially as the binding energies for neutrons are low in any case in view of the excess of neutrons in the fragment as formed. For example, suppose that U^{235} is made to fission by a neutron and that two fragments with $A = 118$ and $Z = 46$ appear. Using the formula for $M(A, Z)$, one can calculate the binding energies of a neutron to the various nuclei having $A = 118$. The values are set out in the following table:

Z	44	45	46	47	48	49	50
Binding energy (MeV)	2.5	6.8	3.6	7.8	4.7	9.0	5.8

$Z = 50$ is the stable species. We see from this that neutrons are lightly bound to the fragments, and whenever the neutron emission is possible energetically it is also likely, because of the absence of a Coulomb barrier. One would conclude, in fact, that the excitation in the fragments is considerable, since the value of ν (ν = average number of neutrons per fission) for U^{235}, to take a typical example, is 2.45. This represents more than one neutron per fragment on the average.

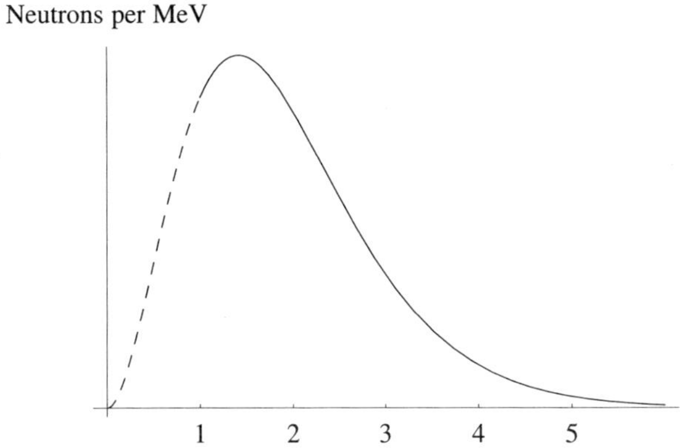

Fig. 12.8

Fig. 12.8 shows the distribution in number and energy of the fission neutrons. In the centre of mass system of neutron and fragment the energy distribution for the neutron would be a kind of Maxwellian with a temperature corresponding to the energy of excitation of the fragment. To get a theoretical curve for the laboratory system one would have to take account of the motion of the fragment and the dependence of emission probability on neutron energy. The observed distribution as shown in Fig. 12.8 agrees roughly with expectations, with the qualification that no accurate measurements have been made below 1 MeV.

Delayed neutrons constitute about 1% of the total. They can be accounted for in terms of an initial β-particle emission from the fragment, leading to a new nucleus which may be formed in some excited level above the binding energy of a neutron. This is depicted in Fig. 12.9. The fission fragment may decay directly by the transition β_1 to the ground state of the neighbouring nucleus,or it may take the transition β_2 to an excited state, and thence to the ground state of an isotope of the same neighbouring nucleus. Just as for the original fragment, one would expect neutron emission to occur immediately after the formation of the excited state. This means that the decay periods for the delayed neutrons correspond to the periods of the β-decays preceding them. The yields and half-lives of the delayed neutrons are shown in the accompanying table. The first two periods are well established; the shorter ones are not so certain.

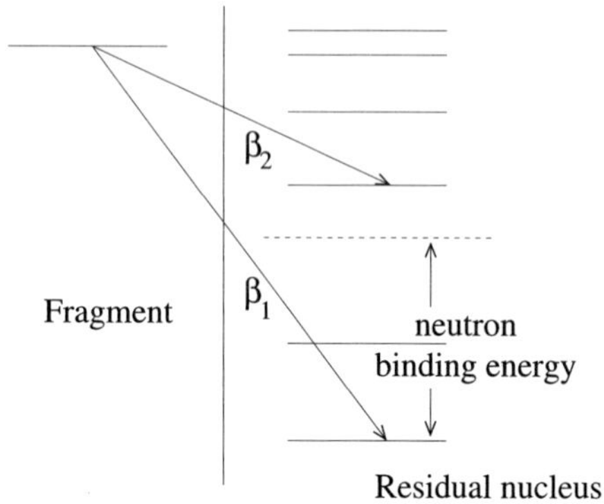

Fig. 12.9

Half-life (sec)	Delayed neuts per prompt
55.6	0.00021
22.0	0.00139
4.51	0.00178
1.53	0.00201
0.42	0.00071

12.5 Cross Sections for Fission and Other Processes

In U^{238}, the processes other than scattering are essentially confined to (n, γ) and (n, f). In Fig. 12.10 the cross sections for these processes are plotted. It may be seen that capture is more likely than fission up to about 1 MeV or so. Figs. 12.11 show the fission cross section of U^{235} in the low and high energy regions respectively. (See next page). At high energy the (n, γ) process in U^{235} is very improbable by comparison with fission. At low energies it follows the general shape of the (n, f) curve, but lies below it; $\sigma_{\mathrm{th}}(n, \gamma) \sim \frac{1}{5}\sigma_{\mathrm{th}}(n, \mathrm{f})$. It also has its own resonances at about 2 eV and 5 eV.

For Pu^{239}, $\sigma_{\mathrm{th}}(n, \mathrm{f}) = 705$ barns, $\sigma_{\mathrm{th}}(n, \gamma) = 345$ barns, so that the total cross section is ~ 1050 barns. The fission cross section as a function of energy is given in the following table. Capture becomes relatively less important at higher energies.

Energy (eV)	0.025 (thermal)	0.1-0.2	0.3	10^4	$3 \cdot 10^5$	Levels off
σ_f (barns)	703	$1/v$ law	6000 (resonance)	3	2	1.96 .

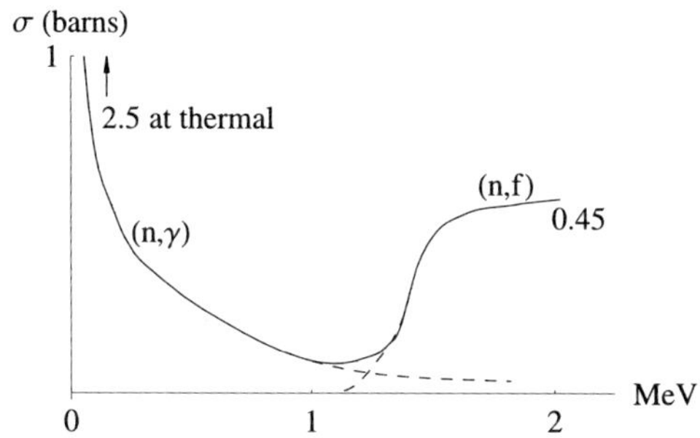

Fig. 12.10

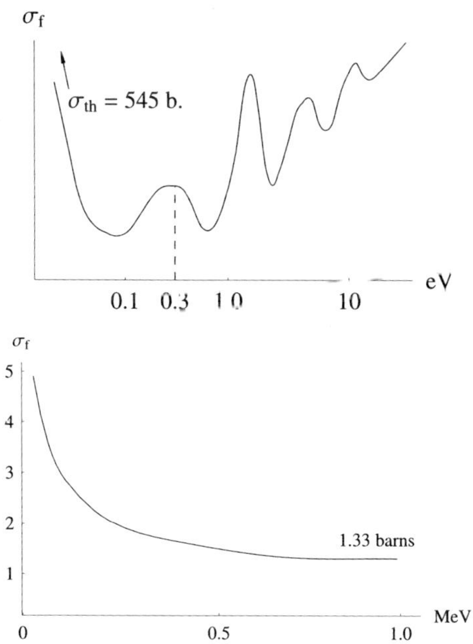

Fig. 12.11

Chapter 13

The Possibility of a Chain Reaction

13.1 The Properties of Natural Uranium

It will be shown here that a chain reaction cannot maintain itself in natural uranium. The persistence of a chain reaction depends on the relative probabilities of the (n, f) and (n, γ) processes. A fission in U produces on the average about 2.4 neutrons, but the (n, γ) cross sections over the total fission neutron spectrum remove most of these and make a self-sustaining chain impossible.

We have, in natural U, which is an isotopic mixture, 1 part in 140 of U^{235}. Let us put $\sigma_{\text{effective}}(235) = \dfrac{1}{140}\,\sigma_{\text{true}}(235)$, and then plot on a single curve the cross sections for the various processes. This is done in Fig. 13.1. If we make observations on natural uranium, we do not of course detect the contributions of the different isotopes as separate effects. We are therefore interested only in the total fission cross section and the total radiative capture cross section of the isotopic mixture. A plot of these quantities is given in Fig. 13.2. It may be observed that σ_{f} is greater than σ_{c} in two regions:

a) For $E > 1.4$ MeV.
b) For $E \ll 1$ eV.

From 1 eV to 1 MeV σ_{c} is about 15 to 20 times bigger than σ_{f}. Thus with unseparated U we could only hope to sustain a chain reaction with thermal neutrons or with neutrons of energy greater than 1 MeV. We will consider these regions in turn.

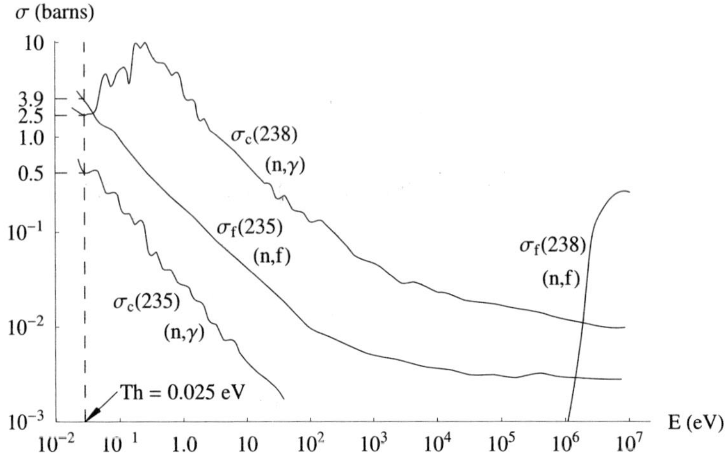

Fig. 13.1

13.1.1 *The High Energy Region*

There are two alternative processes to fission. These are:

a) Elastic scattering, which is not important, since it scarcely affects the neutron energy.
b) Inelastic scattering, which pushes the neutrons down below 1 MeV into the forbidden region where fission is very unlikely.

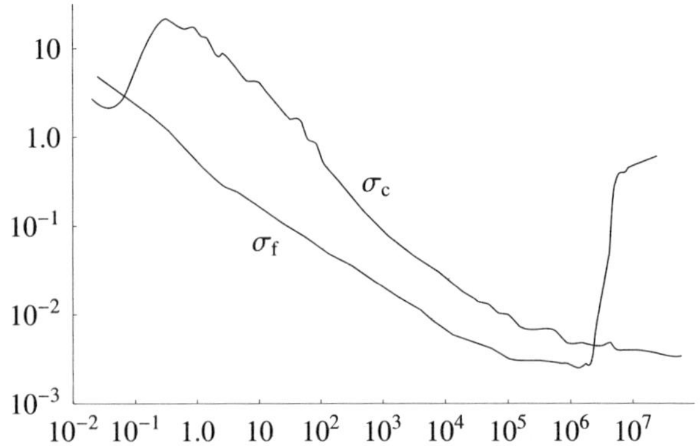

Fig. 13.2

Suppose that we have a large mass of U, and that one fission occurs in the middle. This gives about 2.4÷2.5 neutrons. The fission spectrum is such that about 1.5 of these neutrons are above the fission threshold ($\sim$ 1.5 MeV, effectively), and about 1 neutron below it. This low energy neutron is probably absorbed without causing fission - the probability of fission is only about 0.05. The 1.5 neutrons above 1.5 MeV are also subject to capture to a certain extent, but, more important, they are liable to be inelastically scattered. In terms of cross sections, $\sigma_{\text{inelastic}} = 2.3$ barns, $\sigma_{\text{f}} \sim 0.4$ barns, and $\sigma_{\text{c}} \sim 0.1$ barns. Thus, of the neutrons available, only a fraction $\frac{0.4}{2.3 + 0.4 + 0.1} \sim \frac{1}{7}$ can give fissions. This is about 0.2. Together with the 0.03 neutrons which give rise to fissions after being slowed down below the U^{238} fission threshold, we therefore have 0.25 fissioning neutrons. We need at least 1.0 for maintaining a chain, so that a fast neutron chain in natural U is impossible. The figure 0.25 is the reproduction factor in this case.

13.1.2 *The Thermal Region*

If we have $E = 0.025$ eV,

$$\left.\begin{aligned} \sigma_{\text{f}} &= 3.9 \text{ barns} \\ \sigma_{\text{c}} &= 3.2 \text{ barns} \end{aligned}\right\} \qquad \sigma_{\text{total}} = 7.1 \text{ barns.}$$

All these cross sections go as $1/v$, so the ratio remains nearly constant. Thus, given one fission with its 2.4÷2.5 neutrons, and supposing that we can slow them all down to thermal without absorption, the reproduction factor will be $2.4 \cdot \frac{3.9}{7.1} = 1.32$. This means a self-sustaining chain. But the slowing down has to be accomplished with not more than about 25% loss, if the reaction is to work in a non-infinite system.

13.2 Moderators

To achieve the slowing down with little chance of high energy fission, the U must be spread thinly in a non-absorbing scatterer. He^4 could be contemplated, since He^3 has a negative binding energy for neutrons. But the volume of such a system would be excessive - about 1 cubic mile in fact. Compression, or the use of liquid He, would help, although this is not too practicable. The result would be what is called a pile. The principle of compression, although it cannot actually be applied here, is very useful.

Suppose the linear dimensions of a system to be reduced of a factor 2, and the density to be increased by a factor 2. Then the mean free path is halved, and the whole system is scaled down to work exactly as before. But the mass is reduced by a factor 2^2, and the volume by a factor 2^3.

More suitable substances than He^4 are C (graphite) and D_2O. It is possible that H_2O and Be can also work with natural U. There is some capture for all these substances, but it is small.

13.3 Homogeneous and Lumped Graphite Piles

Consider a graphite pile, and assume it to be so large that surface leakage of neutrons can be neglected. Suppose that the U is uniformly spread through the pile. For graphite at thermal energies,

$$\sigma_{\text{sc}} = 4.8 \text{ barns}$$
$$\sigma_{\text{c}} = 0.0048 \text{ barns}.$$

We want as much graphite as possible relative to the U so as to slow the neutrons efficiently, but there is an upper limit to the ratio R of the numbers of C and U atoms, set by the finite capture by the C. In fact, if we suppose 2.4 neutrons per fission, we must have

$$2.4 \cdot \frac{3.9}{7.1 + 0.0048R} \geq 1.0$$

$$R \leq 470.$$

This means about 25 times as much graphite by weight. A consideration of the losses during slowing down, and of the fact that in actual piles the U is not quite uniform (leading to a higher neutron density in the C than in the U, which is a sink for neutrons) reduce this upper limit for R. (The non-uniformity effectively increases $\sigma_{\text{c}}(\text{C})$, as may readily be seen. There must exist also a lower limit for R, since we have seen that the pile will not go without any graphite (or similar moderator). Evidently, if a working pile is to be possible, the lower limit for R must be smaller than the upper limit.

The problem of capture during the slowing down process must be investigated. Suppose that the slowing down density q (= neutrons per cm^3 per seconds slowed from above to below E, where E is any energy in the slowing region, Fig. 13.3) is uniform in space. This implies a uniform medium and a uniformly spread source.

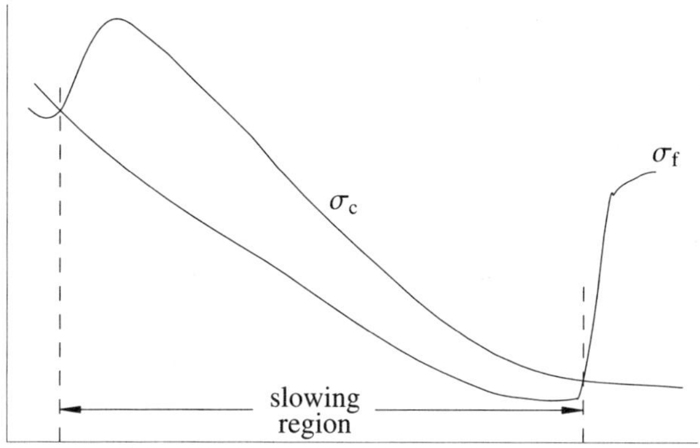

Fig. 13.3

From the discussion in Chapter 8 we have seen that the velocity of drift for a neutron down the ϵ-axis ($\epsilon = \log E$) is $\xi \dfrac{v(\epsilon)}{\lambda}$, where $\xi = \overline{\log \dfrac{E_1}{E_0}}$. Hence

$$q = n(\epsilon) \cdot \xi \frac{v(\epsilon)}{\lambda(\epsilon)} \,.$$

Thus the number of neutrons per cm^3 in the logarithmic energy interval dϵ at ϵ is:

$$n(\epsilon)\, \mathrm{d}\epsilon = q \frac{\lambda}{\xi} \cdot \frac{1}{v} \cdot \frac{\mathrm{d}E}{E} \,.$$

Now consider a U atom in the pile. The number of neutrons in this energy interval which this atom absorbs per second is equal to $n(\epsilon)\, \mathrm{d}\epsilon\, v\, \sigma(\epsilon)$, and therefore the total number of neutrons, irrespective of energy, which the atom absorbs per second is given by

$$q \frac{\lambda}{\xi} \int \sigma(\epsilon) \frac{\mathrm{d}E}{E} \,.$$

The setting of the upper limit is not critical, owing to the rapid decrease of the integrand with increase of E. The lower limit is usually taken as the Cd absorption edge i.e. about 0.3 eV. Then for U,

$$\int_{0.3}^{E_{\max}} \sigma(\epsilon) \frac{\mathrm{d}E}{E} = 240 \text{ barns.}$$

This figure is determined experimentally.

Since q neutrons are produced per cm^3 per second, and since each atom absorbs $q\,\frac{\lambda}{\xi}\int_{0.3}\sigma\,\frac{\mathrm{d}E}{E}$ neutrons per second, each atom can be associated with an absorption volume equal to $\frac{\lambda}{\xi}\int_{0.3}\sigma\,\frac{\mathrm{d}E}{E}$. Then we can say (although this in no way represents the physical reality) that each atom will absorb all neutrons produced in the absorption volume surrounding it, and no others. For U, the volume is $4\cdot10^{-21}$ cm^3, which is much greater than the geometrical volume of the nucleus - about 10^{-35} cm^3. The total number of neutrons absorbed per cm^3 per second in the pile would then be equal to $q\cdot4\cdot10^{-21}$ (atoms U/cm^3). If we tale the value of R to be about 500, the number of atoms of U per cm^3 is $1.6\cdot10^{20}$, and inserting this value we find that $0.64\,q$ neutrons/cm^3/second are absorbed.

The above value for the number of neutrons captured is in fact an overestimate, because of the existence of resonances in the capture cross section of U, shown very crudely and inaccurately on Fig. 13.4. The absorption of n atoms of U is less than n times the absorption of one. The reason is this: The high capture probability for a neutron of one of the resonance energies tends to remove this kind of neutron from the spectrum, so that the number/energy curve for the neutrons in the pile will look like Fig. 13.5. Thus there is a shortage of the kind of neutrons that the U nuclei capture most efficiently, and the mean cross section of a U atom in a large mass such as a pile is appreciably lower than for an isolated atom. This at one suggests that one ought to put the U in a pile in lumps rather than diffusely, since the lump has no moderator to spread out the neutron spectrum and continually put neutrons into the resonance regions. We must, however, investigate the absorption of thermal neutrons by U and C as affected by lumping.

We must in this case assume different neutron densities in the U and the graphite; let these densities (thermal neutrons) be n_{U} and n_{C} respectively. Moreover, let N_{U} and N_{C} be the numbers of U and C atoms per cm^3. Then the ratio of the carbon and uranium absorptions is given by

$$\frac{\sigma_{\mathrm{C}}n_{\mathrm{C}}N_{\mathrm{C}}}{\sigma_{\mathrm{U}}n_{\mathrm{U}}N_{\mathrm{U}}}\,.$$

Now since the U in lumps is a neutron sink, we have $n_{\mathrm{U}} < n_{\mathrm{C}}$. It follows from this that lumping is unfavourable from the point of view of obtaining thermal neutron-induced fissions. This suggests that there must be an optimum size of lump. The matter is discussed a little further in the next chapter.

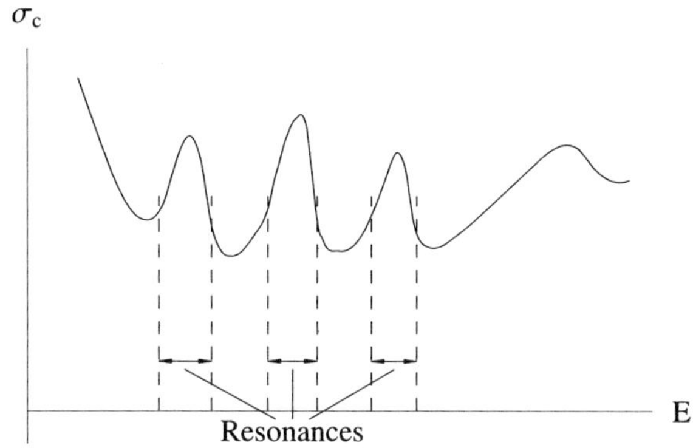

Fig. 13.4

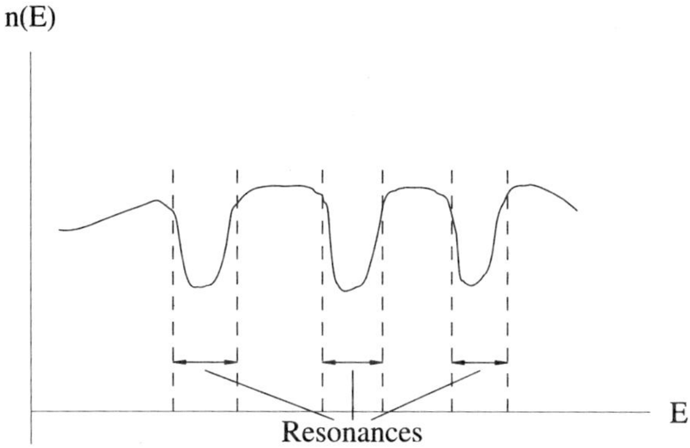

Fig. 13.5

There are certain empirical relations concerning the self-absorption in lumped and mixed piles. For a homogenous U-C mixture,

$$\left[\int \sigma \frac{\mathrm{d}E}{E}\right]_{\text{effective}} = 2.95\left[\sigma_{\text{scattering}}\right]^{0.47} \qquad \text{for } \sigma_{\text{sc}} < 1000.$$

In this formula,

$$\sigma_{\text{scattering}} = \sigma_{\text{sc}}(1 \text{ atom U}) + \sigma_{\text{sc}}(1 \text{ atom C}) \times \frac{\text{No. of C atoms}}{\text{No. of U atoms}}.$$

This is a markedly different result from that obtained from the simple addition of atomic absorptions. Thus, for the U-C mixture that we have already considered the formula gives

$$\left[\int \sigma \frac{\mathrm{d}E}{E}\right]_{\text{effective}} = 124 \text{ barns},$$

compared with our calculated value of 240 barns. Thus, even for a uniform distribution (homogeneous pile) the self-absorption effect is about a factor 2, i.e. the absorption volume per U nucleus is only about half what it would be in the absence of resonances.

For lumps the difference is much more strongly marked. The empirical formula for lumps of U in graphite is

$$\text{Absorption Volume} = (0.385 \times \text{Mass of U in gm.}) + (1.05 \times \text{Surface Area of U in cm}^2).$$

Applying this to a U sphere of 3 cm radius, we find:

$$\text{Actual volume (geometrical)} = 112\,\text{cm}^3$$
$$\text{Empirical absorption volume} = 990\,\text{cm}^3$$
$$\text{Calculated absorption volume} = 23000\,\text{cm}^3.$$

(neglecting self-screening and self-absorption)

We see, therefore, that the calculated value is more than twenty times greater than the observed, the advantage to be gained from lumping the U is therefore apparent. We shall end this chapter by investigating whether the homogeneous pile will work at all.

13.4 The Possibility of a Homogeneous Pile

Let us suppose that the atomic ratio of C to U atoms is 200, the two kinds of atoms thoroughly mixed. We will take the value of ν (ν = no. of neutrons per fission) to be 2.4. Then we find ($\sigma_{\rm sc}(C) = 4.8$ barns, $\sigma_{\rm sc}(U) = 10$ barns)

$$\left[\int \sigma \frac{\mathrm{d}E}{E}\right]_{\text{effective}} = 75 \text{ barns.}$$

The total absorption per cm^3 in slowing down is given by

$$q \cdot \frac{\lambda}{\xi}\left[\int \sigma \frac{\mathrm{d}E}{E}\right]_{\text{effective}} \cdot N_{\rm U} = q \times \frac{2.6}{0.158} \times 75 \cdot 10^{-24} \times 4.0 \cdot 10^{20}$$
$$\sim 0.5\, q \text{ neutrons per second.}$$

Thus 1.2 neutrons per fission reach thermal energies. Of these, the number captured by U nuclei is equal to

$$1.2 \times \frac{7.1}{200 \cdot 0.0048 + 7.1} = 1.06.$$

But of the total U absorption cross section of 7.1 barns, only 3.9 corresponds to fission, so that the reproduction factor is only $\frac{3.9}{7.1} \times 1.06 = \underline{0.58}$.

The pile will not go. By leaving the value of R (the C/U atomic ratio) undetermined, we can try to solve for it, assuming that the reproduction factor is unity. We find, however, that it is impossible to get an answer for positive R. The conclusion to be drawn is that no homogeneous pile of this type will work, and we must therefore devote our attention (if we are considering only the U-graphite combination) to heterogeneous piles.

Chapter 14

The Heterogeneous Pile

14.1 The Design of a Lumped Pile

Two possible geometries, sketched in Figs. 14.1a and b, are composed of spherical lumps or of rods. The design using a lattice of spherical lumps is slightly more efficient than the other, but rods are easier to cool, and this is an important consideration in a pile which is to develop high power.

Let us calculate the reproduction factor for a pile composed of spherical lumps. We shall suppose that we start with a single fast neutron in one of the spheres. Since the neutron has a small but finite chance of producing a fission while it is still fast, a number ϵ of neutrons will emerge from the sphere, where ϵ is slightly greater than unity. Let f_R be the probability that a neutron is swallowed up either in graphite or by resonance capture in U. Then the number of neutrons reaching thermal energies is given by $\epsilon(1-f_R)$. Let f_T be the probability that a thermal neutron is absorbed by U rather than by graphite. Then the number of thermal neutrons captured by U, per fast neutron emitted, is equal to $f_T\,\epsilon(1-f_R)$. Let us further define a quantity η, which is the number of fission neutrons produced as a result of the capture of a thermal neutron by U. Evidently,

$$\eta = \nu\,\frac{\sigma_{\rm f}(\text{th})}{\sigma_{\rm f}(\text{th})+\sigma_{\rm c}(\text{th})} = \frac{3.9}{7.1}\,\nu\,.$$

Then the reproduction factor is k, where

$$\boxed{k = \eta\, f_T\,(1-f_R)\,\epsilon\,.}$$

The design of a pile involves the calculation of the various quantities in the above equation for k. η is a basic property of the fissioning nucleus, but all the other factors depend on the composition and geometry of the pile. In the next section we shall consider how the calculations are made.

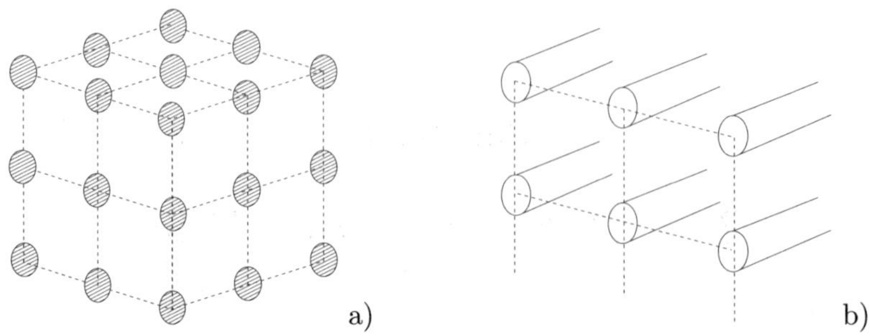

Fig. 14.1

14.2 The Determination of Pile Constants

14.2.1 *The Magnitude of* ϵ

The average path of a neutron escaping from a spherical lump is equal to the radius R of the lump. We shall assume that there is only a 60% chance for a neutron to lie at the high energy end of the fission spectrum ($E > 1.4$ MeV). The cross section for fission at these energies is 0.45 barns, and corresponding to this cross section there is a mean free path for fission production, given by

$$\lambda_\mathrm{f} = \frac{1}{N\sigma_\mathrm{f}},$$

where N is the number of U nuclei per cm^3. Inserting the numerical value of N - about $4 \cdot 10^{22}$ - one finds that λ_f is about 50 cm. Since the number of extra neutrons produced in a fission is $(\nu - 1)$, we have

$$\begin{aligned}\epsilon &\sim 1 + 0.6 \cdot \frac{R}{50} \cdot 1.4 \\ &\sim 1 + 0.017\,R.\end{aligned}$$

This is not, of course, anything more than a rough approximation to the answer, but it gives the order of magnitude.

14.2.2 $(1 - f_R)$

Consider a unit cell of the lattice of the pile structure, as shown in Fig. 14.2. The sphere of U is shaded in the figure. We use the empirical formula for the absorption volume of a lump of U:

$$V_\mathrm{U} = 0.383 \times \text{Mass} + 1.03 \times \text{Surface}.$$

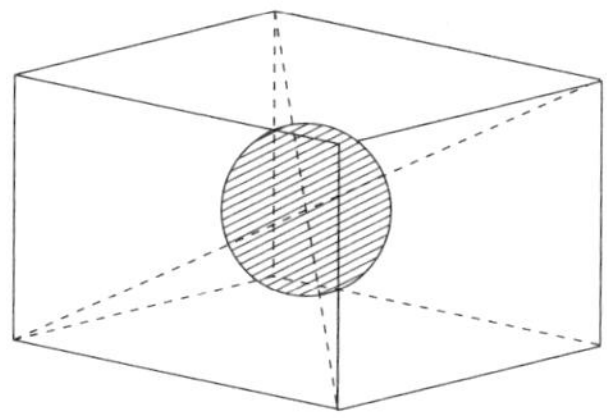

Fig. 14.2

If the slowing down density is q, then qV_{U} neutrons are absorbed each second by the lump. But if V_{C} is the volume of the cubical unit of the lattice, then qV_{C} neutrons per second are produced by the lump. It follows then that

$$f_R = \frac{V_{\mathrm{U}}}{V_{\mathrm{C}}}$$

approximately. A better calculation gives

$$f_R = 1 - \mathrm{e}^{-V_{\mathrm{U}}/V_{\mathrm{C}}},$$

and hence

$$\boxed{1 - f_R = \mathrm{e}^{-V_{\mathrm{U}}/V_{\mathrm{C}}}.}$$

14.2.3 *Calculation of f_T*

If the lattice is effectively infinite, $(\operatorname{grad} n)$, where n is the neutron density, must vanish on any plane of symmetry. The boundaries of a unit cell, such as that shown in Fig. 14.2, are symmetry planes, but it simplifies the problem to replace the cubical cell with one having spherical symmetry throughout, as indicated in Fig. 14.3. Imposing this boundary condition, one then finds

$$f_T = \frac{3\alpha^2}{(\beta^3 - \alpha^3)} \cdot \frac{(1-\alpha)(1+\beta)\mathrm{e}^{-(\beta-\alpha)} - (1+\alpha)(1-\beta)\mathrm{e}^{(\beta-\alpha)}}{(\alpha + s - s\alpha)(1+\beta)\mathrm{e}^{-(\beta-\alpha)} - (\alpha + s + s\alpha)(1-\beta)\mathrm{e}^{(\beta-\alpha)}},$$

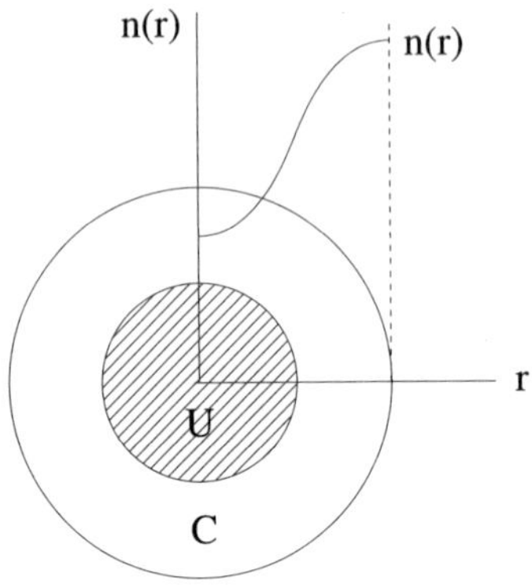

Fig. 14.3

where

$$\alpha = \frac{\text{Radius of U lump}}{\text{Diffusion length in C}}$$

$$\beta = \frac{\text{Radius of cell}}{\text{Diffusion length in C}}$$

$$s = \frac{1+\gamma}{1-\gamma}\cdot\frac{1}{\sqrt{N}}, \qquad \text{where} \begin{cases} \gamma = \text{albedo of the sphere} \\ \qquad \text{(looking from the outside)} \\ N = \dfrac{\sigma_{\text{total}}}{\sigma_{\text{absorption}}} \sim 1100 \text{ for C.} \end{cases}$$

The fraction $\frac{1-\gamma}{1+\gamma}$, whose reciprocal appears in the formula, can be expressed in terms of the diffusion length ℓ_i and the quantity N_i, which is equal to the ratio $\frac{\sigma_{\text{total}}}{\sigma_{\text{abs}}}$, for U. The equation is:

$$\frac{1-\gamma}{1+\gamma} = \frac{\sqrt{3}\ell_i}{\lambda_i N_i}\left\{\frac{e^{2R/\ell_i}+1}{e^{2R/\ell_i}-1} - \frac{\ell_i}{R}\right\}.$$

In uranium, $\ell_i = \frac{24.0}{\rho}$, $\lambda_i = \frac{21.1}{\rho}$, $N_i = 2.79$. (ρ = density). A remarkable sort of lattice on which to base a numerical evaluation of these formulas could have a cubical unit cell of side 20 cm, with a spherical lump of U of weight 3000 g in the middle ($\rho_U = 18$).

We have now seen how to calculate the relevant quantities. In a typical case one might have $\eta = 1.32$, $f_T = 0.88$, $f_R = 0.12$, $\epsilon = 1.05$. For such a

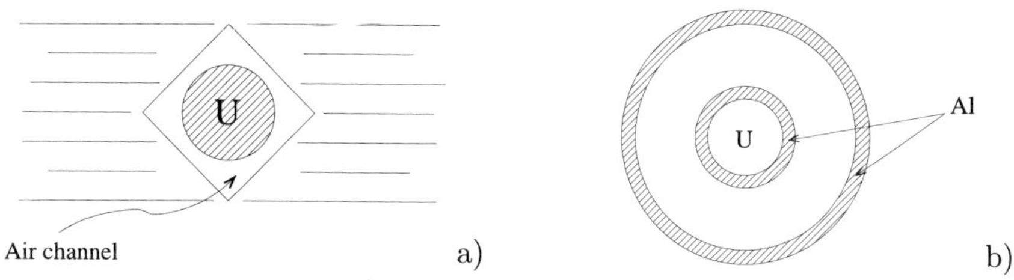

Fig. 14.4

lattice, $k = 1.07$. (The best possible value of k for a U-C system is about 1.10).

In a practical pile, if it is to be run at a high power level, one is bound to sacrifice a certain amount of k in putting in provision for cooling. This will be true in any industrial pile, or in a laboratory pile run at a high flux. The Argonne pile of the University of Chicago is severely handicapped because of the lack of cooling, and cannot be run at a power more than 2 kilowatts. (It is a 20 ft cube in size). Figs. 14.4 a) and b) give some idea of the amount of space taken up by an efficient cooling system. Fig. 14.4 a) shows the type of cooling used at Clinton. The lattice is composed of rods, and the cooling is performed by a blurt of air. At a power of 5 megawatts, the air emerges at a temperature of about 100°C. Other piles are cooled as in Fig. 14.4 b), by water flowing through an annular jacket. The U must in this case be protected with Al, as it is corroded by hot water. The actual size of the loss of k entailed by these systems is very considerable, and may reduce it from 1.10 down to about 1.01.

14.3 Reproduction Factor and Critical Size

A pile is said to be critical if the value of k is exactly equal to 1, and in this condition it just coasts along, maintaining a constant power and a constant neutron population. This corresponds to the actual working of a pile in practice; the neutron flux is allowed to build up to any desired value and is then held constant. Now suppose that we have calculated the value of k for a given lattice, and that it turns out to be in excess of 1 - say 1.05 - so that the pile would be supercritical if infinite. We then are interested in the question, how large must the pile be in order to become just critical? We shall try to answer this question in what follows.

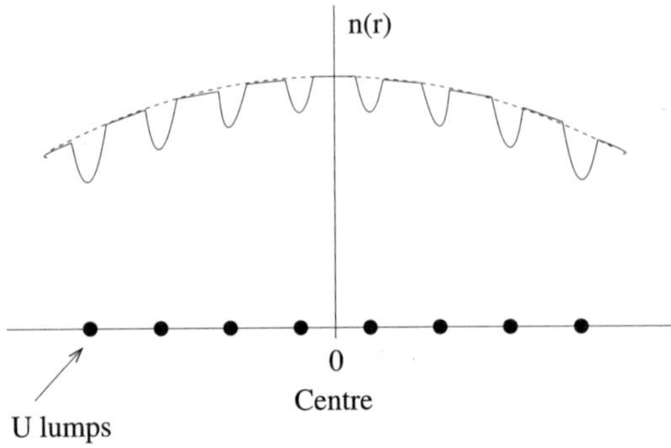

Fig. 14.5

The actual neutron density in a finite pile is somewhat as shown in Fig. 14.5. $n(r)$ has dips at the positions of the U lumps, and suffers a general falling off towards the sides. We shall suppose that the lattice cells are very numerous, so that the dips are smoothed out. Then the diffusion equation is written[1]

$$\frac{\lambda v}{3} \cdot \Delta n - \frac{v}{\Lambda} n + q_T = 0\,.$$

Λ is the absorption mean free path, and represents a combined average for U and C. For pure C, $\lambda \sim 25$ metres, and for a normal mixture in a pile is about 2.5 metres. We have to consider the fast neutrons from fission. We shall suppose that for these also the space distribution is smooth. They obey the slowing down equation

$$\Delta q = \frac{\partial q}{\partial \tau}\,.$$

q and n are of course related. A number $\frac{v}{\Lambda} n$ thermal neutrons are absorbed per cm^3 per second. Of these, $f_T \frac{v}{\Lambda} n$ are absorbed by U, thus giving rise to $\eta f_T \frac{v}{\Lambda} n$ fast neutrons from fission. If one now equates this rate of production of fission neutrons to the value of $q(\tau)$ at $\tau = 0$, as one strictly ought to do, then one has to reconcile the continual absorption of neutrons during the slowing down period with the use of the age equation, which contains no absorption term. If, however, we put for q the number of fast neutrons which are going to reach thermal energies, then the age equation

[1]@ In the original manuscript, the quantity q_T has an incorrect minus sign.

can legitimately be used, and we write

$$q(\tau)|_{\tau=0} = \epsilon\,(1 - f_R)\,\eta\,f_T\,\frac{v}{\Lambda}\,n = k\,\frac{v}{\Lambda}\,n\,,$$

which is the desired relation.

Suppose now that the pile is a cube of side a. We know that $n(r)$ does not become zero at the edge of the pile, but at a distance 0.7λ beyond (Chapter 9). However, this small correction need not affect us here. We shall put[2]

$$n(x,y,z) = \sum_{i,j,k} n_{i,j,k}\,\sin\frac{i\pi x}{a}\,\sin\frac{j\pi y}{a}\,\sin\frac{k\pi z}{a}\,.$$

We shall consider the (1,1,1) term for our solution. Then we have

$$n = n'\,\sin\frac{\pi x}{a}\,\sin\frac{\pi y}{a}\,\sin\frac{\pi z}{a}$$

$$q(0) = q'(0)\,\sin\frac{\pi x}{a}\,\sin\frac{\pi y}{a}\,\sin\frac{\pi z}{a}\,.$$

These equations must simultaneously satisfy the diffusion equation, the age equation and the equation for $q(0)$. Then we must have

$$-\frac{\lambda v}{3}\cdot\frac{3\pi^2}{a^2}\,n' - \frac{v}{\Lambda}\,n' + q' = 0 \tag{14.1}$$

$$-\frac{3\pi^2}{a^2}\,q' = \frac{\partial q'}{\partial\tau} \tag{14.2}$$

$$k\,\frac{v}{\Lambda}\,n' = q'(0)\,. \tag{14.3}$$

From (14.2), we find

$$n'\cdot\left\{-\frac{\lambda v}{3}\cdot\frac{3\pi^2}{a^2} - \frac{v}{\Lambda} + k\,\frac{v}{\Lambda}\,\mathrm{e}^{-3\pi^2\tau/a^2}\right\} = 0,$$

i.e.

$$k = \mathrm{e}^{3\pi^2\tau/a^2}\left\{1 + \frac{3\pi^2}{a^2}\cdot\frac{\lambda\Lambda}{3}\right\}.$$

This solution gives the k value for an infinitely extended pile such that a pile of finite size a^3 will be non-multiplicative i.e. $k(a) = 1$. This is evidently so, since we have solved the equations to the stationary state of the system, and the quantity k in the relation represents the reproduction factor for a

[2]@ Here and in the following the reader should be aware of the fact that the author uses the same symbol k for the reproduction factor and for the mode number along the z direction. However, it seems that no possible confusion arises.

system in which leakage effects can be ignored - this latter fact comes from the equation which defines k, viz.

$$k = \epsilon\,(1 - f_R)\,\eta\, f_T\,.$$

We may note in passing that our solution gives $k = 1$ if we put $a = \infty$.

Approximating the solution, and putting $k(a) = 1$, we find

$$k(\infty) = 1 + \frac{3\pi^2}{a^2}\left(\tau + \frac{\lambda\Lambda}{3}\right).$$

Now τ is, of course, the age at which a fission neutron becomes thermal, and is ~ 300 cm^2. Moreover, $\dfrac{\lambda\Lambda}{3}$ is about 350 cm^2. Then, if we express the length of the side of the pile in metres, we have

$$k(\infty) \sim 1 + \frac{2}{a_{\rm m}^2}\,.$$

Putting $k(\infty) = 1.1$, $a_{\rm m} \sim 4.5$ metres; and for $k(\infty) = 1.03$, $a_{\rm m} = 8$ metres approximately. In general,

$$a_{\rm m} \sim \frac{\sqrt{2}}{\sqrt{k-1}}\,,$$

and the volume of the pile is

$$V_{\rm m} \sim \left(\frac{2}{k-1}\right)^{3/2} \quad \text{cubic metres.}$$

In the case of a rectangular parallelepiped,

$$k(\infty) = 1 + \pi^2\left(\frac{1}{a^2} + \frac{1}{b^2} + \frac{1}{c^2}\right)\left(\tau + \frac{\lambda\Lambda}{3}\right).$$

If we solve the equations of the pile for criticality with respect to one of the higher harmonics, then we find that the above equation for $k(\infty)$ must be replaced by

$$k(\infty) = 1 + \pi^2\left(\frac{i}{a^2} + \frac{j}{b^2} + \frac{k}{c^2}\right)\left(\tau + \frac{\lambda\Lambda}{3}\right).$$

This fact justifies our neglecting all but the fundamental mode of the pile in determining the critical conditions. For, if we suppose that the pile is made just critical for the fundamental, then it is clear, from inspection of the above equations, that it is sub-critical for all other modes. Physically, this means that if the pile is started up with the value of k appropriate to the fundamental, then all other modes will quickly die out after being formed. This predominance of the fundamental holds for any shape of pile.

Chapter 15

The Time Dependence of a Pile

15.1 The Time-Dependent Equations

We shall suppose that all the neutrons produced in the pile are prompt - although in fact the delayed neutrons (p. 101) play an important role in running a pile, and may be decisive in making it work. We shall assume also that the time spent in slowing down a neutron is negligible compared with the time it spends as a thermal neutron; the ratio is in fact $\sim 1:100$. Then the slow neutron diffusion equation is

$$\frac{\lambda v}{3} \cdot \Delta n - \frac{v}{\Lambda} n + q = \frac{\partial n}{\partial t},$$

and, as before,

$$\Delta q = \frac{\partial q}{\partial \tau};$$

$$q(0) = k \frac{v}{\Lambda} n$$

- this condition is valid since it means that $q(0)$ is always proportional to n, which is so if there are no delayed neutrons and if also there is no delay in the slowing down of the prompts.

For a general shape of pile, the problem is to solve a differential equation of the form $\Delta\varphi + \omega\,\varphi = 0$, the ω's being eigenvalues, with the condition $\varphi = 0$ at the boundaries. The time-dependent solution is to be written as a product of functions of space and time.

Put

$$\begin{aligned} n &= f(t) \cdot \varphi(x, y, z) \\ q &= q(t, \tau) \cdot \varphi(x, y, z) \\ q(0) &= q(0, t) \cdot \varphi(x, y, z)\,. \end{aligned}$$

Then the age equation becomes

$$\Delta q = q(t,\tau)\,\Delta\varphi = -\omega\, q(t,\tau)\,\varphi\,.$$

But[1]

$$\Delta q = \frac{\partial q}{\partial \tau} = \varphi\,\frac{\partial q}{\partial \tau}\,,$$

$$\frac{\partial q}{\partial \tau} = -\omega\, q\,,$$

and

$$\boxed{q = q(0)\cdot \mathrm{e}^{-\omega\tau}\,.}$$

Since $q(0) = k\,\dfrac{v}{\Lambda}\,n$, this can also be written in the form

$$\boxed{q = k\,\frac{v}{\Lambda}\,f(t)\cdot\varphi(x,y,z)\cdot \mathrm{e}^{-\omega\tau}\,.}$$

We now solve for n:

$$\Delta n = f(t)\cdot\Delta\varphi = -\omega\, f(t)\,\varphi = -\omega\, n\,.$$

Hence, in the diffusion equation, we have

$$-\frac{\lambda v}{3}\cdot\omega\, n - \frac{v}{\Lambda}\,n + \frac{kv}{\Lambda}\cdot \mathrm{e}^{-\omega\tau}\,n = \frac{\partial n}{\partial t}\,.$$

(Remember that in this equation τ is the age of the thermal neutrons). The solution of the equation is

$$n = n(0)\cdot\exp\left[t\left\{\frac{kv}{\Lambda}\cdot \mathrm{e}^{-\omega\tau} - \frac{v}{\Lambda} - \frac{\lambda v}{3}\,\omega\right\}\right].$$

The exponent decreases as ω increases. Now ω is a constant representing some harmonic of the stationary solution, so we have

$$n = n(t)\cdot\sum_i \mathrm{e}^{-t\cdot F_i(\omega_i)}\times(\mathrm{const.})_i\,.$$

The larger the value of ω_i, the more rapidly this term dies out, so that, starting with any arbitrary distribution of harmonics in the pile, the fundamental soon becomes the most important term, as we noted in the last chapter. The condition for a time-independent solution is:

$$\frac{kv}{\Lambda}\cdot \mathrm{e}^{-\omega_1\tau} - \frac{v}{\Lambda} - \frac{\lambda v}{3}\,\omega_1 = 0\,,$$

[1]@ More appropriately the following equation should be written as

$$\Delta q = \frac{\partial q}{\partial \tau} = \varphi\,\frac{\partial q(t,\tau)}{\partial \tau}\,.$$

which is just the generalization of the particular case of a cube, also discussed in the last chapter. Solving for k, we find

$$k(\infty) = \mathrm{e}^{\omega_1 \tau} \cdot \left(1 + \frac{\lambda\Lambda}{3} \cdot \omega_1\right) .$$

For a cube, $\omega_1 = \dfrac{3\pi^2}{a^2}$; for a sphere, $\omega_1 = \dfrac{\pi^2}{R^2}$, and so on. Taking ω to mean always ω_1, we re-write the exponent in the general solution for n as follows:

$$k\,\frac{v}{\Lambda}\,\mathrm{e}^{-\omega\tau}\left[1 - \frac{\mathrm{e}^{\omega\tau}}{k}\left(1 + \frac{\lambda\Lambda}{3} \cdot \omega\right)\right] .$$

If the reproduction factor were

$$k(\infty) = \mathrm{e}^{\omega\tau} \cdot \left(1 + \frac{\lambda\Lambda}{3} \cdot \omega\right) ,$$

the pile would be just going. Let us write k_0 for this value of k. Then the exponent in the formula for n becomes

$$k\,\frac{v}{\Lambda}\,\mathrm{e}^{-\omega\tau}\left[1 - \frac{k_0}{k}\right] .$$

The time dependence of n is thus given by

$$\boxed{n(t) = n(0) \cdot \exp\left[\frac{v}{\Lambda}\,\mathrm{e}^{-\omega\tau} \cdot (k - k_0)t\right] .}$$

Now, in practice $\mathrm{e}^{-\omega\tau} \sim 1$, so that the expression for $n(t)$ simplifies to

$$\boxed{n(t) = n(0) \cdot \exp\left[\frac{v}{\Lambda}\,(k - k_0)t\right] = n(0)\,\mathrm{e}^{t/T} ,}$$

where T is a time that is characteristic of the system, and is known as the period of the pile.

15.2 Evaluation of the Period

From the previous section we have

$$T = \frac{\Lambda/v}{k - k_0} = \frac{\text{Lifetime of neutron in the pile}}{k - k_0} .$$

We shall now take account of the delayed neutrons from fission in order to obtain a more accurate expression for T. The diffusion equation and the age equation remain the same as before, i.e.

$$\frac{\lambda v}{3} \cdot \Delta n - \frac{v}{\Lambda}\,n + q = \frac{\partial n}{\partial t} ;$$

$$\Delta q = \frac{\partial q}{\partial \tau}\,.$$

The third equation, however, giving the value of $q(0)$, must be revised. For if a neutron is absorbed at the time $t = 0$, 99% of the fission neutrons are prompt, and about 1% come later. That is, out of any number of neutrons delivered at some given time, about 1% come from fissions produced by earlier captures. Let p be the fraction of delayed neutrons; then the number of neutrons made per second in an equilibrium pile[2] is[3]

$$q(0) = \underbrace{k\,\frac{v}{\Lambda}(1-p)\,n}_{\text{prompt}} + \underbrace{p\,k\,\frac{v}{\Lambda}\,n}_{\text{delayed}} = k\,\frac{v}{\Lambda}\,n$$

- the result is just the same as for the case when all the neutrons are prompt, because n is not a function of time. In a non-equilibrium pile, on the other hand, one must take account of the different values of n for the prompt neutrons and the various groups of delayed neutrons. If p_i denotes the fraction of the neutrons which are emitted according to an exponential decay law with mean life θ_i, the fraction of fission neutron appearing in a time dt at time t after neutron capture is given by

$$\mathrm{d}t \cdot \sum_{i=1}^{\sim 10} \frac{p_i}{\theta_i}\,\mathrm{e}^{-t/\theta_i}\,.$$

If, now, we suppose that we are at the time t , and that we look back to the state of affairs at a time $(t - t')$, we observe that the number of neutrons absorbed in a short interval dt' was $\frac{v}{\Lambda}\,n(t-t')$. As a result of this, the number of neutrons appearing at the time t is:

$$k\,\frac{v}{\Lambda}\,n(t-t')\,\mathrm{d}t' \sum_i \frac{p_i}{\theta_i}\,\mathrm{e}^{-t'/\theta_i}\,.$$

It follows, then, that $q(0)$ is now given by

$$q(0) = k\,\frac{v}{\Lambda}\,n(t)\cdot(1-p) + k\,\frac{v}{\Lambda}\int_0^\infty n(t-t')\,\mathrm{d}t' \sum_i \frac{p_i}{\theta_i}\,\mathrm{e}^{-t'/\theta_i}\,.$$

Let us try a solution of the type

$$n = \text{const.} \times \varphi(x,y,z)\cdot \mathrm{e}^{t/T}\,.$$

Inserting this in the equation for $q(0)$, we get

$$q(0) = k\,\frac{v}{\Lambda}\,(1-p)\,n(t) + k\,\frac{v}{\Lambda}\text{const.}\left(\varphi\cdot \mathrm{e}^{t/T}\right)\int_0^\infty \mathrm{e}^{-t'/T}\,\mathrm{d}t' \sum_i \frac{p_i}{\theta_i}\,\mathrm{e}^{-t'/\theta_i}\,.$$

---

[2]@ That is, in a pile just critical.

[3]@ In the original manuscript, the n factor in the delayed term is missing.

This gives

$$q(0) = k\,\frac{v}{\Lambda}\,n(t)\left[(1-p) + \sum_i \left(\frac{p_i}{\theta_i/T+1}\right)\right] \equiv A\cdot n(t)\,.$$

Now, as before,

$$q = q(0)\cdot \mathrm{e}^{-\omega\tau} = A\cdot n(t)\cdot \mathrm{e}^{-\omega\tau}\,.$$

Hence, in the diffusion equation, we have

$$\frac{\lambda v}{3}\cdot \Delta n - \frac{v}{\Lambda}\,n + A\cdot n(t)\,\mathrm{e}^{-\omega\tau} = \frac{\mathrm{d}n}{\mathrm{d}t} = \frac{n}{T}\,.$$

Also, by definition of the solution (cfr. p. 123), we have

$$\Delta n = -\omega\, n\,.$$

We thus arrive at the equation

$$-\frac{\lambda v}{3}\cdot\omega - \frac{v}{\Lambda} + A\,\mathrm{e}^{-\omega\tau} = \frac{1}{T}\,,$$

whence we obtain

$$T = \left\{-\frac{\lambda v}{3}\cdot\omega - \frac{v}{\Lambda} + k\,\frac{v}{\Lambda}\left[(1-p) + \sum_i\left(\frac{p_i}{\theta_i/T+1}\right)\right]\mathrm{e}^{-\omega\tau}\right\}^{-1}.$$

Re-arranging,

$$T = \left\{-\frac{\lambda v}{3}\cdot\omega - \frac{v}{\Lambda} + k\,\frac{v}{\Lambda}\left[1 - \sum_i\left(\frac{p_i\theta_i}{\theta_i+T}\right)\right]\mathrm{e}^{-\omega\tau}\right\}^{-1}.$$

Solving for k,

$$k = \frac{\mathrm{e}^{\omega\tau}\left[\dfrac{\Lambda}{v}\cdot\dfrac{1}{T} + 1 + \dfrac{\lambda\Lambda}{3}\,\omega\right]}{1 - \displaystyle\sum_i\left(\frac{p_i\theta_i}{\theta_i+T}\right)}\,.$$

Now we have seen in the previous section that the reproduction factor k_0 for a pile in equilibrium is given by

$$k_0 = \mathrm{e}^{\omega\tau}\left[1 + \frac{\lambda\Lambda}{3}\,\omega\right],$$

so that we can write

$$k = \frac{k_0 + \dfrac{\Lambda}{v}\cdot\dfrac{\mathrm{e}^{\omega\tau}}{T}}{1 - \displaystyle\sum_i\left(\frac{p_i\theta_i}{\theta_i+T}\right)}\,,$$

and this tends to k_0 as T tends to infinity - i.e. to the case of a time-independent pile. By approximation we can put the equation for k into a simpler form. The numerical values of the quantities involved are as follows:

$$\Lambda \sim 300 \text{ cm}, \qquad v \sim 2 \cdot 10^5 \text{ cm/s}, \qquad \mathrm{e}^{\omega\tau} \sim 1, \qquad T \sim 1 \text{ s},$$

for the second term of the numerator; and in the denominator, we can put

$$\sum_i \left(\frac{p_i \theta_i}{\theta_i + T}\right) < \sum_i p_i < 10^{-2}\,.$$

It may be seen that the second term of both numerator and denominator are $\ll 1$, so that, to a good degree of accuracy,

$$\boxed{k = k_0 \left[1 + \frac{\Lambda\,\mathrm{e}^{\omega\tau}}{v\,k_0\,T} + \sum_i \left(\frac{p_i\theta_i}{\theta_i + T}\right)\right]}\,.$$

The relative excess reactivity is defined by the equation

$$\frac{k - k_0}{k_0} = \frac{\Theta_0}{T} + \sum_i \left(\frac{p_i\theta_i}{\theta_i + T}\right),$$

where

$$\Theta_0 = \frac{\Lambda\,\mathrm{e}^{\omega\tau}}{v\,k_0}\,.$$

Θ_0 is very nearly equal to Λ/v, i.e. to the lifetime of a thermal neutron. Now, for T large ($T \gg \theta_i$), the excess reactivity can be written

$$\frac{k - k_0}{k_0} = \frac{\Theta_0 + \sum_i p_i\theta_i}{T},$$

The numerator of this expression has an important physical significance - it represents the mean lifetime $\overline{\theta}$ of a generation of neutrons. This is easily shown, for the generation time of a prompt neutron is Θ_0, and the generation time of a delayed neutron of type i is $\Theta_0 + \theta_i$. Thus,

$$\begin{aligned}\overline{\theta} &= \Theta_0 \cdot (1 - p) + \sum_i p_i \cdot (\Theta_0 + \theta_i) \\ &= \Theta_0 + \sum_i p_i\theta_i\,.\end{aligned}$$

We therefore have

$$T = \frac{k_0}{k - k_0}\,\overline{\theta}\,.$$

p_i	θ_i
0.00077	0.6 seconds
0.00220	2.2
0.00194	6.5
0.00152	31.7
0.00023	80.2

In the accompanying table we have listed the most important of the mean lives characterizing delayed neutron emission. From these, we find

$$\sum_i p_i\theta_i \sim 0.083 \text{ seconds.}$$

The value of Θ_0 is equal to 0.0015 seconds. We thus see that the delayed neutrons, although they comprise less than 1% of the whole, contribute nearly 60 times more to the generation time than do the prompt neutrons. In the absence of delayed neutrons, it is clear that T would be very short even for a small excess k, and the pile would be a rather dangerous and unstable instrument.

If T is small ($T \ll \theta_i$), we have

$$\frac{k - k_0}{k} = \frac{\Theta_0}{T} + p\,,$$

and the excess k depends chiefly on the prompt neutrons. This condition is not often met with, however, and most piles made up to date have a long period which is governed mainly by the delayed neutrons.

Chapter 16

Practical Aspects of Pile Physics

16.1 The Determination of k

The most important quantities involved in the design of a pile are the dimensions and the value of k. Since the dimensions depend on the value of $(k-1)$, where k itself is very slightly in excess of unity in most cases, a small error in k makes a big error in the required dimensions of the pile we deduce from it.

One determines k experimentally by making a small lattice (say 1/10 full size) whose unit cell has the same size as that of the final pile. A constant source of neutrons (e.g. Ra-Be) is placed at the centre of one of the end faces, as shown in Fig. 16.1. A steady state is established, governed by the usual equations

$$\frac{\lambda v}{3}\Delta n - \frac{v}{\Lambda}n + q = 0\,,$$

$$\Delta q = \frac{\partial q}{\partial \tau}\,,$$

$$q(0) = k\,\frac{v}{\Lambda}\,n\,,$$

and we assume, as in the last Chapter, that we have a space distribution $\varphi(x, y, z)$ such that $\Delta\varphi + \omega\varphi = 0$. We know that the neutron distribution falls off to zero at the sides of the lattice in the x and y directions, and that it decreases from the source upwards in the z direction. We thus put

$$\varphi(x, y, z) = \sin\frac{\pi x}{a}\,\sin\frac{\pi y}{a}\,\mathrm{e}^{-z/b}$$

(a is known and b is measured). It follows that we must have the identity

$$\omega = \frac{2\pi^2}{a^2} - \frac{1}{b^2}\,,$$

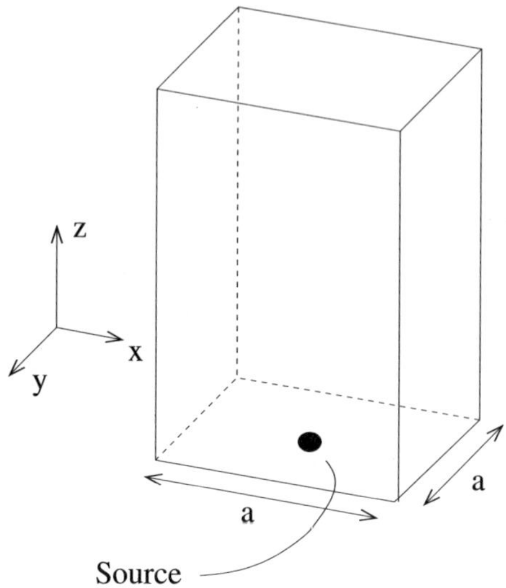

Fig. 16.1

and as before we write

$$k = e^{\omega\tau}\left(1 + \frac{\lambda\Lambda}{3}\,\omega\right).$$

ω gives the size of the pile in equilibrium, since, comparing its value obtained from this "exponential experiment" to that for a cubical pile of side A with the same unit cell, we have

$$\boxed{\frac{2\pi^2}{A^2} = \frac{2\pi^2}{a^2} - \frac{1}{b^2}\,.}$$

16.2 The Study of Pile Materials

Absorption measurements can be made to find the properties of various samples of pile materials, using a working pile. All piles have safety devices, usually in the form of Cd rods, which can be moved in and out to make the pile critical or subcritical as desired. To study the effect of a piece of U, graphite, etc., one can place it in a pile, and exactly compensate its effect by a suitable movement of one of the control rods. The experimental arrangement is sketched in Fig. 16.2.

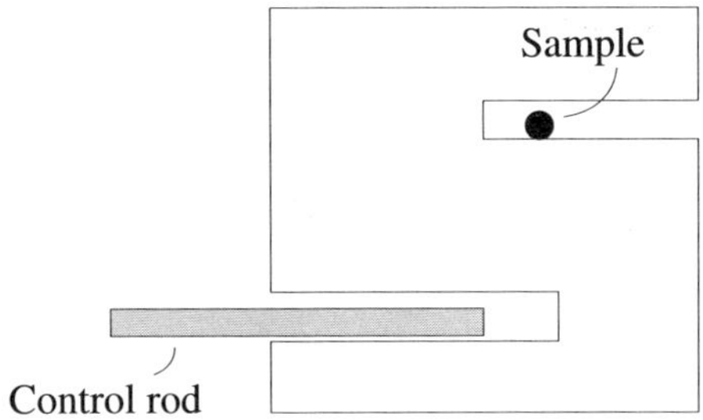

Fig. 16.2

Now, if an absorber is inserted at a certain point in pile, its effect is proportional to

a) the neutron density at the point,
b) the effectiveness of those neutrons in the pile. Peripheral neutrons are likely to be lost, and are therefore less useful than central ones.

It can be shown that both effects are proportional to the neutron density n, so that the total effect is proportional to n^2. If we consider the insertion of the compensating Cd control rod into a cubical pile of side a, its absorptivity as a function of the depth of insertion d takes the form showed in Fig. 16.3; analitically it is given by

$$\text{Absorptivity} = \int_0^d \sin^2 \frac{\pi x}{a}\, dx\,.$$

Very accurate comparisons of the absorption of different samples can be made by this means. For U samples, for instance, the answer can be given correct to 1 part to 10^5.

16.3 Energy and Radiation Production

The number of fissions per cm^3 per second in a pile is given by the expression

$$f_T\, n\, \frac{v}{\Lambda} \cdot \frac{3.9}{7.1}$$

(f_T = fraction of thermals absorbed by U; Λ = mean free path for absorption of a thermal neutron). Supposing that each fission liberates 200 MeV ($= 3.2 \cdot 10^{-4}$ ergs), the energy release is given by

$$\text{ergs/cm}^3\text{/s} = f_T \cdot \frac{1}{\Lambda} \cdot \frac{3.9}{7.1} \cdot 3.2 \cdot 10^{-4} \cdot (n\, v)\,.$$

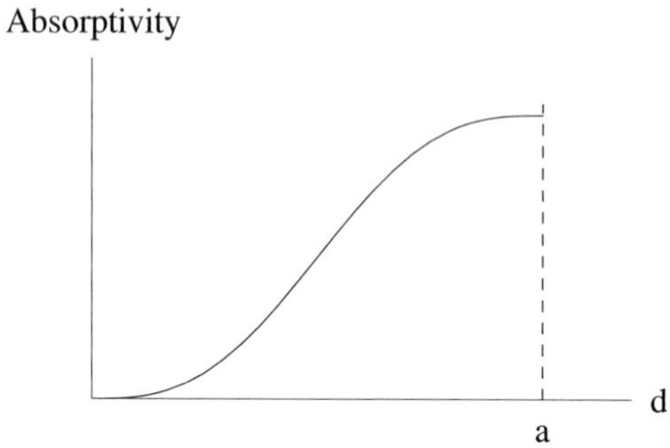

Fig. 16.3

The quantity $(n\,v)$ is the flux of neutrons at any point. For a cubical pile, if $(n_0\,v)$ is the flux at the centre,

$$\overline{(n\,v)} = \frac{8}{\pi^3}\,(n_0\,v)\,.$$

If V is the volume of the pile, the total energy production in kilowatts is given by

$$\begin{aligned} P &= n_0\,v \cdot \frac{8}{\pi^3} \cdot V\,\frac{f_T}{\Lambda} \cdot \frac{3.9 \times 3.2}{7.1} \cdot 10^{-4} \\ &\sim 4 \cdot 10^{-9}\,n_0\,v \text{ kilowatts for actual cases.} \end{aligned}$$

In Clinton, one has $P = 5000$ kW $\longrightarrow n_0\,v \sim 10^{12}/\text{cm}^2/\text{s}$.
In Hanford, $n_0\,v \sim 10^{13}/\text{cm}^2/\text{s}$.

16.4 Shielding

In a pile working at 1 kilowatt there are about $3 \cdot 10^{13}$ fissions per second. Assuming 20 MeV of γ-radiation per fission, of which 5% is not absorbed within the pile, the total γ-radiation is equivalent to about $3 \cdot 10^{13}$ quanta of 1 MeV per kilowatt of power. The surface area of a pile is about $4\pi(500)^2$, so that a person going near the pile, and stay there all day, would receive on each square cm of his body an amount of radiation given by

$$\frac{K \cdot 3 \cdot 10^{13}}{4\pi(500)^2} \cdot 10^5 \sim 10^{12} K \text{ 1 MeV photons,}$$

where K is the power of the pile in kilowatts. (1 day $\sim 10^5$ seconds). A daily dose (i.e. the tolerance dose which can be accepted by the body

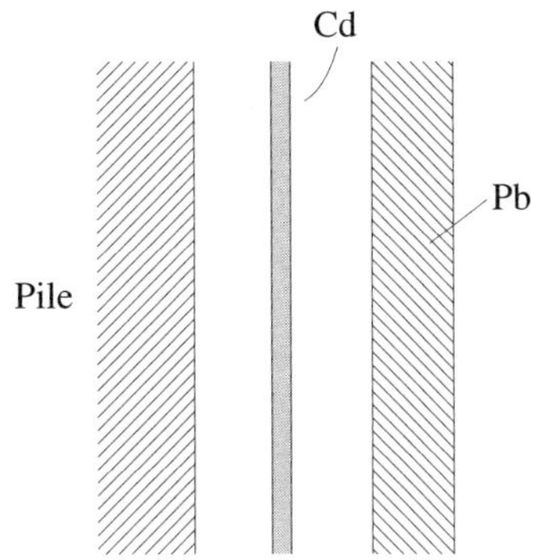

Fig. 16.4

each day for indefinite periods without hurt) is about 10^8 protons/cm^2 (approximately 1R = 1 Röntgen unit). The person standing by the pile would thus receive $\sim 10^4 K$ daily doses per day. If K is of the order 5000, the shielding must reduce the radiation by a factor of 10^9 at least.

One also has to reckon with the neutrons coming out. These are cut off by Cd, which must be inside the lead used to absorbs the γ-radiation (Fig. 16.4), because the (n, γ) capture process in Cd itself gives off a lot of γ-rays.

Cd, however, is good only for the slow neutrons. To cut down the fast neutrons one can use concrete; or, better, since it is less bulky, an arrangement of alternate layers of Fe and paraffin wax. 4 feet of this reduces the γ-radiation by a factor of about 10^9 and also deals with most of the neutrons.

Let us try to estimate the minimum bulk of the radiation shield. The γ-rays are attenuated by a factor 1/e in a thickness of matter of about 25 g/cm^2. We thus need at least 500 g/cm^2 of shielding to attenuate the γ-rays alone by a factor of 10^8-10^9. To deal with the neutrons as well one needs a total shielding of at least 1000 g/cm^2. The lightest portable pile of this type would then weigh about 50 tons.

The escape of radiation via the cooling system must also be considered. In Clinton there is an air cooling system, and the argon in the air becomes radioactive. The air is made to pass through a 100 feet stack and is harmless on emergence. At Hanford the cooling is by water, and is made safe by dilution in the Columbia river - there has been no perceptible mortality among the fish!

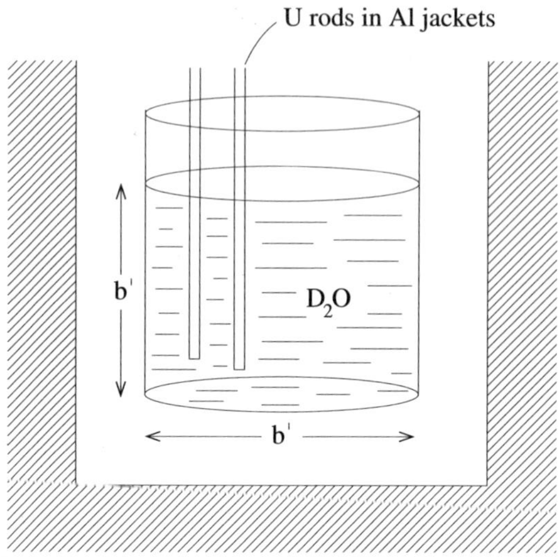

Fig. 16.5

16.5 Other Types of Piles

So far we have only considered the type of pile which works on U with graphite. But this is by no means the only useful sort of pile. At the Argonne laboratory there is a pile which is shown diagrammatically in Fig. 16.5. The moderator is D_2O, and the pile works up to 300 kW. The central flux $n_0 v$ is about 10^{12} - the same as for the 5000 kW Clinton pile. The greater power of the latter is due simply to the large amount of U. D_2O makes a pile easier to cool at low power than if one uses graphite, but it is expensive and is easily decomposed. In a homogeneous U+D_2O pile, 2 mg D_2O per kilowatt per second would be decomposed. This would necessitate recovery of the decomposition gases.

Different moderators can be used if enriched rather than natural U is employed. In natural uranium, for every 3.9 neutrons absorbed in U^{235} to give fission, 0.7 and 2.5 neutrons are lost by the (n, γ) process in U^{235} and U^{238} respectively. If one could use pure U^{235}, one would get $\frac{3.9}{3.9 + 0.7} \cdot 2.4$ neutrons per thermal neutron absorbed. This works out to 2.05 instead of the 1.32 for normal U. Even a small enrichment makes a big difference,

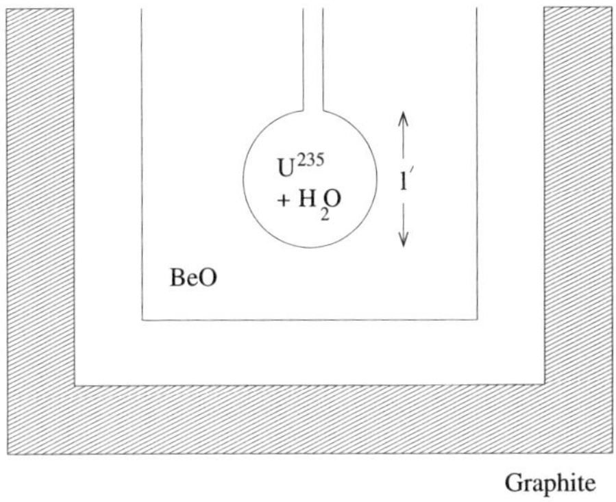

Fig. 16.6

and the water boiler at Los Alamos, sketched in Fig. 16.6, works on 14% enriched material and ordinary water. The power is about 5 kW. A problem here is the disposal of radioactive fission product. Such piles as this are small, and can conveniently be made as experimental neutron sources.

Chapter 17

Fast Reactors

17.1 Elementary Considerations

We have discussed the mechanism of thermal neutron chain reactions. The question now arises how to produce a nuclear explosion. In Chapter 15 we have seen ([first of the highlighted] equation at p. 125) that the growth of neutron density is given by

$$n(t) = n(0) \cdot \exp\left[\frac{v}{\Lambda}\,\mathrm{e}^{-\omega\tau} \cdot (k - k_0)t\right] .$$

We can put $\mathrm{e}^{-\omega\tau} \sim 1$, and then the e-folding time (= time for neutron population to increase by a factor e) is

$$\frac{\Lambda}{v(k - k_0)} .$$

For a thermal pile $\frac{\Lambda}{v} \sim 10^{-3}$ seconds and $(k - k_0) \sim 0.1$. Thus the e-folding time is $\sim 10^{-2}$ seconds or more. In the water boiler (p. 136) $\Lambda \sim 50$ cm, $v \sim 10^5$ cm/s and $k \sim 2$, giving an e-folding time of $\sim 10^{-4}$ seconds. For a fast reactor, of U^{235} or Pu, $\Lambda \sim 10$ cm, $v \sim 10^9$ cm/s and $k \sim 2$, giving an e-folding time of 10^{-8} seconds. These times are all minima, corresponding to the most rapid possible growth.

In 1 kilogram of fissionable material there are $2 \cdot 10^{24}$ atoms, approximately. If we denote the reciprocal of the e-folding time by α, the reaction could continue for at most t seconds, where

$$10^{24} = \mathrm{e}^{\alpha t} .$$

This gives the following values of t:

Graphite pile	0.5 seconds
Water boiler	0.005 seconds
Fast reactor	0.5 microseconds .

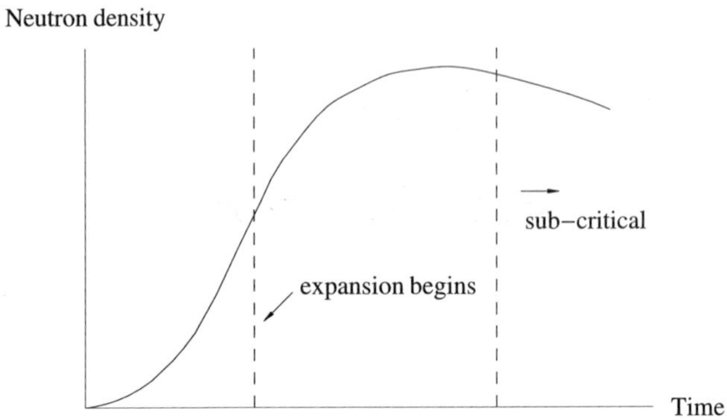

Fig. 17.1

In practice, mechanical effects begin 5 or so generations before the end; the energy release causes pressures which expands the reactor. Eventually expansion and leakage make the assembly sub-critical. The time-dependence of the process is indicated in Fig. 17.1.

Let us suppose that the pressure goes as

$$P(t) = P_0 \cdot \mathrm{e}^{\alpha t} ,$$

and let us suppose also that the reactor is spherical. Then essentially the process of expansion consists in the outward acceleration of a surface shell of the reactor, of mass M and initial radius R_0. This is indicated in Fig. 17.2. The acceleration is given by

$$\ddot{R} \sim \frac{P\,R^2}{M} = \frac{P_0 R^2\,\mathrm{e}^{\alpha t}}{M} .$$

$\mathrm{e}^{\alpha t}$ changes more rapidly than R, and the displacement of the surface as a function of time is given by

$$\delta R \sim \frac{P_0 R^2}{M} \cdot \frac{\mathrm{e}^{\alpha t}}{\alpha^2} = \frac{P\,R^2}{M\alpha^2} ,$$

where $P = P(t)$. If R_{crit} is the critical radius, at which multiplication ceases, the final value of δR is given approximately by $(R_{\mathrm{crit}} - R_0)$, and the energy released up to this time is roughly

$$W = P_{\max} \cdot \frac{4\pi R^3}{3} \sim M\,\alpha^2\,R \cdot (R_{\mathrm{crit}} - R_0) .$$

Thus for the maximum efficiency we must make α as large as possible - i.e., the e-folding time must be as short as possible.

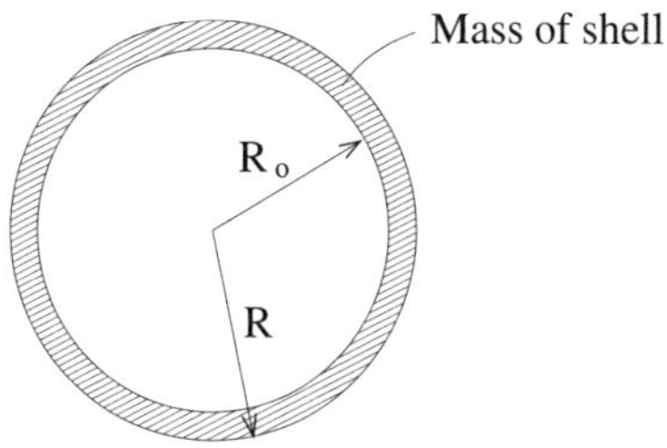

Fig. 17.2

17.2 The Integral Equation to the Neutron Distribution

Let us calculate the critical properties of a fast reactor with a core of U^{235} and a tamper (neutron reflector) of U, as shown in Fig. 17.3. In such a reactor the size of the active material is $\sim$ 10 cm, and λ is also about 10 cm. Thus differential diffusion theory cannot be used (cfr. p. 83). We shall make use of the concept of the transport cross section. This is the scattering cross section modified to take account of the anisotropy of elastic scattering, and it is defined through the equation

$$\sigma_{\mathrm{tr}} = \int_0^\pi \sigma_{\mathrm{c}}(\beta)\,(1 - \cos\beta)\,\mathrm{d}\Omega\,,$$

where $\sigma_{\mathrm{c}}(\beta)$ is the differential cross section for scattering into unit solid angle in the direction β. This may be put in the alternative form

$$\sigma_{\mathrm{tr}} = \sigma_{\mathrm{sc}} \cdot \left(1 - \overline{\cos\beta}\right)\,,$$

where σ_{sc} is the total elastic scattering cross section, and $\overline{\cos\beta}$ is the value of $\cos\beta$ averaged over all collisions. Corresponding to the transport cross section there is also a transport mean free path. It is related to the true mean free path by the equation

$$\lambda_{\mathrm{tr}} = \frac{\lambda}{1 - \overline{\cos\beta}}\,.$$

This quantity is familiar to us; it appears in the diffusion coefficient in our derivation of the age equation (Chapter 8), and it is from its use in the study of these transport phenomena that it takes its name.

We shall define a quantity σ, which is to be the sum of the transport cross section of all the atoms in 1 cm^3 of material:

$$\sigma = \sum_{i=1}^{N} \sigma_i \cdot \left(1 - \overline{\cos\beta_i}\right)\,,$$

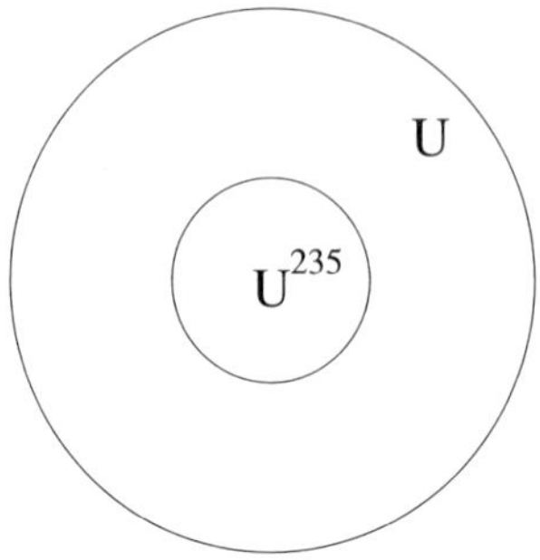

Fig. 17.3

where N is the total number of atoms of all kinds per cm^3. We also define λ such that

$$\sigma\lambda = 1\,.$$

The processes other than transport (= elastic scattering) are as follows:

	No. of neutrons hitting nucleus	No. of neutrons leaving nucleus
Capture	1	0
Inelastic scattering	1	1
Fission	1	ν

Suppose that we get $(1+f)$ neutrons out for every neutron that goes into a nucleus. Then

$$1+f = \frac{0\cdot\sigma_{\text{abs}} + 1\cdot\sigma_{\text{sc}} + \nu\cdot\sigma_{\text{f}}}{\sigma_{\text{total}}}$$

$$f = \frac{(\nu-1)\cdot\sigma_{\text{f}} - \sigma_{\text{abs}}}{\sigma_{\text{total}}}\,.$$

In a non-fissionable material,

$$f = \frac{-\sigma_{\text{abs}}}{\sigma_{\text{sc}}+\sigma_{\text{abs}}}\,,$$

and $|f| < 1$. For a chain reaction to occur we must have $f > 0$ in at least some part of the system.

An exact calculation of the fast reactor would treat all neutron energies, but we shall here assume some average energy - this is called 'one group theory' - and establish the integral equation for the neutron density in terms of f and σ.

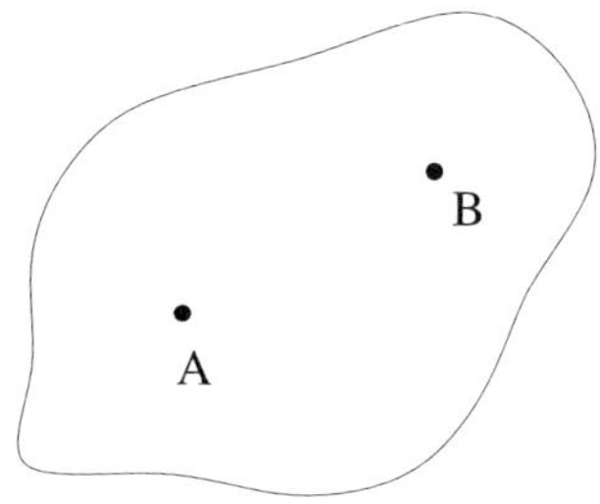

Fig. 17.4

The first question we have to ask ourselves is the following: If at some point one neutron is produced per second, what is the neutron density at some distance r from the point if the neutron emission is isotropic? The time spent by a neutron in traversing a spherical shell of volume $4\pi r^2 \mathrm{d}r$ is $\frac{\mathrm{d}r}{v}$. The average number of neutrons found in the shell is equal to the number emitted per second multiplied by the time spent in the shell. With our assumptions that is just $\frac{\mathrm{d}r}{v}$. The neutron density at r is thus $\frac{1}{4\pi r^2 \mathrm{d}r \cdot v}$. In obtaining this result we have ignored the possibility of absorption.

Now let us suppose that we want to know the neutron density n_A at a point A (Fig. 17.4) in a critical reactor. We need the contributions to the volume element at A from the volume elements at all points such as B in the figure. In the volume $\mathrm{d}B$ there are

$$n(B)\, v \cdot \sigma(B)\, \mathrm{d}B$$

collisions per second, and the number of neutrons escaping from $\mathrm{d}B$ per second is given by

$$n(B)\, v \cdot \sigma(B)\, \mathrm{d}B \cdot [1 + f(B)]\ .$$

At A, if $BA = r$, the neutron density is therefore

$$\frac{n(B)\, \mathrm{d}B\, v \cdot \sigma(B) \cdot [1 + f(B)]}{4\pi r^2 v}\,,$$

in the absence of absorption, and with isotropic scattering. We can take care of the absorption by introducing a factor $\exp\left[-\int_B^A \sigma\, \mathrm{d}r\right]$. Then, adding up the contributions from all the elements $\mathrm{d}B$ we find

$$n(A) = \int_{\text{all space}} \frac{n(B) \cdot \sigma(B) \cdot [1 + f(B)] \cdot \exp\left[-\int_B^A \sigma\, \mathrm{d}r\right]}{4\pi r_{AB}^2} \cdot \mathrm{d}B\,.$$

This is the integral equation required. For a system in equilibrium, i.e. just critical,

$$v \cdot \int_{\text{all space}} n(A)\,\mathrm{d}A\,\sigma(A)\,f(A) = 0\,.$$

17.3 The Critical Size for a Fast Reactor

The conditions

$$n(A) = \int_{\text{all space}} \frac{n(B)\cdot\sigma(B)\cdot[1+f(B)]\cdot\exp\left[-\int_B^A \sigma\,\mathrm{d}r\right]}{4\pi r_{AB}^2}\cdot\mathrm{d}B$$

and

$$0 = v \cdot \int_{\text{all space}} n(A)\,\mathrm{d}A\,\sigma(A)\,f(A)$$

are the equilibrium conditions for a fast reactor, analogous to the differential equations for a pile. They can be used to determine the critical size, but we shall use differential theory first to get approximate solutions.

If N is the number of scattering collisions per capture collision,

$$N = \frac{3\ell^2}{\lambda^2}\,,$$

where ℓ = diffusion length, λ = mean free path. (See p. 80). Thus,

$$\ell = \lambda\sqrt{\frac{N}{3}}\,,$$

with $N = \dfrac{\sigma_{\text{total}}}{\sigma_{\text{abs}}}$ = a large number. This condition on N appeared because we assumed that the neutron flux was proportional to the neutron density gradient, neglecting higher derivatives. This amounts to the same as assuming that $n(x)$ goes as $\mathrm{e}^{x/\ell}$, and for the curvature of this function to be small over the distance λ, the quotient λ/ℓ must be small i.e. N must be large.

Let us try first of all to modify the differential diffusion equation so that it holds good for small values of N. Suppose that the gradient of neutron density is in the direction of the x-axis. By virtue of the gradient, the density of neutrons travelling in a direction θ with respect to the x-axis is not the same for all θ at a given point, and is a maximum for neutrons travelling in the direction of x. Let the density of neutrons at x moving in the direction θ be $n(x,\theta)$. Then the number in an element of volume $\mathrm{d}\tau$ whose directions lie between θ and $\theta+\mathrm{d}\theta$ is

$$n(x,\theta)\,\mathrm{d}\tau\cdot 2\pi\sin\theta\,\mathrm{d}\theta = n(x,\xi)\,\mathrm{d}\tau\cdot 2\pi\,\mathrm{d}\xi\,,$$

say, where $\xi=\cos\theta$. We will now examine all the factors which tend to make $n(x,\xi)$ vary with time.

(i) We assume only one neutron velocity; therefore for neutrons of direction ξ_0 the x-component of velocity is $v\cdot\xi_0$. If n varies with x, the change in $n(x,\xi_0)$ at x in time $\mathrm{d}t$ will be

$$n(x-v\,\xi_0\,\mathrm{d}t)-n(x) = -v\,\xi_0\,\mathrm{d}t\,\frac{\partial n}{\partial x}\,.$$

If we consider a finite range of directions, within the element of solid angle $\mathrm{d}\Omega=2\pi\mathrm{d}\xi$, the number of neutrons disappearing from this range per cm^3 per second by this effect of bodily transfer is given by

$$-\delta n_1 = -v\,\mathrm{d}t\,\mathrm{d}\xi\cdot 2\pi\xi\,\frac{\partial n}{\partial x}\,,$$

the ξ now being any arbitrary direction.

(ii) Another source of loss of neutrons from the range $\mathrm{d}\xi$ is by collisions, which deflect neutrons out of the original direction ξ. In a distance $\mathrm{d}x$, the number of collisions is $n\,\frac{\mathrm{d}x}{\lambda}$; if we again take into account the size of the solid angle element, we find that the number scattered out is

$$\begin{aligned}-\delta n_2 &= -\frac{n}{\lambda}\,2\pi\mathrm{d}\xi\cdot\mathrm{d}x\\ &= -\frac{n}{\lambda}\,2\pi\mathrm{d}\xi\cdot v\mathrm{d}t\,.\end{aligned}$$

(iii) How many neutrons are scattered <u>into</u> $\mathrm{d}\xi$ by collisions? This will evidently be an integral of some sort. For various directions ξ' at x, the result of the last paragraph (ii) tells us that the numbers of neutrons scattered out of those directions will be represented by terms such as

$$\frac{n(x,\xi')}{\lambda}\cdot 2\pi\mathrm{d}\xi'\cdot v\,\mathrm{d}t\,.$$

Of these neutrons, a fraction $\frac{N-1}{N}$ will have escaped capture, and will be moving in other directions. The number scattered into the element of solid angle at ξ is thus given by

$$\Delta\,(+\delta n_3) = \frac{n(x,\xi')}{\lambda}\cdot 2\pi\mathrm{d}\xi'\cdot v\,\mathrm{d}t\cdot\frac{N-1}{N}\cdot\frac{2\pi\mathrm{d}\xi}{4\pi}\,.$$

Integrating this with respect to $\mathrm{d}\xi'$,

$$+\delta n_3 = \frac{N-1}{N} \cdot v\,\mathrm{d}t \cdot \frac{2\pi}{4\pi} \cdot \frac{2\pi \mathrm{d}\xi}{\lambda} \cdot \int_{-1}^{1} n(x,\xi')\,\mathrm{d}\xi' .$$

In the equilibrium state, the sum of these three contributions must be zero. We therefore have

$$-2\pi\xi \frac{\partial n}{\partial x} - \frac{n}{\lambda} 2\pi + \frac{N-1}{N} \cdot \frac{\pi}{\lambda} \cdot \int_{-1}^{1} n(x,\xi')\,\mathrm{d}\xi' = 0 ,$$

or

$$\boxed{-\xi \frac{\partial n}{\partial x} - \frac{n}{\lambda} + \frac{N-1}{N} \cdot \frac{1}{2\lambda} \cdot \int_{-1}^{1} n(x,\xi')\,\mathrm{d}\xi' = 0 .}$$

Let us assume that the solution is separable, so that we may write

$$n(x,\xi) = n(x) \cdot f(\xi) .$$

And let us assume further, on the basis of past experience, that $n(x)$ is of the form

$$n(x) = \mathrm{e}^{x/\ell} ,$$

with ℓ to be determined. Then by substituting in the above integro-differential equation, we find[1]

$$\frac{1}{\lambda} f(\xi) + \frac{\xi}{\ell} f(\xi) = \frac{N-1}{2N\,\lambda} \cdot \int_{-1}^{1} f(\xi')\,\mathrm{d}\xi',$$

i.e.

$$\left(\frac{\lambda\xi}{\ell} + 1\right) \cdot f(\xi) = \frac{N-1}{2N} \cdot \int_{-1}^{1} f(\xi')\,\mathrm{d}\xi' = \text{ constant.}$$

Thus we must have

$$f(\xi) = \frac{C}{\dfrac{\lambda\xi}{\ell} + 1} ,$$

and we must establish that this has some values of ℓ which give solutions. This means that we must investigate whether our form for $f(\xi)$ holds for all ℓ, for some ℓ, or for none. Substituting back into the equation, we have the identity

$$1 = \frac{N-1}{2N} \cdot \int_{-1}^{1} \frac{\mathrm{d}\xi'}{1 + \dfrac{\lambda}{\ell}\xi} = \frac{N-1}{2N} \cdot \frac{\ell}{\lambda} \log\left(\frac{\ell+\lambda}{\ell-\lambda}\right) .$$

[1] @ In the original manuscript, a factor λ is missing in the denominator of factor $\dfrac{N-1}{2N}$ appearing in the following equation. However, the subsequent calculations are correct.

It is this equation which defines the permitted values of ℓ.

We shall suppose for the moment that N is still fairly large. Then we can put[2]

$$\frac{2N}{N-1} = 2\cdot\left(1-\frac{1}{N}\right)^{-1} = 2\left(1+\frac{1}{N}+\frac{1}{N^2}+\ldots\right).$$

Putting $\frac{\lambda}{\ell} = \epsilon$, and assuming ϵ to be small, we obtain the following simple expansion for $\frac{\ell}{\lambda}\log\left(\frac{\ell+\lambda}{\ell-\lambda}\right)$:

$$\frac{\ell}{\lambda}\log\left(\frac{\ell+\lambda}{\ell-\lambda}\right) = \frac{1}{\epsilon}\cdot\left\{\left(\epsilon-\frac{\epsilon^2}{2}+\frac{\epsilon^3}{3}-\frac{\epsilon^4}{4}+\ldots\right)+\left(\epsilon+\frac{\epsilon^2}{2}+\frac{\epsilon^3}{3}+\ldots\right)\right\}$$
$$= 2\cdot\left[1+\frac{\epsilon^2}{3}+\frac{\epsilon^4}{5}+\ldots\right].$$

Hence we find (through recourse to the identity):

$$1+\frac{1}{N}+\frac{1}{N^2}+\cdots \equiv 1+\frac{\epsilon^2}{3}+\frac{\epsilon^4}{5}+\ldots.$$

To a first approximation, therefore,

$$\epsilon^2 = \frac{3}{N},$$

i.e.

$$\ell = \lambda\sqrt{\frac{N}{3}},$$

which is exactly the result of a solution by differential diffusion theory. To get the next approximation we insert the result of the first approximation in the equation; i.e. we put

$$\epsilon^4 = \frac{9}{N^2}.$$

We accordingly find

$$\frac{1}{N}+\frac{1}{N^2} = \frac{\epsilon^2}{3}+\frac{9}{5}\cdot\frac{1}{N^2},$$

$$\frac{1}{N}\left(1-\frac{4}{5N}\right) = \frac{\epsilon^2}{3}.$$

[2]@ In the original manuscript, in the expansion on the right hand side the incorrect variable λ replaces the quantity N.

giving

$$\frac{\lambda}{\ell} = \sqrt{\frac{3}{N}\left(1 - \frac{4}{5}\cdot\frac{1}{N}\right)}\,.$$

This is a good enough approximation with which to return to the integral equations set out in full at the beginning of this section.

If there is no fission, but only scattering and capture, we have (cfr. p. 142):

$$f = -\frac{1}{N}\,; \qquad\qquad \sigma = \frac{1}{\lambda}\,.$$

Substituting these in the relation we have just obtained between λ, ℓ and N, we get

$$\frac{1}{\ell} = \sigma\sqrt{-3f\cdot\left(1 + \frac{4}{5}f\right)}\,.$$

It can be shown that this continues to hold even with f positive (e.g. in U^{235}), in which case the density goes as $\exp\left(\pm\frac{i\cdot x}{|\ell|}\right)$, and is therefore sinusoidal in form. The diffusion equation using the integral theory is simply

$$\Delta n - \frac{n}{\ell^2} = 0\,.$$

The source term q disappears, being taken care of in the new value of ℓ - which comes from assuming that n goes as $e^{x/\ell}$ and is a function of x only. A 3-dimensional treatment leads to similar results (the work was done by R. Serber), and we shall apply them to the case of a fast reactor with spherical symmetry. The conditions are:

$f > 0$	within the spherical core of radius a,
$f < 0$	outside the core.

We shall use the corrected values of ℓ throughout. Inside the core, $\ell^2 < 0$, and the solutions for $n(r)$ are sinusoidal. Outside the core, $\ell^2 > 0$ and the solutions are exponentials. We will pause at this point to collect the essential result obtained so far; we also introduce two new quantities h and H, which are defined below:

Inside core ($r < a$)	Outside core ($r > a$)
$\Delta n - \frac{1}{\ell^2}\,n = 0\,, \qquad \sigma = \frac{1}{\lambda}\,,$	$\Delta n - \frac{1}{\ell'^2}\,n = 0\,, \qquad \sigma' = \frac{1}{\lambda'}\,,$
$h = \frac{1}{\sigma\ell}\,; \quad H = -i\,h = \frac{-i}{\sigma\ell}\,,$	$h' = \frac{\lambda'}{\ell'}\,; \quad \frac{N}{N-1} = \frac{1}{2h'}\log\frac{1+h'}{1-h'}\,,$

whence

$$\frac{1}{1+f} = \frac{1}{2i\,H} \log \frac{1+i\,H}{1-i\,H} \qquad\qquad \frac{1}{\ell'} = \sigma' \sqrt{-3f' \cdot \left(1 + \frac{4}{5}\, f'\right)} .$$

$$= \frac{1}{H} \tan^{-1} H .$$

We can write

$$\Delta n(r) = \frac{\mathrm{d}^2 n}{\mathrm{d}r^2} + \frac{2}{r} \cdot \frac{\mathrm{d}n}{\mathrm{d}r} = \frac{1}{r} \cdot \frac{\mathrm{d}^2}{\mathrm{d}r^2}\,(n\,r) .$$

Thus, for both regions,

$$\frac{\mathrm{d}^2}{\mathrm{d}r^2}\,(n\,r) = \frac{1}{\ell^2}\,(n\,r) .$$

The solutions are:

Outside core: $$(n\,r) = A\,\mathrm{e}^{-r/\ell'} + B\,\mathrm{e}^{+r/\ell'} ,$$

Inside core: $$(n\,r) = C\,\sin(\sigma\,H\,r) + D\,\cos(\sigma\,H\,r) ,$$

putting $\ell = \dfrac{-i}{\sigma\,H}$. We cannot admit a solution which permits the neutron density to tend to infinity; this means that the constants B and D must be zero. Otherwise, the neutron density outside the core would increase indefinitely with increasing r, and the neutron density inside the core would become infinite at the origin. If we normalize to $n(0) = 1$, we have finally

$$n_{r<a} = \frac{\sin(\sigma\,H\,r)}{\sigma\,H\,r}$$

$$n_{r>a} = B\,\frac{\mathrm{e}^{-\sigma' h' r}}{r} .$$

B can be found by appealing to our initial integral equation; we will write down once again, this time in terms of points P' and P (Fig. 17.5) rather than of A and B as before, so as to avoid confusion. We then have

$$n(P') = \int \frac{n(P)\,\sigma(P) \cdot [1 + f(P)] \exp\left[-\int_P^{P'} \sigma\,\mathrm{d}r\right]}{4\pi r_{PP'}^2} \cdot \mathrm{d}P$$

$$0 = \int n(P)\,\sigma(P)\,f(P)\,\mathrm{d}P .$$

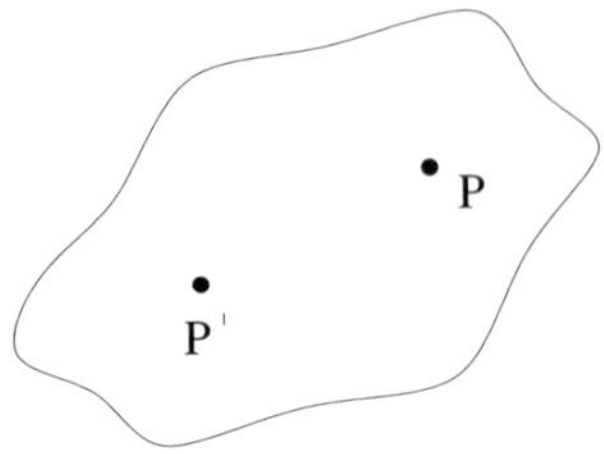

Fig. 17.5

We cannot satisfy both equations exactly; we fulfil the second and then satisfy the first only at the centre. It is the second equation which determines B. We substitute the value thus obtained in the first equation; if it does not check for the particular values of f and f' (cfr. p. 142) characteristic of the core and tamper, we must try a different radius. A correct check implies a just critical assembly, corresponding to our condition

$$\Delta n - \frac{n}{\ell^2} = \frac{\mathrm{d}n}{\mathrm{d}t} = 0\,.$$

We start, then, by putting our solutions for $n(r)$ into the second integral equation. This gives

$$4\pi \int_0^a \sigma\, f \cdot \frac{\sin(\sigma\, H\, r)}{\sigma\, H\, r} \cdot r^2\, \mathrm{d}r + 4\pi \int_a^\infty \sigma'\, f' \cdot \frac{B \mathrm{e}^{-\sigma'\, h'\, r}}{r} \cdot r^2\, \mathrm{d}r = 0\,.$$

f', in the non-multiplying tamper, is negative; we therefore put

$$\sigma\, f \cdot \int_0^a \frac{\sin(\sigma\, H\, r)}{\sigma\, H\, r} \cdot r^2\, \mathrm{d}r = \sigma'\, |f'|\, B \cdot \int_a^\infty \frac{\mathrm{e}^{-\sigma'\, h'\, r}}{r} \cdot r^2\, \mathrm{d}r\,.$$

From this equation we find

$$B = \frac{h'^2\, \sigma'\, f}{H^3\, \sigma^2\, (-f')} \cdot \frac{\sin(H\, \sigma\, a) - H\, \sigma\, a\, \cos(H\, \sigma\, a)}{1 + h'\, \sigma'\, a} \cdot \mathrm{e}^{h'\sigma' a}\,.$$

We now substitute in the first integral equation, letting P' refer to $r = 0$. With our normalization to $n(0) = 1$, this gives

$$1 = \int_0^a \frac{\sin(\sigma\, H\, r)}{\sigma\, H\, r} \cdot \sigma\, (1+f) \cdot \mathrm{e}^{\sigma r}\, \mathrm{d}r + \int_a^\infty B\, \frac{\mathrm{e}^{-\sigma'\, h'\, r}}{r} \cdot \sigma'\, (1+f') \cdot \mathrm{e}^{-\sigma a - \sigma'(r-a)}\, \mathrm{d}r\,.$$

The final result, inserting the value of B, is as follows:

$$\frac{1+f}{f}\, H^2\, (\sigma\, a)^2 \cdot \mathrm{e}^{\sigma a} \cdot \frac{\tan^{-1} H - \displaystyle\int_0^{\sigma a} \frac{\sin H\, x \cdot \mathrm{e}^{-x}}{x}\, \mathrm{d}x}{\sin(H\, \sigma\, a) - H\, \sigma\, a\, \cos(H\, \sigma\, a)}$$

$$= \frac{1+f'}{-f'}\, h^2\, (\sigma'\, a)^2 \cdot \mathrm{e}^{-\sigma' a (1+h')} \cdot \frac{\displaystyle\int_{(1+h')\sigma' a}^{\infty} \frac{\mathrm{e}^{-x}}{x}\, \mathrm{d}x}{1 + h'\, \sigma'\, a}\,.$$

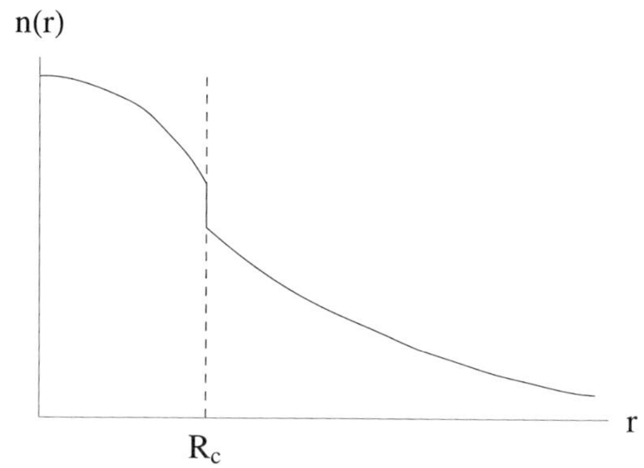

Fig. 17.6

The left-hand side of the equation contains only quantities related to the core, and the right-hand side is in terms only of the properties of the tamper. We substitute the values of f and f' for the materials considered, and plot the left- and right-hand sides as functions of a. The curves will interact at some point, thus defining the critical radius.

Given the critical radius R_c we can find B and plot n as a function of r. The general shape is as shown in Fig. 17.6. The curves exhibit a discontinuity at R_c, but this is not from the truth, as $\dfrac{\mathrm{d}n}{\mathrm{d}x}$ is very steep at the boundary between core and tamper.

17.4 Supercritical Reactors

With some modifications the critical theory can be extended to supercritical reactors. Let us refer again to our integral equation for $n(P')$. We have

$$n(P') = \int \frac{n(P)\,\sigma(P)\cdot[1+f(P)]\exp\left[-\int_{P'}^{P}\sigma\,\mathrm{d}r\right]}{4\pi r_{PP'}^2}\cdot \mathrm{d}P\,.$$

This says that all neutrons in P' at time t come from points where they suffered a last collision at a time $(t-t')$ earlier. For a system in equilibrium the times are of no importance, since $n(P,t)$ is constant. But for a multiplying system we must write

$$n(P,t) = n(P,t')\cdot \mathrm{e}^{\alpha(t-t')}\,;\qquad t > t'.$$

Moreover, $(t - t')$ is equal to r/v. The integral equation thus becomes:

$$n(P') = \int \frac{e^{-\alpha r/v} \cdot n(P)\,\sigma(P) \cdot [1 + f(P)] \exp\left[-\int_{P'}^{P} \sigma\, dr\right]}{4\pi r_{PP'}^2} \cdot dP\,.$$

Using this equation, it is possible to solve for α in the following way. Suppose that we have a core of radius 20 cm, the radius for an equilibrium system being 10 cm. We proceed by trial and error until we find a value for a which, when substituted into the time-dependent equation for $n(P')$, makes the equation hold at the centre for $a = 20$. In this way one can calculate the e-folding time for a fast reactor, and its value thus found will be valid until mechanical effects set in - these having to be known before the efficiency etc. of the bomb can be estimated.

Problems and Exercises

1. Verify the conversion factor for mass to MeV.

2. 1 g of Ra in the form of a sphere is surrounded by a hollow sphere of Be of outer radius 3 cm, inner radius 1 cm. How many neutrons/second does this source emit?

3. If 1 mA of 500 kEV deuterons strikes a D_2O target, how many neutrons per second are emitted?

4. A Cu plate 1 cm thick reduces the intensity of a neutron beam by a factor 0.36. Find σ_{total} in barns.

5. Find an example of each of the following processes: (n, n) elastic, (n, n) inelastic, (n, α), (n, p), (n, γ).

6. A boron layer contains 0.5 g/cm^2 trough which a beam of 1 eV neutrons passes. What is the reduction of intensity?

7. Calculate the average life of thermal neutrons in
   (a) Li of normal density,
   (b) BF_3 gas at N.T.P. ,
   (c) air at N.T.P. .

8. Find five examples each of the (n, α), (n, p) and (n, γ) processes, and give the lifetimes of decay of the products.

9. Find the binding energy of a neutron to: H^1, H^2, Li^6, Li^7, Be^9,

B^{10}, B^{11}, C^{12}, C^{13}.

10. Find the average of the binding energies for nuclei in the neighbourhoods of Fe, Ag, Pb and U.

11. Find the resonance levels of In and Au in the literature.

12. Find from the literature the first resonances for nuclei having $A = 100$ to $A = 150$. Estimate the level spacing here.

13. Make a table of level densities for the following nuclei at excitations of 4, 6, 8, 10 MeV, using the neutron-proton gas model of the nucleus: Be, Fe, Ag, Au.

14. Work in some detail the wave-mechanics of the $n - p$ collision, taking into account the singlet and triplet states of the system separately. Assume a square well potential of radius $0.282 \cdot 10^{-12}$ cm $(= e^2/mc^2)$ for both, the depth of the singlet $(\uparrow\downarrow)$ being 10.8 MeV, and for the triplet $(\uparrow\uparrow)$ being 19.7 MeV.
(a) Are there any bound states?
(b) Find a from which σ_{sc} is derived.
(c) Find σ_{sc} for average collisions with arbitrary spin orientations.

15. Design a crystal spectrometer with NaCl, considering only the (001) planes. Make a table of θ vs. λ (1st and 2nd orders) and the corresponding energies in eV. Assuming a Maxwell distribution at 300°K for the incident beam, and 100% reflection, find the intensities of the 1st and 2nd orders.

16. Prove that, if one measures σ_{sc} for a Maxwellian distribution of thermal neutrons on crystal when the crystals are sufficiently small (in fact see M of p. 44), the order of magnitude of σ_{apparent} is about equal to σ_{sc} for isolated atoms. (Hint: The width of the diffraction maximum for M' diffracting elements is $\delta\lambda/\lambda \sim 1/M'$. Neglect all numerical constants!).

17. Find ξ for elastic collisions of neutrons with D, He, Be, O, U. Find the number of collisions in these elements to reduce the neutron energy from 1 MeV to 1 eV.

18. From the following figures of Rh detector activity as a function of distance from a Ra-Be source in water, find the value of $\overline{r^2}$. (To get sufficient accuracy, make a logarithmic extrapolation to larger distances for $A(r)$):

r(cm)	0	1	2	3	4	5	6	7	8
$A(r)$	1000	960	890	770	630	500	390	290	210

r(cm)	9	10	11	12	13	14	15	16	17
$A(r)$	150	110	88	67	51	39	29	23	19

r(cm)	18	19	20
$A(r)$	16	12	10

19. Find the relation between the 'age' of a neutron and the time from its birth in seconds. Assume λ = constant, and suppose the energy loss per collision is small.

20. Find $\overline{r^2}$ from the Gaussian which represents the solution of the age equation for a point source and spherical symmetry, and show that the result agrees with that obtained by different methods in Chapter 7.

21. Work out in detail the integral for N at p. 73.

22. σ_c for H at a neutron temperature of 300°K ($v = 2.2 \cdot 10^5$ cm/s) is 0.51 barns. Find $\overline{v}$, and hence $\overline{\sigma_c}$, for a Maxwellian distribution of this temperature, assuming the $1/v$ law for σ_c. Find Λ and thence λ from this.

23. Compare the formula for Z_A (p. 90) with the curve of stable isotopes (Dempster), by plotting both on one piece of graph paper.

24. Compare the theoretical and experimental (Dempster) curves of packing fractions. Neglect δ in the formula.

25. Calculate the binding energy of a neutron for various places in the periodic table, e.g. Th^{232}, Au^{197}, Sm^{149}, In^{115}, Mn^{55}.

26. Derive the energy distribution curve for fission neutrons, assuming a velocity V for the fission fragment and a Maxwellian distribution of neutron velocities in the C.G. system.[3] Assume that the probability of escape is proportional to the neutron velocity.

27. From the following table, and assuming a fission spectrum where 40% of the neutrons are below the U^{238} fission threshold, find the reproduction factor for an infinite mass of natural U:

	Threshold	
	below	above
$\sigma_{\text{inelastic}}$ (above to below)	/	2.2
Average σ_f	0.0096	0.47
Avcrage σ_c	0.17	0.12

28. Show that no homogeneous U-C pile will go, whatever the U/C atomic ratio.

29. Suppose a 3000 g spherical lump of U ($\rho = 18$), in cubical cells of graphite of side 20 cm. Calculate f_T, using the formula on p. 118. Use high accuracy, as numbers tend to cancel.

30. A U-graphite lumped pile is in the form of a rectangular parallelepiped, with $a = 6$ cm, $b = 7$ cm, $c = 8$ cm. Find $k(\infty)$ to make the pile critical for the modes (1,1,1), (2,1,1), (1,2,1), (1,1,2).

31. Calculate the relation between the height and the radius of a cylindrical pile in equilibrium, given $k(\infty)$.

32. Given $\Theta_0 = 0.0015$, plot $(k - k_0)/k_0$ as a function of $1/T$, using the complete formula

$$\frac{k - k_0}{k_0} = \frac{\Theta_0}{T} + \sum_i \frac{p_i \theta_i}{\theta_i + T} \cdot$$

(see p. 129).

---

[3]@ That is, the centre of gravity system.

The following questions were set as a final examination at the end of the lecture course:

1. In the reaction

$$N^{14} + He^4 \rightarrow F^{17} + n^1 ,$$

the masses are N^{14}=14.00750, He^4=4.00390, F^{17}=17.00760, n^1=1.00896.
(a) Is the reaction endo- or exo-thermic?
(b) What is the Q in MeV?
(c) What is the reaction threshold for the α-particle?

2. Assuming the Cd cross section for thermal neutrons is 2500 barns, find the thickness of Cd required to reduce the intensity of a thermal neutron beam by a factor 10^6. ($\rho = 8.65$ g/cm^3, atomic weight of Cd = 112.4).

3. A neutron of 1.5 MeV collides against a proton at rest. Assuming that the neutron is scattered by 30° in the laboratory system, find:
(a) the angle made by the proton trajectory with the original neutron path.
(b) the energy acquired by the proton in the collision.

4. A neutron of initial energy E_0 has two collisions with H atoms. What is the probability that after the collisions the neutron energy is between E and $E + \mathrm{d}E$?

5. A square graphite column of density 1.6 has sides of 150 cm. The intensity of thermal neutrons diffusing up the column decreases by a factor e each 28 cm. Find the absorption cross section for C, assuming the scattering cross section to be 4.8 barns. (A.W.[4] of C = 12).

6. An atom has $Z = 50$. What is the minimum value of the atomic weight A for which the atom emits neutron spontaneously? (Use the full formula for $M(A, Z)$ on p. 90). Neutron mass = 1.00893.

---

[4]@That is, atomic weight.

PART 2
Nuclear Reactors

Chapter 1

Process for the Production of Radioactive Substances

Enrico Fermi, Edoardo Amaldi, Bruno Pontecorvo, Franco Rasetti, and Emilio Segré, Rome, Italy, assignors to G.M. Giannini & Co., Inc., New York, N.Y., a corporation of New York

Application October 3, 1935, Serial No. 43,462
In Italy October 26, 1934

U.S. Patent No. 2,206,634 - Patented July 2, 1940

In this Patent, a very detailed description of the experimental results, obtained by studying the radioactivity induced in a number of chemical elements by irradiation with slow neutrons, is reported, along with a corresponding theoretical interpretation.

It is here emphasized, in order to achieve better efficiencies, the use of neutrons instead of charged particles, as considered in previous works on nuclear reactions. Moreover, since neutrons produced by bombardment of atomic nuclei with artificially accelerated particles have high average energies, a method for slowing down fast neutrons is described, by passing the neutrons through a screen of hydrogenous materials, like water or paraffin. The reduction of the energy of the neutrons is interpreted as due to their collisions against the nuclei or the entire atoms of energy reducing materials. An interpretation is provided for the experimental observations: in the case of a strong induced radio-activity following the absorption of the slow neutrons, the formation of an unstable isotope is assumed, while the formation of a stable nucleus is assumed in case no activation or, at least, no strong activation follows an anomalously large absorption.

Particularly interesting is the mention of the possible discovery of "transuranic" elements given in the present Patent. Even here, some caution was adopted about its interpretation, as well as the theoretical interpretation of the effects induced by slow neutrons considered in the paper: "The theoretical statements and explanations are, of course, not conclusive and our invention is in no way dependent upon their correctness. We have found them helpful and give them for the aid of others, but our invention will be equally useful if it should prove that our theoretical conclusions are not altogether correct."[1]

The original Patent application, Metodo per accrescere il rendimento dei procedimenti per la produzione di radioattività artificiali mediante il bombardamento con neutroni *(Method for increasing the efficiency of the processes for the production of artificial radioactivities by neutron bombardment), was submitted in Italy just after the achievement (on October 22, 1934) of the first experimental results, and later extended to U.S.A. and other countries. The intriguing story about this Patent (seemingly without reference to its content), which resulted to be of fundamental relevance for the subsequent development of the atomic energy, is well described in the literature.*[2]

The reference article[3] *for the material here contained is of February 15, 1935, to which we refer the reader for further details. However, at least in part, specific results discussed here are somewhat different from those in the article mentioned.*

This invention relates to the production of isotopes of elements from other isotopes of the same or different elements by reaction with neutrons, and especially to the production of artificial radio-activity by formation of unstable isotopes.

It has been known for many years that, although each chemical element has always the same atomic number or charge, it may exist in different forms having different atomic weights. These forms of the elements are referred to as isotopes.

It has also been known that the radio-active elements, by disintegration or break down occurring in their nuclei are spontaneously converted into various isotopes of other elements. Thus, for example, the radio-active element uranium may be converted into lead of atomic weight 206, while the element thorium may be converted into a different isotope of atomic weight 208.

It has long been known that such spontaneous disintegration of radio-active elements is accompanied by emission of alpha, beta, and gamma ray, that is to say, of the helium nucleus, electrons, and electromagnetic radiation of extremely short wave length.

---

[1]Similar sentences appear also in other Patents for evident legal reasons, but here the dubious "theoretical correctness" is particularly pointed out.

[2]See, for example, Segrè E. (1970), *E*nrico Fermi, physicist (The University of Chicago Press, Chicago).

[3]Amaldi, E., D'Agostino, O., Fermi, E., Pontecorvo, B., Rasetti, F. and Segrè, E. (1935), *P*roc. Roy. Soc. London **A149**, pp. 522–538.

In more recent years it has been demonstrated that isotopes of various elements could be converted into other isotopes of the same or different elements by bombardment with alpha particles, diplons,[4] protons or gamma rays of very high energy, and that the isotopes thus produced may be unstable so as to decompose with features similar to those of the naturally radio-active bodies. That is to say, radio-active isotopes may, in this way, be artificially produced.

Accordingly, it is an object of the present invention to provide a method and apparatus by which nuclear reactions can be carried on with high efficiency and with the heavier as well as with the lighter elements. A more specific object of the invention is to provide a method and apparatus for artificially producing radio-active substances with efficiency such that their cost may be brought below that of natural radio-active materials.

Our invention is based upon the use of neutrons instead of charged particles for the bombardment and transformation of the isotopes.

All of the prior work on nuclear reactions has been done with high energy particles and every effort has been bent toward increasing the energy of the particles as the means of extending and making more efficient the nuclear reactions. We have now discovered that effort in this direction is sound only when charged particles are used which require tremendous energy to break through the potential barrier surrounding the nucleus; and that if, instead of charged particles, neutrons are used for the nuclear reactions, the greatest efficiencies are in same cases attained with low energy or "slow" neutrons, e.g., of the order of a few hundred electron volts, or even much less down to a small fraction of an electron volts.

Neutrons when produced in any ordinary manner, e.g., by the action of radon on beryllium or of polonium on beryllium or by bombardment of atomic nuclei with artificially accelerated particles, might have a very wide range of energies but high average energy. These energies range up to several million volts. It is necessary, therefore, if the greatest efficiency of reaction is to be attained, to reduce by artificial means the energy of these neutrons. We describe below a method for slowing down fast neutrons.

We have demonstrated that the absorption of slow neutrons is anomalously large as compared with that of the faster or higher energy neutrons. The simplest explanation for most cases is to admit that the neutron is captured by the nucleus with formation of an isotope heavier by one mass unit. If this heavier isotope is unstable a strong induced radio-activity may

---

[4]@ That is, deuterons.

be expected. This occurs, for example, with silver and iridium which go over into radio-active isotopes. In other cases it is found that no activation, or at least no strong activation, follows an anomalously large absorption. This is the case with many elements, e.g., yttrium and cadmium. In these cases the formation of a stable nucleus upon the capture of the neutrons is to be expected.

In some cases the absorption of the slow neutrons results in the emission of a relatively strong gamma-radiation with energy corresponding to the binding energy of the neutron. This gives a reliable source of very hard gamma rays, even harder than the naturally produced gamma rays, e.g., from radium.

In view of these considerations it is obviously desirable to convert as many as possible of the available neutrons into the slow or low energy condition in which they may be readily captured by the nuclei of the substance being reacted. We have found that it is possible to achieve the desired results by passing the neutron radiation against or through a screen of a suitable material.

The materials which have been found best suited to this purpose are those containing hydrogen (including all its isotopes, but the light isotope which predominates in natural occurrence being most efficient) and especially water and the hydrocarbons, such as paraffin for example. Other materials, as for example beryllium, carbon, silicon, lead, show this effect to a lesser degree. Other materials, of which iron is an example, do not produce a similar effect to any practical extent, probably because of a relatively large absorption of the neutrons when their energy is reduced.

The increase in activity of the neutrons by such substances is apparently due to two effects both resulting from collisions of the neutrons. In the first place it is readily shown that an impact of a neutron against a proton reduces, on the average, the neutron energy by a factor $1/e$. From this it follows that ten impacts reduce the energy to about 1/20000 of its original value. Assuming the initial energy to be $4 \cdot 10^6$ electron volts, the energy after ten impacts would be about 200 electron volts and 20 impacts would reduce the energy of the neutron down to a value corresponding to thermal agitation. Thus the first important effect is probably the reduction of the energy of the faster neutrons by impact and the efficiency of hydrogen for this purpose is probably due to the low mass of the hydrogen nucleus. Although we refer to the nucleus, because almost the entire mass is represented by the nucleus, it will be understood that the impact for slowing down may be, and for reasons of economy ordinarily will be, with atoms, i.e., combined nuclei.

The second probable effect is the scattering and reflection of the neutrons.

Hydrogen is so much more effective than any of the other elements for reducing the energy of neutrons that it will ordinarily be used. It must not be overlooked, however, that the elements having a lesser effect offer possibilities for control of the neutron energy. Where neutrons of initially lower energy are used or where their use requires a higher energy than in the reactions with which we are here especially concerned, the less effective elements may be used singly or combined with elements of different energy reducing power.

The density of the energy reducing or scattering substance has also been found to be an important factor. This follows, likewise, from the theoretical explanation given above. If the energy reduction and scattering of the neutrons is due to impact with atomic nuclei, the probable frequency of such impacts will be directly dependent upon the number of atoms in a given space. For this reason such substances ordinarily should be used in liquid or solid (i.e., non-gaseous) form and, so far as is practicable, substances will be chosen having as high as possible a proportion of hydrogen in the molecule. Here again, the gaseous state under various pressures, and substances having less hydrogen offer the possibility for accurate control if less the maximum slowing of neutrons is desirable.

It will be readily understood from what has been said above that the greatest effect is attained if the source of neutrons and the substance being irradiated are both surrounded by the energy reducing and scattering material. This could easily be accomplished in many cases by immersing the neutron source in a solution or emulsion of the substance being bombarded. We have illustrated diagrammatically in Figs. 1.1 and 1.2 arrangements by which this may be accomplished.

In Fig. 1.1 a cylindrical paraffin block **10** is provided with a hole **11** into which is inserted a source of neutrons, e.g., a tube containing radon and beryllium. The material being irradiated is placed above the source on the paraffin block as shown at **12** and is covered by a second paraffin block **13** having a central opening **14** to accommodate the material being irradiated. For the treatment of small amounts of materials the block **10** may be, for example, about 24 centimeters in diameter and about 14 centimeters in height with the neutron source about 2 centimeters under the upper surface. It will be observed that these dimensions give radial thickness of the material surrounding the neutron source approximately equal to the

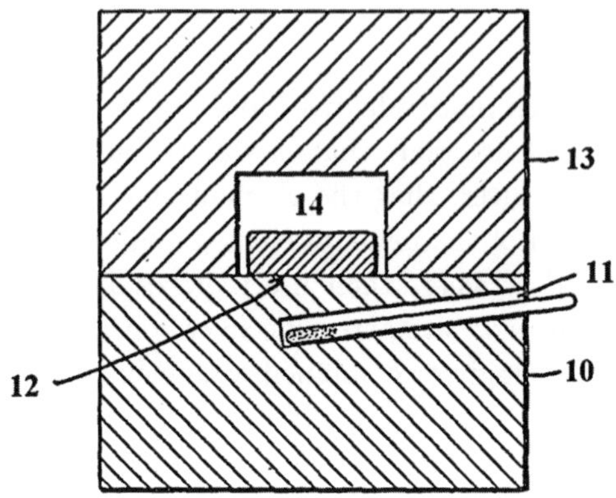

Fig. 1.1

mean free path in the substance of the high energy neutrons.

Where the substances to be irradiated are soluble in or can be suspended in water or a hydrocarbon or other energy reducing or dispersing substance, etc., the substances irradiated therein may be formed by immersing the neutrons source directly into the solution etc. (See Fig. 1.2.)

The hydrogen which serves to reduce the energy of the neutrons may also be in chemical combination with the substance being irradiated.

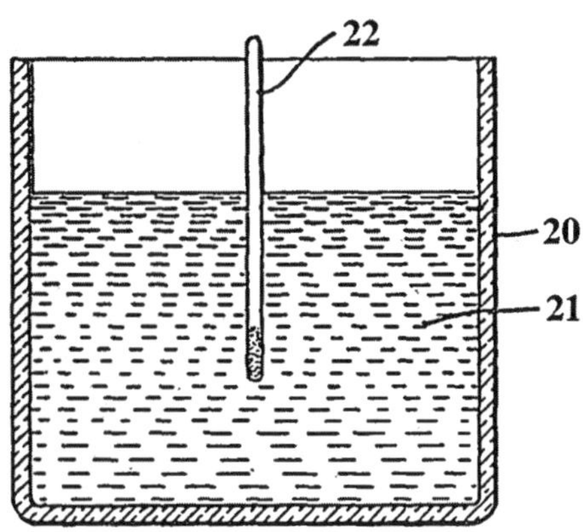

Fig. 1.2

Fig. 1.2 is a diagrammatic illustration of a neutron irradiating device in which the substance being irradiated is dissolved or dispersed in the energy reducing or dispersing material. A suitable vessel **20** is provided for holding the solution or dispersion **21** and into this is immersed the neutron source **22**.

Instead of the radon beryllium source, any other source of neutrons may be used, as for example neutron tubes as developed by Oliphant and as more recently developed by laboratories of the General Electric Company and the Westinghouse Electric and Manufacturing Company or cyclotrons as developed by Lawrence. In such tubes deuteron oxide (heavy water), which may be separated by known methods from naturally occurring water, is bombarded with deuterons accelerated in an electric field produced by a grid tube. The deuteron nucleus is disintegrated with the emission of neutrons.

Obviously the apparatus can be endlessly modified, the essential being the combination of the energy reducing substance near the substance being irradiated, and a suitable source of neutron radiation.

In the following we have tabulated the results of various irradiations which we have carried out:

1. *Hydrogen.* – No activity could be detected either in water or in paraffin irradiated in a large can of water with 500 millicuries Rn+Be for several days.

3. *Lithium.* – Lithium hydroxide was found to be inactive after irradiation with slow neutrons (14 hours, 400 millicuries). Although lithium remains inactive, it strongly absorbs the slow neutrons; half-value thickness $\delta = 0.05$ g/cm^2. This absorption is not accompanied by a γ-radiation. It was shown independently by Chadwick and Goldhaber[5] and by us that when the slow neutrons are absorbed, heavy charged particles are emitted. According to Chadwick and Goldhaber, the nuclear process is represented by the following reaction,

$$^6\text{Li} + n = {}^4\text{He} + {}^3\text{H}.$$

4. *Beryllium.* – Metallic beryllium (purity 99%), strongly irradiated with slow neutrons, showed only an extremely weak activity possibly due to impurities. Owing to the very strong activation of several elements when irradiated under water, impurities might easily be misleading.

5. *Boron.* – Metallic boron irradiated 14 hours under water with 500 millicuries was found inactive. Boron has the highest absorption coeffi-

[5] *Nature*, vol. 135, pag. 65 (1935).

cient as yet found for slow neutrons, $\delta = 0.004$ g/cm^2, corresponding to a cross-section of about $3 \cdot 10^{-21}$ cm^2. No γ-rays have been found to accompany this absorption: instead of a γ-radiation in this case as well as for lithium, α-particles are emitted, as was shown by Chadwick and Goldhaber (*Nature*, vol. 135, 1935, pag. 65) and by us. This effect can be easily detected by the strong discharge in an ionization chamber filled with boron trifluoride surrounded by paraffin and irradiated with a Po+Be neutron source. Screening the ionization chamber with a thin cadmium foil in order to absorb the slow neutrons, reduces considerably the ionization current. The same effect was observed with the ionization chamber filled with air, some boron being spread on its floor. The emission of α-particles was also detected with a small ionization chamber connected to a linear amplifier, either spreading some boron on its walls or filling it with boron trifluoride. In order to explain these phenomena we have proposed the nuclear reaction,

$$^{10}\mathrm{B} + n = {}^{7}\mathrm{Li} + {}^{3}\mathrm{He}$$

6. *Carbon.* – No activity; see hydrogen.

7. *Nitrogen.* – Ammonium nitrate irradiated 12 hours with 600 millicuries under water showed no activity.

8. *Oxygen.* – No activity; see hydrogen.

9. *Fluorine.* – Both activities of this element (periods 9 seconds and 40 seconds) are not sensitive to hydrogen containing substances.

11. *Sodium.* – This element has two activities: one of these (period 40 seconds) is not sensitive to hydrogen-containing substances. A very weak activity with a long period was reported by Bjerge and Westcott. As this activity is strongly enhanced by water, we were able to measure its period with reasonable accuracy and found it to be 15 hours. Owing to the theoretical importance of this activity, we compared very carefully its decay curve with that of the long period of aluminium in order to check their identity. For a chemical investigation of the active substance we irradiated pure sodium carbonate. We dissolved the irradiated substance in hydrochloric acid and added aluminium and magnesium chlorides. A precipitate of the hydroxides of the latter elements obtained by adding ammonia, was found inactive. Afterwards we added some sodium fluoride to the solution, and precipitated the fluorine as barium fluoride; this precipitate was also found inactive. The solution containing the original sodium was then evaporated and ignited gently, in order to eliminate neon, where an active isotope of this element would have been formed. The activity was found in the dried sodium salt. We conclude that the active product is an isotope of sodium,

^{24}Na. The same isotope was produced by us last year by bombarding magnesium or aluminium with neutrons. ^{24}Na has also been produced recently in considerable amounts and studied very completely by Lawrence bombarding several elements with artificially accelerated particles.

12. *Magnesium.* – Pure magnesium oxide, especially tested by us in order to ensure that it was aluminium free, was irradiated under water. The substance was kept at some distance from the source in order to prevent the activation of the periods not sensitive to water. A new very weak activity with a period of about 10 minutes was found. As this period coincides with the 10-minute period of aluminium, which is known to be due to ^{27}Mg (see aluminium), it is very likely that it is due to the same isotope formed by the capture of a neutron by ^{26}Mg present in an amount of 11% in ordinary magnesium.

The 40-second period is insensitive to water.

13. *Aluminium.* – Aluminium irradiated in water shows a fairly strong new activity decaying with a period of 2.3 minutes (measured with an ionization chamber). Irradiated outside of water, this activity is extremely weak. As the period of the new activity coincides with the 2.3 minute period of silicon, which is due to ^{28}Al, we assume that this activity is also due to the same isotope formed by capture of a neutron from ^{27}Al.

The second period of aluminium has been measured with the ionization chamber and found to be 10 minutes instead of 12. This activity is insensitive to water. A chemical separation of the carrier of this activity has been performed. Irradiated metallic aluminium was dissolved in a caustic soda solution and magnesium chloride was added. The precipitate of magnesium hydroxide carried the 10-minute activity. We assume that the active isotope is ^{27}Mg formed according to the reaction,

$$^{27}\mathrm{Al} + n = {}^{27}\mathrm{Mg} + {}^{1}\mathrm{H}$$

14. *Silicon.* – We have determined with the ionization chamber the short period of this element, finding it to be 2.3 minutes. This activity is insensitive to water.

Besides this activity, we found a new, longer period of some hours in fused silica irradiated in water. This activity is very weak and very sensitive to water. We think probably that its carrier is ^{31}Si which is obtained by irradiated phosphorus and has a period of 2.4 hours. ^{31}Si could be formed by capture of a neutron from ^{30}Si present in an amount of 3%.

15. *Phosphorus.* – The short-period activity of this element (2.3 minutes) is not enhanced by water. Curie, Joliot and Preiswerk ascribe this

period to ^{28}Al. A chemical test in favor of this hypothesis is the following: we irradiated phosphoric acid, neutralized the solution with sodium carbonate, and added aluminium chloride; the activity was found to be concentrated in the precipitated aluminium.

We have observed, with the aid of the ionization chamber, the decay-curve of the longer period of phosphorus. Its period is 2.4 hours instead of 3 as given before. We have also measured, with the ionization chamber, the half-value thickness of the corresponding β-rays and found it to be 0.15 g/cm^2 Al.

16. *Sulphur.* – We have determined in the ionization chamber the period of phosphorus extracted from irradiated sulphur. We found: period 14 days, half-value thickness of the β-rays 0.10 g/cm^2 Al.

17. *Chlorine.* – Chlorine irradiated under water showed a new period of 35 minutes measured electrometrically.

Chlorine absorbs fairly strongly the slow neutrons (half-value thickness $\delta = 0.3$ g/cm^2). The process of absorption is accompanied by emission of γ-rays.

19. *Potassium.* – We have found in irradiated potassium an induced activity strongly sensitive to water, decaying with a period of 16 hours. A chemical investigation of the carrier of the activity, performed by the same method described for sodium, excluded the elements Cl, Al, Ca. We conclude therefrom that the activity is probably carried by an isotope of potassium. According to v. Hevesy, this isotope is to be identified with a ^{42}K, that was obtained by him by neutron bombardment of scandium, and has the same decay period.

20. *Calcium.* – No activity was found in calcium fluoride irradiated 14 hours in water with a 600 millicuries source. A very weak activity sensitive to water has been demonstrated.

23. *Vanadium.* – The decay of the activity induced in vanadium has been measured in the ionization chamber with the following results: half-value period 3.75 minutes; half-value thickness of the β-rays 0.17 g/cm^2 Al. The β-rays are accompanied by a γ-radiation. The activation of vanadium is strongly sensitive to hydrogenated substances.

24. *Chromium.* – The activity of chromium is insensitive to water.

25. *Manganese.* – The activity with short period (3.75 minutes) is insensitive to water. On the other hand, the activity with longer period (2.5 hours measured in the ionization chamber) is strongly enhanced by water. Half-value thickness of β-rays measured electrometrically is 0.14 g/cm^2 Al; the disintegration is accompanied by γ-rays. The 2.5-hour prod-

uct is known to be an isotope of manganese. In order to get new evidence in favor of the fact that the active product is really an isotope of manganese, we first concentrated the activity obtained in irradiated manganese permanganate by a precipitation of manganese carbonate. The carbonate containing the activity was then dissolved in hydrochloric acid, and large amounts of chromium, vanadium and iron salts were added to the solution. Afterwards the manganese was separated once again as dioxide, with nitric acid and sodium chlorate. The manganese precipitate carried the activity, while the fractions containing chromium, vanadium and iron were found to be inactive.

26. *Iron.* – The activity of this element (period 2.5 hours) is insensitive to water. Half-value thickness for the absorption of slow neutrons 8 g/cm^2.

27. *Cobalt.* – This element absorbs fairly strongly the slow neutrons; half-value thickness 0.7 g/cm^2. The absorption is accompanied by the emission of a γ-radiation.

28. *Nickel.* – Strongly irradiated nickel showed only a dubious trace of activity.

29. *Copper.* – Both induced activities of this element (period 5 minutes, measured electrometrically, and 10 hours) are strongly enhanced by water. Copper absorbs the slow neutrons with a half-value thickness of about 3 g/cm^2; this absorption is accompanied by a weak γ-radiation.

Irradiated metallic copper was dissolved in hydrochloric acid, and small quantities of cobalt, nickel and zinc salts were added. Copper sulphide was precipitated from the acid solution and found to be active. The precipitates of the zinc, cobalt and nickel sulphides, obtained by neutralizing the solution and adding ammonium sulphide, were inactive. As the time employed for this test was rather long, the test refers only to the longer period. The carrier of this activity can then be assumed to be an isotope of copper.

30. *Zinc.* The activity of the short period of zinc is not enhanced by water. The longer period was measured electrometrically and found to be 10 hours. The carrier of this activity has been investigated by means of the following test: irradiated metallic zinc was dissolved in hydrochloric acid, and a small quantity of copper, nickel and cobalt salts added. Copper was precipitated partially by reduction on small traces of undissolved metallic zinc and partially as sulphide in the acid solution. The collected copper was strongly active. Neutralizing the solution and adding ammonium sulphide, the other elements were precipitated and found to be inactive.

31. *Gallium.* – The 20-minute period (measured electrometrically) is not very sensitive to water. Half-value thickness of the corresponding β-

rays is 0.17 g/cm^2 Al. The carrier of this activity is probably an isotope of gallium. In order to test this point, we irradiated gallium nitrate and afterwards added to the solution traces of copper and zinc. Copper was separated as a metallic deposit on zinc powder and zinc as zinc mercury sulphocyanate after adding mercury sulphocyanate. Both elements were found to be inactive.

Besides this 20-minute activity, we have also found, irradiating under water, a new activity which is accompanied by a rather strong γ-radiation; it decays with a period of 23 hours (measured electrometrically).

33. *Arsenic.* – The activity of this element is strongly sensitive to water. We have measured electrometrically its period (26 hours) and its half-value thickness of the β-rays (0.16 g/cm^2 Al).

34. *Selenium.* – The activity if this element (period 35 minutes) is sensitive to water. Irradiated selenious anhydride was dissolved in 30% hydrochloric acid and some arsenious anhydride added to the solution. We precipitated metallic selenium by reduction with gaseous sulphurous anhydride, and found it strongly active. We precipitated from the solution arsenic sulphide and found it inactive. This test seems to rule out also germanium, and we conclude that the activity is due to an isotope of selenium.

35. *Bromine.* – Both activities of this element are sensitive to water. The periods have been measured electrometrically; they are 18 minutes and 4.2 hours. The half-value thickness of the γ-rays is for both activities 1.12 g/cm^2 Al, and both are accompanied by γ-rays.

38. *Strontium.* – No activity was found after a long and strong irradiation under water.

39. *Yttrium.* – Strongly irradiated yttrium oxide showed only a very weak activity possibly due to impurities. Yttrium absorbs very intensively the slow neutrons (half-value thickness $\delta = 0.015$ g/cm^2). This absorption is accompanied by γ-rays.

40. *Zirconium.* – Strongly irradiated zirconium nitrate showed only a very weak activity probably due to impurities.

41. *Niobium.* – The same as zirconium.

43. *Rhodium.* – The short period activity is sensitive to water. Period and half-value thickness of the β-rays have been determined electrometrically (44 seconds; 0.15 g/cm^2 Al). We also made a more accurate measurement in the ionization chamber of the longer period and found it to be 3.9 minutes. The activity is accompanied by a weak γ-radiation. Rhodium absorbs fairly strongly the slow neutrons (half-value thickness 0.3 g/cm^2): the absorption probably corresponds to the formation of the active isotopes.

46. *Palladium.* – Also the activities of this elements are sensitive to water. We find at least two periods: a short one of about a quarter of an hour and one of about 12 hours.

47. *Silver.* – The two periods have been redetermined with the ionization chamber. They are 22 seconds and 2.3 minutes. They are both very sensitive to water. To the strong activation of this element corresponds a considerable absorption for slow neutrons (half-value thickness $\delta = 1.2$ g/cm^2).

We added palladium nitrate and rhodium chloride to a solution of irradiated silver nitrate. Adding hydrochloric acid, we precipitated silver which was found active. From the filtered solution we precipitated palladium with dimethylglioxime and rhodium by reduction. Both were inactive. This test is valid only for the longer period, owing to the time employed, and shows that the carrier of the activity is probably an isotope of silver.

48. *Cadmium.* – Cadmium irradiation under different conditions showed several weak activities with various periods not yet identified. Cadmium absorbs with great intensity the slow neutrons. (Half-value thickness 0.013 g/cm^2. The corresponding cross-section is the largest as yet found for slow neutrons ($\sigma = 10^{20}$ cm^2). The absorption is accompanied by an intensive γ-radiation and probably corresponds to the transformation of a stable isotope of cadmium into another stable isotope of the same element.

49. *Indium.* – The activity induced in indium shows three periods: the shortest period (13 seconds) has an activity highly sensitive to water. Also the second period (54 minutes, measured electrometrically) is very sensitive to water. Magnetic deflection experiments show that the corresponding electrons are negative; their half-value thickness is 0.045 g/cm^2 Al. A still longer period of some hours is recorded by Szilard and Chalmers; this last activity is either insensitive to water or is only moderately sensitive.

Chemical tests have been made in order to identify the carriers of the last two activities. To a solution of irradiated indium nitrate, silver was added and precipitated as silver chloride; the precipitate was inactive. Afterwards we added to the solution tin, antimony and cadmium and precipitated them as sulphides with sulphuretted hydrogen. The acidity of the solution must be adjusted in such a way as to leave the indium in solution while precipitating the other metals. This precipitate also was inactive; neutralizing the solution, we precipitated the indium sulphide which carried the activity.

Corresponding to the strong activation of indium, it is found that this element has a considerable absorption power for the slow neutrons: half-value thickness $\delta = 0.3$ g/cm^2.

50. *Tin.* – Tin strongly irradiated under water showed no activity.

51. *Antimony.* – We have found an induced activity in this element, decaying with a period of 2.5 days; the activation is sensitive to hydrogenated substances. The half-value thickness for the emitted β-rays is 0.09 g/cm^2 Al. The following chemical test indicates that the carrier of this activity is probably an isotope of antimony. We dissolved metallic irradiated antimony in aqua regia and added some tin to the solution; after separation of tin as a sulphide according to Clarke, we found the activity in a precipitate of sulphide of antimony. The antimony sulphide was then dissolved again; indium was added to the solution and antimony separated as a sulphide in a moderately acid solution; the solution was neutralized and indium precipitated and found to be inactive. To a new solution of the antimony we added tellurium and iodine and separated the first by reduction and the second as a silver iodide. Both were inactive.

52. *Tellurium.* – Shows a weak activity sensitive to water; the period resulted 45 minutes.

53. *Iodine.* – Period and half-value thickness of the β-rays were determined electrometrically: period 25 minutes; half-value thickness 0.11 g/cm^2 Al. The activation is moderately sensitive to water.

56. *Barium.* – A new activity sensitive to water with a period of 80 minutes has been found. The following chemical test is in favor of the assumption that the carrier of this activity is an isotope of barium. We dissolved irradiated barium hydroxide in hydrochloric acid, and added a small quantity of sodium chloride and precipitated barium sulphate. The activity was carried by the precipitate; we evaporated the solution and found the residual sodium to be inactive.

57. *Lanthanum.* – No activity was found after strong irradiation under water.

58. *Cerium.* – Same as lanthanum.

59. *Praseodymium.* – The short-period activity (5 minutes) is insensitive to water. Irradiating under water we have found a new water-sensitive activity decaying with a period of 19 hours; half-value thickness of the corresponding β-rays 0.12 g/cm^2 Al (both measured electrometrically).

72. *Hafnium.* – v. Hevesy has found an activity having a long period of several months which is sensitive to water.

64. *Gadolinium.* – We irradiated under water a very pure sample of gadolinium oxide. We found an activity, decaying with a period of 8 hours.

73. *Tantalum.* – Only a dubious activity was found after 12 hours' irradiation under water with 500 millicuries.

74. *Tungsten.* – Metallic tungsten was irradiated under water and showed an activity decaying with a period of about 1 day.

We irradiated tungstic anhydride, dissolved it in caustic soda and then added and separated tantalum pentoxide which was found to be inactive. To the tungstic solution we added a nitric rhenium solution and precipitated the tungstic anhydride adding hydrochloric acid. The precipitate carried the activity, while the rhenium, precipitated from the filtrate as sulphide, was inactive. As we have no hafnium, we have made the following experiment in order to exclude an isotope of this element as carrier of the activity. From a solution of irradiated tungstic anhydride in ammonia, we precipitated zirconium hydroxide. The precipitate was inactive. We conclude that the activity of tungsten is probably due to an isotope of this element.

75. *Rhenium.* – We irradiated pure metallic rhenium under water; its activity is enhanced by water and decays with a period of about 20 hours. The half-value thickness of the electrons is 0.12 g/cm^2 Al. The activity is probably carried by an isotope of rhenium. Irradiated rhenium was dissolved in nitric acid; we added tantalum and tungsten and separated them as tantalum pentoxide and tungstic anhydride. Both were inactive, while rhenium conserved the activity.

77. *Iridium.* – The activity induced in this element is strongly sensitive to water. Period and half-value thickness of the β-rays have been measured in the ionization chamber; period 19 hours, half-value thickness 0.12 g/cm^2 Al. To the strong activation of iridium corresponds a strong absorption of the slow neutrons; half-value thickness 0.3 g/cm^2; the absorption is accompanied by the emission of γ-rays.

78. *Platinum.* – Very pure metallic platinum irradiated under water showed an activity decaying with a period of about 50 minutes.

79. *Gold.* – The activity of this element is sensitive to water; its period has been measured electrometrically and is 2.7 hours. The β-rays were magnetically deflected and found to be negative. They have a very small penetrating power: half-value thickness 0.04 g/cm^2 Al. Strong γ-radiation is omitted during bombardment with slow neutrons.

80. *Mercury.* – No activity was found after strong irradiation. This element absorbs intensely the slow neutrons, half-value thickness 0.2 g/cm^2. γ-rays are emitted during the absorption.

81. *Thallium.* – No activity was found after strong irradiation.

82. *Lead.* – The same as thallium.

83. *Bismuth.* – The same as thallium.

90. *Thorium.* – The 1-minute and 24-minute (measured electrometrically) periods are scarcely sensitive to water.

92. *Uranium.* – We have also studied the influence of hydrogenated substances on the induced activities of this element. (Periods 15 seconds, 40 seconds, 13 minutes, 100 minutes.) The results was that while the activities corresponding to the first, third and fourth period are slightly increased by water, no increase was found for the activity corresponding to the 40-second period.

Chemical evidence seems to indicate that the carriers of the 13- and the 100-minute activities were not isotopes of any of the known heaviest elements, and that they were probably due to transuranic elements.

The precipitation of the activity with a sulphide was repeated, precipitating several metals (silver, copper, lead, mercury); the acidity of the solution (hydrochloric acid) was about 20%; sometimes slightly varied in order to facilitate the precipitation of the sulphide of the metal used. The yield in activity of the precipitate was generally good – about 50% – and varied according to the condition of the precipitation. Nitric acid lowers the yield of the reaction very much. The usual high yield of the sulphide reaction is also obtained in presence of a hydrofluoric solution of tantalum. We also made a test in order to see whether the induced activities presented a reaction which is given by von Grosse as the most characteristic of protoactinium.

We dissolved in a 25% hydrochloric acid solution uranium oxide which had been purified and irradiated; we added to the solution zirconium nitrate and phosphoric acid; the precipitate of zirconium phosphate was inactive. After the separation of zirconium we precipitated a sulphide from the filtered solution, and collected the activity in the sulphide with the usual yield. According to von Grosse and Agruss, this reaction must be considered a proof by the non-identity of the carrier of the activity with a protoactinium isotope. The 15-second, 13-minute and 100-minute activities are probably chain products, with atomic number 92, 93 and 94 respectively and atomic weight 239.

From the above tabulation it is apparent that the increase in activities by the hydrogen containing substances, etc., is particularly applicable to those nuclear reactions in which the neutron is captured with the formation of a heavier isotope of the same element; and the present invention makes possible numerous reactions of this type which could not be appreciably carried out without the use of our invention.

With these procedures the isotopes produced by the nuclear reactions are ordinarily mixed with other substances, although in much higher concentrations than was heretofore possible. We have utilized for the separation of these isotopes and especially of the radio-active isotopes the method of Szilard and Chalmers (*Nature*, vol. 134, pag. 462, 1934) and extended their procedure to cover other cases. This and further examples and theoretical discussion of our invention are set forth in our publication: Fermi, Amaldi, D'Agostino, Rasetti, Segre, *Proc. Roy. Soc.* A, vol. 146, pag. 483 (1934); A, vol. 149, pag. 522; Fermi, Amaldi, Pontecorvo, Rasetti, *Ric. Scient.*, vol. 2, pag. 380 (1934); Fermi, Pontecorvo, Rasetti, *Ric. Scient.*, vol. 2, pag. 380 (1934); Amaldi, D'Agostino, Segre, *Ric. Scient.*, vol. 2, pag. 381 (1934); Amaldi, D'Agostino, Fermi, Pontecorvo, Rasetti, Segre, *Ric. Scient.*, vol. 2, pag. 467 (1934); vol. 1, pag. 123 (1935).

Claims directed to the broad method of producing radio-active substances by a neutron reaction and to beta-emissive substances so produced are being presented in the copending application Serial No. 57,325 to Enrico Fermi, filed January 2, 1936.[6]

Although we have herein described our invention in detail and specified particular examples of apparatus and processes and various modifications thereof, and have proposed various theoretical explanations, it is to be understood that these are not binding nor exhaustive but are intended rather for the assistance of others skilled in the art to enable them more easily to apply our invention under widely varying conditions encountered in actual practice and the change and modify the particular embodiments and examples herein described as may be necessary or desirable under such varying conditions. The theoretical statements and explanations are, of course, not conclusive and our invention is in no way dependent upon their correctness. We have found them helpful and give them for the aid of others, but our invention will be equally useful if it should prove that our theoretical conclusions are not altogether correct.

We claim:

1. The process for the production of radio-active isotopes, which comprises generating neutrons having a high average electron voltage, slowing down and scattering said neutrons by projecting them through a screen of an element of the class consisting of hydrogen, helium, beryllium, carbon, silicon and lead which screen is of such thickness that the neutrons are

[6]@ It is remarkable that no such an application is present in the archive of the U.S. Patent Office.

slowed down to an average energy of not more than a few hundred electron volts, then passing said neutrons into a mass of an element of the groups having atomic numbers 11, 12, 13, 14, 17, 19, 23, 25, 29, 31, 33, 34, 35, 43, 47, 48, 49, 51, 52, 53, 56, 64, 72, 74, 75, 77, 78, 79 and 92, and thereby producing from the latter element a radio-active isotope capable of emitting beta rays.

2. The process of producing radio-active isotopes which comprises generating neutrons having a high average energy, slowing down said neutrons so that they have an average energy of not more than a few hundred electron volts by projecting them through an energy-reducing screen of an element of the class consisting of hydrogen, helium, beryllium, carbon, silicon and lead, and passing said neutrons of reduced energy into a mass of an element of the group having atomic numbers 11, 12, 13, 14, 17, 19, 23, 25, 29, 31, 33, 34, 35, 43, 47, 48, 49, 51, 52, 53, 56, 64, 72, 74, 75, 77, 78, 79 and 92 to thereby produce from the latter element a radio-active isotope capable of emitting beta rays.

3. The process of producing radio-active isotopes which comprises, generating neutrons having a high average energy, slowing down said neutrons so that they have an average energy of not more than a few hundred electron volts by projecting them through a hydrocarbon, and passing said neutrons of reduced energy into a mass of an element of the group having atomic numbers 11, 12, 13, 14, 17, 19, 23, 25, 29, 31, 33, 34, 35, 43, 47, 48, 49, 51, 52, 53, 56, 64, 72, 74, 75, 77, 78, 79 and 92 to thereby produce from said latter element a radio-active isotope capable of emitting beta rays.

4. The process of producing radio-active isotopes which comprises, generating neutrons having a high average energy, slowing down said neutrons so that they have an average energy of not more than a few hundred electron volts by projecting them through paraffin, and passing said neutrons of reduced energy into a mass of an element of the group having atomic numbers 11, 12, 13, 14, 17, 19, 23, 25, 29, 31, 33, 34, 35, 43, 47, 48, 49, 51, 52, 53, 56, 64, 72, 74, 75, 77, 78, 79 and 92 to thereby produce from the latter element a radio-active isotope capable of emitting beta rays.

5. The process of producing radio-active isotopes which comprises, generating neutrons having a high average energy, slowing down said neutrons so that they have an average energy of not more than a few hundred electron volts by projecting them through water, and passing said neutrons of reduced energy into a mass of an element of the group having atomic numbers 11, 12, 13, 14, 17, 19, 23, 25, 29, 31, 33, 34, 35, 43, 47, 48, 49, 51, 52, 53, 56, 64, 72, 74, 75, 77, 78, 79 and 92 to thereby produce from the

latter element a radio-active isotope capable of emitting beta rays.

6. The process of producing a radio-active isotope from an element of the group having atomic numbers 11, 12, 13, 14, 17, 19, 23, 25, 29, 31, 33, 34, 35, 43, 47, 48, 49, 51, 52, 53, 56, 64, 72, 74, 75, 77, 78, 79 and 92, comprising the steps distributing the element throughout an energy-reducing screen of an element of the class consisting of hydrogen, helium, beryllium, carbon, silicon and lead, generating neutrons having a high average energy, and projecting the high-energy neutrons through said energy-reducing screen to slow down the neutrons to an average energy of not more than a few hundred electron volts and thereby produce from said first element by the reaction of the neutrons of reduced energy a radio-active isotope capable of emitting beta rays.

7. The process of producing a radio-active isotope from an element of the group having atomic numbers 11, 12, 13, 14, 17, 19, 23, 25, 29, 31, 33, 34, 35, 43, 47, 48, 49, 51, 52, 53, 56, 64, 72, 74, 75, 77, 78, 79 and 92, comprising the steps distributing the element throughout an energy-reducing screen of an element of the class consisting of hydrogen, helium, beryllium, carbon, silicon and lead, generating at a point substantially surrounded by said screen neutrons having a high average energy, and projecting the high-energy neutrons through said energy-reducing screen to slow down the neutrons to an average energy of not more than a few hundred electron volts and thereby produce from said first element by the reaction of the neutrons of reduced energy a radio-active isotope capable of emitting beta rays.

Emilio Segré
Enrico Fermi
Edoardo Amaldi
Bruno Pontecorvo
Franco Rasetti

Chapter 2

Test Exponential Pile

Enrico Fermi, Chicago, Ill., assignor to the United States of America as represented by the United States Atomic Energy Commission

Application May 4, 1944, Serial No. 534,129

U.S. Patent No. 2,780,595 - Patented Feb. 5, 1957

The Patent contains an extremely detailed description of an atomic pile employing natural uranium as fissile material and graphite as moderator. It starts with the discussion of the theory of the intervening phenomena, in particular the evaluation of the reproduction or multiplication factor, K, that is the ratio of the number of fast neutrons produced in one generation by the fissions to the original number of fast neutrons, in a system of infinite size. The possibility of having a self-maintaining chain reaction in a system of finite size depends both on the facts that K is greater than unity and the overall size of the system is sufficiently large to minimize the percentage of neutrons escaping from the system. After the description of a possible realization of such a pile (with many detailed drawings), the various kinds of neutron losses in a pile are depicted.

Particularly relevant is the reported "invention" of the exponential experiment: since theoretical calculations can determine whether or not a chain reaction will occur in a give system, but can be invalidated by uncertainties in the parameters of the problem, an experimental test of the pile is proposed, aimed at ascertaining if the pile under construction would be divergent (i.e. with a neutron multiplication factor K greater than 1) by making measurements on a smaller pile. The idea is to measure, by a detector containing an indium foil, the exponential decrease of the neutron density along the length of a column of uranium-graphite lattice, where a neutron source is placed near its base. Such an exponential decrease is greater or less than that expected due to leakage, according to whether the K factor is less or greater than 1, so that this experiment is able to test the criticality of the pile, its accuracy increasing with the size of the column. In order to perform this measure a mathematical description of the effect of neutron production, diffusion, and absorption on the neutron density in the structure is *given. In particular, a mathematical formula is given for the neutron density distribution at various points throughout a pile with a parallelepiped geometry. This expression is discussed in details, with regard to its possible approximate form and the corrections needed when the approximations are not valid. From the discussion, a definition of the so-called "critical size" emerges, that is the size for which the total surface to volume ratio of the entire pile is such that the rate of neutron loss is reduced to a value less than the rate of neutron production; the expressions of the critical radius for various geometrical structures are given. Finally, it is reported in detail a measurement of the neutron density in a pile built with a rectangular geometrical arrangement of uranium boxes in graphite blocks. As a result of this "invention", it was proposed a structure with an actual K factor of 1.054: this structure was built and operated at various rates of power production in the form of heat.*

For the present Patent, there is no "reference" published article, although some material appears also in [Anderson (1942b)] of March 26, 1942. More in general, some results are as well present in several papers of Volume II of the Fermi Collected Papers [Fermi (1962)] but many details (including several figures) are reported only in the present Patent.

My invention relates to the general subject of nuclear fission and more particularly to a means and method for creating and measuring a chain reaction obtained by nuclear fission of natural uranium having a U^{235} isotope content of approximately 1/139.

When it became known that the isotope U^{235} in natural uranium could be split or fissioned by bombardment with thermal neutrons – i.e., neutrons in thermal equilibrium with the surrounding medium – many predictions were made as to the possibility of obtaining a self-sustaining chain reaction system, i.e., a system where the fission neutrons produced give rise to new fission neutrons in sufficiently large numbers to overcome the neutron losses in the system. Since the result of the fission of the uranium nucleus is the production of two lighter elements with great kinetic energy plus about 2 neutrons on the average for each fission, together with beta and gamma radiation, a large amount of power could be made available if a self-sustaining system could be built.

However, the neutrons arising from the fission process are set free with the very high energy of above one million electron volts average and are therefore not in condition to be utilized most efficiently to create new fissions in U^{235}. The fast fission neutrons, after they are created, must be slowed down to thermal energies before they are most effective to produce fresh fission by bombardment of additional U^{235} atoms. If a system can be made in which neutrons are slowed down without much absorption until they reach thermal energies and then mostly enter into uranium rather than into any other element, a nuclear chain reaction might be obtained. Light elements, such as deuterium, beryllium, or carbon, the latter preferably in the form of graphite, can be used as slowing agents. Hydrogen would be most advantageous were it not for the fact that there is a high probability of neutron capture by the hydrogen nucleus. A special advantage of the use of the light elements mentioned for slowing down fast fission neutrons is that fewer collision are required for slowing than in the case with heavier elements. Both beryllium and deuterium are expensive and not readily available in large quantities. However, carbon in the form of graphite is a relatively inexpensive, practical, and readily available agent for slowing fast neutrons to thermal energies. Recently, beryllium has been made available in sufficiently large quantities for test as to suitability for use as a neutron slowing material in a system of the type to be described. It has been found to be in every way as satisfactory as carbon.

This invention is concerned with a chain reacting system based on the combination of a fissionable material and a slowing medium in such a manner that the neutron reproduction in a system of practical size is greater than the number of neutrons lost in the system.

The earliest attempts to provide a structure capable of sustaining a chain reaction, using natural uranium, involved the use of fine uranium particles such as uranium oxide powder, dispersed in water as the slowing agent. However, these attempts were not successful, and analysis of experiments made has indicated that the neutron losses in such a system will prevent a chain reaction from being sustained irrespective of the size of the system. A chain reaction, however, will take place in such a system but the chain started by any number n of neutrons will die out after producing a number of fissions. Such a chain is known as a convergent chain reaction. It is only when the chain started by a number n neutrons continues with more than n neutrons being produced at each fission cycle that the chain can be said to be divergent. Even when the chain may be said to be divergent, neutron losses still may be sufficiently high to prevent the sustaining of the chain reaction in a system of practical size. It is therefore important to know where such neutron losses occur.

During the interchange of neutrons between uranium particles or bodies dispersed in a slow medium, neutrons may be lost in four ways: by absorption in the uranium metal or compound in which the neutrons originated or in which they finally may enter without producing fission, by absorption in the slowing down material, by capture by impurities present in the system, and by escape from the structure by diffusion.

Natural uranium, particularly by reason of its large U^{238} content, has an especially strong absorbing power for neutrons when they have been slowed down to a moderate energy, this absorption occurring at the so-called resonance energies. The absorption in uranium at these energies is termed the uranium resonance absorption or capture; it is caused by the isotope U^{238} and does not result in fission but leads to the production of the relatively stable nucleus 94^{239}. It is not to be confused with absorption or capture of neutrons by impurities, referred to later.

Neutron resonance absorption in uranium may take place either on the surface of the uranium bodies, in which case the absorption is known as surface resonance absorption, or it may take place further in the interior of the uranium body, in which case the absorption is known as volume resonance absorption. Volume resonance absorption is due to the fact that neutrons make collisions inside the uranium body and may thus reach resonance energies inside the body and be absorbed at once. After successfully reaching thermal velocities, the neutrons may also be subject to capture by U^{238} without fissions, leading to the production of 94^{239}.

The thermal neutrons are also subject to capture by the slowing material. While carbon, beryllium and deuterium have very small capture cross-sections for thermal neutrons, an appreciable fraction of thermal neutrons is lost by capture in the slowing material during diffusion therethrough before a uranium body is reached. It is therefore desirable to have the neutrons, after reaching thermal energy, promptly enter uranium.

In addition to the losses which are inherently a part of the nuclear chain reaction process, impurities present in both the slowing material and the uranium add a very important neutron loss factor in the chain. The effectiveness of various elements as neutron absorbers varies tremendously. Certain elements such as gadolinium, boron, cadmium, and some others, if present even in a few parts per million, could very likely prevent a self-sustaining chain reaction from taking place. It is highly important, therefore, to be able to evaluate in some manner the effect of losses due to such impurities, and as far as possible to remove from both the slowing material and the uranium all impurities capturing neutrons to the detriment of the chain reaction.

The ratio of the number of fast neutrons produced in one generation by the fissions, to the original number of fast neutrons, in a system of infinite size, is called the reproduction or multiplication factor of the system and is denoted by the symbol K. As it disregards finite size of the system, K therefore is a factor determinable from the geometry of the system, i.e. the slowing material employed and the manner of disposition of the uranium or other fissionable material in the slowing material, together with the effect of the fixed impurities in the materials. If K can be determined to be greater than unity, indicating that fewer neutrons are parasitically absorbed than are gained by fission, the uranium will support a self-maintaining chain reaction provided the overall size of the system is sufficiently large to prevent too large a percentage of the generated neutrons from escaping from the system. Thus the overall size is important, when K is greater than unity.

The overall size required for the system will vary, depending upon the neutron losses in the chain. If the structure is too small, with a large surface to volume ratio, there will be a high rate of loss of neutrons from the structure by leakage through the outer surfaces which may reduce the number of neutrons available for the fission chain inside the structure to a point where the chain reaction will not be self-maintaining. For each value for the reproduction factor K greater than unity, there is thus required a minimum overall size of the structure known as the critical mass or size, above which the rate of loss of neutrons by diffusion through the periphery of the structure is less than the net rate of production of neutrons within the system, thus making the system self-sustaining. The rate of diffusion of neutrons away from a large structure in which they are being liberated, can be treated by mathematical analysis.

However, K can still be greater than unity but so close to unity that no system of practical size will sustain a chain reaction as the rate of exterior neutron loss will always be too high, irrespective of any practical size to which the structure is built. It is important therefore, that all losses be kept to the minimum to provide a K factor sufficiently large to enable a system capable of sustaining a chain reaction to be of practical size.

Fig. 2.1 is a diagram or chart showing what happens to the neutrons during the chain reaction;

Fig. 2.2 is a front elevational view of an intermediate pile forming part of the subject matter of the present invention;

Fig. 2.3 is a horizontal sectional view taken on line **2–2** of Fig. 2.2;

Fig. 2.4 is a vertical sectional view taken through the device on line **3–3** of Fig. 2.5;

Fig. 2.5 is a horizontal sectional view taken on the line **4–4** of Fig. 2.4;

Fig. 2.6 is a vertical sectional view taken on the line **5–5** of Fig. 2.2;

Fig. 2.7 is a perspective view of a portion of the intermediate pile, showing the block construction;

Fig. 2.8 is an enlarged perspective view of the detector unit employed for making measurements in the device, the view showing the pans or trays spaced apart for purposes of clarity;

Fig. 2.9 is a longitudinal sectional view through the plate portion of the detector unit shown in Fig. 2.8;

Fig. 2.10 is a perspective view of a rod containing the neutron source shown in Fig. 2.6; and

Fig. 2.11 is an enlarged, fragmentary, detailed sectional view of the rod and neutron source shown in Figs. 2.6 and 2.10, the view being taken on the line **9–9** of Fig. 2.10.

To illustrate the importance of the various factors entering into a chain reaction, I shall next describe the chain reaction process, as it is presently understood to occur in any system of finite size utilizing natural uranium dispersed in graphite. For better explanation, reference is here made to the diagram constituting Fig. 2.1 of the accompanying drawings.

In Fig. 2.1:

A represents a uranium body of any size from which fast neutrons are set free as a result of the fission process.

B represents a fast neutron loss due to the leakage from the system.

C represents a uranium body of any size in which both volume and surface resonance absorption of neutrons by U^{238} takes place, at resonance energies above thermal energy, to form 94^{239}.

D represents the number of neutrons reaching thermal energy.

E represents a thermal neutron loss by diffusion of thermal neutrons from the system.

F represents a neutron loss caused by capture of neutrons by impurities in both uranium and graphite.

G represents a neutron loss due to capture of thermal neutrons by the graphite as the thermal neutrons diffuse therethrough before entering uranium.

H represents the number of thermal neutrons entering a uranium body.

I represents a uranium body of any size in which part of the thermal neutrons entering the body are absorbed by U^{238} to form 94^{239}, the remaining thermal neutrons causing new fissions in U^{235}, thereby producing fast neutrons, a few of which produce additional fast neutrons by fission of U^{238} atoms.

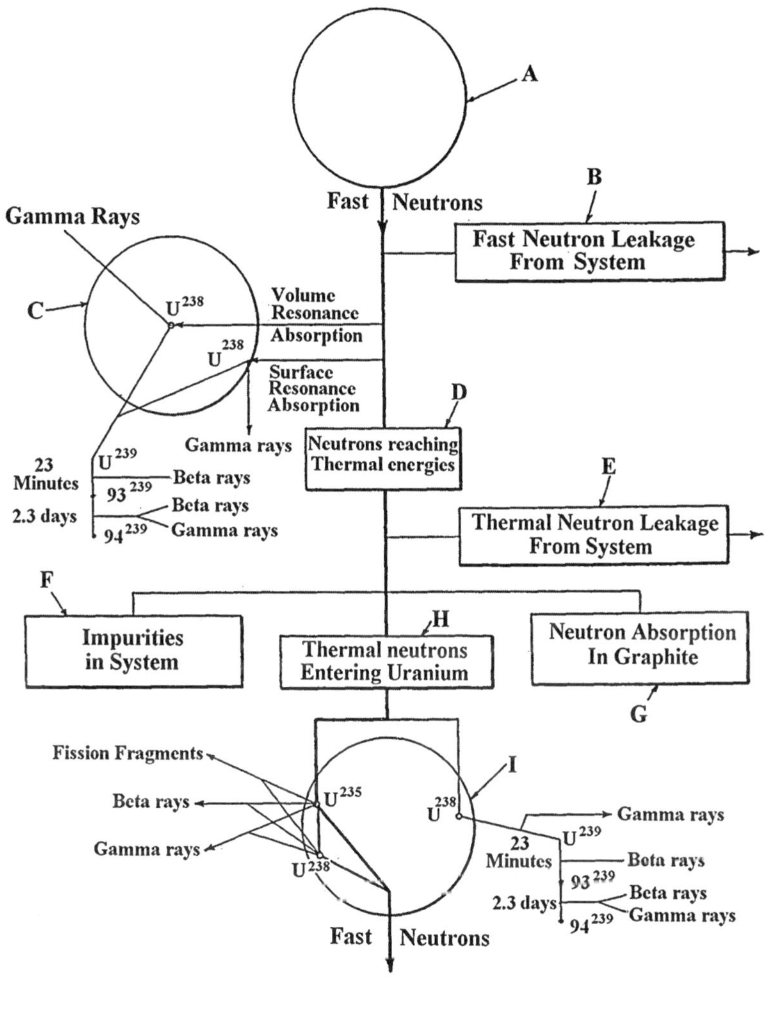

Fig. 2.1

The four neutron losses from the chain reaction referred to above are represented in Fig. 2.1, where the resonance absorption at **C** and the fraction of thermal neutrons absorbed by U^{238} at **I** represent the uranium absorption losses. Losses due to impurities are represented at **F**, whose due to absorption in the slowing material at **G**, and the leakage losses due to the finite size of the system at **B** and **E**.

It has been determined from previously performed experiments that a mixture of natural uranium oxide powder dispersed in water will not sustain a chain reaction even if used in a system infinite in size, because K has been determined to be less than unity for such a combination. It is necessary, therefore, in order to increase the probability of a system attaining a K greater than unity, to either increase the ratio of fission neutrons produced per cycle, or to reduce the neutron losses in a proposed system, over those obtaining in the mixture of uranium powder and water referred to.

The use of uranium enriched by supplementing the natural content of fissionable material therein will increase the probability of an absorbed thermal neutron producing fission rather than being absorbed by U^{238} to produce 94^{239}, and the number of fast neutrons per cycle will be increased. However, if natural uranium alone is to be used, the only way to increase the probability of K becoming greater than unity is to reduce the parasitic neutron losses in the system from those existing in the uranium oxide powder-water system. The slowing medium can be changed from water to heavy water, graphite or beryllium to decrease neutron capture by the slowing medium. The effect of impurities can be reduced by purifying both the uranium and the slowing material to the maximum practicable extent. All of these procedures tend to reduce the neutron losses in the chain, and to increase the possibility of K becoming grater than unity.

In addition, in order for the premise to be fulfilled that the fast fission neutrons be slowed to thermal energies in a slowing medium before again entering uranium atoms without too large an absorption in uranium, a change in the physical arrangement of the uranium in the system from that obtaining in a mixture of uranium powder and a slowing material can be utilized for more efficient reproduction of neutrons.

I have found, for example, as will be hereinafter proved, that it is possible by proper physical arrangement of the materials to substantially reduce the resonance absorption loss of neutrons in a system below the uranium resonance absorption loss occurring in a mixture of fine uranium oxide powder in water having the same uranium to slowing materials volume ratio. By the use of the light elements as described above for slowing materials, fewer collisions are required to slow the neutrons to thermal energies, with resultant large increments of energy loss between collisions, thus decreasing the probability of a neutron being at any resonance energy as it enters a uranium atom. During the slowing process, however, neutrons are diffusing through the slowing medium over random paths and distances so that the uranium is exposed not only to thermal neutrons but also to neutrons of energies varying between thermal energy and the high initial energy of fission evolved neutrons. Neutrons at uranium resonance energies will, if they enter uranium at these energies, be absorbed on the surface of a uranium body, whatever its size, giving rise to surface absorption. Any substantial overall reduction of surface of the same amount of uranium will therefore reduce surface absorption, and any such reduction in surface absorption will release neutrons to enter directly into the chain reaction.

For a given ratio of natural uranium to slowing material, surface resonance absorption losses of neutrons in the uranium can be substantially reduced by a large and significant factor from the losses occurring in a structure comprising a distribution of fine uranium oxide powder throughout a slowing medium by aggregating the uranium into a plurality of substantial masses in which the mean spatial diameter is at least 1 cm. In graphite, for example, an appreciable advantage is obtained. An important gain thus can be made in the number of neutrons made directly available for the chain reaction. Graphite is an efficient slowing medium, is readily obtainable, and is preferred for most structures. Consequently, I prefer to place the uranium in the system in the form of spaced uranium masses or bodies of substantial size preferably either of uranium metal, oxide, carbide, or combination thereof, instead of using the uranium in the form of fine particles. The uranium bodies can be in the form of layers, rods or cylinders, cube or spheres, or approximate shapes, dispersed throughout the graphite preferably in some geometric pattern. As graphite can be obtained in the form of blocks, and structures built by stacking or piling the blocks, a structure called a pile can readily be built for test purposes.

Theoretical calculations can be made to determine whether or not a chain reaction having a K factor greater than unity would be possible in various types of geometrical combinations of uranium and slowing material. Such calculations, however, of necessity must assume more or less arbitrary values for the fission factors and for the various neutron absorption effects of the uranium, the carbon, and the impurities set forth in **C**, **F**, **G** and **I** of Fig. 2.1 including the effect of the physical arrangement of the materials, as accurate values for the constants representing these effects are not available. It would be impossible by such theoretical calculations to predict accurately, either that a self-sustaining chain reaction was possible in a given system, or to predict the type of structure, critical size or design of a system that could be built with any assurance that a self-sustaining chain reaction capable of producing power would be obtained. The errors due to the imperfectly known constants become cumulative during calculation of the reproduction factor K, and the conclusions arrived at from such calculations can be little more than mere guesses at approximate conditions that might obtain if the assumed values of the constants were eventually found to be close to their real or proper values.

Progressive empirical enlargement of any proposed system for which the factor K is not accurately known, in an attempt to attain the overall size of the structure above which the rate of loss of neutrons by diffusion through the periphery of the structure is less than the rate of production of neutrons in the system leads only to an expensive gamble with no assurance of success. The fact that K is greater than unity and the fact that the critical size is within practical limits must be known rather accurately in advance, as otherwise a proposed structure having a K factor less than unity, or even a K factor greater than but close to unity, might not sustain a chain reaction even if all of the uranium in the world were included.

In order, therefore, that any practical self-sustaining chain reacting system be built, the following factors must be accurately known prior to building the system:

1. Whether K is above or below unity, and
2. If K is above unity, what will be the minimum critical size required for a self-sustaining chain reacting system built up of the materials chosen, and with the geometrical arrangement proposed.

As the effect of any attempt, such as mentioned above, to increase the reproduction factor K of a system, cannot, as pointed out above, be accurately foretold from calculations alone, it has been necessary to provide a means and method of determining the value of the reproduction factor in any proposed system. As all of the steps in the chain reaction outlined above contribute to the value of the reproduction factor, if the value of K can be directly determined as to its relation to unity by direct measurement of neutron density existing in structures having specific geometries and materials, then values of the individual contributing constants do not have to be known.

It follows that the main objects of my invention is to provide a means and method of determining, by means of a single series of neutron density measurements which can be performed with precision or an experimental structure whether the reproduction factor of any proposed chain reacting structure is greater or less than one, and then determining from the same measurements the critical size of a self-sustaining system, if K is found to be greater than one, without the necessity of knowing the values of individual factors entering into the chain process.

In addition, fairly accurate evaluations of the reproduction factor for various systems can be deduced by including two other quantities, the so-called "range" of the fission neutrons in being slowed down and the "diffusion length" of thermal neutrons in the structures. Although there is at present an inaccuracy of 5 percent to 10 percent attached to the values of these quantities, they enter only into the determination of the departure of the reproduction factor form unity. Since the departure of K from unity in structures of the types set forth herein is usually not greater than 10 percent, the value of K can be found within an accuracy of 1 percent or less by using the values presently known.

Thus, a further object of my invention is to provide a method of determining the value of the reproduction factor K of a given design within a 1 percent error for specific sizes, shapes, and spacings of uranium bodies in the slowing down material and for the particular purities of the elements making up the device.

Inasmuch as the same nuclear fission mechanism is acting in all portions of a structure where the same geometry of uranium and slowing material exists, the only difference in the chain reaction in a small but representative portion of the structure from the chain reaction in the entire structure is that the rate of neutron loss by exterior leakage is greater in small structures than in larger structures. I have found that by measuring the neutron densities in a small structure and mathematically correlating the density measurements made with the dimensions of the small structure, I can determine the chain reaction conditions that would exist in a much larger structure having the same geometrical arrangement of the same materials.

The present invention is therefore valuable as a relatively inexpensive device for measuring the action of the nuclear chain mechanism in a relatively small structure, and particularly for determining a quantity Δ (defined below), for specific conditions, and for studying the effect on this and related quantities of changes in these conditions. The factor K will be greater than unity if Δ is found to be negative. An accurate calculation can be made as to the critical size required for a self-sustaining chain reaction structure providing the value of Δ is ascertained to be negative and its value determined, as I have found that it may be, in a small but representative portion of any proposed chain reacting structure. There is, therefore, no need of actually building a structure of supposed critical size based solely on calculations using assumed constant values, in order to make a trial of the geometry and purity of materials to be used. Thus the purchase of the complete amounts of materials for the proposed structure need be made only after the desired and accurate specifications for the structure have been experimentally determined by means of my invention.

It is, therefore, another object of my invention to be able to determine in advance of building a large scale structure, whether or not the system proposed will operate to sustain a chain reaction, by measurement of the neutron activity in a much smaller but representative section of the proposed system, this latter section by itself being too small under any conditions to operate to sustain a chain reaction, because of excessive loss of neutrons from the exterior surface of the smaller structure.

It therefore follows that, having attained the foregoing objects, it is a still further object of my invention to test means for decreasing the parasitic neutron losses obtaining in a mixture of uranium and slowing down material so that K factors more favorable to the attainment of a self-sustaining chain reaction can be obtained in a system of finite size.

In a presently preferred form, as to apparatus, my invention relates to a test exponential pile comprising a column of rectangular parallelepiped shape consisting of a desired arrangement of uranium containing bodies of sub-critical mass embedded in a slowing material such as heavy water, graphite or beryllium. The column may be built up or piled on a suitable base layer by layer. Valuable material can be saved by constructing the column on a base made of slowing down materials such as, for example, graphite or paraffin. A source of neutrons may be placed near the bottom of the column. The source may be placed inside the base so that the neutrons are slowed down partially before entering the column to be investigated. Slots are preferably provided in the column at varying distances from the neutron source, and neutron detectors inserted therein as desired for measurements of neutron intensity. This structure can be considered as a small but representative portion of a proposed self-sustaining chain reacting structure. According to the type of experiment being performed, the weight of the structure can vary widely, but in many instances 25 to 100 tons of materials will suffice for the system. In the case of the heterogeneous structure, it should have a sufficient number of repetitions of cells comprising the uranium bodies and their individual complements of slowing material, so that the variations in neutron density through the cells along any chosen axis are minor variations in comparison with the overall neutron density variation along the same axis. The amount of material required for this purpose is substantially smaller than any structure capable of a self-

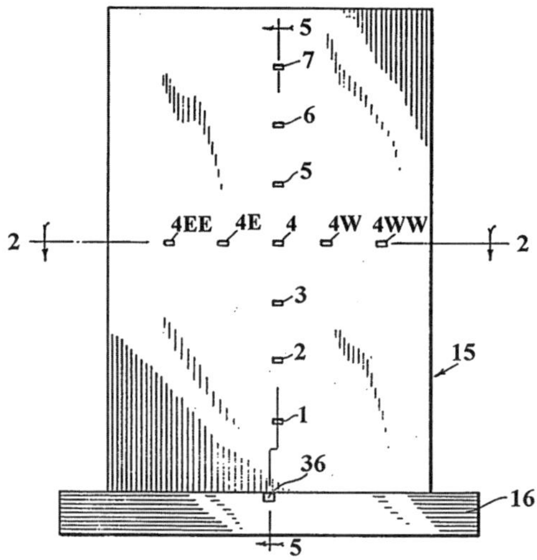

Fig. 2.2

sustaining chain reaction utilizing natural uranium, and experiments have proven that accurate measurements can be made on structures 1/10 of the critical volume when K is greater than unity.

The foregoing constitute some of the principal objects and advantages of the present invention, others of which will become apparent from the following description and the drawings which illustrate one example of the invention.

A specific embodiment of the present invention is selected for purposes of disclosure, it being recognized that many modifications may be made without departing from the intended scope of the invention.

Referring to the figures, the structure or pile, generally indicated at **15** in Fig. 2.2 representing, for example, a portion only of a proposed self-sustaining chain reaction system, is a column 11 feet high having a square cross-section with a dimension of about 8 feet on a side. The column rests on a paraffin base **16**, which is about 10 feet square and 1 foot thick. The base may be also made of a layer of about 3 feet (thick) of graphite, if desired.

As shown in Figs. 2.4 and 2.5, the device comprises a plurality of uranium bodies in the form of cubes **17** of about 8 inches on a side, arranged

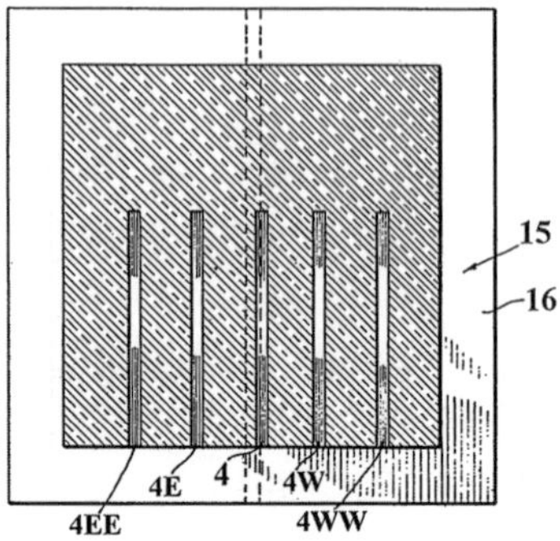

Fig. 2.3

in graphite **18**. The distance between the uranium bodies is 8 inches. Between the paraffin base **16** and the first layer of uranium cubes **17** is a layer of graphite 8 inches thick.

Referring to Fig. 2.4, the structure as shown consists of a plurality of cell units **19**, each cell comprising one uranium body and a mass of graphite, and, as shown in Fig. 2.4, each cell has a side equal in length to twice that of the uranium body. In accordance with the general meaning, a cell is that unit which is repeated throughout the mass to form the entire structure.

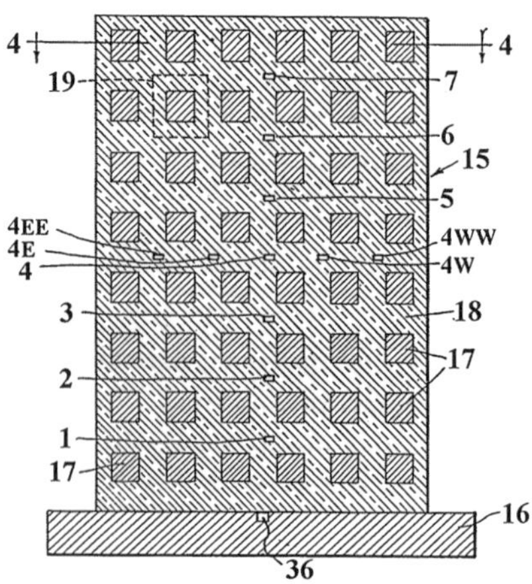

Fig. 2.4

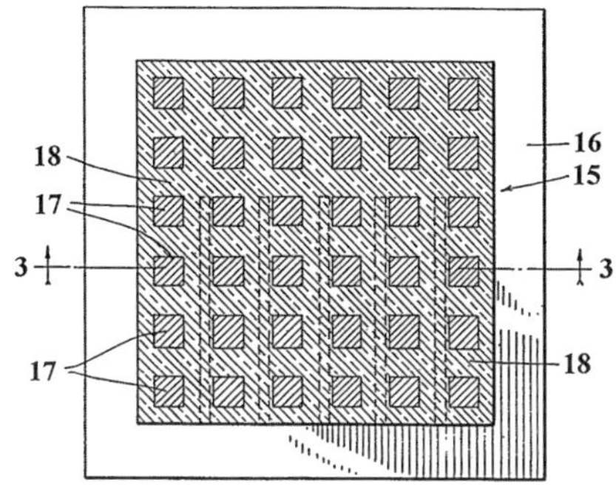

Fig. 2.5

Through specific dimensions for the reacting structure have been given, it is recognized that certain variations may be made. As a general proposition, it can be stated that reliable readings may be taken in structures of smaller sizes, but as the size of the structure decreases the readings will become less accurate. Conversely, as the size of the structure increases over and above that specified, the readings become more accurate. Very satisfactory results within 1 percent error can be obtained, however, from the specific size of structure illustrated. If fewer cell units are used, the error will be increased above 1 percent. In the example illustrated, 288 cells were used. The overall size should be sufficiently large so that the variations in neutron density within one cell are small compared with the average neutron density along the measured axis. An appreciable number of fast neutrons must be slowed to thermal energy in the graphite in order to accomplish this result.

A plurality of measuring slots **1** to **7**, inclusive, are spaced vertically in the column **15**, and additional slots **4E**, **4EE**, **4W**, and **4WW** are spaced horizontally at the level corresponding to slot **4**. These slots are all five feet long and have a cross-section measurement of 2 inches by 3/8 of an inch. Slots **1** to **7**, inclusive, extend through the graphite to points near, but slightly to one side of the vertical axis of the column.

A detector device, shown generally in Fig. 2.6 at **30**, is insertable in any one of the detecting slots **1** to **7**, inclusive, and includes a relatively thin graphite tray **31** hollowed out to receive a pair of nesting aluminum or cadmium trays **32** and **32A**. The size of the graphite tray **31** is slightly

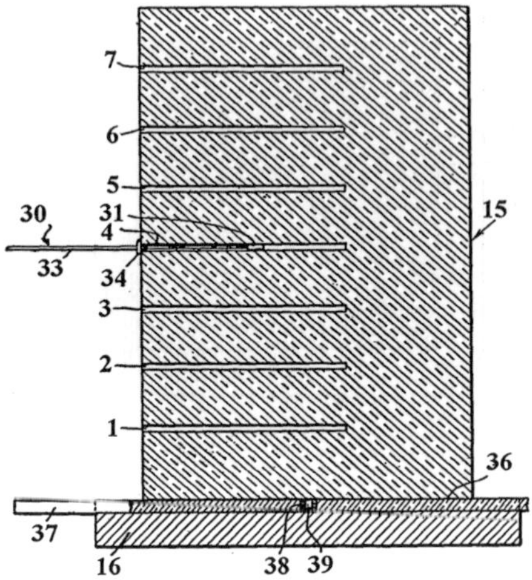

Fig. 2.6

less than 2 inches by 3/8 of an inch, so that it may readily be passed in and out of the detector slots. An aluminum rod or handle **33** is fastened to one end of the detector tray **31** and is of sufficient length to permit moving the graphite tray **31** throughout the entire length of any one of the slots **1** to **7**, inclusive. An adjustable collar **34** is movable lengthwise on the handle **33** and is provided with a set screw **35**, so that the collar may be fixed at any desired position on the handle. The purpose of this collar is to facilitate the positioning of the detector tray **31** at any desired location in one of the detecting slots.

A detector foil **32B** of indium is placed between the nesting trays **32** and **32A** for the purpose of making measurements of neutron density in the column. The foils are rectangular, having dimensions of 4 by 6.35 centimeters, and are between 5.5 and 6 mm in thickness. All foils have been previously standardized for their radioactivity response to neutron bombardment by comparison with a standard foil. The trays **32** and **32A** are of two kinds, i.e., either both of aluminum or both of cadmium. The aluminum permits thermal neutrons to pass through to the detector foil, whereas the cadmium absorbs the thermal neutrons before they can reach the foil. It is possible to determine the relative density of thermal neutrons present in the region of the detector element by exposing the indium foil to neutron bombardment when held between aluminum trays, and taking the difference between this induced radioactivity, and the radioactivity induced when the foil is exposed to neutron bombardment while held between cadmium trays.

The method of measuring the neutron density is as follows:

Assume that a source of neutrons has been placed at the bottom of the pile, aligned with the vertical axis. An indium foil **32B** of the dimensions specified is placed between aluminum nesting pans **32** and **32A** of the detector device **30**. The detector device is then inserted in one of the detecting slots, as, for example, slot **1**, and exposed to the neutrons present at a point near the vertical axis. In one mode of measurement the indium foil is then allowed to remain in the slot for a predetermined period of time in order to come to a known fraction of its saturation activity – say, 15 minutes – during which time is bombarded by the neutrons inside the structure. At the end of this time interval, the detector device **30** is withdrawn from the slot, and after a predetermined lapse of time – say three minutes – during which time the short-lived radioactivity of the indium foil disappears, the indium foil is then wrapped around the cylinder of a standardized Geiger counter and the beta rays emanating from the the foil due to the remaining and longer lived radioactivity of the foil are counted for a given time interval, say 30 minutes. From the number of counts during this time interval, the initial saturation radioactivity at the time of removal is computed. The indium foils can be reused after several days, as the induced radioactivity has then decayed to an insignificant value. Measurements are made in the same manner and in the same positions with the same time factors to determine the cadmium ratio, later referred to.

However, for convenience and also because a given exposure time may be too long in regions of high neutron density, yielding too many counts per minute for the counter to handle accurately or may be too short in very low density regions, yielding too few counts for accurate representations, the time schedules may be changed to compensate. In this case, however, the saturation activity A_s in counts per minute should be calculated, as is well known in the art. The following equation has been used under these circumstances:

$$A_s = \frac{fC}{e^{-f\,t_w}(1 - e^{-f\,t_x})(1 - e^{-f\,t_c})},$$

where C is the observed number of counts in time t_c, and in minutes, t_x, t_w and t_c are respectively the periods of irradiation, waiting and counting, and where f is the fraction of the excited (radioactive) atoms that disintegrate in a unit of time, and for indium equals 0.012836 per minute.

The detector device **30** is then inserted into the slot **2** and a similar reading is taken close to the vertical axis in this slot, corresponding to that in slot **1**. Likewise, readings are taken close to the vertical axis of the column in all of the other slots **3** to **7**, inclusive. In this manner, the radioactivity of the indium caused by the slow neutron bombardment at points close to the vertical axis of the reacting structure may be determined accurately, and expressed in terms of counts per minute of the Geiger counter. From this information, the ratios of the counts measured in adjacent horizontal planes in the structure can be obtained. Other readings are taken on the top and sides of the structure, as will be pointed out later.

As there are minor variations in neutron density along any given axis due to the periodically recurring cells in a structure wherein the uranium is aggregated, the neutron density can be considered as averaged over the individual cell. In order that measurements along an axis may be properly compared, the measurements are made in the same relative position in the various cells along the axis being investigated.

The measurements made close to the vertical axis express the decay of the chain reaction started by the source, and when the position readings are corrected for the finite size of the structure, can be used to find the value of Δ.

To explain the relation of Δ and the reproduction factor K to the observable quantities as measured, it is desirable to first consider a mathematical description of the effect of diffusion, absorption, and neutron production on the neutron density at some point in the structure.

In the following mathematical discussion, the symbols listed below will be used with the meanings as indicated:

X, Y, Z are the rectangular coordinates of a point in a coordinate system with its origin at the neutron source at the base of the pile, whose X and Y axes are parallel to the sides of the structure, and whose Z axis is the central vertical axis of the structure;

a is the effective length of side of the structure;

a_0 is the actual length of side of the structure;

n is the number of thermal neutrons per cc. in the neighborhood of the point X, Y, Z;

v is the mean velocity of the thermal neutrons;

ds is an increment of distance traversed by a neutron;

dt is the increment of time corresponding to ds;

Λ is the mean total distance a thermal neutron travels;

D is the diffusion coefficient for thermal neutrons;

Δn or Δq is an abbreviation for the sum of the three second derivatives of n or q with respect to the three variables X, Y, Z. Δn is also referred to as the Laplacian of n;

Δ standing alone and not immediately followed by n or q signifies a number given by the ratio of Δn to n. This number is found to be constant throughout any structure utilizing given geometries and materials;

λ is the neutron mean free path between collisions;

r is the radius of a sphere or cylinder;

r_0 is the range, or the mean distance the neutrons reach from their point of origin during the process of slowing down to thermal energy. For chain reacting structures the values is calculated to be about 35.5 cm for uranium and graphite, for uranium and beryllium about 27 cm, and for uranium and deuterium (employed as heavy water) about 22 cm;

τ is a variable called the age of the neutrons which is a function of their energy during slowing down. For thermal neutrons τ becomes $1/4r_0^2$;

q is the number of neutrons destined to escape resonance capture by uranium, slowed down per sec. per cc. at X, Y, Z, to a given energy E or below, from higher energies, and is to be regarded as a function of the age corresponding to the energy E;

q_0 is the number of fast fission neutrons destined to escape resonance capture by uranium produced per sec. per cc.;

L is the diffusion length of thermal neutrons in the structure. For uranium and graphite chain reacting structures the value is calculated to be about 18.5 cm, and for uranium and beryllium about 25.3 cm. For uranium and deuterium the value is between 100 and 200 cm;

K is the reproduction factor;

Σ signifies a sum and is used to indicate a series of which a typical term follows the symbol;

C_{jl} is one typical constant in this series; j and l are two odd integers characterizing the terms of the series;

$b_{1,1}$ is the value of the quantity occurring in the expression

$$e^{-\frac{Z}{b_{1,1}}}$$

which is part of the first term in the series giving the value of n for different heights Z;

$$C_{1,1}\cos\pi\frac{X}{a}\cos\frac{Y}{a}\left(e^{-\frac{Z}{b_{1,1}}}\right)$$

is the first term in a series, and is referred to as the first harmonic term, the following terms being referred to as the third, fifth, etc., harmonic terms;

$$\frac{\delta q}{\delta \tau}$$

is the partial derivative of q with respect to τ.

With these symbols in mind, the various factors governing the variation in the thermal neutron density n at a point in the graphite may be taken into account as follows: The number of thermal neutrons produced per second per cc. at X, Y, Z in consequence of the reduction of more energetic neutrons to the thermal condition is given by the value of q for thermal energy. The contribution to n by diffusion from the surrounding medium is $D\Delta n$. The chance of a neutron being absorbed in traversing a distance ds is

$$\frac{ds}{\Lambda}.$$

The number of thermal neutrons absorbed per second is

$$\frac{nv}{\Lambda} \text{ since } \frac{ds}{dt}$$

is the velocity v. Consequently, in a steady state:

$$D\Delta n - \frac{nv}{\Lambda} + q = 0. \tag{2.1}$$

The number q is given by a solution of the slowing down equation

$$\Delta q = \frac{\delta q}{\delta t}$$

obtained by consideration of the diffusion of the neutrons during the slowing-down process and by suitably fixing the dependence of the age variable τ on the energy E. Initially

$$\text{(i.e. for } \tau = 0) \quad q = K\frac{nv}{\Lambda},$$

since K is defined as the number of fast neutrons destined to escape resonance capture, produced per thermal neutron absorbed. A solution is

$$q = K\frac{nv}{\Lambda}\, e^{\tau\Delta}$$

since, as stated above

$$\Delta n/n$$

is observed to be a constant whose value is designated as Δ. If this is substituted in (2.1) using the kinetic theory formula $D = 1/3\lambda\nu$ and putting

$L^2 = 1/3\lambda\Lambda$ the following equation governing the neutron density is obtained:

$$\Delta n - \frac{1}{L^2}n + K\frac{n}{L^2}\ e^{\tau\Delta} = 0 \tag{2.2}$$

and by rearranging terms and putting

$$\Delta n/n = \Delta \tag{2.3}$$

$$K = (1 - L^2\Delta)e^{-\tau\Delta}. \tag{2.4}$$

In most cases in which this invention is used, the value of $K - 1$ is less than 0.1 so that the exponential term may be replaced by the first two terms of the expansion in series. Thus, since $\tau = 1/4r_0^2$,

$$K - 1 = -\Delta\left(L^2 + \frac{1}{4}r_0^2\right). \tag{2.5}$$

Equations (2.4) and (2.5) clearly show that the value of Δ depends only on K, L and r_0 and that the same value of Δ will hold for a large structure as well as a small one having the same internal construction and materials. Substituting

$$\Delta n/n$$

for Δ in (2.5), the differential equation (2.2) becomes

$$\Delta n = -\frac{K-1}{L^2 + \frac{1}{4}r_0^2}\ n. \tag{2.6}$$

Equations (2.4) and (2.5) show that if the value of Δ as found by measurements on the pile comes out negative, then the reproduction factor K is greater than unity, and if the value of Δ is positive then K is less than unity. It will be seen later that the numerical value of Δ also fixes the overall size of a self-maintaining chain reaction structure when Δ is negative.

It is also to be understood that Δ is a value independent of the external shape of the intermediate pile from which it is ascertained, and of the external shape of the operating pile to be designed, so that, for example, Δ determined by measurements of a parallelopiped-shaped intermediate pile may be used to compute critical size for a self-sustaining pile of any shape, such as a sphere, cylinder or cube.

Theoretically from (2.4) the quantity Δ could be found from neutron density measurements made on a structure of any size and shape. However, certain simple regular structures have great advantages for convenience and accuracy in calculation. A sphere or a cylinder could be used, but the preferred form of my invention is a rectangular parallelopiped, more particularly a long vertical parallelopiped with a rectangular cross-section. For this case a mathematical formula can be given for the neutron density distribution at various points throughout the pile when a source of neutrons is used at an end thereof. Having this formula available with measurements of the density distribution and with the proper constants chosen to fit the measurements, the value of Δ can be very readily deduced for the geometry and materials used in the structure.

For the case of a long vertical rectangular parallelopiped with square cross-section of side length a, this formula is the double Fourier series:

$$n = \sum C_{jl} \cos\left(j\pi \frac{X}{a}\right) \cos\left(l\pi \frac{Y}{a}\right) e^{-\frac{Z}{b_{jl}}}, \tag{2.7}$$

in which

$$\frac{1}{b_{jl}^2} = \frac{1}{b_{1,1}^2} + \frac{(j^2 + l^2 - 2)\pi^2}{a^2}. \tag{2.8}$$

Here each term satisfies (2.6), so that the sum is also a solution. This is a series in which the various terms are given by putting j and l equal to (1,1), (1,3), (3,1), (3,3), etc. in succession. The adequacy of the formula is tested by measuring the neutron densities in a horizontal plane at some constant height Z. These measurements also serve to determine the constants $C_{1,1}$, $C_{1,3}$, $C_{3,1}$, $C_{3,3}$, etc., which multiply the succession terms of the series (2.7).

However, the neutron density distribution across the pile at considerable distances from the sources approaches a true cosine curve and is given more and more accurately by the first term alone of formula (2.7) as the height of the pile increases and the plane of measurement is more and more removed from the influence of the source. This simplifies formula (2.7) to:

$$n = C_{1,1} \cos\left(\pi \frac{X}{a}\right) \cos\left(\pi \frac{Y}{a}\right) e^{-\frac{Z}{b_{1,1}}} \tag{2.9}$$

since $b_{1,1}$ is the largest b, and the amplitudes of all the higher harmonic terms decrease at a faster rate with increasing distances from the source.

When measurements are confined to the regions away from the source where formula (2.9) is applicable, the value of Δ can be found very easily. By differentiating n as given by formula (2.9) twice with regard to X, Y, Z, adding and dividing by n, it is found that

$$\frac{\Delta n}{n} = \Delta = \frac{-2\pi^2}{a^2} + \frac{1}{b_{1,1}^2}. \tag{2.10}$$

This value of Δ applies to a structure having a square cross-section. For one having a rectangular cross-section with the sides a_1 and a_2, the formula for Δ becomes:

$$\Delta = \frac{1}{b_{1,1}^2} - \frac{\pi^2}{a_1^2} - \frac{\pi^2}{a_2^2}. \tag{2.11}$$

To find a numerical value for Δ, it is necessary to substitute the value for a, the length of the side of the structure, and the value of $b_{1,1}$, which is given by the decrease of the neutron density along the central axis of the structure. For $X = 0$, $Y = 0$ formula (2.9) gives

$$n = C_{1,1}e^{-\frac{Z}{b_{1,1}}}. \tag{2.12}$$

From two observed values of n (n_1 and n_2) in adjacent horizontal planes Z_1 and Z_2 of the structure, we have

$$log_e \frac{n_1}{n_2} = \frac{Z_2 - Z_1}{b_{1,1}}. \tag{2.13}$$

The value of $b_{1,1}$ can thus be obtained. With the value of $b_{1,1}$ found, the value of Δ can be found from formula (2.10) for a structure with a square cross-section or formula (2.11) for a structure with a rectangular cross-section.

It is to be noted that $b_{1,1}$ is the coefficient of only the first term of the series in formula (2.7). If the planes at which measurements are made are sufficiently far from the source, it has been found that the value of $b_{1,1}$ does not need correction for the small second and third terms, which are only large near the source. The effect of the higher harmonic terms can be checked by measuring the neutron density across a Z plane to see if the measurements indicate the presence of the higher harmonics. If the measurements show a significant departure from a cosine curve, then coefficients of the higher harmonic terms can be found to suit the measurements made. Since $b_{1,1}$ is to be found, the values of the harmonic terms in (2.7), if found to be significant, can be computed from (2.8) with the value of $b_{1,1}$ as found by formula (2.13), as a correction to the observed value of n, in a formula where

$$n - (\text{higher harmonic terms}) = \text{first term} \tag{2.14}$$

a well known mathematical expedient in dealing with an infinite series.

Slots **4EE**, **4E** and **4W**, **4WW** in the structure described are used to determine whether harmonic corrections are required. This is done by measuring the neutron distribution across the pile by using indium foils as above described to see whether this distribution fits formula (2.9) using the values of $b_{1,1}$ as obtained by formula (2.13). If not, harmonic corrections are applied to the observed values n which then agree with formula (2.9). The corrections can then be computed for lower planes if measurements in these planes are to be utilized for Δ determinations.

However, as such harmonic corrections 3 feet or more away from the source approach the experimental error in measuring the neutron density, good values of Δ can be obtained by disregarding measurements made in planes near the source, and by computing the Δ value found for a higher plane, when the density readings on these planes are corrected for the side and end effects. These latter corrections are: a correction required because the effective side of the structure is larger than the actual side and a correction required by the fact that the height of the column is finite. These boundary corrections are needed because the neutron density is not zero either at the sides or top of the structure due to neutron leakage out of the structure, whereas formula (2.7) assumes the neutron density to be zero at these boundaries. These corrections will be discussed in turn.

The neutron density according to formula (2.7) should reach zero at $X = a/2$ and $Y = a/2$ for a structure having a square cross-section. It is found that the observed neutron densities are not zero at the surface but do agree with formula (2.7) where a is larger than the geometrical side a_0 by a few centimeters. The proper value of a to use in (2.7) may be found by two observations of the neutron density n_0 at the center and $n(a_0)$ at the surface of the pile between the levels of the Z planes measured, then

$$n(a_0) = n_0 \cos\left(\frac{a_0}{a}\frac{\pi}{2}\right) \tag{2.15}$$

the value a, rather than a_0, should be used in computing Δ from formula (2.10).

The second correction is due to the finite height of the column, known as the end or top correction. Close to the end, or top, the neutron leakage is increased, and to obtain corrected neutron densities in planes, Z_0, Z, near the top, the values given by formula (2.12) must be multiplied by

$$1 - e^{-\frac{2}{b_{1,1}}(Z_0 - Z)} \tag{2.16}$$

to give the observed neutron densities where Z_0 is the actual value for Z for the top. If conversely, we take the observed neutron densities which have been reduced by nearness to the top and divide them by (2.16), we obtain increased values which we may call corrected neutron density values which will then agree with formula (2.12) and which may be used to compute $b_{1,1}$ from formula (2.13) from measurements made in planes near the top.

In case a spherical structure is desired in place of the parallelopiped just described, it is also possible to find a mathematical description for the decay of neutron density along a radius of the sphere when the source is at the center. The differential equation (2.6) in polar coordinates is

$$\frac{d^2(nr)}{dr^2} = -\frac{K-1}{L^2 + \frac{1}{4}r_0^2}\, n\, r. \tag{2.17}$$

If the source is placed at the center of the sphere and if the density at the outside is maintained at zero by surrounding the structure with a sheet of cadmium that absorbs all neutrons reaching the periphery, a solution that agrees with the observation is

$$n = \frac{1}{r}\left(Ae^{\sqrt{\Delta}\, r} - Be^{-\sqrt{\Delta}\, r}\right), \tag{2.18}$$

where

$$\Delta = \frac{1-K}{L^2 + \frac{1}{4}r_0^2},$$

and where A and B are two constants chosen to make the solution suit the boundary conditions. The value of Δ may then be deduced from two neutron density observations at different positions along the radius.

However, if K is greater than unity, the constants A and B will be chosen as complex numbers. In such a case an alternative equation can be given in terms of trigonometric functions:

$$n = \frac{1}{r}\left(E\cos(\sqrt{-\Delta}\, r) + F\sin(\sqrt{-\Delta}\, r)\right), \tag{2.19}$$

where E and F are real numbers.

Other solutions may also be found for cylindrical and other structural shapes, but as these are more complicated, they are not to be recommended in comparison with the two structures described above. The parallelopiped structure is preferred from a structural stand-point.

Two important deductions may be reached at once from the numerical value of Δ. The first is that the value of the reproduction factor K is greater or less than unity depending on whether the value of Δ comes out negative or positive, respectively. This fact, which follows from equation (2.5), does not require a knowledge of the value of the quantities in the bracket of that equation.

The second deduction that may be drawn from the magnitude of Δ is the critical size to which the system must be built to maintain a nuclear chain reaction. This size is the size where the total surface to volume ratio of the entire pile is such that the rate of neutron loss is reduced to a value less than the rate of neutron production, the latter being expressed by the numerical value of Δ. As Δ is a function of the geometrical arrangement of the materials and the impurities in the materials, the system to be self-sustaining, when Δ is found to be negative (determining K to be greater than unity), can be built to critical size in a number of shapes. In case the self-sustaining chain reacting system is to be built in the form of a large spherical structure, the critical radius r_c is given by

$$r_c = \frac{\pi}{\sqrt{-\Delta}}. \tag{2.20}$$

In case the structure is to be a rectangular parallelopiped with sides a_1, a_2 and a_3, the critical size s given by the formula

$$-\Delta = \frac{\pi^2}{a_1^2} + \frac{\pi^2}{a_2^2} + \frac{\pi^2}{a_3^2}. \tag{2.21}$$

In case the structure is to be built up as a cylinder of height H and radius r, the critical values of these quantities may be computed from the formula

$$-\Delta = \frac{\pi^2}{H^2} + \frac{(2.405)^2}{r^2}. \tag{2.22}$$

It will thus be seen that for a given amount of material in the system the spherical overall shape is to be preferred. Any deviations from the spherical shape must be compensated for by the use of additional material. These computations do not require knowledge of the values in the bracket of equation (2.5).

In the manner stated above, from a single series of measurements, the fact that K is greater or less than unity can be determined from the sign of Δ, and if K is determined to be greater than unity, the critical size required for a self-sustaining chain reaction can be calculated, using the numerical value obtained for Δ. The value of K will hold for the materials and geometries used, no matter what the overall shape of the system may be. From the value of Δ and by using the presently known values of L and r_0 the numerical value of K can be computed within a 1 percent error from formula (2.5).

Broadly stated, therefore, as to the method of my invention, I provide a small but representative portion of a structure proposed as having a possibility of sustaining a chain reaction, and then measure the neutron distribution at least along one dimension of said structure, and from the measurements so made compute the ratio of the Laplacian of the neutron density to the neutron density found in the structure. This ratio is known as Δ and has, for chain reacting structures of the type described, the direct relationship to K, so that the value of K can also be determined, if desired, using presently known values of L and r_0.

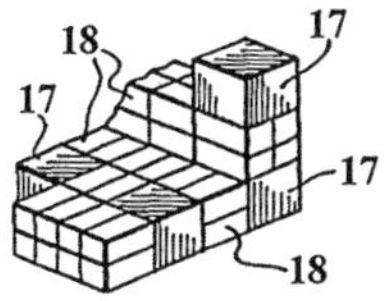

Fig. 2.7

As a specific example of the uses of the exponential pile, and the described method of measurement for a specific geometry and for specific materials, a pile has been built wherein the uranium bodies **17** (Fig. 2.4) consisted of 288 boxes in the shape of cans 8 inches on the side, made of thin tinned iron sheets each weighting about 500 g. Each box was filled with 60 pounds of commercially obtained uranium oxide (U_3O_8) powder. The boxes were uniformly disposed throughout a pile of graphite blocks 4 inches×4 inches×8 inches, in cubical spaced relationship, with the boxes spaced 16 inches apart in alternate layers measuring from center to center, as shown in Fig. 2.7. The graphite was similar to that used as anodes in electrolytic smelting. The total weight of the uranium oxide was 17,280 pounds, or approximately $8\frac{2}{3}$ tons. The graphite used had a density of 1.63 and a total weight of about 26 tons. The units were tightly packed to eliminate air spaces as far as possible, and the materials were known to be impure.

A neutron source was disposed at the bottom of the pile. For this purpose, two grams of radium mixed with beryllium were placed in a slot **36** disposed in the paraffin base **16** (Fig. 2.2). A graphite rod **37** (Figs. 2.6 and 2.10) may be placed in the slot **36**, in which case the rod is slidable in the slot. The neutron source may then be disposed in the cylindrical slot **38** in the graphite rod **37**, a suitable capsule **39** (Fig. 2.11) made for

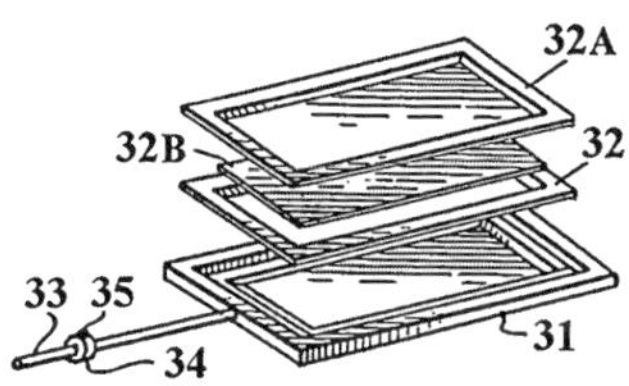

Fig. 2.8

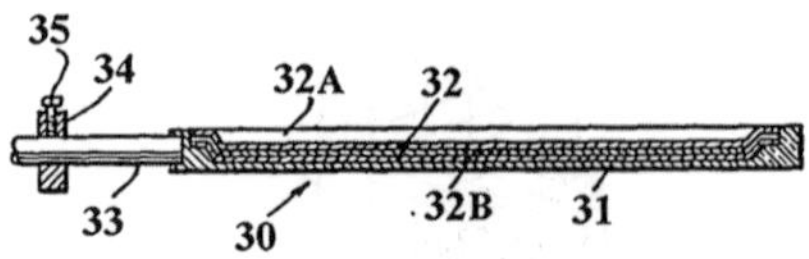

Fig. 2.9

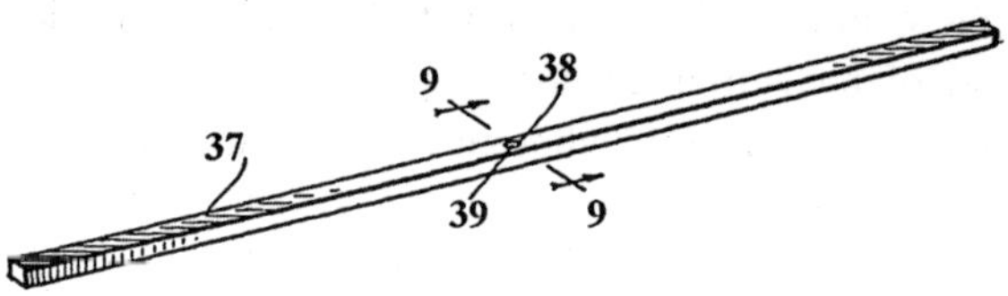

Fig. 2.10

example, of beryllium, being adapted to contain the neutron source. Geiger-Muller counter readings of the neutron induced indium foil radioactivity were made for the seven slots shown in Fig. 2.2 and identified as slots **1** to **7**, inclusive, at points close to the vertical axis of the column, and on top of the pile. Measurements were made both with and without the artificial source disposed in the slot **36**.

The following Table 2.1 gives the values of measurements made, the values representing neutron density in terms of counts per minute obtained from the exposed indium foils:

The first column of Table 2.1 represents the position of the indium foil in the column, that is, for which of the seven slots the readings shown in the remaining columns were made. Columns 2 and 3 in the table give the indium foil counts recorded with the source of neutrons in place in the slot **36**. Column 2 gives the readings where no cadmium surrounded the indium foil. Column 3 is a tabulation of count readings with the foil surrounded

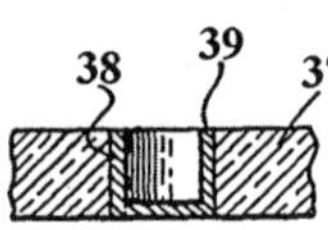

Fig. 2.11

Table 2.1

[Measurements near Z axis]

1	2	3	4	5	6
Position	Counts with source		Cd ratio	Counts without source, No Cd	Net effect of chain reaction started by source neutrons
	No Cd	With Cd			
1	234.8	44.4	5.29	0.252	234.5
2	69.37	12.00	5.78	0.320	69.05
3	23.64	4.143	5.71	0.365	23.28
4	8.93	1.610	5.54	0.365	8.57
5	3.594	0.638	5.63	0.394	3.200
6	1.572	0.286	5.50	0.324	1.248
7	0.684	0.112	6.11	0.235	0.449
(8, top)	0.070	–	–	–	–

with cadmium. Column 4 gives the ratio of counts between the values in column 2 and column 3; that is, the ratio between the counts with the cadmium and without cadmium surrounding the detector. The fact that this ratio, referred to as the cadmium ratio, is substantially the same for all positions after the number 1 position is due to the fact that the neutron energy spectrum beyond the first position has reached a value which is no longer affected appreciably by the proximity of the source. Although in this particular case, the cadmium ratio is somewhat higher at position **7**, the departure is not sufficient to prevent the achievement of reasonably accurate results when reading **7** is used with others, for the computation of Δ.

Beyond the second slot, identified as slot **2** in Fig. 2.2, the source no longer directly contributes thermal neutrons to the chain reaction. The original neutrons from the source disposed in the slot **36** are all reduced to thermal energies before distances beyond slot **2** are reached. These thermal neutrons are absorbed partly in the uranium and give rise to nuclear fissions, producing fast fission neutrons. These fast fission neutrons are in turn slowed down to thermal energies in the graphite to produce further fast fission neutrons, by entering uranium, etc. Thus, as indicated by the substantial uniformity of the cadmium ratio and as required by the analyses upon which the actual computation of Δ, below is based, the chain fission reaction is responsible for the neutron density at slots **2** to **7**, inclusive, this chain reaction being initiated in the uranium of the lowest uranium bearing

Table 2.2

1 Position	2 Counts on surface	3 Counts on axis	4 Actual side a_0 in cm	5 Effective side a in cm
5 1/2	0.170	2.36	243.84	254.5
5 1/2	0.129	1.03	243.84	262.5

layer of column **15** by the primary neutrons emanating from the source in the slot **36**.

Column 5 in Table 2.1 gives the indium foil counts measured at the various positions in the column without the presence of the source in the slot **36** and without cadmium surrounding the indium foil. The count values set forth in column 5 are due to neutrons from spontaneous fission of the uranium in the structure.

It is to be noted that the chain reaction is also present due to this natural fission. However, the count values are low, and a more complicated solution would be required to find Δ from the observed values representing the neutron density without the source, which is quite different from that existing when a source is utilized to start the chain reaction at one end of the structure. The use of the source provides larger density values, and an exponential density decay curve along one dimension of the structure, from which Δ can readily be computed, as pointed out above.

Column 6 of Table 2.1 gives the mathematical difference between the values in columns 2 and 5, these values indicating the net effect of the chain reaction initiated by the source neutrons. Because the cadmium ratio is substantially constant, these values in column 6 can be taken to represent the thermal neutron densities, as well as the total neutron densities, resulting from the reaction.

Indium foil measurements made in slots **4WW**, **4W** and **4E**, **4EE** indicated that the neutron density distribution across the pile at that level was very little affected by the source, and that good values of Δ could be obtained by using the first term (which includes $b_{1,1}$) alone in higher planes when the readings in these higher planes are corrected for top effect and the proper value for a is found in each case. Consequently, readings taken in positions 1 to 4, inclusive, may be discarded.

To find the effective sides a at various levels, formula (2.15) is used. Measurements were made of indium foil counts due to the neutron densities on the outer surface of the column at various heights, and are tabulated in

Table 2.2 for the top portion of the pile. The detector foil positions were between the positions of the measuring slots, on the exterior surface of the column. The axis density counts in column 3 are extrapolated for interplane positions corresponding to the positions of the foils on the exterior surface.

The values in Table 2.2, column 5, can be used for the quantity a in computing the value of Δ.

The end or top corrections are found by the formula (2.16), using the approximate values of $b_{1,1}$ obtained from the readings in Table 2.1, column 6 by the use of formula (2.13), where Z_1 and Z_2 are the distances from the source of the positions in which readings were taken.

The correction factors are as follows:

Table 2.3

1 Position	2 Correction factor
5	0.997
6	0.98
7	0.87

It is to be noted that the end correction increases as the end is approached, and is very small at position 5.

The measurements of Table 2.1, column 6, are now ready for correction for top effect, and for use in computation with the values found for the effective sides at various Z planes to ascertain the final and more accurate values of $b_{1,1}$, and then to find Δ at two positions in the pile where the effect of the source is minimized. The two values found for Δ are then averaged to find the average Δ for the upper portion of the pile. The results are set forth in Table 2.4, below.

Table 2.4

1 Position	2 Net counts	3 With top correction	4 n_1/n_2	5 Effective side a in cm	6 $b_{1,1}$	7 Δ
5	3.200	3.209	2.5208	254.5	48.948	$2.127 \cdot 10^{-4}$
5 1/2 ..						
6	1.248	1.273	2.5108	262.5	44.63	$2.156 \cdot 10^{-4}$
6 1/2 ..						
7	0.449	0.507				
Average				$\Delta = 2.141 \cdot 10^{-4}$		

In the above Table, column 2 gives the observed counts due to the chain reaction taken from column 6, Table 2.1. In column 3 the count readings have been multiplied by the end correction factors given in Table 2.3. Column 4 gives the count ratios n_1/n_2 of readings in adjacent planes. In column 5 are set forth the effective sides a as given by Table 2.2. In column 6 are given the values of $b_{1,1}$ as computed from the formula (2.13).

In the case of the structure $Z_1 - Z_2$ measured 16 inches $= 40.64$ cm between all positions. In column 7 the values of Δ are given as found from the use of the computed values of $b_{1,1}$ and the effective side a by use of the formula (2.10).

The mean value of Δ computed from readings in positions 5, 6 and 7, is 2.141×10^{-4} and is positive in sign. It follows that the value of the reproduction factor K of the structure measured was less than unity. Consequently, this particular geometry and materials would never provide a self-maintaining chain reaction no matter how large a structure was used.

For some purposes, particularly to study variations in the reproduction factor K as brought about by changes in the geometry of the uranium bodies and the purity of the uranium and the slowing medium in the system, it is desired to find the numerical value of the reproduction factor K. For this purpose, formula (2.5) is useful by including the best presently known values of L and r_0 and using the value of Δ as computed, as follows. Substituting the value of $\Delta = 2.141 \times 10^{-4}$ in the structure measured and using the values of $L = 18.5$ cm and $r_0 = 35.5$ cm as previously given for a structure of uranium and graphite, K is found as follows: substituting in (2.5) $K-1 = -2.141 \times 10^{-4}(18.5^2 + 1/4\,35.5^2)$ solving: $K = 0.86$.

The value of the reproduction factor K is thus found to be 0.86.

While the calculated values of $L = 18.5$ cm and $r_0 = 35.5$ cm are probably not accurate within a 5 percent to 10 percent error when combined, it will be noted that these values only enter into the departure of K from unity. As the departure of K from unity for structures such as described is only on the order of 10 percent, the use of these L and r_0 values introduces an error of 1 percent or less in the final value of K.

As stated above, the best previously known figures for the K factor of a mixture of theoretically pure uranium powder and water is $K = 0.79$. Similarly the best K factor for a mixture of theoretically pure uranium powder and graphite has been computed to be $K = 0.785$. The value of $K = 0.86$ obtained as above set forth for uranium oxide bodies of substantial size spaced in graphite, both known to have impurities therein, indicated that a saving in resonance absorption losses must have been made by aggregating the uranium into substantial masses, but that other losses were too large to provide chain reaction capable of being maintained in a system of finite size.

Auxiliary measurements were made, and it was found that part of the loss in the K factor was due to neutron absorption in the tined iron cans holding the uranium oxide. It was further found that the purity of the oxide could be improved, and that the effect of the uranium could be still further improved by compression of the oxide or use of the uranium in the form of metal.

A number of intermediate piles were then built and measured by the method set forth herein, using compressed uranium oxide and graphite and uranium of improved purity, in structures having improved geometries of the uranium bodies in the graphite. The first design found to give a K factor greater than unity, when measured by the the means and method described herein, attained a Δ of -1.06×10^{-5} indicating a K factor of 1.007. As the critical size required for this particular type of structure to become self-sustaining, was found to require approximately 10,000 tons of graphite and 3,000 tons of uranium oxide in accordance with formula (2.20), further experiments were made to increase the K factor by including uranium metal in the structure and also by utilizing UO_2 in place of U_3O_8 together with graphite of still better purity, the uranium bodies being again rearranged to a more favorable geometry. A structure was built and measured that indicated that a chain fission reaction could be produced and sustained in a system of practical size.

As an example of the accuracy of the measurements that can be obtained by utilizing the means and method described herein, a self-sustaining chain reacting structure was proposed, utilizing both uranium metal bodies and compressed UO_2 bodies spaced in graphite of several different impurities. The K factor for each of the combinations of uranium and carbon proposed was measured by means of the intermediate pile and method as described herein. K factors were found which indicated that a self-sustaining chain reacting structure could be built having an average K factor of 1.050 utilizing about 480 tons of graphite and 50 tons of uranium. The proposed structure was built and was found to be capable of sustaining a chain reaction by virtue of nuclear fission at a very slightly smaller size than originally computed, with an actual K factor of 1.054. This self-sustaining chain reacting system has been operated at various rates of power production in the form of heat.

Dissipation of large amounts of power in a self-sustaining chain reacting system such as described immediately above, requires the insertion within the structure of a coolant, tubes and passageways for the coolant, and, in many cases requires modification of the shape of the uranium bodies to provide for efficient cooling thereof. The procedure as outlined herein is ideally suited to determine effect of these foreign bodies in the structure, as it is impossible to find the K factor by reliance on theory alone. By the use of the intermediate pile, any proposed cooling system can be incorporated therein, the Δ measured and the K determined.

It should be pointed out here, with respect to the successful attainment of a self-sustaining chain reacting system, that the number of neutrons made directly available to the chain reaction by aggregating the uranium into separate bodies spaced through the slowing medium was found to be a critical factor in obtaining a self-sustaining chain reaction when utilizing natural uranium and graphite. Utilizing theoretically pure carbon and theoretically pure uranium, it has been computed that the highest K obtainable is about 1.1 whereas the same volume ratio of uranium powder and graphite, with the powder distributed uniformly throughout the graphite, gives a K less than unity. Actual K factors as high as 1.08 have been obtained using aggregation and as pure materials as is presently possible to obtain, combined in the best known geometry. It is thus clearly apparent that the aggregation of the uranium into masses separated in the slowing material is one of the most important, if not the most important factor entering into the successful construction of a self-sustaining chain reacting system utilizing relatively pure materials in the best geometry at present known. The presently described structure and its method of measurement determined what the effect of such aggregation would be on the reproduction factor K of the system and thereby proved that my invention incorporating aggregation of the uranium into masses of substantial size was the principal factor in the successful construction of a self-sustaining chain reacting system utilizing natural uranium.

It will be understood that unless otherwise specified, references in the claims to measurement of neutron density or gradient or of neutron density distribution are intended generically to include measurements which may actually be no more than measurements of relative neutron densities, since an important advantage of the invention, when practiced in the presently preferred manner, is that actual neutron densities need not be determined, i.e., the readings need only be relative, as by noting the counts per minute obtained from irradiated foils.

It is also to be noted that while my invention has been described herein as applied to and utilizing a structure having a solid slowing agent such as graphite incorporating uranium aggregated into separate masses, the invention can equally well be used in conjunction with intermediate structures having liquid slowing agents, such as heavy water, with the uranium in the form of separate large masses therein; or in the form of fine particles mixed with the liquid in a form commonly called a slurry. In such cases neutron density measurements are made in the liquid contained in a tank of the desired size and shape in the same manner as described herein.

While the theory of the nuclear chain fission mechanism in uranium set forth herein is based on the best presently known experimental evidence, I do not wish to be bound thereby, as additional experimental data later discovered may modify the theory disclosed. Any such modification of theory, however, will in no way affect the results to be obtained in the practice of the invention herein described and claimed.

I claim:

A test exponential pile comprising a column of graphite of rectangular parallelepiped shape, a plurality of bodies containing natural uranium disposed therein in a geometrical pattern of cell units, each cell unit being composed of uranium embedded in graphite, said pile being of sub-critical mass, a neutron source located adjacent the bottom face of said column of graphite at the center thereof, there being a plurality of spaced vertically aligned measuring slots extending into the column to points near the vertical axis of the column and a plurality of spaced horizontally aligned measuring slots located approximately at the vertical midplane of the column, and an indium foil within at least one of the slots for use in determining the neutron densities within the aforesaid slots.

References cited in the file of this Patent

UNITED STATES PATENTS

2,127,193	Toulmin	Aug. 16, 1938
2,206,634	Fermi et al.	July 2, 1940

FOREIGN PATENTS

114,150	Australia	May 2, 1940

114,151 Australia ....May 3, 1940

OTHER REFERENCES

Anderson at al.: Phy. Rev. 56, 284-6 (1939).
Naturwissenschaften, vol. 27 (1939), pages 402, 403, 406-410.
Rev. Physical Chem. (Japan), vol. 13 (1939), pp. 145-150.
Naturwissenschaften, vol. 27 (1939), pages 404, 405.
Power, July 1940, page 58.
Smyth Report, "Atomic Energy for Military Purposes", August 1945.
Anderson et al.: MDDC-880, U.S.A.E.C., date of manuscript Mar. 21, 1942, date declassified, January 28, 1947, 5 pages.
Kelly et al.: Phy. Rev. 73, 1135-9 (1948).

Chapter 3

Method of Operating a Neutronic Reactor

Enrico Fermi and Leo Szilard, Chicago, Ill., assignors to the United States of America as represented by the United States Atomic Energy Commission

Application December 1, 1952, Serial No. 323,452

U.S. Patent No. 2,798,847 (2,708,656) - Patented July 9, 1957

This Patent is a later,[1] *almost faithful, copy of Patent No. 2,708,656 (which is then not reported in the present volume). This revised version was probably prepared (by the authors) in order to correct several misprints of the previous version.*

As emphasized in The New York Times *of May 19, 1955, Patent No. 2,708,656, an "historic Patent, covering the first nuclear reactor", is the first one on this topic issued by the U.S. Patent Office, and served as a reference for the subsequent Patents on the same subject. In this long Patent, the theory, experimental data and principles of construction and operation of "any" type of nuclear reactor known at that time are discussed in an extremely detailed way. Various possible fission fragments produced by the reactor, several forms of the uranium employed (metal, oxide and so on, grouped in different geometrical forms), various materials adopted as moderators, several cooling systems, different geometries of the reactors, etc. are considered accurately.*

The theoretical description, centered around the achievement of a self-sustaining chain reaction, is exhaustive, and great attention is devoted to any possible cause of neutron loss, to the resonance capture of neutrons and to the effect of the presence of relevant impurities in the reactor. The chain production of neutrons in the pile is described in great detail, along with the theoretical arguments underlying the exponential experiment.

The problem of the variation of the multiplication factor due to the production of radioactive elements, such as xenon, is discussed extensively. In particular it is pointed out that, although the initial production of xenon lowers the multiplication

---

[1] Note that the application for the present Patent was filed on the tenth anniversary of the operation of the first chain reacting pile at Chicago, on December 2, 1942.

factor K due to its relevant neutron absorption, it subsequently increases again due to the decay of xenon into another isotope which absorbs fewer neutrons.

The building up of reactors with solid (graphite) or liquid (heavy water) moderators is discussed, as well as other possible moderators such as light water or beryllium. In particular, the ratio is given of the absorption cross section to the scattering cross section for several moderators.

Procedures for the purification of uranium are described as well. Several methods (i.e., the exponential pile or the "shotgun" method; see Patent No. 2.969,307) are reported for testing the purity against neutron absorption of different materials. The effect of the boron and vanadium impurities in the graphite and light water in the heavy water are considered.

Different cooling systems for the reactors are considered and compared in the Patent, based on the circulation of a gas (typically, air) or a liquid (light or heavy water, diphenyl, etc.).

The principles and practice for the construction, functioning and control of several kinds of reactors are reported in detail.

One reactor considered in the present Patent is a low power uranium-graphite one without cooling system, where the active part consists in (small) cylinders of metallic uranium or pseudo-spheres of uranium oxide (or cylinders of U_3O_8). The control rods are made of steel with boron inserts, while limitation and safety rods are made of cadmium.

In addition, an uranium-graphite pile cooled by air or even by water or diphenyl is considered. It is pointed out that dyphenil should usually be preferred with respect to water, due to its lower absorption of neutrons and to its higher boiling temperature, but the disadvantage related to its use is mainly due to the closed pumping system required and to the possible occurrence of polymerization which makes the fluid viscous.

Another kind of reactor described in detail is made of uranium (vertical) bars immersed in heavy water. When, during the operation, heavy water is dissociated into D_2 and O_2, these two gaseous elements are carried by an inert gas (helium) into a recombination device. The control and safety rods are made of cadmium.

Hybrid reactors composed of different lattices in the same neutronic reactor, in order to increase the multiplication factor K, are considered as well.

A description of the possible uses of nuclear reactors, other than as power supplies, including the production of collimated beams of fast neutrons, the production of plutonium (a fissionable material usable in other reactors) or several other radioactive isotopes (for possible utilization in medicine) is as well given.

As it results clear, no published reference article behind the present Patent exists. Some partial results may be found in several papers[2] *of Volume II of [Fermi (1962)] (see, for example, [Fermi (1952)]), but here very many technical data*

[2] Just to cite some of them, we mention [Fermi (1941)] for the use of uranium spheres or in lumps, [Anderson (1941c)] for the use of indium foils to measure slow neutron density, [Fermi (1942a)] for the introduction of danger coefficients, [Fermi (1942g)] for the methods of cooling, [Fermi (1942h)] and [Fermi (1952)] for the discussion about the location of uranium and control rods in the pile, [Fermi (1943f)] for the use of heavy water as moderator, and so on.

and some information of historic interest (mainly on the experiments performed in order to obtain the data reported) are given.

The most "relevant" change of Patent No. 2,798,847 with respect to the original Patent No. 2,708,656 is the replacement of the 8 claims of the previous one by the following only one claim, which well summarizes the work done:

"A method of operating a neutronic reactor including an active portion having a neutron reproduction ratio substantially in excess of unity in the absence of high neutron absorbing bodies, said method comprising the steps of inserting in the active portion a shim member consisting essentially of a high neutron absorbing body in an amount to reduce the neutron reproduction ratio to a value slightly higher than unit to prevent a dangerous reactivity level, controlling the reaction by moving a control member consisting essentially of a second high neutron absorbing body inwardly and outwardly in response to variations in neutron density, to maintain the neutron reproduction ratio substantially at unity, and withdrawing successive portions of the shim member to the extent necessary to enable the reactor to be controlled by movement of the control member after the neutron reproduction value has been lowered to the point where the outward movement of the control member is insufficient to maintain the neutron reproduction ratio at the desired point, and thus to maintain the range of control effected by such movement of the control member substantially constant despite diminution of neutron reproduction ratio caused by operation of the reactor, the active portion being substantially free of high neutron absorber other than the control member and the shim member."

The present invention relates to the general subject of nuclear fission and particularly to the establishment of self-sustaining neutron chain fission reactions in systems embodying uranium having a natural isotopic content.

Experiments by Hahn and Strassman, the results of which were published in Naturwissenshaften, vol. 27, page 11, in January 1939, led to the conclusion that nuclear bombardment of natural uranium by slow neutrons causes explosion or fission of the nucleus, which splits into particles of smaller charge and mass with energy being released in the process. Later it was found that neutrons were emitted during the process and that the fission was principally confined to the uranium isotope U^{235} present as 1/139 part of the natural uranium.

When it became known that the isotope U^{235} in natural uranium could be split or fissioned by bombardment with thermal neutrons, i.e., neutrons at or near thermal equilibrium with the surrounding medium, many predictions were made as to the possibility of obtaining a self-sustaining chain reacting system operating at high neutron densities. In such a system, the fission neutrons produced give rise to new fission neutrons in sufficiently large numbers to overcome the neutron losses in the system. Since the result of the fission of the uranium nucleus is the production of two lighter elements with great kinetic energy, plus approximately 2 fast neutrons on the average for each fission along with beta and gamma radiation, a large amount of power could be made available if a self-sustaining system could be built.

In order to attain such a self-sustaining chain reaction in a system of practical size, the ratio of the number of neutrons produced in one generation by the fissions, to the original number of neutrons initiating the fissions, must be known to be greater than unity after all neutron losses are deducted, and this ratio is, of course, dependent upon the values of the pertinent constants.

In the co-pending application of Enrico Fermi, Serial No. 534,129, filed May 4, 1944, now Patent No. 2,780,595, dated February 5, 1957, and entitled "Nuclear Chain Reacting Systems",[3] there is described and claimed a means and method of determining the neutron reproduction ratio for any type of uranium containing structure, directly as a result of a simple measurement which can be performed with precision. Accurate values for all of the pertinent nuclear constants need not be known.

We have discovered certain essential principles required for the successful construction and operation of self-sustaining neutron chain reacting systems (known as neutronic reactors) with the production of power in the form of heat. These principles have been confirmed with the aid of measurements made in accordance with the means and method set forth in the above-identified application, and neutronic reactors have been constructed and operated at various power outputs, in accordance with these principles, as will be more fully brought out hereinafter.

In a self-sustaining chain reaction of uranium with slow neutrons, as presently understood, 92^{238} is converted by neutron capture to the isotope 92^{239}. The latter is converted by beta decay to 93^{239} and this 93^{239} in turn is converted by beta decay to 94^{239}. Other isotopes of 93 and 94 may be formed in small quantities. By slow or thermal neutron capture, 92^{235} on the other hand, can undergo nuclear fission to release energy appearing as heat, gamma and beta radiation, together with the formation of fission fragments appearing as radioactive isotopes of elements of lower mass numbers, and with the release of secondary neutrons.

[3]@ This corresponds to the previous Patent (see Chapter 2), though the title is different.

The secondary neutrons thus produced by the fissioning of the 92^{235} nuclei have a high average energy, and must be slowed down to thermal energies in order to be in condition to cause slow neutron fission in other 92^{235} nuclei. This slowing down, or moderation of the neutron energy, is accomplished by passing the neutrons through a material where the neutrons are slowed by collision. Such a material is known as a moderator. While some of the secondary neutrons are absorbed by the uranium isotope 92^{238} leading to the production of element 94, and by other materials such as the moderator, enough neutrons can remain to sustain the chain reaction, when proper conditions are maintained.

Under these proper conditions, the chain reaction will supply not only the neutrons necessary for maintaining the neutronic reaction, but also will supply the neutrons for capture by the isotope 92^{238} leading to the production of 94, and excess neutrons for use as desired.

As 94 is a transuranic element, it can be separated from the unconverted uranium by chemical methods, and as it is fissionable by slow neutrons in a manner similar to the isotope 92^{235}, it is valuable, for example, for enriching natural uranium for use in other chain reacting systems of smaller overall size. The fission fragments are also valuable as sources of radioactivity.

The ratio of the fast neutrons produced in one generation by the fissions to the original number of fast neutrons in a theoretical system of infinite size where there can be no external loss of neutrons is called the reproduction or multiplication factor or constant of the system, and is denoted by the symbol K. For any finite system, some neutrons will escape from the periphery of the system. Consequently a system of finite size may be said to have a K constant, even though the value thereof would only exist if the system as built were extended to infinity without change of geometry or materials. Thus when K is referred to herein as a constant of a system of practical size, it always refers to what would exist in the same type of system of infinite size. If K can be made sufficiently greater than unity to indicate a net gain in neutrons in the theoretical system of infinite size, and then an actual system is built to be sufficiently large so that this gain is not entirely lost by leakage from the exterior surface of the system, then a self-sustaining chain reacting system of finite and practical size can be built to produce power and related by-products by nuclear fission of natural uranium. The neutron reproduction factor in a system of finite size therefore, differs from K by the external leakage factor, and by a factor due to the neutron absorption by localized neutron absorber, and the reproduction ratio must still be sufficiently greater than unity to permit the neutron density to rise exponentially with time in the system as built.

Progressive empirical enlargement of any proposed system for which the factor K is not accurately known, in an attempt to attain the overall size of a structure of finite size above which the rate of loss of neutrons by diffusion through the periphery of the structure is less than the rate of production of neutrons in the system leads only to an expensive gamble with no assurance of success. The fact that K is greater than unity and the fact that the critical size is within practical limits must be known rather accurately in advance, as otherwise a proposed structure having a K factor less than unity, or even a K factor greater than but close to unity, might not sustain a chain reaction even if all of the uranium in the world were included.

The earliest attempts to predict a structure capable of sustaining a chain reaction, using natural uranium, involved the use of fine uranium particles such as uranium oxide powder, dispersed in hydrogen in combined form as the slowing agent. However, these attempts were not successful, and analysis of experiments made has indicated that the neutron losses in such a system when natural uranium is used, will prevent a chain reaction from being sustained, irrespective of the size of the system.

However, in considering such experiments, it was found that even when hydrogen was used as a moderator, as for example, as water, that resonance losses in uranium could be very substantially reduced by aggregating the uranium into bodies of substantial dimensions as compared to the uranium powder-hydrogen mixture previously suggested. It was also found that such aggregation will reduce resonance losses when a moderator such as graphite is used. This gain in neutrons, saved for use in the chain, has proved to be one of the major factors in obtaining a sufficiently low over-all neutron loss as to make possible the attainment of a self-sustaining chain reaction in various moderators when other losses are also controlled.

During the interchange of neutrons in a system of finite size, comprising bodies of any size disposed in a neutron moderator, neutrons may be lost to the chain reaction in four ways:

1. By absorption or capture in the uranium content of the bodies without producing fission,
2. By absorption or capture in the moderator material itself,
3. By absorption or capture by the impurities present on both the uranium bodies and the moderator,
4. By leakage out of the system through the periphery thereof.

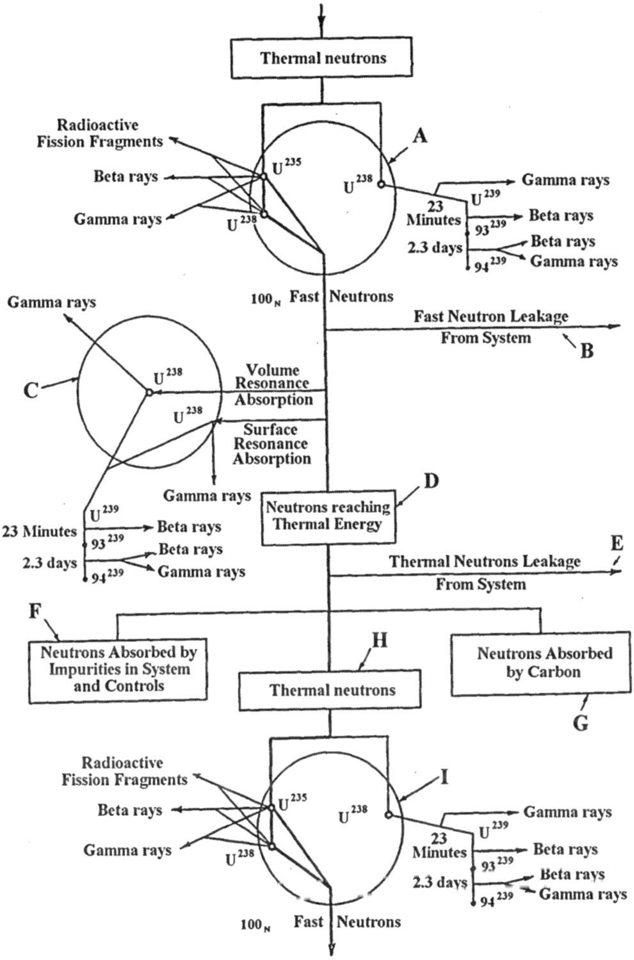

Fig. 3.1

3.1 The Chain Fission Reaction

To illustrate the importance of the various factors entering into a chain reaction, we next describe the chain reaction process for example, as it is presently understood to occur in any system of finite size utilizing natural uranium bodies dispersed in a graphite moderator at some position in the reactor where the neutron density is substantially constant. For better explanation, reference is here made to the diagram constituting Fig. 3.1 of the accompanying drawings; description of the remaining figures being more conveniently set forth in a subsequent part of this specification.

In Fig. 3.1, the letter

A represents a uranium bodies of any size form which fast neutrons are set free as a result of the fission process.

B represents a fast neutron loss due to leakage out of the system.

C represents a uranium body of any size in which both volume and surface resonance absorption of neutrons by U^{238} takes place, at resonance energies above thermal energy, leading to the formation of element 94.

D represents the number of neutrons reaching thermal energy.

E represents a thermal neutron loss by diffusion of thermal neutrons out of the system.

F represents a neutron loss caused by capture of neutrons by impurities in uranium, graphite, and controls.

G represents a neutron loss due to capture of thermal neutrons by the graphite as the thermal neutrons diffuse therethrough before entering uranium.

H represents the number of thermal neutrons entering the uranium body.

I represents a uranium body of any size in which part of the thermal neutrons entering the body are absorbed by U^{238} leading to the formation of 94^{239}, the remaining thermal neutrons causing new fissions in U^{235} thereby producing fast neutrons, a few of which produce additional fast neutrons by fission of U^{238} atoms in the same body.

We will first consider the condition obtaining where thermal neutrons enter uranium body A. Some of these thermal neutrons will cause fission in the U^{235} content of the uranium body A to produce fast neutrons, the yield being at an average rate of about 2 neutrons per fission. As a result of this fission, fission fragments are released together with beta and gamma rays, thereby producing energy which, in the system, is manifested mostly by the heating of the uranium bodies with only a slight release of heat in the graphite. The actual average yield of fast neutrons by fission of U^{235} is slightly higher, e.g., by a few percent, than the average of 2 mentioned above. Some of the fast neutrons released in the fission of U^{235} by the thermal neutrons of this example almost immediately produce fast fission of U^{238} in the same uranium body, with the production of additional fast neutrons.

The fast neutrons leaving the uranium body, for example 100 neutrons, enter the mass of moderator, travel therethrough, and through the uranium bodies over paths long in comparison with the spacing of the uranium bodies, to undergo successive collisions that slow them down. A substantial proportion of the fast neutrons are thus destined to be reduced, by about 100 elastic collisions apiece in the case of graphite and mostly in the moderator, to thermal energy. During this travel, before the neutrons arrive at thermal energies, a small percentage of the higher energy neutrons on the average may leak out of the system because of the finite size of the reactor, and be lost to the chain reaction. Furthermore, during the extremely irregular path of the neutrons while they are being slowed down by elastic collisions in the graphite, some of the neutrons will reach a uranium resonance absorption energy as they are about to enter a uranium body, such as C, and are absorbed immediately on or close to the surface of the uranium body. In addition some neutrons are reduced to resonance energy after entering the uranium body by an elastic collision with the uranium, and are therefore immediately absorbed within the uranium body. Irrespective of whether the neutron resonance absorption in U^{238} is on the surface, or in the volume of the uranium body, element 94 is produced by the resonance absorption according to the following process:

$$_{92}U^{238}+n \rightarrow {}_{92}U^{239}+[\text{6 MeV of } \gamma \text{ rays, not necessarily all of one frequency}]$$

$$_{92}U^{239} \xrightarrow{\text{23 min.}} {}_{93}Np^{239} + \beta^- \; [\text{1 MeV } \beta^-\text{, no } \gamma \text{ rays}]$$

$$_{93}Np^{239} \xrightarrow{\text{2.3 days}} {}_{94}Pu^{239} + \beta^- \left[\begin{array}{l} \text{600 keV upper } \beta^- \text{ energy limit. Also} \\ 2\ \gamma \text{ rays, 400 keV and 270 keV, about} \\ \text{1/2 converted to electrons} \end{array} \right]$$

A small amount of 94^{240} may also be found, formed by addition of a neutron to 94^{239}. Capture of thermal neutrons by U^{238}, as indicated in bodies A and C of Fig. 3.1, also results in production of element 94 by the same process.

The predominant isotope produced, 94^{239}, is a long lived radioactive product with a half life of about 20,000 years.

A large percentage of the original fast neutrons escape resonance capture and fast neutron leakage, and are reduced to thermal energy within the system. Of these thermal neutrons, a small number on the average may leak by diffusion out of the system and be lost from the chain reaction, leaving the remainder of the thermal neutrons diffusing through the moderator in condition to produce fission if they promptly enter U^{235} or element 94 without being captured by any other material.

The fission reaction is as follows:

$$_{92}\mathrm{U}^{235} + \text{neutron} = \mathrm{A} + \mathrm{B} + n \text{ neutrons (average)}$$

A=Light fission fragment, e.g., Br, Kr, Rb, Sr, Y, Zr, Cb, Mo, 43, Ru, Rh. Atomic mass, 83-99, inclusive. Atomic number, 35-45, inclusive.

B=Heavy fission fragment, e.g., Sb, Te, I, Xe, Cs, Ba, La, Ce, Pr, Nd. Atomic mass, 127-141, inclusive. Atomic number, 51-60, inclusive.

In any practical system, impurities will be present in both the moderator and the uranium. In the chain described, a small fraction of the neutrons can be captured and absorbed by impurities in the system without the reproduction factor of the system falling below unity. Thus for example in Fig. 3.1, if impurities necessarily present in the materials do not consume too many neutrons, some excess neutrons are available to be captured by "impurities" intentionally introduced for control purposes, i.e., by a con trol rod, later to be described. Furthermore, since many of the thermal neutrons diffusing through the moderator are not in a position to promptly enter a uranium mass when they reach thermal energy, these thermal neutrons must continue to diffuse through the moderator until they do reach a uranium body. During this diffusion, a small percentage of the neutrons are absorbed by the moderator, leaving sufficient thermal neutrons to enter a uranium body to produce new fast neutrons by fission, to repeat the cycle. In the uranium-graphite system about 72 thermal neutrons enter the uranium body to produce 100 new fast neutrons, i.e., a survival of about 72 percent of the original 100 fast neutrons during the slowing process.

The four neutron losses from the chain reaction referred to above are represented in Fig. 3.1, where the resonance absorption at C and the fraction of thermal neutrons absorbed by U^{238} at I represent the uranium absorption losses. Losses due to impurities are represented at F, those due to absorption in the moderator at G, and the leakage losses due to the finite size of the system at B and E.

These losses will be considered in detail in the order named, as any one of these losses, or their total if too high, can prevent a self-sustaining chain reaction from being attained in a system of any size.

3.1.1 *Neutron Loss by Absorption in Uranium*

It is possible by proper physical arrangement of the materials substantially to reduce uranium resonance absorption, as will be shown later. By the use of light elements for moderators, fewer collisions are required to slow the neutrons to thermal energies with large increments of energy loss per collision, thus decreasing the probability of a neutron being at a resonance energy as it enters a uranium atom. During the moderation, however, neutrons are moving through the slowing medium over random paths and distances so that the uranium is not only exposed to thermal neutrons but also to neutrons of energies varying between the energy of fission and thermal energy. Neutrons at uranium resonance energies will, if they enter uranium at these energies, be absorbed on the surface of a uranium body whatever its size, giving rise to surface absorption. Any substantial reduction of overall surface of the same amount of uranium will reduce surface absorption, and any such reduction in surface absorption will release neutrons to enter directly into the chain reaction.

For a given ratio of moderator to uranium, surface resonance absorption losses of neutrons in the uranium can be substantially reduced by a large factor when the uranium is aggregated into substantial masses in which the mean spatial diameter is at least about 0.5 centimeter for natural uranium metal and somewhat larger when the bodies are of a uranium compound, as hereinafter more fully discussed. For example with UO_2 the minimum radius is larger and with other uranium compounds a similar variation from metallic uranium may be observed. The degree of this variation is dependent upon the density of the uranium compound, its bulk density, and the absorption coefficient of other elements therein for neutrons. In any event the uranium may be placed in the system in the form of geometrically spaced uranium masses or bodies of substantial size, preferably either of metal, oxide, carbide, or combinations thereof. The term geometric is used to mean any pattern or arrangement wherein the uranium bodies are distributed in the moderator with at least a roughly uniform spacing and are roughly uniform in size and shape, or are systematic in variations of size, shape or spacing to produce a volume pattern conforming to a generally symmetrical system. If the pattern is a repeating or rather exactly regular one, the structure may be conveniently described as a lattice. The uranium bodies can be in the form of layers, rods, or cylinders, cubes or spheres, or approximate shapes, dispersed throughout the moderator. Optimum conditions are obtained with natural uranium by using metal spheres.

The resonance losses in uranium constitute one of the critical factors in coordinating the total losses permissible in a neutronic reactor. Proper sizes and shapes of the uranium bodies and volume ratios of uranium to moderator must be fairly accurately known in order that optimum geometry be approached, or if the use of near-optimum geometry is not desirable, then the permissible ranges of departure from the optimum should be determined, so that a reproduction ratio greater than unity can be maintained in a reactor of practical size.

The K constant of a mixture of fine uranium oxide particles in a light element such as graphite, found to be satisfactory as a neutron moderator, assuming both of them to be theoretically pure, would only be about 0.785. Actual K constants as high as about 1.04 have been obtained using aggregation of natural uranium oxide in graphite, and with as pure materials as is presently possible to obtain showing a substantial gain due solely to reduction of resonance loss.

Assuming theoretically pure graphite, and theoretically pure natural uranium metal, with the presently obtainable densities of 1.65 and 18 g/cm^3, respectively, the maximum possible K constant theoretically obtainable is about 1.1 when the uranium metal of density 18 g/cm^3 is aggregated with optimum geometry in the moderator. When heavy water (D_2O) is used as a moderator, higher K constants approaching 1.3 are obtainable. Still higher K constants can be obtained in uranium having more than the naturally occurring content of fissionable elements. Adding such fissionable material is termed enrichment of the uranium.

3.1.2 *Neutron Loss by Absorption in the Moderator*

Neutrons are also subject to capture by the moderator. While carbon and beryllium have very small capture cross sections for thermal neutrons, and deuterium still smaller, a fraction of the thermal neutrons present in the system under best conditions is lost by capture in the moderator during diffusion therethrough. It is therefore desirable to have the neutrons reaching thermal energy enter uranium as promptly as possible. This may be taken care of by using optimum or near optimum geometry where the resonance absorption is substantially equal to absorption in the moderator.

Moderators differ in their ability to slow down neutrons and in their capacity to absorb neutrons. The ability to slow down neutrons may be expressed by what is known as the scattering cross section of the nucleus, whereas the ability to absorb or capture neutrons is expressed by what is known as the capture cross section of the nucleus. The ratios of absorption cross section to scattering cross section for moderators discussed herein are approximately as follows:

Light water (H_2O) 0.00478

Diphenyl	0.00453
Beryllium	0.00127
Graphite	0.00726
Heavy water (D_2O)	0.00017

The choice of moderators therefore will depend on many considerations, as will be apparent from further discussions herein.

3.1.3 *Neutron Loss by Absorption by Impurities in the System*

However, even when resonance and moderator losses are reduced to a practical minimum, no self-sustaining chain reaction can be obtained in any system unless impurities in the material used for the reaction are reduced to such an extent that the loss by parasitic capture by such impurities will not, in combination with the other losses, prevent the reaction from becoming self-sustaining. Impurities present in both the uranium and the moderator consequently constitute a very important neutron loss factor in the chain. The effectiveness of various elements as neutron absorbers varies tremendously.

Certain elements such as boron, cadmium, samarium, gadolinium, and some others, for example, if present even in a few parts per million, could very likely prevent a self-sustaining chain reaction from taking place. It is highly important, therefore, to remove as far as possible all impurities capturing neutrons to the detriment of the chain reaction from both the slowing material and the uranium. If these impurities are present in too great quantity, the self-sustaining chain reaction cannot be attained. The permissible amounts of impurities will vary for each specific geometry, depending upon such considerations as the form in which the uranium is used – that is, whether natural or enriched, whether as metal or oxide. The type of slowing down material used also influences the effect of impurities, as do the weight ratios between the uranium and the slowing down material. Elements such as oxygen may be present, and the uranium may be in form of oxide, such as UO_2 or U_3O_8, a carbide, or fluoride, but the metal is preferred. Nitrogen may be present in the reactor in fairly large amounts, and its effect on the chain reaction is such that the neutron reproduction ratio of the system may be changed by changes in atmospheric pressure. This latter effect may be eliminated by excluding nitrogen from the system, or by sealing the system from the effects of changes of atmospheric pressure.

The effect of impurities on the optimum reproduction factor K may be conveniently evaluated by means of certain constants known as "danger coefficients" which are assigned to the various elements. These danger coefficients for the impurities are each multiplied by the percent by weight of the corresponding impurity, with respect to the weight of uranium in the system, and the total sum of these coefficients gives a value known as the total danger sum. This total danger sum is subtracted from the reproduction constant K as calculated for theoretically pure materials and for the specific geometry under consideration.

The danger coefficients are defined in terms of the ratio of the weight of impurity per unit mass of uranium and are based on the cross section for absorption of thermal neutrons of the various elements. These values may be obtained from physics textbooks on the subject, and by direct measurement, and the danger coefficient computed by the formula[4]

$$\frac{\sigma_i}{\sigma_u}\frac{A_u}{A_i}$$

wherein σ_i and σ_u represent the cross sections for the impurity and the uranium respectively, A_i the atomic weight of the impurity and A_u the atomic weight for uranium. Regardless whether the impurities are in the moderator or in the uranium, they are computed as their percent by weight, of the uranium in the system.

Danger coefficients for some elements are given in the following table, wherein the elements are listed in order of their atomic number.

The sum of the danger coefficients of the impurities in any given composition entering into a reactor as multiplied by the percent by weight of the uranium in the reactor, is known as total danger sum of the composition. This figure is a dimensionless constant like K and accordingly can be directly subtracted from K. It will be noted that the danger coefficients given are related to the neutron absorption value of unity for uranium.

As a specific example, of the use of danger coefficients, if the materials of a system under consideration has 0.01 percent by weight of each of the elements H, Co, and Ag with respect to the weight of the uranium in the system, the total danger sum in K units for such an analysis would be:

$$0.0001 \times 10 + 0.0001 \times 17 + 0.0001 \times 18 = 0.0045$$

This figure can then be subtracted from the K calculated for a particular geometry of theoretically pure materials to give the actual K constant for the materials used. This would be a rather unimportant reduction in the

[4]@ In the original text the symbols T_i, T_u appear instead of σ_i, σ_u.

Table 3.1

Element	Danger Coefficient	Element	Danger Coefficient
H	10	Co	17
D	0.1	Ni	3
He	0	Cu	1.8
Li	310	Zn	0.61
Be	0.04	Ga	∼ 1
B	2.150	As	2
C	0.012	Se	6.3
N	4	Br	2.5
O	0.002	Rh	50
F	0.02	Ag	18
Na	0.65	Cd	870
Mg	0.48	In	54.2
Al	0.30	Sn	0.18
Si	0.26	Sb	1.6
P	0.3	I	1.6
S	0.46	Ba	0.30
Cl	31	Sm	∼ 1.430
K	2.1	Eu	435
Ca	0.37	Gd	∼ 6.320
Ti	3.8	Pb	0.03
V	4	Bi	0.0025
Cr	2	Th	1.1
Mn	7.5		
Fe	1.5		

reproduction factor K unless the reproduction factor K for a given geometry and materials without considering impurities, is very nearly unity. If, on the other hand, the impurities in the uranium are Li, Co, and Rh in the same percentage, the total danger sum would be:

$$0.0310 + 0.0017 + 0.0050 - 0.0377$$

reduction in K due to impurities. This latter reduction in the reproduction factor for a given system would be serious and might well reduce the reproduction factor below unity for certain geometries.

The maximum possible K constants for neutronic reaction systems when natural uranium aggregates in optimum geometry (i.e., best apportionment of resonance and moderator losses) are used, and where the materials used are assumed to be theoretically pure, have been calculated as follows:

In reactors operating at high neutron densities an equilibrium poisoning factor up to 0.024 K can develop and must be taken into consideration, as will be brought out later.

Table 3.2

Materials	K for pure materials
U metal–graphite moderator	1.1
U oxide–graphite moderator	1.07
U metal–beryllium metal moderator	1.11
U metal–beryllium oxide moderator	1.1
U metal–heavy water moderator	About 1.3
U metal–light water moderator	About 1.

It can readily be seen from the above tabulation that the total danger sum for impurities in both the uranium and moderator must be less than about 0.3 in order that the K factor remain equal to or greater than unity with a deuterium moderator, about 0.11 for a beryllium moderator, and about 0.1 for a graphite moderator. Light water can be used as a moderator, at least in part of a reactor, as will be pointed out later.

In the chain reaction outlined in Fig. 3.1 for a natural uranium reactor of practical size, a small percentage of neutrons can be absorbed by impurities without reducing the neutron reproduction ratio below unity. Not all of these neutrons, however, should be absorbed by the residual impurities in the uranium and the moderator, because if this were so the system would always just be self-sustaining and no exponential rise in neutron density could be obtained. Some means must be provided to release additional neutrons to enter the chain.

For example, in Fig. 3.1, it may be considered that only half of the neutrons that can be absorbed by impurities are absorbed by materials actually present as impurities in the uranium and the moderator, and that the other half of the neutrons are absorbed by a strong neutron absorbing material, such as cadmium, for example, that is wholly or partially removable from the system. Under these conditions, with the chain reaction in balance, if the amount of cadmium or other neutron absorbing material is reduced in the system by removal therefrom to a point where less than the number of neutrons that can be spared for impurity absorption are absorbed by both the impurities in the materials and by the remaining cadmium, for example, then the neutron density in the system will rise exponentially when the system is large enough, because the neutron reproduction in each cycle then will exceed 100 fast neutrons for each original 100 fast neutrons.

In order to stabilize the reaction at any desired neutron density within the system, the neutron density is measured as it is rising. When a predetermined neutron density is reached within the system, the cadmium or other neutron absorbing material is reinserted into the system to a point where the total permissible number of neutrons is again absorbed by the total impurities within the system. The chain reaction will again be in balance at the new neutron density. To reduce the neutron density, still more absorbing material is introduced into the system sufficient, for example, to increase the total impurity absorption to the point where less than 100 new fast neutrons are produced per cycle. The neutron density will then decay. The system can then be stabilized when a new desired lower neutron density is reached by decreasing the amount of the absorbing material in the system until only the number of neutrons permissible for balance are again absorbed, and the system will then be balanced at the lower neutron density. The reaction is completely stopped by leaving sufficient absorbers in the system to prevent the reaction from building up or remaining in balance. The neutron density will then drop to a low natural neutron background value, and remain there until the absorbers are again removed to raise the reproduction ratio above unity.

3.1.4 *Exterior Neutron Loss in a Neutronic Reactor of Practical Size*

In any chain reacting system, it is only when the system has infinite size that there will be no exterior leakage of neutrons from the system. For any reactor of finite size exterior neutron losses will occur, and these losses will increase as the size of the reactor decreases.

Losses will occur both of fast neutrons, which during their slowing down by scattering collisions by the nuclei of the moderator may become directed outwardly when near the periphery of the reactor and thus escape before they reach thermal energy, and of slow neutrons. The latter may escape when they are diffusing through the moderator near the periphery of the reactor.

Thus while a system of infinite size can have a reproduction constant K of, for example 1.1, the identical construction made smaller than infinite size, because of this exterior leakage, can non longer have 110 new neutrons produced for each 100 initiating a cycle. The number will be less by the leakage factor and will, as the size of the reactor is decreased reach the point where only 100 new neutrons are produced for each 100 neutrons starting each generation. This size is known as the critical size of the reactor and is, of course, dependent on the K constant of the system, when moderator characteristics are known, or, upon a factor known as the Laplacian (Δ), as found directly in a lattice of small volume as will be shortly brought out. For each value for the reproduction factor K greater than unity as modified by moderator characteristics, there is thus a minimum overall size of the reactor known as the critical size wherein the neutron reproduction ratio is unity. However, if the reproduction ratio is exactly unity no rise in neutron density will occur. The reactor, to be operative at any desired power output, must be capable of providing a reproduction ratio of slightly over unity, and therefore must be made slightly larger than critical size so that the required reproduction ratio can be attained. Once the desired operating power output is established, the reproduction ratio can be reduced to unity by insertion of control absorbers and the reaction maintained thereafter at the desired power output. Critical size and operating size can be determined mathematically when losses are known and coordinated, as brought out later.

Some of the exterior losses however, can be reduced by surrounding the reactor with what is known as a reflector, of a material preferably having a low absorption to scattering cross section ratio. The effect of the reflector on operating size will be shown later.

3.2 Measurement of Neutron Losses

In using the exponential pile to test for neutron losses, a pile structure with a specific geometry of uranium and graphite blocks, for example, is built which is known to be of non-operating dimensions. In other words, it is of sufficiently small size that the exterior leakage from the system would prevent a self-sustaining neutronic chain reaction even if the K constant were to be above unity. In such a small pile, uranium compositions can be arranged in any desired geometry and by placing a neutron source, such as a radium-beryllium source, at the bottom of the pile, a neutron density distribution specific to the geometry and materials used will be created through the pile. By measuring the neutron density in various parts of the pile, it is found that the neutron density declines exponentially with the distance away from the source. This characteristic gives the pile its name of exponential pile. Different types of geometries can be tested in such a pile to determine optimum conditions, for example, or range of conditions for which K is greater than unity.

If, instead of varying the geometry, the geometry and the moderator is maintained constant and uranium compositions of different neutronic purities are substituted, the change in slope of the curve showing the decay in neutron density through the structure away from the source will give an accurate measure of the neutronic purity of the uranium composition, in terms of the K constant if desired. In this way, different uranium compositions can be compared with one another and the effect on the K constant determined. The departure of the K constant as found for a particular composition from the K constant calculated for theoretically pure uranium will thus give the danger sum for neutron absorbing impurities in the uranium, when the K reduction due to moderator impurities is known.

As the structure is customarily built up or piled from moderator blocks when solid moderators are used and having the uranium lumps geometrically arranged therein, the structure is commonly called a "pile", and the geometrical arrangement of the uranium lumps in the moderator called a lattice.

Briefly, the theory of exponential pile measurements is as follows:

Considering a uranium-graphite lattice structure or column of square cross section with sides equal to a, and semi-infinite height, with a source of fast neutrons at the center of the base of the column, then, at points sufficiently far removed from the source, the neutron density due to any chain reaction present will be given by an equation of the following form where x, y, and z are the axes of the structure:

$$n = \sum_{ij} A_{ij}\ e^{\frac{-x}{b_{ij}}} \cos\frac{i\pi y}{a} \cos\frac{j\pi z}{a}, \tag{3.1}$$

where the symbols ij represent the orders and arguments of the Bessel function series, A is a constant which varies with the Bessel functions included in the summation and b is a constant associated with the exponentials in the summation.

The x axis is taken along the vertical axis of the structure, and the $x = 0$ plane coincides with the base of the pile. Thus, for points close to the vertical axis, each harmonic of the neutron density decreases exponentially as follows:

$$n_{ij} = A_{ij}\ e^{\frac{-x}{b_{ij}}}, \tag{3.2}$$

with a relaxation distance or length equal to b_{ij}, the relaxation distance or length being the distance in which the neutron flux is reduced to a fraction of $1/e$ of its original value. At a sufficiently large distance from the source the first harmonic only is important. The relaxation length can then be taken as b, and b taken alone is related to the reproduction factor K through the following equation:

$$K = 1 - \frac{\lambda\Lambda}{3}\left(\frac{1}{b^2} - \frac{2\pi^2}{a^2}\right)\exp\left\{-\frac{r_0^2}{4}\left(\frac{1}{b^2} - \frac{2\pi^2}{a^2}\right)\right\}, \tag{3.3}$$

where

a = length of side of the structure
b = relaxation distance
λ = mean free path of thermal neutrons in graphite
Λ = mean free path for absorption collision
$r_0^2/4$ = the age of nascent thermal neutrons

The quantity

$$\frac{1}{b^2} - \frac{2\pi^2}{a^2} = \Delta.$$

Δ signifies a number given by the ratio of Δn to n where n is the number of thermal neutrons per cubic centimeter at the point x, y, z. Δn is an abbreviation for the sum of the three second derivatives of n with respect to the three variables x, y, z. Δ is found to be constant throughout any structure utilizing given geometries and materials. For the cases where K is close to unity, Δ is small, so that the equation can be written

$$K = 1 - \left(\frac{\lambda\Lambda}{3} + \frac{r_0^2}{4}\right)\left(\frac{1}{b^2} - \frac{2\pi^2}{a^2}\right). \tag{3.4}$$

By defining

$$M^2 = \left(\frac{\lambda\Lambda}{3} + \frac{r_0^2}{4}\right),$$

then M is the migration length of thermal neutrons in the structure, and is roughly proportional to the average distance between the place of birth of a neutron as a fission neutron and its place of death by thermal absorption.

Substituting in (3.4) the quantity Δ for the quantity

$$\frac{1}{b^2} - \frac{2\pi^2}{a^2}$$

and M^2 for

$$\frac{\lambda\Lambda}{3} + \frac{r_0^2}{4}$$

the equation can be written

$$K = 1 - M^2\,\Delta. \tag{3.5}$$

The final equation for K can then be written to include M^2, a and b as follows:

$$K = 1 - M^2 \left(\frac{1}{b^2} - \frac{2\pi^2}{a^2} \right). \tag{3.6}$$

M^2 has been found to be from about 650 cm^2 to 750 cm^2 for chain reacting structures of uranium and graphite, formed by building up or piling graphite moderator blocks with uranium lumps geometrically arranged therein, and can be used in Equation (3.6) to find K for such structures.

The length of a side, a, to be used in calculating K from Equation (3.6) must be that value for which the neutron intensity actually becomes equal to 0. Because of the finite length of the mean free path λ, compared to the dimensions of the pile, the effective side is larger than the physical side. From neutron density measurements made at the outer surface of the pile, the effective value of a can be estimated, for various x planes. Using the quantities found for M and a, a measurement of the relaxation distance b, associated with the first harmonic of the neutron density will then determine, from Equation (3.6), the reproduction factor corresponding to a lattice of infinite dimensions similar in geometry and materials to the structure being tested. This reproduction factor must be modified when used in conjunction with reactors attaining high neutron densities for prolonged time periods, by an operational poisoning factor. This factor can be added into the exponential pile by adding equivalent absorbers to each cell and then finding Δ or K. When K is found without such absorbers this factor can be directly deducted.

To determine the relaxation distance b, thin indium foils (0.0924 g/cm^2), are placed at positions along the axis of the pile for a predetermined time for example and the 54 minute radioactivity induced by neutron bombardment is measured on Geiger-Mueller counters for a predetermined time. For these measurements the indium foil is held in a nickel holder. Thus the activation of the foil (A_{ni}) is due to the absorption of both thermal and indium resonance neutrons. All measurements are corrected to give the foil activity values for infinite times of irradiation. The emission of neutrons by spontaneous fissions of the uranium in the pile produces a small neutron background which must be subtracted form the density measurements.

Because of the finite height of an exponential pile, two corrections may be applied to neutron density measurements. First, a harmonic correction due to the presence of higher harmonics in the neutron density curve in horizontal planes near the source; and second, an end-correction due to the proximity of the top of any practical column to the measuring positions.

Finally after making the harmonic and end-corrections, b is calculated from the relation

$$b = \ln \frac{D}{\dfrac{(A_{ni})_2}{(A_{ni})_1}}, \tag{3.7}$$

where D is the distance between the two positions x_1 and x_2 along the vertical axis at which $(A_{ni})_1$ and $(A_{ni})_2$ are measured and ln the mean logarithm to the base e.

For measurements near the top of the pile the harmonic correction may be ignored. For measurements away from the top the end-correction may be ignored. Thus the best values are obtained from measurements in x planes intermediate between the sources and the top.

Two neutron density measurements made in adjacent positions along the vertical axis of the exponential pile will, therefore, give b and a value Δ or a value for K when the value of M^2 is known. It is customary to average the values obtained by using measurements made in several adjacent and equally spaced positions along the vertical axis to obtain the average Δ or K constant for the entire pile.

The same procedure can be used when liquid moderators are involved by placing the liquid in a tank and suspending the uranium, in the form of rods, for example, so that it enters the moderator. Measurements are made as set forth herein for solid moderators.

The migration length has been described as roughly proportional to the average displacement of a neutron from the point of its origin as a fast neutron in a uranium lump to the point of its disappearance in the pile. More precisely, we define the square of the migration length by the formula

$$M^2 = \frac{r_0^2}{4} + \frac{\lambda\Lambda}{3}, \tag{3.8}$$

where r_0^2 is the mean square distance between production and disappearance of neutrons in the lattice.

In principle, an experiment for the actual measurement of the migration length could be performed as follows: A lattice of a given type is set up, as for the exponential pile. For best results it would be desirable to suppress neutron multiplication in this lattice, which could be done, for example, by using instead of normal uranium, uranium completely depleted in the fissionable isotope, and readjusting the neutron absorption to equal that of normal uranium by the addition, say, of boron as a neutron absorber. Into this prepared lattice introduce a point source of fission neutrons, which might be a lump of spontaneously fissioning material. Then, by the usual foil techniques we could measure the distribution of thermal neutrons through the lattice, and compute the mean square distance by known methods. In principle one would thus obtain the correct value of M^2.

Such experiments have not to date been performed, because the preparation of the material is very expensive, and no proper fission source is presently available. Actually the best existing knowledge of M^2 for the present lattices is obtained by measurements made in an exponential pile using the formula:

$$\frac{K-1}{M^2} = -\Delta, \tag{3.9}$$

which is Formula (3.5) above. The Laplacian Δ can be measured directly in the exponential pile as follows using the formula:

$$\Delta = \frac{1}{b^2} - \frac{2\pi^2}{a^2} \tag{3.10}$$

and by finding the values of a and b as outlined above, the value of Δ may be determined. A neutron absorber of known neutron capture cross section is then introduced into the exponential pile in known amounts, the change in the Laplacian measured, and M^2 calculated from the measurements.

In one specific instance an exponential pile having uranium rods arranged in graphite in such a manner that liquid could be passed over the uranium bodies, borated water in various concentrations of boron was passed through the lattice. It was found that there was a change in Δ of

$$0.0584 \times \frac{10^{-6}}{\text{cm}^2}$$

for one part per million of boron in the water. From this change the value of M^2 was calculated to be about 590 cm^2, accurate within about 10 percent of error. It is to be noted however that M^2 enters into $K-1$ only, so that the error in K would then be only about 1 percent at the most. M^2 in this case is slightly lower than in a reactor without a cooling system and for a uranium metal sphere-graphite lattice M^2 has been found to be about 700 cm^2.

The practical calculations for pile design do not even depend upon this procedure but upon a more theoretical one still. M^2 can be written:

$$M^2 = \tau + L_0^2(1-f), \tag{3.11}$$

where the symbol τ designated the age of nascent neutrons and is essentially the mean square distance that fission neutrons may travel before becoming thermal. This can be directly measured in the moderator used, since the metal has a very small effect on slowing down. The second term (L_0^2) is the diffusion length squared for thermal neutrons in the lattice in question, which is equal simply to the diffusion length in the moderator. L_0 can also be directly measured in the moderator used, and is multiplied by the fraction of neutrons absorbed in the moderator, which is $(1 - f)$, where f is the thermal utilization defined as the fraction of the thermal neutrons absorbed by the uranium (both by simple capture and to produce fission) rather than by the moderator. Such calculations find many objections but are adequate to 10 to 15 percent and are suitable, therefore, for design purposes in finding $K - 1$.

The following values of M^2 have been found by measurements and calculation to be indicative for preliminary design purposes in building reactors.

For water: $M^2 = 40$ cm^2
For D_2O: $M^2 = 230$ cm^2
For beryllium: M^2 on the order of 300 cm^2
For graphite: $M^2 = 600$ cm^2–700 cm^2

By the use of the exponential pile, various sizes and shapes of uranium bodies have been tested and the related K factors found for various moderators.

By testing uranium compositions in the exponential pile, the neutronic purity can be determined in terms of K when the same moderator is used or when the effect of the moderator impurities is known, with geometry unchanged. The test is equally reliable for uranium compounds such as the uranium oxides U_3O_8 and UO_2, uranium carbide, uranium tetrafluoride, uranium hexafluoride, etc., compounds which contain, in addition to traces of elements having high neutron capture cross sections, large amounts of elements such as O, C, and F, all of which have relatively low neutron capture cross section. The test can evaluate the total effect of both types of impurities in terms of K reduction, as well as the effect of changing geometries on K.

When M^2 is known, this factor can be used to determine critical size of the structure for various moderators.

Thus the determination of (1) the proper size, shape and disposition of the uranium bodies in the moderator to reduce resonance losses; the determination of (2) the amounts of neutron absorbing impurities that can be tolerated in addition to other losses before a self-sustaining chain reaction will become impossible in a system of practical size; and determination of (3) the nuclear characteristic of the moderator with respect to requirements of critical size and tolerable exterior losses; has enabled us to provide a means and method of building neutronic reactors capable of sustaining a chain neutron reaction by virtue of nuclear fission, even when individual values for constants entering into the nuclear processes are only imperfectly known.

It is, therefore, the main object of the present invention to provide a means and method of designing and building and operating neutronic reactors capable of sustaining a chain nuclear reaction by virtue of nuclear fission, and to outline the variations that can be tolerated before the reaction will become impossible of attainment in structures of practical sizes.

Other objects and advantages of our invention will be apparent from a description of several operative reactors as shown in the attached drawings, wherein:

Fig. 3.1 is a diagram or chart illustrating the balanced condition of a chain reaction in a system of practical size employing natural uranium in graphite;

Fig. 3.2 is a perspective view of a uranium-graphite reactor completely enclosed in a radiation shield;

Fig. 3.3 is a front end plan view of the reactor shown in Fig. 3.2, a portion of which is shown in central vertical section;

Fig. 3.4 is a side plan view of the reactor a portion of which is shown in central vertical section;

Fig. 3.5 is a top plan view of the reactor a portion of which is shown in central horizontal section;

Fig. 3.6 is a plan view of one of the graphite blocks containing uranium metal with a portion broken away to show in section one of the uranium metal cylinders;

Fig. 3.7 is a longitudinal sectional view taken on the line **12–12** of Fig. 3.6;

Fig. 3.8 is a longitudinal sectional view of a graphite block and showing pseudospheres of uranium oxide in place of the uranium metal;

Fig. 3.9 is a plan view of a graphite block loaded with pseudospheres of uranium oxide, with a portion of the block broken away to show a pseudosphere in a section taken as indicated by line **14–14** in Fig. 3.8;

Fig. 3.10 is a plan view of a dead graphite brick with a portion broken away and shown in section;

Fig. 3.11 is a schematic wiring diagram of a neutron density monitoring circuit;

Fig. 3.12 is a graph showing neutron density values plotted with relation to the number of layers as a cubical reactor is built;

Fig. 3.13 is a diagrammatic side view of a safety rod;

Fig. 3.14 is a diagrammatic side view of a shim or limiting rod;

Fig. 3.15 is a diagrammatic side view of a control rod;

Fig. 3.16 is a diagram showing the outline of a reactor in the shape, roughly, of an ellipsoid;

Fig. 3.17 is a graph on which are plotted neutron density value relations found in the active portion of the system plotted against number of layers of graphite bricks for an ellipsoidal reactor;

Fig. 3.18 is a diagram illustrating the distribution of neutron density in a spherical reactor;

Fig. 3.19 is an enlarged, fragmentary, perspective view of a modified active portion in which the overall shape is in the form of a cube or parallelepiped and the uranium is arranged horizontally in cylinders or rods;

Fig. 3.20 is a second modification of the active portion of the system wherein the overall shape is cylindrical and the uranium is disposed vertically in the form of cylinders or rods;

Fig. 3.21 is a vertical sectional view of a neutronic reactor employing deuterium oxide as the moderator;

Fig. 3.22 is an enlarged fragmentary vertical sectional view through a portion of the reactor showing in particular a uranium rod used in the reactor shown in Fig. 3.21;

Fig. 3.23 is an enlarged vertical sectional view of a portion of a uranium rod equipped with an attached adapter for removing the uranium rod from the reactor;

Fig. 3.24 is a fragmentary detail sectional view corresponding to Fig. 3.22 but showing only a modification of the ball valve seal shown in Fig. 3.22;

Fig. 3.25 is a horizontal sectional view shown partially in elevation, the section being taken on the line **29–29** of Fig. 3.21;

Fig. 3.26 is a graph on which are plotted contour lines representing various reproduction constants K for systems employing uranium metal spheres and graphite;

Fig. 3.27 is a graph similar to that of Fig. 3.26 for cylindrical rods of uranium metal;

Fig. 3.28 is a graph on which are plotted contour lines representing various values for the reproduction constants K for a uranium oxide (UO_2)-

graphite system wherein the oxide is in the form of spheres;

Fig. 3.29 is a graph on which are plotted contour lines representing various reproduction constants K for systems employing uranium oxide (UO_2) and graphite wherein the oxide is in the form of cylindrical rods;

Fig. 3.30 is a graph showing K contour lines for uranium metal rods immersed in D_2O;

Fig. 3.31 is a longitudinal view partly in section and partly in elevation of an air cooled neutronic reactor system;

Fig. 3.32 is a cross sectional view, partly in elevation, taken as indicated by the line **32–32** in Fig. 3.31;

Fig. 3.33 is a plan view of the system shown in Figs. 3.31 and 3.32;

Fig. 3.34 is a longitudinal sectional view partly in elevation, of a jacketed slug;

Fig. 3.35 is a longitudinal sectional view, partly in elevation of a horizontal channel during a loading and unloading operation;

Fig. 3.36 is a cross sectional view taken as indicated by the line **36–36** in Fig. 3.35;

Fig. 3.37 is a vertical sectional view (partly in elevation) of a liquid cooled reactor;

Fig. 3.38 is a vertical section view (partly in elevation) of the reactor shown in Fig. 3.37, and taken as indicated by the line **38–38** in Fig. 3.37;

Fig. 3.39 is a diagrammatic perspective view of a uranium rod and associated coolant channel;

Fig. 3.40 is a diagram showing the statistical weight of concentric lattice portions of uniform K plotted against the extent of the same lattice portions within the structure;

Fig. 3.41 is a diagram showing the effect of reflectors of various thickness on the size of the reactor; and

Fig. 3.42 is a diagram showing change of critical size in U-C reactors with change in K.

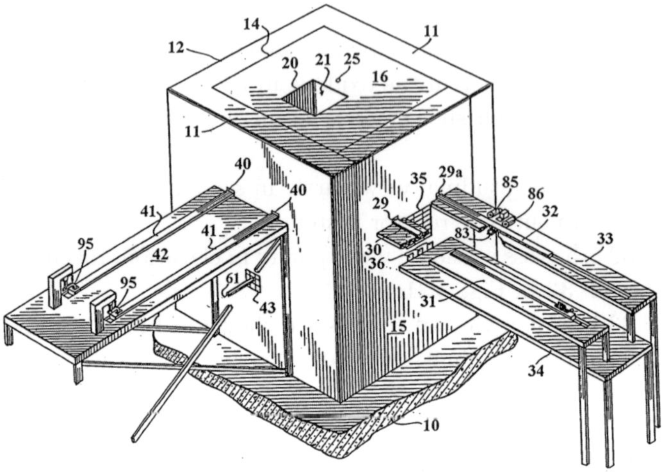

Fig. 3.2

3.3 An Illustrative Neutronic Reactor having a Solid Moderator

One of the simplest ways to accomplish a self-sustaining chain reaction operating by virtue of nuclear fission is to utilize either uranium metal, uranium oxide, or both, aggregated into bodies od substantial size and spaced in a solid moderator such as graphite to form a lattice, and build without the introduction of a cooling system into the reactor. Such a neutronic reactor is shown in Figs. 3.2 to 3.17, inclusive.

Fig. 3.2 shows the neutronic reactor system diagrammatically in perspective and will be first referred to. As the active portion of the reactor loses large quantities of neutrons during operation, and the fission reaction creates gamma radiation, it is desirable to protect operating personnel from the radiations resulting from the chain reaction. In this instance, protection is provided by surrounding substantially all of the reactor with concrete or equivalent shielding.

A heavy concrete foundation **10** is first poured and side walls **11** and connecting backwall **12** are then erected. This provides a vault space **14** (Figs. 3.3, 3.4 and 3.5) in which the chain reacting lattice of uranium and graphite is erected until the vault is filled within about five feet of the top and five feet of the front, as will be later described. The front of the

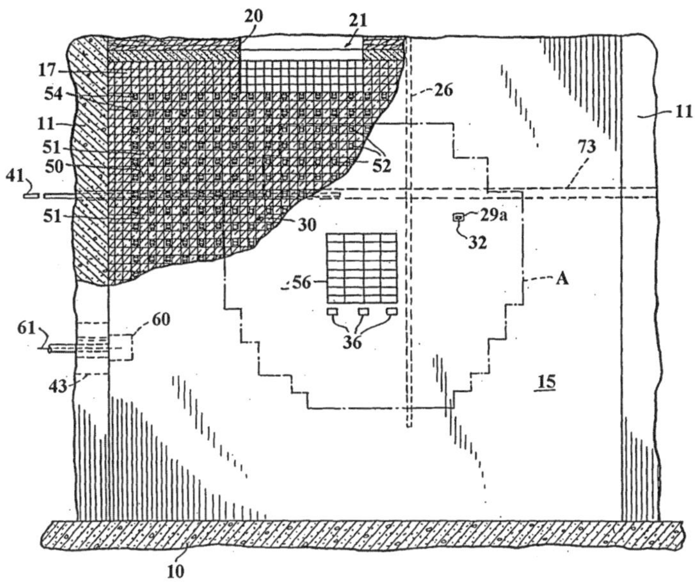

Fig. 3.3

vault is then closed by a front wall **15** formed of concrete, and the top is closed by a top wall **16** which may be of wood and lead layers. The top wall **16** is pierced by a large opening **20**, leading to a well **21** extending inwardly to the peripheral layer of uranium bodies in the internal lattice. A smaller adjacent aperture **25** is the exterior opening of a shaft **26** (Fig. 3.3) extending into the central portion of the reactor.

Front wall **15** is pierced by shim and regulating rod apertures **29** and **29a** respectively, positioned on each side of and slightly above the center of front wall **15**. A "shim" or limiting rod **30** is positioned on a limiting rod platform **31** and is movable to enter aperture **29** in a horizontal plane; and a regulating or control rod **32** is positioned on a control rod platform **33** to enter aperture **29a** in a horizontal plane. Below the plane of these two rod platforms is a removal platform **34** positioned to receive lattice portions that may be removed from the reactor through a removable section channel **35** and from removable stringer channels **36**. Details of the rod mechanisms and use of the platforms will later be described.

One side of the reactor side wall **11** is also pierced by a pair of spaced safety rod apertures **40** through which two safety rods **41** can be horizon-

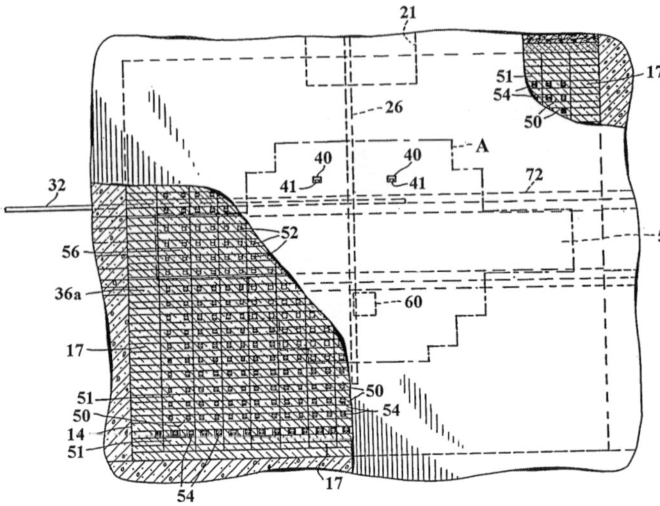

Fig. 3.4

tally inserted into the reactor from safety rod platform **42**. Just below the safety rod apertures is an ionization chamber channel **43**. This completes the description of the exterior of the reactor.

The self-sustaining chain reacting unit to be built up within vault space **14** is designed so to reduce total neutron losses as to make a self-sustaining chain reaction possible, as has been previously outlined, using specific types and sizes of uranium masses, both of metal and oxide in graphite, all of obtainable purity, and spaced with a specific geometry. As will be shown later, an all metal structure can be built, i.e., wherein all of the uranium bodies are of metallic uranium, but the combination of metal and oxide in the present example is utilized for economy.

The basic construction unit used to fill vault space **14** is a graphite block $4\frac{1}{8}$ inches by $4\frac{1}{8}$ inches in cross section, used in a number of lengths. The blocks are carefully planed by woodworking machinery to have smooth rectangular sides and end faces, so that they may be readily piled or stacked to fill the vault space **14** without substantial air spaces. Such construction has led the device to be termed a "pile", but the more generic term "reactor" is preferred.

Vault space **14** is dimensioned, in this instance, to receive horizontal graphite block layers. Two main types of graphite blocks are used as shown

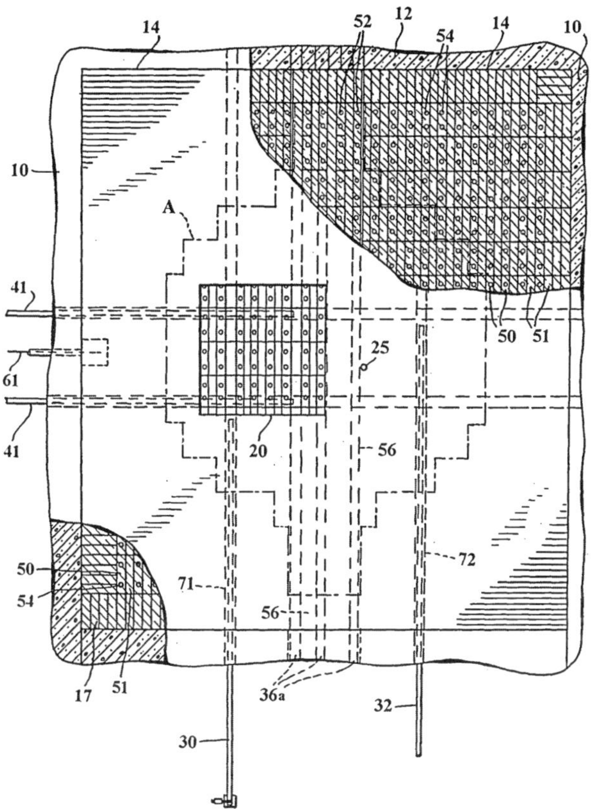

Fig. 3.5

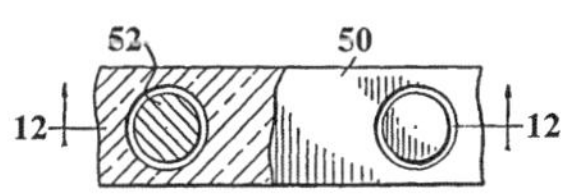

Fig. 3.6

in Figs. 3.6 - 3.10, inclusive. Certain of the blocks **50** are drilled with cylindrical holes spaced $8\frac{1}{4}$ inches center to center to receive the uranium bodies which are placed therein. The uranium bearing graphite blocks **50** are termed live graphite. Other blocks **51**, as shown in Fig. 3.10, contain no

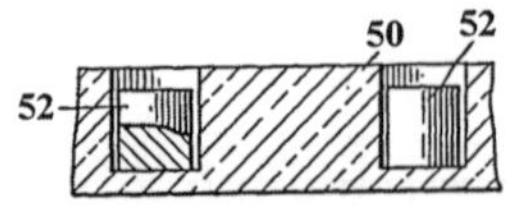

Fig. 3.7

uranium and may be termed dead graphite.

The uranium bodies are in two main forms, one form being cast uranium metal cylinders **52** having a metal density of slightly over 18 grams per cubic centimeter, of several weights as later listed. Other uranium bodies **54** are in the form of pseudospheres of uranium oxide UO_2, similar to the metal cylinders with the exception of a beveling on top and bottom to approximate spherical contours as shown in Figs. 3.8 and 3.9. A few U_3O_8 cylinders are also used. The oxides are compressed to a density of about 6 g/cm^3 and all of the uranium is purified as herein later described.

In any event, the uranium bodies are placed in the holes in blocks **50**, and these live graphite blocks **50**, in conjunction with dead graphite blocks **51** are used to build up the chain reacting system in vault space **14** by assembling the blocks into a uranium lump lattice arrangement to provide an active portion of substantially cubical form, surrounded by several layers of dead graphite to act as a reflector **17**.

To start the building of the reactor in vault space **14**, three bottom layers of dead graphite are laid down on the foundation 22 feet deep and 20 feet wide to start a reflector **17**. For more uniform distribution of weight, alternate layers may have the graphite blocks crossed at right angles. The blocks are closely piled to minimize air spaces.

After three layers of dead graphite are piled, the uranium bearing layers are started with adjacent rows of live graphite cut in length so that the uranium bodies are spaced along the row $8\frac{1}{4}" \times 8\frac{1}{4}"$ center to center, each live graphite row spaced by a row of dead graphite, with the uranium bodies aligned both across and in depth in the vault space. The uranium bearing rows do not begin until 12 inches of dead graphite is laid down next to the

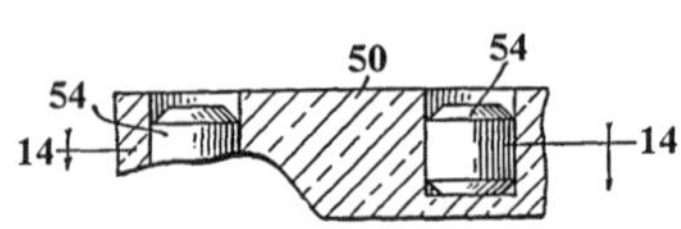

Fig. 3.8

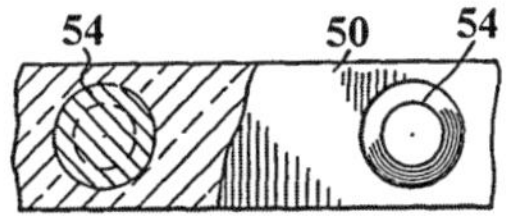

Fig. 3.9

concrete walls of the vault and at the open front, and three sides have 16 inches of dead graphite. Thus the foundation of an active portion having a substantially square base is set up, with the base surrounded on all sides by at least 12 inches of graphite, with the uranium bearing portion of the layer being about $17\frac{1}{2}$ feet wide by $19\frac{1}{2}$ feet deep.

A layer of dead graphite is then laid over the first uranium bearing layer, and the next uranium bearing layer is laid with the uranium bodies substantially aligned vertically. Thus as the reactor is built up, layer by layer of alternate graphite and uranium-graphite layers, the uranium lumps form a cubic lattice with the uranium bodies aligned with the rectangular co-ordinates of the vault space **14**.

As the presently designed reactor is designed to have a central portion where the lumps are of uranium metal, the metal lumps are positioned in stepped relation in the various layers to form a mass about 13 feet wide, 10 feet high and 10 feet deep positioned between the 16th and 48th layer centrally of the reactor as indicated by broken line **A** in Figs. 3.3, 3.4 and 3.5.

Along lines passing close to the center of the reactor, removable stringers, such as indicated by numeral **36a** in Figs. 3.4 and 3.5, of live carbon blocks, are preferably provided so that one or more complete rows of uranium bodies can be removed from close to a central diameter of the system, as may be desired for test purposes, as will later be described. Such stringers are easily provided by making the row of live graphite blocks it is desired to remove, slightly smaller in cross section than the surrounding

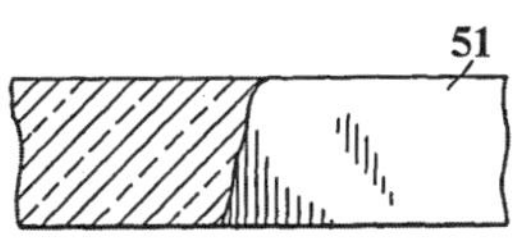

Fig. 3.10

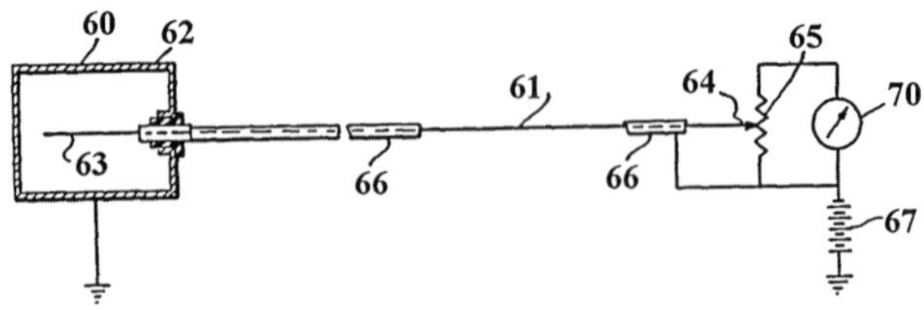

Fig. 3.11

blocks so that the row can readily be pushed out of the pile from outside the shield and reinserted when desired without disturbing the remainder of the structure.

As it may be desirable to be able to remove a larger amount of the uranium from the central portion of the reactor, a horizontal removable section **56** is provided, extending from front to rear of the reactor and through the central portion containing metal. This removable section contains metal uranium bodies extending all the way to the front and back of the active portion of the reactor and is 8 rows wide and 8 rows high.

Matching blocks bored with a vertical $2\frac{5}{8}$" hole between the uranium bodies in the live blocks, and in proper position in the dead blocks, are aligned as layers are added, to provide internal continuity of the shaft **26**.

As the reactor is being built the ionization chamber **60** is installed in channel **43** just inside wall **11**. Wire line **61** is led to the outside and connected to a monitoring circuit, as shown in Fig. 3.11. The opening through the concrete shield is closed with concrete bricks.

Ionization chamber **60** comprises a sealed metal casing **62** containing approximately 18 liters of boron fluoride at one atmosphere pressure into which projects a central electrode **63**. The central electrode is connected through central wire **61** to a contact **64** adjustable across resistor **65**. One end of resistor **64** is connected to an outer wire shield **66** and to a battery **67** of about 450 volts potential, the other end of which is grounded, as is casing **62**. The other end of resistor **65** is connected through galvanometer **70** to the wire shield side of battery **67**. Neutron absorption by the boron in the chamber releases alpha particles by nuclear reaction and causes alpha ray ionization in the chamber, the amount thereof being measured by galvanometer **70**. Only the ionization chamber need be exposed to the neutrons developed in the reactor, with the galvanometer positioned up the 300 feet away. The galvanometer deflection preferably positioned adjacent the control rod is only approximately linear with neutron density, but is reproducible, and readily calibrated in terms of neutron density or power if desired.

As the reactor is built up layer by layer, slots **71** and **72** are provided in a dead graphite layer for entrance of the shim and regulating rods respectively, and at right angles thereto, on a higher level, safety rod slots **73** are provided in a dead graphite layer. All of these slots pass entirely through the reactor with the safety rod slots at right angles to the shim and control rod slots.

Construction is continued with the shim rod, control rod and safety rods fully inserted into the reactor.

Preferably, at least from the halfway point of construction, the natural neutron density in the pile is monitored as layers are added. That is, the relatively constant though small "natural" neutron contribution by spontaneous fission and from other natural sources, causes fission and initiates short chains of fission reaction in the partly built pile. Until the critical size of the pile is reached, such chain reactions are convergent, i.e., are not self-sustaining, but the fissions provide a measurable increase of neutron density in the pile, over that provided by the natural neutrons which traverse the structure. Thus it has been found that by plotting the neutron density within the pile as layers are added thereto, with the control rod and safety rods withdrawn, a prediction can be made in advance as to the size at which the chain reaction in the structure will become just self-sustaining. The personnel building the pile can thus be warned that the critical size is being approached.

In Fig. 3.12 the results of indium foil measurements in the reactor are shown plotted against the number of layers placed on the pile during construction of the reactor, the foil measurements being obtained in the following manner.

A_0 is the so called saturation radioactivity expressed in counts per minute of a standard indium foil activated by the neutrons at the approximate center of the structure as far as constructed.

The indium foils are held, for example, between aluminum trays and exposed to neutron bombardment at the approximate center of the pile for a predetermined period of time, to produce in the indium foil a condition of partial radioactive saturation. Various slots (not shown) extending into the lattice may be provided for insertion of the indium foil as construction proceeds, i.e., so that such access can be add to localities which successively constitute the approximate center of the existing structure.

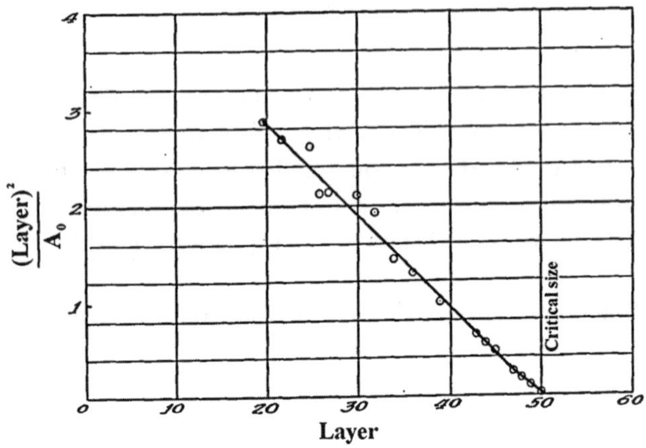

Fig. 3.12

After the predetermined period of exposure to the neutrons, the indium foil is removed from the pile and is allowed to stand for exactly three minutes to permit the short-lived radioactivity produced by the exposure to the neutrons to decay substantially to zero. The remaining induced radioactivity of the indium foil is then determined by utilizing a suitable and previously standardized Geiger counter to count the beta rays emanating from the foil over a predetermined time period. The results thus obtained are then converted into values which would have been obtained if the indium foil had been exposed to the neutrons sufficiently long to produce a state of saturation.

However, a given exposure time may be too long in regions of high neutron density, yielding too many counts per minute for the counter to handle accurately, or may be too short in very low density regions yielding too few counts for accurate representation, in which event the time schedules may be changed to compensate. The activity of the indium foil may then be expressed in terms of counts per minute at saturation for the foil used, although in the latter case and for long or short exposure times, the saturation activity in counts per minute may be determined from the equation

$$A_0 = \frac{f\,c}{e^{-f\,t_x}(1-e^{-f\,t_w})(1-e^{-f\,t_c})}, \tag{3.12}$$

where c is the observed number of counts in time t_c, and in minutes, t_w, t_x and t_c are respectively the period of irradiation, waiting and counting, and where f is the fraction of the excited (radioactive) atoms that disintegrate in a unit time, and for indium equals 0.012836 per minute. The indium foils are preferably 4 cm $\times$ 6.4 cm and have a thickness corresponding to 0.094 g/cm^2.

The values of layers/A_0 are plotted against the number of layers, i.e., for measurements of A_0 and computations of layers made as the illustrated pile was built, each measurements of A_0 being made for, and thus corresponding to, the steady state value of neutron density reached at the particular number of layers to which such measurement corresponds. It should be noted that as the critical size is approached, the steady state values of A_0 approach infinity, and when the critical size is exceeded, A_0 no longer has a steady state value. In place of a horizontal line as would be obtained if K were exactly unity, the curve slopes downwardly, indicating that K is greater than unity, i.e., that the neutron density is increasing more rapidly than R^2_{eff}, and that at some value of R_{eff}, and consequently at some determinable number of layers, indicated by the intercept with the axis, the density becomes or can become infinite. Thus as the plotted points approach the axis a simple extension of the curve clearly indicates in advance the layer at which the system will become chain reacting, with the ability to maintain the reaction, which in this case was slightly above the 50th layer.

With the safety, shim, and control rods fully inserted into the structure, the active portion and the reflector **17** are completed to the final size and shape desired, with layers of dead carbon blocks being positioned over the top thereof to complete the reflector **17**, except for the well **21** and the shaft **26**. The front of the vault was then closed. Except for the upward extension of well **21** the top was closed by adding 40 inches of wood and 6 inches of lead in order that personnel could be present on top of the reactor while it was operating. A structure is thus obtained having a substantially cubical active portion therein, surrounded by a carbon reflector **17**, and enclosed in a shield on all sides.

While the critical size was reached at slightly above the 50th layer, the reactor to be operative must have a size larger than critical size so that the reproduction ratio is greater than unity, and so that a rise in neutron density can occur. In this case, the effective operating size for low, intermittent power was reached by the addition of four additional dead graphite layers, which completed the reflector across the top of the reactor and thus increased the effective size. The effect of the reflecting layer is given below in terms of the time for doubling the neutron density inside the reactor, with all control absorbers removed.

Table 3.3

Layer	Time for doubling in seconds
50.09	∞ (critical size)
51	90.
52	32.9
53	19.0
54	12.5 (operating size)

The shielding around the reactor was found to have a neutron reflecting capability equal to about 10 centimeters of dead graphite.

The concrete walls also serve as the main shield to prevent gamma radiation escaping outside the structure, as carbon, being a light material, absorbs gamma rays only to a relatively small extent. The water in the concrete also serves to slow down and absorb escaping neutrons.

More detailed specifications of the cubical reactor just described are given below:

URANIUM

Total No. of U lumps in reactor	17,640
Total number of U metal lumps mostly 6, 3.202, 7, and 8 lbs. each	20,368 lbs.
Total number of U oxide lumps mostly UO_2, about 6 lbs. each. About 5% U_3O_8 averaging 5 pounds each positioned in outer layers only	14,438
Total weight of U oxide in reactor	84,000 lbs.
Total weight of uranium and uranium oxide in reactor	104,000 lbs.
Total weight of graphite (including reflector)	472 tons.
K of metal section	1.07
K of oxide section	1.035
Average K	About 1.055
Fastest doubling time	12.5 seconds

The graphite used was made from row materials specially selected for minimum neutron absorption.

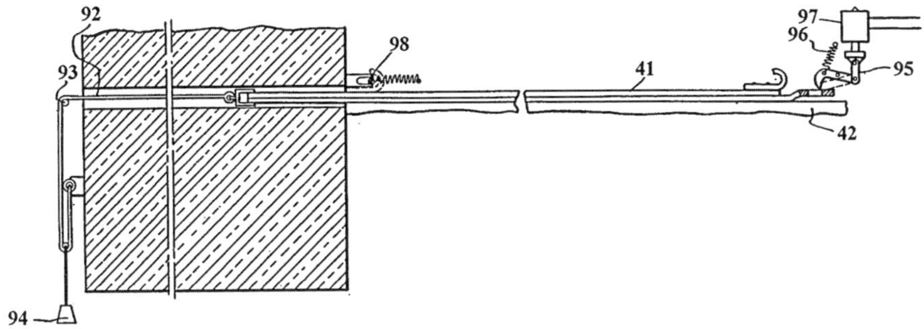

Fig. 3.13

The various rods entering the reactor will next be described, as shown diagrammatically in Figs. 3.13, 3.14, and 3.15.

Referring first to Fig. 3.15 showing the control rod **32** the rod proper is a sheet of 5/16" steel $3\frac{1}{2}$ inches wide and 15 feet long, to which are attached 1/4" blocks of 15 percent boron steel in 4 inch lengths. The rod is mounted on lateral wheels **80** operating on tracks **81** on platform **33** outside of the reactor and on the bottom of slot **72** inside the reactor. The composite rod is provided with a rack **82** engaged by pinion **83**. This pinion is driven by belt **84** from a differential gear box **84a** in opposite directions by in motor **85** and out motor **86**. Limit switches **87** and **87a** are positioned to break the common motor lead at either end of the rod travel by contact of stop **88**. The motors **85** and **86** are under control of the operator, and the action of the control rod is fully described in a subsequent section of this specification. A "selsyn" indicator system **86a** is used with the indicator in view of the control rod operator to tell the operator the exact position of the control rod at all times. The indicator may be calibrated as discussed

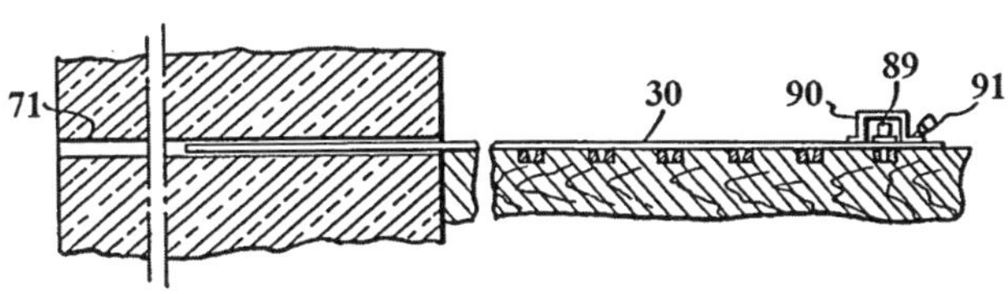

Fig. 3.14

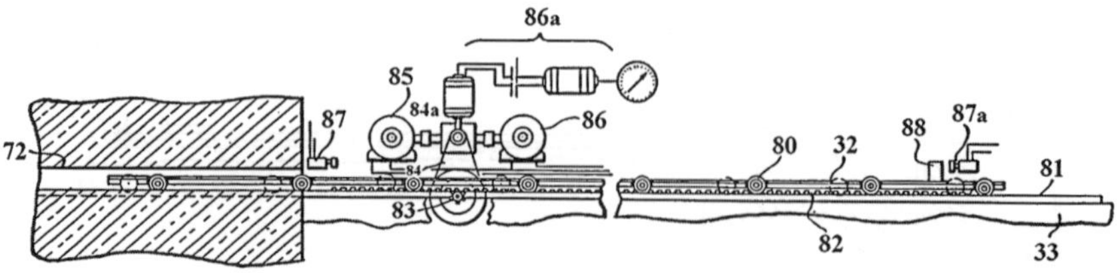

Fig. 3.15

later.

The "shim" or limiting rod **30** is shown in Fig. 3.14. This rod may be simply a cadmium sheet 1/16" by $3\frac{1}{2}$" by 15 feet riveted to a fiber backing, movable by hand into and out of the reactor, held in place by pin **89** and locked by pin guard **90** and padlock **91**. This rod is so positioned in the reactor that when the control rod is completely out of the reactor the maximum reproduction ratio cannot exceed a value dependent upon the position of the shim rod.

The two safety rods **41** are alike and one is shown in Fig. 3.13. They are formed from 1/16" cadmium sheet $3\frac{1}{2}$" wide backed by fiber, and long enough to completely cross the reactor. They are drawn into the reactor from platform **42** by cable **92** passing over pulley **93**, the cable also carrying weight **94**. The safety rods are normally held out of the reactor by latch **95** opened by spring **96** and held in latched position by current passing through solenoid **97**. Accidental or deliberate interruption of current in solenoid **97** will cause the latch to open and the safety rods will be pulled into reactor by gravity to stop the reaction. Spring bumper **98** cushions the rod at the end of its travel. Normally when the reactor is left unattended, all rods are inserted fully into the reactor. The above described reactor is capable of being operated at an output as high as 10,000 kilowatts for short periods. Since the reactor is only conductively cooled, only small powers can be continuously maintained without an appreciable internal temperature rise. However, the reactor is valuable for the manufacture of radioactive elements and 94^{239} and subsequent removal of the irradiated uranium by use of the removable section, for use as an intense source of neutrons available in well **21** and shaft **26** (Fig. 3.3), as a generator of high energy gamma rays, and as a means for testing materials by use of the removable stringers. These uses are more fully described later in the section on uses of neutronic reactors.

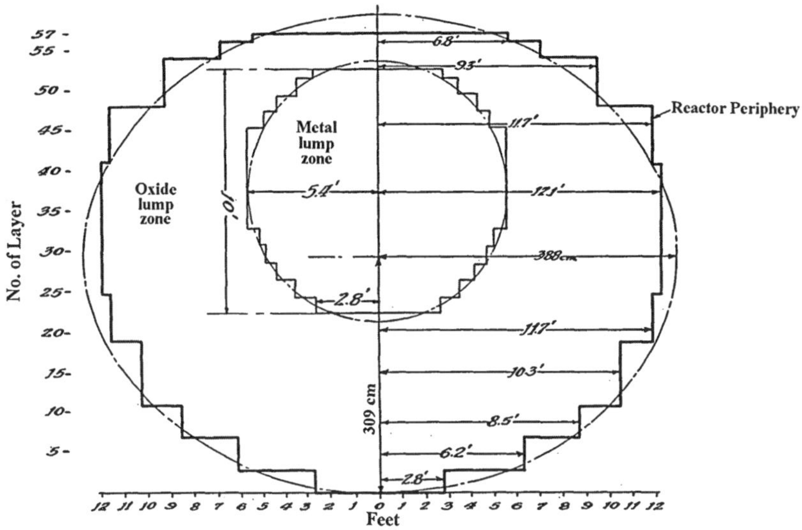

Fig. 3.16

The power produced by the reactor at any attained neutron density may also be calculated from measurements in standard indium foils in locations spaced across the reactor. Again using a symbol A_0 for the saturation radioactivity value computed from the counts per minute obtained in a Geiger counter from the standard indium foils distributed across the pile, and assuming the total energy produced per fission is 200 million electron volts (MeV), equivalent to 3.2×10^{-4} ergs, the power of the pile at the measurements location is given by the following formula:

$$\text{Power} = 2.3\ A_0\ \text{erg/s} = 2.3 \times 10^{-7} \times A_0\ \text{W}.$$

Such indium foil measurements can be used to accurately calibrate galvanometer **70** in terms of watts, if desired. The power can be removed as heat from neutronic reactors by the use of suitable circulating media if desired, as will taken up later.

A prototype of the reactor as above described was built in a slightly non-spherical shape, and successfully operated to create a self-sustaining chain reaction at about 200 watts power. This reactor was then torn down and a large portion thereof incorporated in the reactor just previously described.

As originally operated, the active portion of the reactor was not cubical but was substantially in the shape of a flattened rotational ellipsoid with a polar semi-axis of 309 centimeters and an equatorial semi-axis of 388 centimeters as shown diagrammatically in Fig. 3.16. The effective radius was about 355 centimeters (12.7 feet) and the average K constant was about 1.054. It was surrounded by about 12 inches of graphite and supported by a wooden framework.

The uranium in the reactor was as follows:

Table 3.4

Geometrical Shape	Compound	Weight, lbs.	Density, g/cm^3	Number	Total Weight in Reactor lbs.
$2\frac{1}{4}$" cylinder	Metal	6.0	18	2,060	12,400
$3\frac{1}{4}$" pseudosphere	UO_2	4.72	6.10	14,840	70,000
$3\frac{1}{4}$" pseudosphere	U_3O_8	3.99	5.17	1,200	4,790
3" cylinder	UO_2	4.56	6.14	540	2,460
3" cylinder	U_3O_8	3.97	5.20	840	3,340
				19,480	[1]92,990

[1]Equals 46.5 tons.

Various grades and makes of graphite were used in the reactor, the reflector and a pier extending upwardly for use as a thermal neutron source. The graphite was made from raw materials selected to give a K reduction which averaged about 0.02.

3.4 Graphite in the Reactor

As this reactor was built up the neutron activity was also monitored with indium foil exposures as above described. However, in this case, the changing shape of the reactor must be taken into account.

In this reactor, R_{eff} is the effective radius of the structure at various stages during construction. It is given by the formula

$$\frac{3}{R_{\text{eff}}^2} = \frac{1}{a^2} + \frac{1}{b^2} + \frac{1}{c^2}, \tag{3.13}$$

where a, b, c are the sides of a rectangular parallelepiped which is drawn to conform as closely as possible to the actual shape of the structure in its various stages of construction. If a structure employing a geometry giving K exactly unity is built up gradually maintaining a true spherical shape,

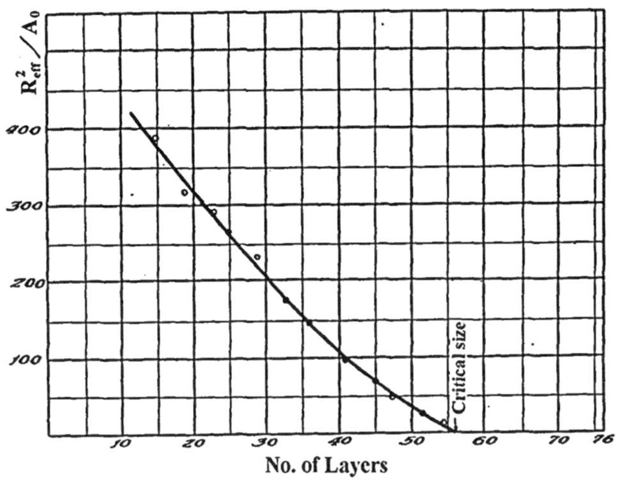

Fig. 3.17

then A_0 increases approximately as R^2, where R is the radius of the sphere at any time.

If it is built with an ellipsoidal shape, then A_0 increases approximately as R^2_{eff}, and in the actual structure that is built, approximate values of a, b, c, to agree with the actual shape at any stage can be estimated, and R^2_{eff} calculated.

The values of R^2_{eff} are then used to plot

$$\frac{R^2_{\text{eff}}}{A_0}$$

against layers to predict the critical layer as shown in Fig. 3.17.

This reactor became chain reacting after the 57th layer was added, this being about one layer beyond critical size. With 57 layers in position, the time for doubling the reaction was found to be about 1 minute. The reactor construction was started to provide a spherical shape. While the K factor of the bulk of the metal and the graphite, and the oxide and graphite, was known from exponential pile measurement, a substantial amount of untested graphite was used in the outer portions of the reactor. This graphite proved to have a lower danger sum than that predicted, giving an average K factor higher than expected. In consequence, measurements showed that the reactor would reach $r = 1$ sooner than expected, and therefore it was finished off without completing the sphere. When the cubical

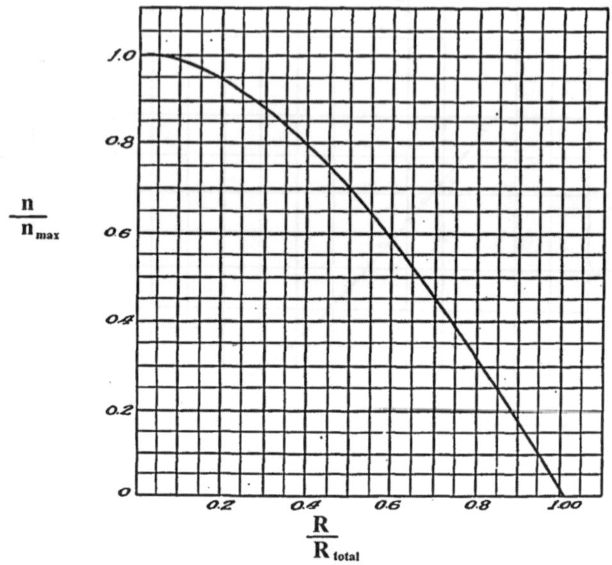

Fig. 3.18

reactor previously described was built the K factors were better known for the various combinations of uranium and graphite and the structure more closely approached the calculated critical size and symmetry.

The neutron density distribution in a spherical reactor is shown in Fig. 3.18 where the ratio of the neutron density to the maximum neutron density in the pile is given for points within the reactor set forth as the ratio of the partial radius at those points, to the total radius. The maximum neutron density occurs at the center of the reactor, falling off rapidly as the periphery is approached with approximately a cosine curve. The curve shows only a very small relative density at the periphery of the reactor, but this density represents a flux of about 4 percent (in graphite) of the total neutrons generated that are continuously leaking out of the reactor. The density, however, of the neutrons leaking from the reactor is low compared to those diffusing through the reactor and thus causes only a slight displacement of the curve shown in Fig. 3.18 past the 1.0 radius line. As the heat released in the reactor is caused by the fissions, the heat distribution curve across the reactor will be similar in shape to the neutron density distribution curve. The neutron density curve across reactors of other shapes

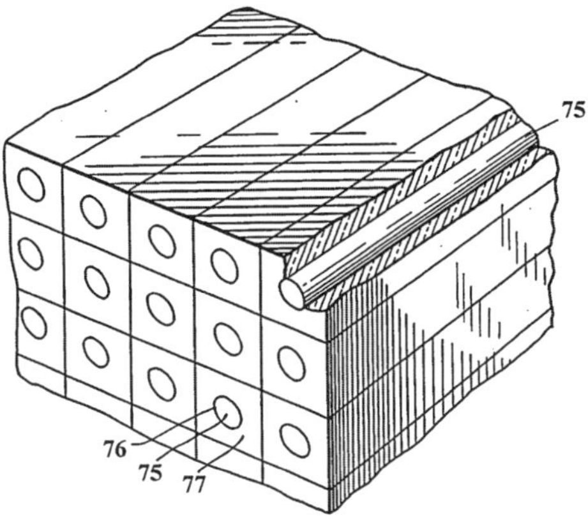

Fig. 3.19

is also substantially a cosine curve.

Reactors of the above types can also be built with rod geometry as shown in Figs. 3.19 and 3.20. In Fig. 3.19 uranium rods **75** are horizontally positioned in bores **76** in live graphite blocks **77** and piled side by side to make for example, a cubical active portion. Similarly as drawn in Fig. 3.20 the uranium rods **75** and the live graphite blocks **77** can be stacked vertically to form a cylindrical active portion. Either shape of active portion

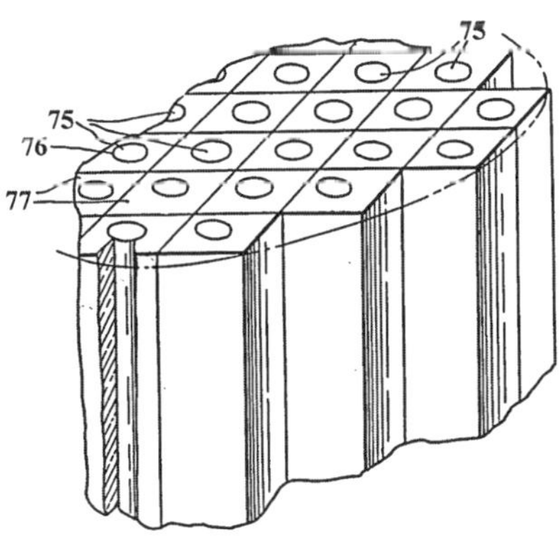

Fig. 3.20

can be used in either of the geometries.

A description of a liquid moderated reactor with the uranium in the form of rods vertically positioned will next be taken up.

3.5 An Illustrative Neutronic Reactor having a Liquid Moderator

A chain reaction can also be maintained in a uranium-D_2O reactor as shown herein, certain details of which are more fully described and claimed in the U. S. Patent No. 2,714,577 to Fermi and Zinn. Referring first to Fig. 3.21 of the drawings, numeral **101** denotes a neutronic reactor tank or container of cylindrical shape and of a material that is relatively non-corrosive at low temperatures and that is relatively non-absorbent with respect to neutrons, such as, for example, aluminum or stainless steel. A suitable size of such tank for a self-sustaining chain reaction when D_2O is used as a moderator is one that is 6 feet in diameter and 7 feet 4 inches high, although other sizes may be used as well. Suspended in tank **101** are 136 rods **102** of uranium metal 1.1 inches diameter sheathed by aluminum about 0.035 inch thick, to prevent extreme radioactivity and contamination of the D_2O moderator by fission products emanating from the uranium. Such coating also prevents corrosion of the uranium by the heavy water. Rods **102**, that will be described in detail hereinafter, extend to about 1/4 inch of the tank bottom. Sufficient deuterium oxide is introduced into tank **101** to obtain a volume of uranium and heavy water slightly over the critical size. Such critical size may be predicted well in advance of attainment thereof. This is done by taking measurements of the neutron density, preferably adjacent to the perimeter of the tank, for example, in the reflector **104** described hereinafter, as the tank is being filled, and by plotting, for instance, the reciprocals of such neutron densities as ordinates against some measure of the overall size of the filled portion of the tank (e.g., the volume of D_2O) as abscissas, as the size is being increased by raising the level of heavy water in the tank in a manner similar to that described for the construction of the uranium-graphite reactor. This will give a curve that, when extrapolated, will indicate or forecast the critical size at the point where the curve crosses the axis of abscissae. When the critical size has been exceeded a self-sustaining nuclear reaction is initiated as described hereinbefore. In the reactor described, the critical size was obtained when the tank **101** was filled to a D_2O level of 122.4 centimeters

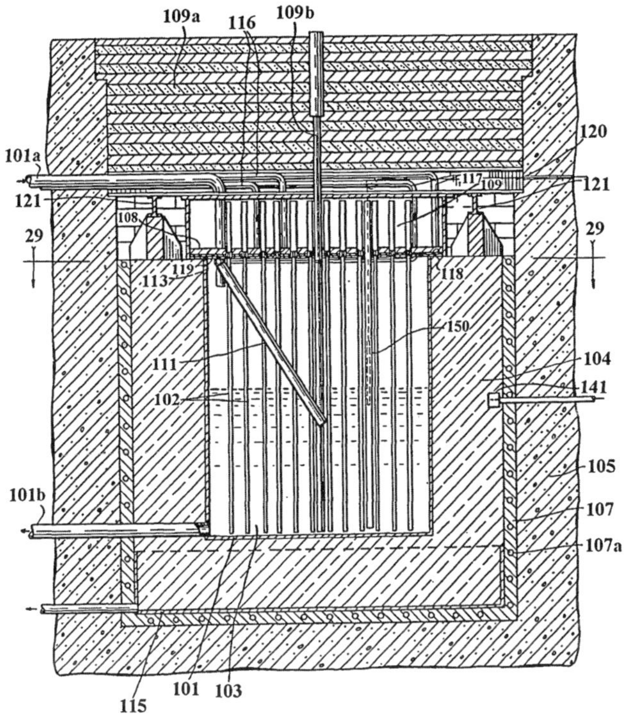

Fig. 3.21

from the bottom, and an operating size giving a neutron density doubling time of 37.6 seconds was obtained at a D_2O level of 123.1 centimeters. At a level of 124.7 the doubling time was 6.52 seconds.

Immediately surrounding tank **101** is a neutron reflector **104** of graphite, for example, having substantially a cup shape. Reflector **104** and tank **101**, together with its contents, are referred to as the neutronic reactor. The moderator, together with the uranium immersed in the moderator, constitutes what may be termed the active portion of the reactor. The thickness of the reflector may be of the order of two or three feet or more depending on the size of the active portion and the degree of neutron scattering required. By using this peripheral layer of scattering or reflecting material, the overall size of the active portion of the reactor may be made somewhat smaller than in a case where no scattering layer is employed, since neutron losses to the exterior ar effectively reduced. Tank **101** and its contents may be built slightly below the diameter that would be required without a reflector so that the addition of reflector **104** with its neutron reflecting action will convert the reactor from one that is not self-sustaining to one that is self-sustaining.

A concrete shield **105** surrounds the graphite reflector **104** and serves to prevent neutrons and gamma radiations from escaping to the outside of the structure. The carbon in reflector **104**, being a relatively light element, absorbs gamma rays only to a relatively small extent. The concrete shield may be of the order of 5 or 10 feet in thickness. The water of crystallization in the concrete absorbs escaping neutrons.

Generally speaking, the higher the atomic weight of an element, the better it serves as a shield to prevent escape of penetrating radiations, such as gamma radiations. Lead, therefore, is an excellent material for a shield for certain purposes while water of the same thickness is only fair. However, thick water shields are sometimes convenient and satisfactory, as will be shown later. Interposed between the concrete shield **105** and graphite reflector **104** is a cooled metal shield **107**, preferably of a lead-cadmium alloy, having cooling tubes **107a** passing therethrough, through which light water, that is, ordinary water, or other coolant may be circulated. Shield **107** is especially adapted not only to cool the outside of the neutronic reactor but to minimize the escape of gamma and other penetrating radiations. A suitable thickness may be 4 inches or more.

A cover plate **108**, for example, of stainless steel, is provided at the top of tank **101** and serves as a thermal shield as well as a support for rods **102** and for control and safety rods pivotally mounted thereunder, as will be described later. Above the cover plate **108** is a space **109** into which the upper ends of rods **102** project, and through which piping can be conducted. Above space **109** is a shield **109a** having alternate layers of iron and Masonite (compressed wood fiber) for preventing the escape of neutrons, gamma rays, and other penetrating radiations from the top of tank **101** to the outside.

A tube or well **109b** of any suitable diameter, for example 4 inches, extends through shields **109a**, space **109**, and plate **108** into the tank **101**, preferably axially of the tank. Objects may be introduced from the exterior through the tube and into the center of the tank so that they may be bombarded by high intensity neutron radiations for the production of radioactive isotopes or for other purposes desired. Aluminum is suitable for tube **109b**.

A pan **115** of stainless steel or other suitable material is located at the bottom of reflector **104** for collecting any heavy water that may leak from tank **101**, inasmuch as heavy water, at present, is relatively expensive. The collected heavy water is drained as indicated by the arrow.

Helium at substantially atmospheric pressure is introduced through pipes **116** into the top of reactor tank **101**, that is, above the level of the heavy water, and thence is circulated to the exterior of the tank **101** through pipes **117**. As the result of high neutron densities and heat developed during the operation of the chain reaction in tank **101** some of the deuterium oxide will decompose into D_2 and O_2. These uncombined gases will collect at the top of tank **101**. In order to remove such uncombined gases, recombine and recondense them without explosive effect, a gas circulating system of any well-known type (not shown) may be used for circulating the helium together with the uncombined gases into a recombiner of any well-known type (not shown) such as a hot grid or platinum-charcoal catalyst, or both, for effecting recombination of the D_2 and O_2 into heavy water, and for returning the condensed D_2O into tank **101**. Helium may be circulated, for example, at the rate of 3 cubic feet per minute. A suitable ratio of D_2 to helium may be 1 to 150 although other dilutions may be used instead.

Space **109** also carries D_2O inlet pipe **101a** by which D_2O can be supplied to the top of reactor tank **101**. Outlet pipe **101b** is positioned at the bottom of tank **101**. If desired, the D_2O can be circulated through external heat exchangers through pipes **101a** and **101b** to cool the moderator and thereby cool rods **102**.

Immediately below coverplate **108** there is provided a cadmium sheet **118** to act as a shield for minimizing the escape of slow neutrons. Immediately below cadmium sheet **118** there is provided an aluminum sheet **119** which is useful primarily to prevent electrolytic action between otherwise dissimilar metals (cadmium and aluminum) in the interior of tank **101**.

Shield **109a** is supported by two pairs of crossed I beams **120** and **121**. Both pairs of I beams have their extremities supported by the concrete shield **105**.

Referring to Fig. 3.22 numeral **102** denotes one of the composite uranium containing rods that is suspended in the deuterium oxide moderator as shown in Fig. 3.21. Rod **102** comprises a cylindrical rod of uranium or uranium containing material **122** 6 feet long that is screw-threaded into a supporting rod **123** of non-fissionable material such as, for example, aluminum. A thin tubing of aluminum **124** is drawn on to the outer surface of uranium rod **122** and supporting rod **123** by any well-known drawing

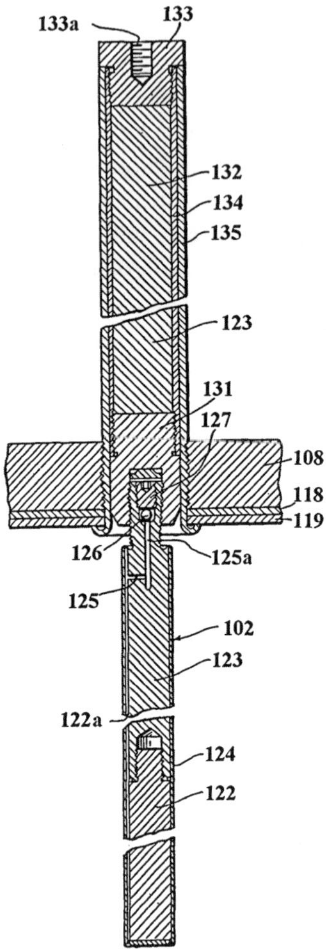

Fig. 3.22

process by screwing on an attachment **137** (see Fig. 3.23) that is subsequently removed. Thereafter, the joints formed at the top of supporting member **123** and at the bottom of rod **122** are welded so as to form an airtight seal in tubing **124**, thus protecting the uranium rod **122** from the effects of the D_2O. A narrow longitudinal groove **122a** is provided between the connected elements **122–123** and tubing **124**. An axial groove **125** and

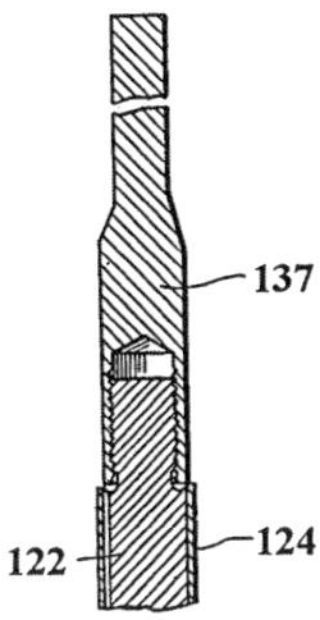

Fig. 3.23

communicating radial groove **125a** are provided at the top of supporting rod **123** that communicates with space **122a** thereby making it possible to evacuate the air space and thereby test for possible leaks in the top and bottom welded joints in tubing **124** by noting the pressure change interiorly of the air space after a predetermined evacuation.

A ball valve **126** is provided to seal the space **122a** from the atmosphere. A set screw **127** is screwed downwardly of supporting rod **123** so as to firmly seat the ball valve **126**. A shank **131** is provided and supporting member **123** are screw-threaded thereto. An upwardly extending aluminum sleeve **134** is screwed to shank **131** and closed by an upper flanged member **133**. A lead rod **132** is positioned between the lower shank **131** and upper flanged member **133** for the purpose of reducing gamma ray escape axially of rods **102**. The flanges of member **133** are supported on the top of a tube **135** screwed to cover plate **108**. Flanged member **133** is provided with an internal thread **133a** by which the rod assembly can be attached to a lifting crane, for example, for installation and removal.

Fig. 3.24 shows an alternate form of seal which may be used in place of the ball valve **126** shown in Fig. 3.22. The seal comprises a stud **126a**

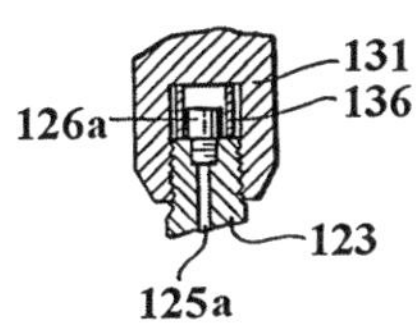

Fig. 3.24

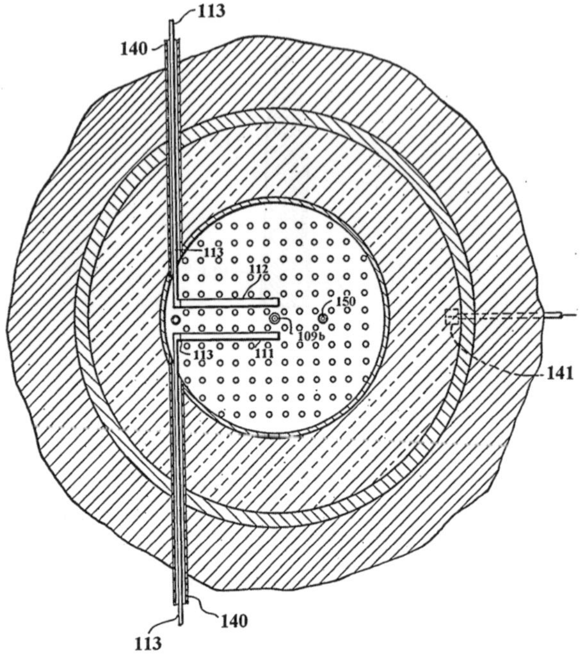

Fig. 3.25

screw-threaded into bore **125a** in supporting member **123**. A compression gasket or sealing ring **136** of lead, for example, is then provided between elements **123** and **131** so that as the latter-mentioned elements are screw-threaded together, gasket **136** is compressed thereby forming an airtight seal between supporting rod **123** and shank **131**.

The neutron chain reaction may be prevented merely by immersing into the heavy water one or more rods, such as hollow rods **111** and **112** (Fig. 3.21 and Fig. 3.25), 4 inches in diameter, containing a 1/16 inch layer of cadmium sandwiched between aluminum walls, and pivotally supported by the cover plate **108**. Rod **111**, for example, may be operated as a control or regulating rod, being immersed to greater or less extent in the body of heavy water, as desired. Rod **112** may be considered as a safety rod which normally is held out of contact with the body of heavy water in tank **101** and is immersed into the heavy water only for emergency purposes, that is, when control rod **111** *per se*, is insufficient for immediate stoppage of the chain reaction. Rod **111** may be pivotally mounted and rigidly secured to a shaft **113**, extending through the tank **101** through sealed bearings **140**. A method of automatic control is to make rotation of shaft **113** responsive to the neutron density at a peripheral portion of reflector **104**, for example, as indicated by ionization chambers such as chamber **141** having suitable amplifiers (not shown) so as to be effective to control the neutron density of the reactor and keep it substantially constant. Such means of automatic control, however, forms no part of the present invention when manual control is used the response of ionization chamber **141** is used to monitor reactor power when operating.

In utilizing the output of the reactor, well **109b** plays an important role. It extends through the center of the reactor where the highest neutron density exists, and intense neutron bombardment of materials inserted into this well will take place, even at relatively low reactor powers.

From the above description it will be seen that U-D_2O reactors are, in general, smaller than U-graphite reactors. This means that both the neutron leakage and the central neutron density are higher than in U-graphite reactors for a given total power output. Such a reactor as just described has been operated continuously at 250 kilowatts when filled to higher levels and properly shimmed by shim rod **150** to compensate for operational poisoning, later discussed.

3.6 Neutronic Reactors with Other Moderators

While we have illustrated our invention as including moderators of graphite and D_2O, other moderators can also be used. Beryllium, for example, either as metal or oxide, can also be used. When Be or BeO is used, the constructional details can be substantially as described, for example, for the uranium-graphite reactor, using the moderator in the form of bricks.

The following tables set forth constants for representative beryllium-uranium reactors, as presently known.

With an efficient reflector, critical amounts of Be and U can be reduced a few percent.

Sphere and rod geometry as shown therein can be used with light water to give K factors around unity even with natural uranium. For example, a K constant of slightly over 1 has been obtained by the use of uranium rods of 1.5 centimeters diameter placed parallel in light water with a volume ratio of water to U metal of 1.65. Diphenyl can also be used as a moderator

Table 3.5

[Be metal density 1.85 g/cm^3]

	U Sphere	U Rod	U Slab
Radius of uranium bodies	5.0 cm	3.5 cm	1.5 cm (thickness)
Critical cylinder	168 x 309.1 cm	165.7 x 304.9 cm	179 x 343.8 cm
Amount Be	51.5 tons	48.9 tons	63.7 tons
Amount U	43.9 tons	47.3 tons	69.2 tons
K constant	1.0968	1.0982	1.0842

Table 3.6

[Be oxide density 2 g/cm^3]

	U Sphere	U Rod
Radius of uranium bodies	3.0 cm	1.5 cm
Critical cylinder	194.2 x 358 cm	199.3 x 368 cm
Amount BeO	134 tons	145 tons
Amount U (tons)	40.4	35.8
K	1.0670	1.0628

and closely resembles light water giving a gain of from 0.2 to 0.4 percent in K. With either, a slight enrichment of the uranium with one of the fissionable isotopes such as, for example, U^{233}, U^{235}, 94^{239}, will provide a K sufficiently greater than unity, to enable the construction of operating reactors.

However, the water or diphenyl lattice can also be used as part of the reactor, with for example a seed, or portion having a higher K, in the center of the reactor so that the average K will be sufficiently above unity to provide a reactor of practical size. A heavy water lattice, for example, can be made to provide the higher K factor for the center of the composite device and the average K, and hence the critical size computed as set forth elsewhere herein. Water lattices ar also useful as reflectors around other reactors, and as neutron reproduction takes place therein, they are very efficient.

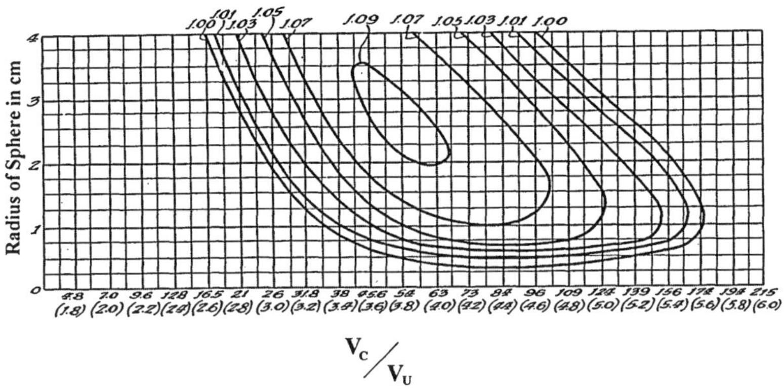

Fig. 3.26

3.7 Reduction of Losses Due to Resonance Capture

Having described several specific embodiments of operative reactors, limit curves for theoretically pure natural uranium metal spheres and rods and oxide spheres and rods will next be described as shown in Figs. 3.26, 3.27, 3.28, 3.29 and 3.30, respectively, when used in various moderators.

The shapes and extents of the curves are based on the fact that K is proportional to the product of three factors: p, f and ϵ where p is the probability of a fast fission neutron escaping resonance capture and becoming a thermal neutron; f is the fraction of the thermal neutrons absorbed by uranium (both by simple capture and to produce fission) rather than by the carbon; and ϵ is the factor by which the number of neutrons is increased because of the additional neutrons due to fission produced by the fast fission neutrons before leaving the lump of uranium. Each of these factors may be computed separately by methods known to physicists, using experimentally determined constants appropriate to these phenomena. The proportionality factor required to obtain the values of K for these curves from the product of these three factors however has been determined from the measured values of K in certain actual measurements of lattice arrangements, such as, for example, the results obtained by the measurements of pile structures too small to support a self-sustaining chain reaction, as set forth herein and in the Fermi application referred to previously, together with results obtained by measurements made in operating reactors. The K values are therefore accurate within the limits of error of this type of

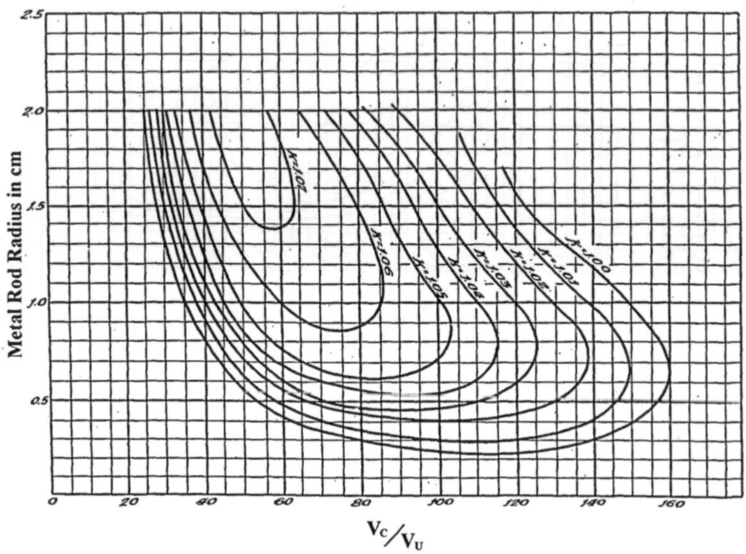

Fig. 3.27

measurements. However, if structures are built well within the limits of the curves shown, using materials giving favorable danger sums and the proper critical sizes, a self-sustaining chain reacting system will result.

Though K contour lines for natural uranium metal spheres, uranium oxide (UO_2) spheres, U metal and UO_2 cylindrical rods only have been shown for graphite moderators and U metal rods for a heavy water moderator, it is to be realized that similar curves can be made for other geometrical shapes of the uranium bodies and for other uranium compounds with proper allowance for the nature of the compound, the uranium content, bulk density and neutron absorption of the moderator.

The highest values for the reproduction factor K are obtained where both the neutron resonance absorption in the uranium and the neutron absorption in the moderator are about equal. As the uranium bodies are enlarged, for any given volume ratio, surface resonance absorption will decrease but moderator absorption will increase. The increase in moderator absorption in this case is due to the fact that the larger the uranium bodies are, for any given volume ratio, the greater will be the distance between

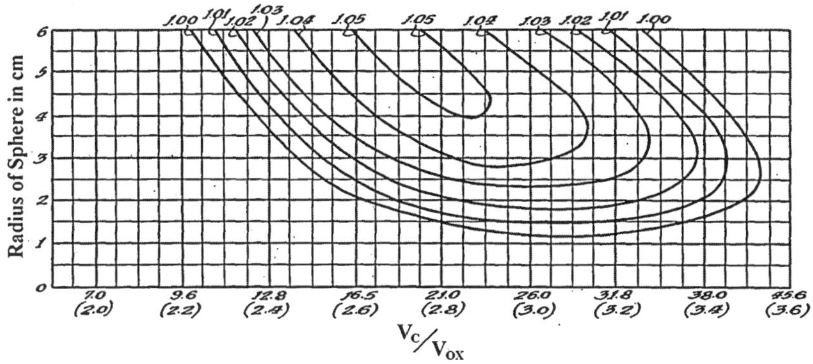

Fig. 3.28

the bodies. The neutrons reduced to thermal energies will then have to diffuse in the moderator over a longer path, thus increasing the probability of capture by the moderator before entering the uranium. Likewise, as the size of the uranium bodies is decreased, uranium resonance absorption increases, and in fact, increase faster than the decrease of moderator absorption. Consequently, with all other factors remaining constant, as the size of the uranium bodies and the ratio of the volume of graphite to uranium depart from the optimum, the reproduction constant K will decrease in value.

In Fig. 3.26 contour lines have been plotted for theoretically pure spherical uranium metal of density at least about 18 g/cm^3, imbedded in graphite. In Fig. 3.28 contour lines have been plotted for spherical lumps of UO_2 of density 6 g/cm^3 imbedded in graphite. In Figs. 3.27 and 3.29, K contour lines have been plotted for cylindrical rods of metallic uranium, and rods of uranium oxide (UO_2) of density 6 g/cm^3, respectively, extending through the reactor. Along the ordinates of the graphs are plotted the radii of the spheres or rods. Along the abscissae of Figs. 3.26 and 3.27 are plotted volume ratios of graphite to uranium. Beneath the volume ratio numerals are numerals in the parentheses representing the linear divisions of the graphs in terms of the cube roots of the ratios of the total volume of the graphite and the uranium to the volume of the uranium. These parenthetical values represent the ratio of the radius of the unit cell to that of the uranium body, both the unit cell (carbon and uranium) and the uranium being considered

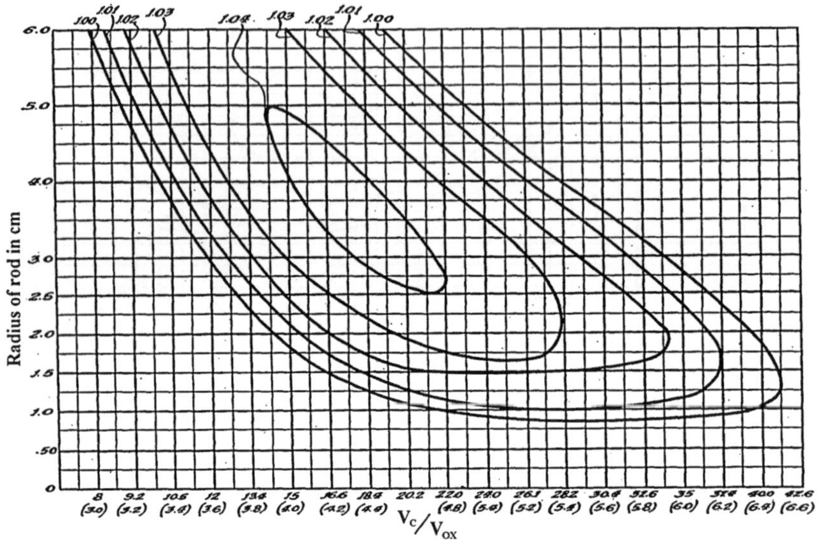

Fig. 3.29

as spheres. The values not in parentheses give the volume ratio of carbon to uranium for the cell and for the structure, as the lattice is a mere repetition of cells.

In Figs. 3.27 and 3.29 the unit cell is expressed in cylindrical terms, and along the abscissae are plotted volume ratios. In Fig. 3.29 the parenthetical numerals represent the ratio of the radius of the unit cell (carbon and uranium) and the uranium, both being considered as cylinders.

Referring first to Fig. 3.26, it can been seen that if the radii of the metallic uranium spheres are less than about 0.3 centimeter, the value of the reproduction constant K is less than unity for all volume ratios, so that for sphere sizes of natural uranium metal much less than this value it would be impossible to build a self-sustaining chain reacting system irrespective of the overall size of the structure. For spheres of greater size than 0.3 centimeter, it is possible to obtain values for the reproduction factor K greater than unity, providing the ratio between graphite volume and the volume of the uranium is within certain limits as shown on the graph. From a consideration of Fig. 3.27 it will be apparent that the limiting radius may be somewhat smaller where the uranium metal aggregates are

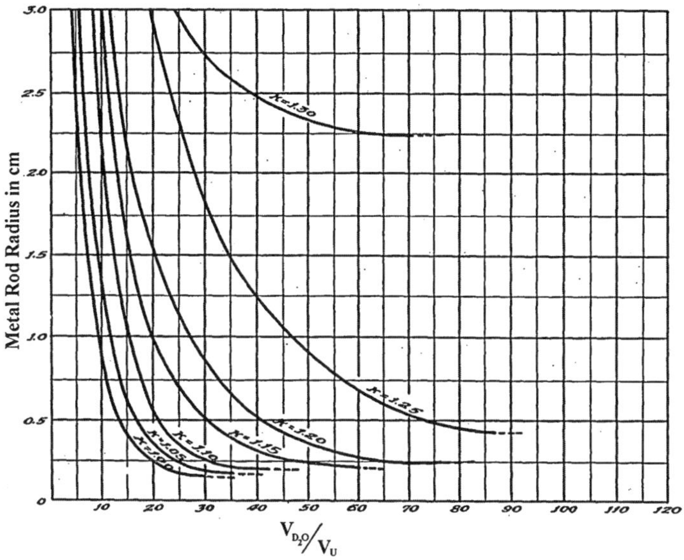

Fig. 3.30

rod-like in shape and as shown, a K of unity may be secured when the radius of the rods is as low as about 0.25 centimeter. The innermost closed contour shown in Fig. 3.26 represents a value for the reproduction constant K of about 1.09. At approximately the center of this contour the highest possible reproduction factor K of about 1.10 would be found. This highest value is for optimum conditions with theoretically pure spherical uranium lumps of about 2.75 centimeters in radius, and a volume ratio of about 54 carbon to 1 uranium. Similarly in Fig. 3.28 it will be seen that if the radii of the uranium oxide spheres are less than about 1.2 centimeters no chain reaction will take place with any volume ratio or size. For oxide spheres of greater than 1.2 centimeters radius, K factors greater than unity are possible within wide limits of volume ratio, with the optimum K of about 1.06 at the center of the innermost contour line obtained by using oxide spheres of about 5.75 centimeters radius and a volume ratio of 18.7 carbon to 1 uranium.

As shown in Fig. 3.29, the minimum oxide rod radius for K greater than unity is about 0.75 centimeter. The optimum K factor of over 1.04 is obtained within the innermost contour at around 3.75 centimeters radius, and a volume ratio of about 17.5 carbon to 1 uranium. It will thus be seen that while rod geometry gives somewhat smaller values of K than sphere geometry, the neutron saving due to aggregation is still large enough to provide a self-sustaining chain reaction in a system of practical size even when uranium oxide is used.

While the optimum conditions are found in a system of lumped uranium and graphite when uranium metal spheres are used, it will be obvious that in many cases other shapes of uranium bodies, such as rods will be preferable. Foe example, if 94^{239} and various fission products are to be recovered from the uranium after operation, any large numbers of spheres or short cylinders are not easily removed without tearing down the reactor. The metal or oxide bodies from the removable stringers described as extending through the reactor can be removed, but such procedure will only remove a small portion of the total material. Rods, or rods made up of short slugs in end to end relation, however, are easily removed, leaving the moderator undisturbed, as will be shown later.

Rod geometry also assumes practical importance when reactors of large power output are built, as rods lend themselves readily to incorporation in fluid heat absorbing systems, as will be taken up in conjunction with methods of cooling the reactors.

K curves for uranium metal rods in a D_2O moderator have also been made and are shown in Fig. 3.30, where the ordinates are given as rod radii and the abscissae as volume ratios of D_2O to U. It will be noted that the curves resemble the graphite curves, except that K constants are higher and only the lower volume ratios are shown. These are the volume ratios where the amount of element 94 produced will be the greatest and where the relative amount of D_2O will be the smallest, as D_2O is presently more expensive than uranium. The curves clearly indicate that optimum K constants of about 1.3 can be obtained with rods of about 2.25 to 2.5 centimeters radius immersed in D_2O at volume ratios of from 40 to 80 D_2O to 1 uranium. The ability to obtain such high K factors by properly aggregating the uranium in D_2O, together with the favorable ratio of scattering cross section to absorption cross section, and with M^2 being considerably less than that of carbon, leads to an operating reactor considerably smaller than can be obtained with graphite or beryllium. In addition the range of volume ratios at which K factors will be greater than unity is very wide, and the upper limits of the curves are not presently well-known. However, for optimum and near optimum geometries at the lower volume ratios the curves shown are sufficiently accurate for use in designing practical D_2O reactors, such as that described herein in detail.

It can be observed from Figs. 3.26, 3.27, 3.28, 3.29 and 3.30 that for a given size of the uranium lumps, either spheres or rods, the value of K will diminish from the maximum, representing optimum conditions, as the volume ratio either increases or decreases. The same effect is produced by increasing or decreasing the size of the uranium bodies from that representing optimum conditions. It is evident, therefore, that for every value of K there is a range of uranium body sizes and volume ratios which can be selected as desired. From the standpoint of economics, the more costly material, when it is uranium, can be saved by selecting the uranium body size and volume ratio at the most extreme righthand position on the contour line representing the K factor required for the system. This saving in proportion of uranium may counterbalance in cost, the increased overall size required.

On the other hand, if one desires to obtain the greatest possible yield of 94^{239} or to reduce the amount of moderator when D_2O is used, rather than reduce the initial cost of the uranium, one should select a point near the lefthand extreme of the contour in question, and furthermore should choose the lowest value of K consistent with purity of available materials and limitations on the overall size of the structure. The geometry of the system desired, therefore, can be selected in accordance with the desired balance of economics and engineering factors.

The curves shown in Figs. 3.26, 3.27, 3.28 and 3.29 have only been carried upwardly to 4 centimeters spheres of metal, 6 centimeters spheres of oxide, 2 centimeters for U metal rods, and 6 centimeters radius for oxide rods. It is only in the regions shown that economical structures can be built with natural uranium because the proportion of uranium required for the same reproduction factor becomes inordinately large as the size of the lumps increases, and cooling difficulties are introduced by using large uranium bodies. However, even using much larger bodies of uranium than those set forth in the curves, the chain reaction will take place with K greater than unity within the areas enclosed by extrapolations of the curves as shown, in systems above critical size. We do not, therefore, desire to be limited to the preferred region shown on the curves where only conditions surrounding and including the optimum are set forth.

It is obvious, from the description given above, that aggregation of the uranium also makes it possible to increase K values obtained when natural uranium is enriched by addition thereof of fissionable material such as U^{233},

U^{235}, or 94^{239}. Enrichment and the use of more efficient slowing media provide a greater number of neutrons per cycle, and thus increase K over the values obtained for given geometries and volume ratios using natural uranium and graphite. The increased K constants provided by aggregation in these instances, permit the overall sizes for the systems to be reduced. The shape of the curves shown in Figs. 3.26, 3.27, 3.28, 3.29 and 3.30 will not be greatly changed by the use of enriched uranium, but will merely take on a higher value of K in accordance with the neutrons gained for the chain by the enrichment. The $K = 1$ curve will then be outside of the curves shown, indicating that enrichment, for example, widens the limits of the volume ratios within which the chain reaction can be sustained, but does not eliminate the need for aggregation of the uranium if maximum K constants are to be obtained.

The curves just described take into account resonance and moderator losses only. To find a true K constant for presently available materials, impurity losses must be taken into account.

3.8 Reduction of Neutron Losses Due to Impurities in the Material

Uranium and its compounds can be produced in a condition substantially free from neutron absorbing impurities within the requirements of neutronic reactor operation with various moderators. A composition that is substantially free from neutron absorbing impurities may be said to have a high neutronic purity. It should be noted that neutronic purity has no necessary connection whatever with chemical purity, i.e., a composition having high neutronic purity is one which is substantially free from particular elements having a high danger sum. Such a composition may be far from chemically pure in that it may contain many foreign elements having low danger sums. For example, substantial amounts of oxygen, fluorine, carbon, or beryllium and many others, all of which have low danger coefficients may be present as impurities, and yet the composition still may have high neutronic purity.

In a well-known process of producing uranium oxide from pitch-blende ore, the last step may be a hydrochloric acid leach. This ordinarily produces an almost chemically pure (better than 99.5 percent uranium oxide) but not a neutronically pure uranium composition, as many elements having exceedingly high neutron capture cross sections may still be present in amounts on the order of parts per million. The raw material that is treated as herein described is material which is neither chemically or neutronically pure, but is ordinarily the end product of a uranium recovery process from uranium ore.

Uranium compositions of high neutronic purity, i.e., those having a danger sum in K units of less than 0.3 and preferably less than 0.01, are suited for use in neutronic reactors. Such compositions can be produced by various purification procedures.

One illustrative procedure involves the steps of forming an ether solution of uranyl nitrate, washing the impurities from the solution with small volumes of water and thereafter recovering the purified uranyl nitrate from which uranium metal or other uranium compositions suitable for use in the neutronic reactor may be prepared.

Neutronically impure uranium oxide may be reacted with nitric acid to obtain a solution containing uranyl nitrate with soluble and insoluble foreign matter. This composition is filtered to obtain a solution of uranyl nitrate and soluble impurities. The solution is heated to boiling to convert the uranyl nitrate to uranyl nitrate hexahydrate, and the water evaporated to obtain a composition consisting essentially of uranyl nitrate hexahydrate and impurities. This composition is then treated with ether to form a solution thereof and the ether solution so obtained is extracted or washed with relatively small quantities of water. In such an extraction procedure the impurities having high neutron absorptive capacity or neutron capture cross section are more readily dissolved in the water than in the ether. Although the uranium compound is also more soluble in water than in ether, the loss of uranium is kept very low by two expedients, namely, by using a quantity of water to extract the impurities small in relation to the amount of ether solution, as for example, from one-half to 5 percent water by volume; and by using as extraction portions water already saturated with uranyl nitrate. The term water extraction as used broadly herein includes the water solutions of uranyl nitrate.

As a result of extraction of the ether solution by successive water portions, a new composition can be produced from the remaining uranyl nitrate having extremely high neutronic purity, although not necessarily of high chemical purity. The remaining purified uranyl nitrate can be recovered in two ways, by evaporation of the ether or by extraction of the uranyl nitrate from the ether solution by pure or substantially pure water which is substantially free from impurities having a high danger sum. The purified uranyl nitrate may then be converted to other uranium compositions suitable for use in a neutronic reactor.

In the large scale production of uranium compositions, it has been found that neutronically pure uranium compositions can be prepared from neutronically impure uranium oxide, for example, by conversion of the oxide to uranyl nitrate hexahydrate, dissolving the uranyl nitrate once in ether, and making several water extractions of the ether solution to remove impurities and then making a water extraction of the ether solution to remove the bulk of the purified uranyl nitrate, or by obtaining the uranium by evaporation of the ether. In such production, generally only one ether solution is required while the number of water extractions may be varied to suit the amount of purification required by the impure oxide. The final water extraction is of substantially pure water to remove the bulk of the uranyl nitrate from the ether. The uranyl nitrate may then be converted to U_3O_8, UO_2, the tetra or hexafluoride, or to the metal or carbide.

As a preferred method applied for large scale operations, the purification may consist essentially of dissolving the uranium oxide in nitric acid, filtering to remove the insoluble residue, and evaporating the solution in order to crystallize the uranyl nitrate hexahydrate. These crystals are then dissolved in ether to form a saturated solution and the impurities extracted therefrom by permitting several batches of small amounts of saturated aqueous solutions of uranyl nitrate to settle through the ether solution. Following this purification sufficient water of high purity, preferably distilled water, to dissolve the bulk of the purified uranyl nitrate out of the ether solution is added, so as to extract the grater part of the purified uranyl nitrate from the ether. The amount of water required for this purpose is large in comparison to the relatively small portions used for washing and frequently the volume of water exceeds or at least is equal to the volume of ether solution. The ether is then ready for the next charge of uranyl nitrate hexahydrate. The aqueous solution of purified uranyl nitrate is evaporated and the nitrate calcined to the oxide, the nitrous oxide fumes evolved being recovered by appropriate recovery means. In this manner it is not necessary to evaporate the ether to obtain the purified uranyl nitrate.

The efficiency of the ether solution and water extraction process has been shown in a practical manner by applying the process to the treatment of crude uranium oxide derived from ores from three widely geographical sources, namely, Canada, Colorado, and the Belgian Congo. These oxides originally obviously contain widely different impurities. After being subjected to the ether solution process the resultant products are neutronically indistinguishable from each other, and are all of such high neutronic purity so that they can be used in a self-sustaining chain neutron reacting system.

The oxide produced by calcining the purified uranyl nitrate according to the process described above is ordinarily UO_3. This may be reduced to UO_2 by heating in a furnace in an atmosphere of hydrogen.

The UO_2 so produced is sufficiently pure neutronically to be used directly in a self-sustaining neutron reacting system in spite of its high oxygen content. However, the use of such oxide has certain disadvantages in that the effective uranium density is below that of metal. For example, when neutronically pure UO_2 is used in a neutron reacting system the critical size is larger than for metal, necessitating the use of larger quantities of uranium and moderator than would be necessary in a neutronic reactor utilizing uranium metal.

It is therefore advantageous to be able to convert the uranium oxide resulting from the purification procedure of the present invention, to massive metal in order that the size of the reactor be reduced, and that the reactor be capable of operating at higher temperatures, as UO_2 forms U_3O_8 when heated, and U_3O_8 is not as efficient as UO_2. Uranium carbide, uranium tetrafluoride and uranium hexafluoride will also support a chain reaction when combined with the proper moderator and with appropriate consideration for the changes in bulk density.

It has been found that a convenient method of obtaining neutronically pure metal is by treating the neutronically pure uranium oxide (UO_2) with fluorine to convert the oxide to uranium tetrafluoride, without introducing additional impurities other than fluorine. This uranium tetrafluride, a solid, may then be mixed with finely divided magnesium and placed in a calcium oxide-lined iron bomb where it is subjected to heat. The magnesium reduces the uranium tetrafluoride to uranium metal during an exothermic reaction. This metal collects at the bottom of the bomb, and acquires only neutronically negligible amounts of the magnesium, calcium and oxygen and iron necessarily present during the reaction. Massive billets of neutronically pure uranium can be obtained weighting from 10 to 200 pounds, and these billets can be recast as desired.

In practice it has been found best to recast the massive uranium metal obtained from magnesium reduction, in graphite crucibles in the absence of air. By this recasting a still more complete separation of the uranium metal from any acquired volatile impurities is obtained, as the volatiles boil off during the recasting process. The uranium is thus produced in massive form suitable for use in a chain reacting system, and can be machined or otherwise worked into rods, tubes or other forms that may be desired. By this procedure whatever small quantities of impurities are permanently introduced into the bodies during conversion to metal, are materials having relatively low neutron capture cross sections and accordingly are only those causing a minimum of difficulty in the neutron reacting system.

In order to determine the efficiency of the purifying process and to determine whether a sufficient number of water extractions of the ether solution have been made, it is desirable to have a practical means of determining the neutronic purity of the resulting composition. The method of chemical analysis for various neutron absorbing elements remaining in the product other than the H, N, and O of the solution, in combination with calculation of the K reduction by the use of the danger coefficients described above for the elements found, can be used, but has practical limitations which make it inconvenient and tedious, as uranium is extremely active chemically and contains many elements in various forms.

The exponential pile described above is one of the best ways to test uranium compositions in terms of purity. Using the same geometry in the exponential pile and the same moderator, or another moderator the characteristics of which are known, the uranium compositions can be substituted one for another and the efficiency thereof determined directly in terms of K. Such a test is particularly valuable after a uranium composition has been determined by simpler procedures to be suitable for use in a neutronic reactor. This test also gives the effect of decreased uranium density when uranium compositions are used.

A somewhat simpler test for neutronic purity is a so-called shotgun test. In this test a thin neutron detector, e.g., a piece of indium foil placed near a neutron source inside a block of paraffin, is made radioactive by an amount proportional to the density of thermal neutrons absorbed thereby. This radioactivity can be measured. A neutron absorbing pellet, such as a standard amount of boron, placed close to the detector foil, decreases the thermal neutron density in the neighborhood of the foil and lowers the induced radioactivity of the detector foil. By replacing the standard boron absorbing pellet, with a corresponding pellet containing the impurities removed from a known amount of uranium composition to be tested by a super-extraction procedure, and again measuring the radioactivity of the neutron detector, a direct comparison is obtained between the absorption caused by the unknown composition and the standard boron absorber. From this comparison the danger sum of the impurities in the uranium composition can be calculated in terms of boron equivalent. From the danger coefficient of boron the K reduction can be calculated.

In order to use the shotgun test so as to determine the efficiency of the ether purification process, a practical procedure is to take a representative sample of material having, for example, 10 kilograms of uranium content after normal ether-water purification. The sample is submitted to an additional extremely exhaustive ether solution purification on a laboratory scale. In this way, practically all of the impurities left in the uranium composition to be tested can be removed and incorporated into the pellet to be tested.

The results of the shotgun test are usually reported as percent absorption which is equal to

$$\frac{\text{Absorption of impurities in pellet expressed in equivalent milligrams of boron}}{\text{Absorption of 10 kilograms of uranium expressed in equivalent milligrams of boron}}. \tag{3.14}$$

The absorption of 10 kilograms of uranium in terms of boron can be readily calculated from the danger coefficients given above as equal to 4.560 milligrams of boron. Thus, by measuring the absorption by impurities in the pellet, and expressing the results in terms of equivalent boron absorption, a close approximation of the decrease in K for the impurities found can be computed.

Therefore, as a close approximation, absorption ratio from shotgun test = change in K.

Following are exhaustive analyses of residual impurities in metallic uranium produced and tested by the above outlined procedure, starting with UO_2 neutronically purified by the above-described process, for impurities having higher absorption cross sections.

This results shows that when proper precautions are taken to avoid contamination of neutronically pure uranium oxide by introduction of neutron absorbing materials during conversion to metal, the neutronic advantages of the original ether solution purification process are carried over into the uranium metal. As a result of obtaining such high neutronic purity of uranium metal, combined with the reduction of resonance losses by aggregation of the uranium and by using suitable moderators, it has been possible to construct neutronic chain reacting systems of practical size.

From the above discussion it will be seen that the extreme importance of certain "dangerous" impurities in uranium compositions has been recognized and controlled by proper purification of raw materials and care to avoid contamination during manufacture. These precautions have led to final products of an unusual degree of overall purity.

Table 3.7

Impurity	Average Amount Parts per million	Danger Product (K units)
Ag	< 1	< 0.000023
B	0.5	0.001075
Ca	Trace	0.000050
Cd	0.86	0.000877
Cl	< 20	< 0.000640
Co	21	0.000370
Cu	3.5	0.000909
Fe	50	0.000080
H	50	0.000600
Mg	< 50	< 0.000025
Mn	< 200	< 0.001340
Na	< 50	< 0.000050
N	30	0.000120
Ni	0.8	0.000003
Si	55	0.000040
Total	...	0.003-0.0053 K units

However, on a weight basis, for example, roughly ten times as much moderator as uranium is used in neutronic reactors such as the uranium-graphite reactor (with reflector), for example. Consequently, the relative effect of the impurities is increased by approximately this weighting factor. The following table gives the result of an analysis of one batch of graphite for the most important impurities therein, the danger coefficients being weighted by the factor of 10.

Table 3.8

Material	Dangerous Impurity	Average Amt. p.p.m.	Weighted Danger Coeff.	loss in K units
Graphite	B	0.55	21,500	0.0118
	V	64	50	0.0032
	Ti	19	47	0.0009
	Fe	45	16	0.0007
	Ca	207	5	0.0010
Total	...	...	...	0.0176

Graphite for use in neutronic reactors is produced by impregnating calcined petroleum coke with pitch and then graphitizing under heat. Several pitch impregnations may be made to increase density of the graphite. However, as petroleum coke and pitch are not readily subject to chemical purification, it is important that such raw materials be chosen with great care as to their impurity content, particularly as to boron and vanadium. The impurity reduction in K for graphite can be limited to from 0.01 to 0.015 by careful selection of the raw materials entering into the manufacture of graphite for use in neutronic reactors.

Impurities in D_2O are easier to control. D_2O is produced in quantity with a purity of about 99.8 percent, the main impurity being light water. This small amount of light water does not affect the efficiency of moderation and with some small changes in geometry of the lattice, 95 percent D_2O, 5 percent H_2O can be used satisfactorily. However, the main effort in manufacturing D_2O is in the early states of concentration, and no significant saving in cost can be made by stopping at 95 percent purity with respect to H_2O. Other impurities in D_2O are small to start with and are relatively unimportant. They come mostly from containers in which the D_2O is handled, and with care such contamination is minimized. Contamination during use due to corrosion of tank walls, uranium rod sheaths, etc., can be removed by distillation of the D_2O if it is found that such impurities are impairing the operation of the reactor. Be and BeO are susceptible to chemical purification.

In any moderator, neutron bombardment during use in a reactor tends to purify the moderator. For example, boron, an element having a high capture cross section is converted to lithium by neutron capture and subsequent alpha emission, and lithium has a much smaller capture cross section than boron. In D_2O, light water contamination is reduced by conversion of the light water to heavy water by neutron capture. Thus moderators improve with use, if not re-contaminated.

The introduction of losses due to neutron absorbing materials formed in the uranium during operation at high neutron density will be later discussed.

3.9 Effect of a Cooling System in a Neutronic Reactor

Reactors conductively cooled by dissipation of the heat of the reaction through the exterior of the structures can only be operated at low powers continuously or at high powers for short periods, for otherwise heat will accumulate in the reactor. In some instances continuous operation at high power outputs is desirable. Under these circumstances, a coolant can be circulated through the reactor provided proper precautions are taken to maintain the proper neutronic conditions therein.

The sources of heat generated in a uranium-graphite reactor, for example, can be summarized as follows, all figures being approximate:

1. Summary by type

Table 3.9

	MeV/fission	Percent
Gamma Radiation	23	11
Beta Radiation	11	6
Kinetic Energy of Fission fragments	159	79
Kinetic Energy of neutrons	7	4
	200	100

2. Summary by locale where heat is generated

Table 3.10

	MeV/fission	Percent
In uranium	184	92
In Carbon	12	6
Outside Pile	4	2
	200	100

3. Summary by type and locale

Table 3.11

	MeV per fission	Percent in U	Percent in C	Percent Outside
Kinetic energy of fission fragments	159	100	....	....
Kinetic energy of neutrons	7	....	90	10
Gamma radiation from fission fragments	5	70	25	5
Beta radiation from fission fragments	11	100	....	....
Nuclear affinity of neutrons (gamma radiation)	18	60	25	15

In D_2O reactors, the operating sizes are smaller and exterior neutron losses are larger. Consequently, more heat is developed outside the reactor. At first glance, it might appear obvious to circulate a cooling fluid through a neutronic reactor to remove heat therefrom. However, it must be kept in

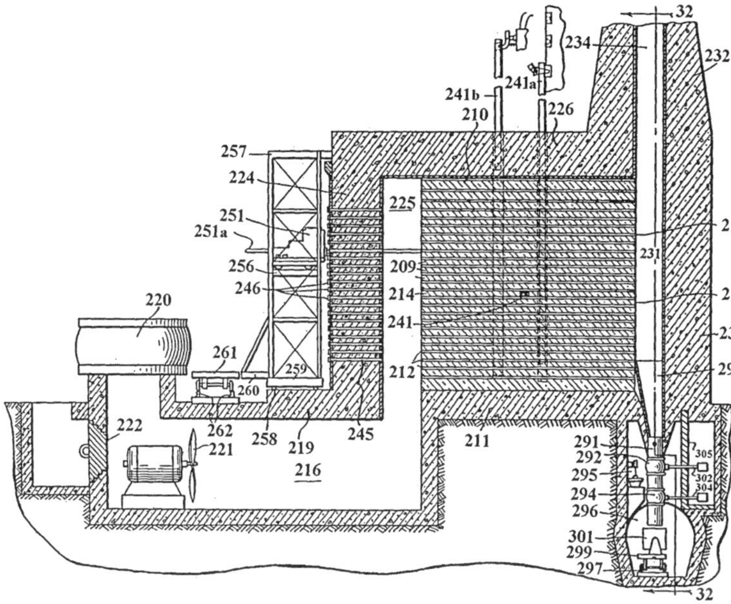

Fig. 3.31

mind that the coolant itself, with few exceptions, will be a neutron absorber, and that tubes or pipes, if used for coolant direction in the active portion of the reactor, will also be neutron absorbing.

The coolant and tubes, if used, can be passed through the reactor in heat exchange relation to the moderator, to the uranium bodies, or to both. A number of methods of cooling are available. One of the simplest is to run aluminum tubes through the moderator and then pass water through the tubes. However, most of the heat generated by the reaction is released in the uranium, and as most moderators are relatively poor heat conductors, cooling the moderator alone, for example, is limited to reactors operating around 1000 kW. For higher power reactors, cooling of the uranium directly may be desirable.

However, uranium is chemically very active, and it is usually desirable to protect the uranium itself form direct contact with the cooling medium. The uranium may need to be protected from chemical reaction with the coolant, and fission fragments from nuclear fissions originating on or near the surfaces of the uranium bodies should, in most cases, be kept from entering the coolant stream, as these fragments are highly radioactive. If the fission fragments should be allowed unrestrained entrance to the coolant, every part of the cooling system exteriorly of the reactor would have to be heavily shielded for protection of operating personnel, and circulating machinery and piping might become inaccessible for repair for long periods of time after reactor shut down.

Air has been used to cool a uranium-graphite reactor operating continuously up to 3000 kw with a construction as shown in Figs. 3.31 to 3.36 inclusive. Such a reactor will next be described.

3.10 An Illustrative Gas-Cooled Neutronic Reactor

One such structure broadly comprises a mass of graphite blocks **209** closely piled or stacked into a cube **210** shown in Figs. 3.31 and 3.32. This structure is more fully described and claimed in the application of Fermi and Szilard, Serial No. 596,465, filed May 29, 1945, copending.[5] This graphite cube may be, for example, 24 to 26 feet on a side and rest on a concrete foundation **211**. The graphite cube **210** is pierced with horizontal air channels **212** of square cross section, with one of the diagonals vertical, as shown in Fig. 3.36. The channels may be readily made by grooving adjacent blocks. The channels are 1.75 inches on a side and extend completely through the reactor, from an inlet face **214** to an outlet face **215**. About 2000 channels may be provided, and as will be later brought out, any unused channel can be plugged. Only a few of the channels are shown in the drawing for the sake of clarity.

Adjacent the inlet face **214** of the cube, the foundation is continued downwardly to form the floor of an inlet air duct **216** extending outwardly. The inlet air duct **216** is completed by concrete side walls and top **219**.

At some distance away from the graphite cube **210** the inlet duct is turned upwardly to terminate in an air filter **220**, relatively close to the surface of the ground. A fan or blower **221**, here illustrated as electrically drive, is installed on the floor of the inlet duct just below the air filter, access to the fan being conveniently obtained through duct door **222**, behind the fan.

The concrete top **219** of the inlet air duct is continued upwardly as inlet

[5]@ The authors refer to E. Fermi and L. Szilard, *Air cooled neutronic reactor*, U.S. Patent No. 2,836,554, May 27, 1958 (see page 409).

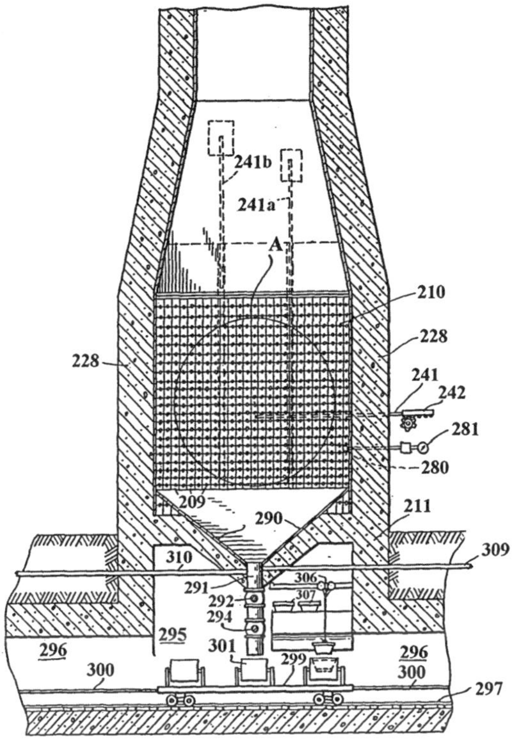

Fig. 3.32

shield **224** positioned parallel to but spaced away from inlet face **214** of the cube **210** to form an inlet chamber **225** communicating with the air channels **212**.

Above the inlet chamber **225** and the cube **210** the concrete is continued horizontally to form a top shield **226**, and side shields **228** are built up from the foundation **211** to enclose cube **210**. Shields **226** and **228** closely approach the top and side faces of the cube, to minimize air flow around the outside of the cube. A small amount of air circulation, however, may be desirable over the top and side faces to cool these faces.

At the outlet face **215**, an outlet end shield **230** of concrete is provided. End shield **230** is parallel to and spaced from the outlet face **215** of the graphite cube to form an outlet chamber **231** communicating above with the

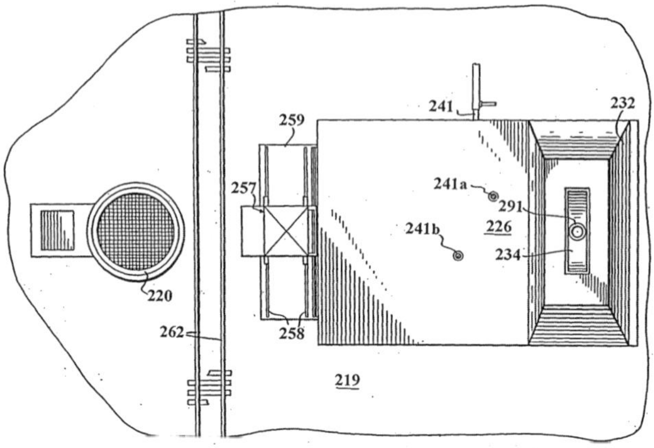

Fig. 3.33

base **232** of a stack **234**, projecting upwardly and formed as a continuation of the concrete top, side, and outlet end shields. Thus, the cube **210** is completely enclosed by concrete shields, with a duct system operating by virtue of pressure provided by fan **221** to conduct air from close to ground level through channels **212** into the stack and then into the atmosphere well above ground level at the top of the stack. The concrete shields may be from five to twenty feet thick in accordance with the maximum desired operating power of the reactor, and serve as shields to reduce escape of neutrons and gamma radiation.

As a neutronic reaction will take place when uranium bodies are prop-

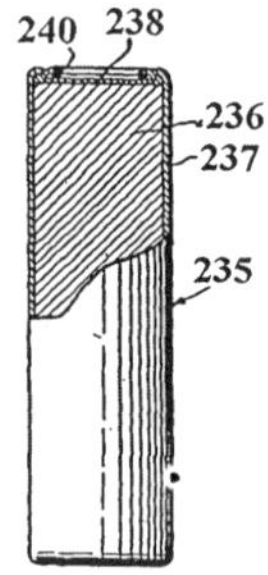

Fig. 3.34

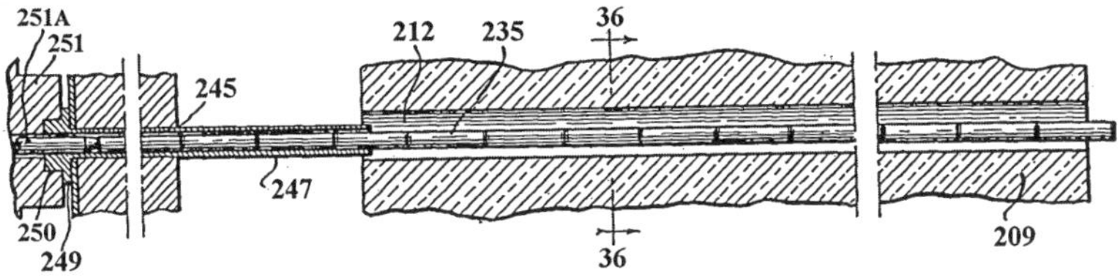

Fig. 3.35

erly spaced in a moderator mass of a certain finite size, the above-described device can be made chain reacting by placing uranium bodies in the horizontal channels in such a manner and in such an amount that a neutron reproduction ratio of slightly over unity is obtained, exclusive of all neutron losses within the reactor and from the exterior of the reactor. The neutron activity is checked during loading, as has been previously described for building other type reactors.

Using the graphite mass **210** as the moderator to slow fast neutrons to energies where they again are able to create fission in 92^{235}, the device as described will have a reproduction ratio of unity when approximately 700 of the channels **212** in the graphite cube are each loaded with 68 aluminum jacketed uranium slugs **235** lying end to end, with a channel spacing of 7 inches measured center to center, and with the loaded channels roughly defining a cylindrical active portion as indicated by line **A** in Fig. 3.32. Both graphite and uranium should be of highest possible purity. Presently obtainable materials have impurities therein reducing K by a total of about 0.015 to 0.02 unit from the base K for pure materials and specific geometry.

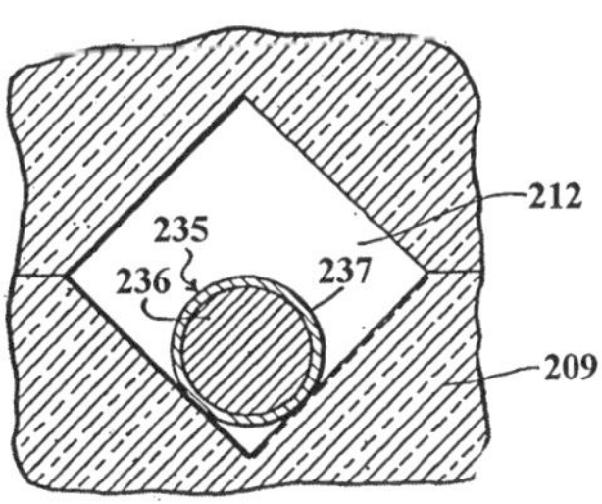

Fig. 3.36

However, more than a unity reproduction ratio is required, as when the reproduction ratio is exactly unity no rise in neutron density will occur as has been previously brought out. Under such conditions the device will not develop high neutron density or power in the form of heat. By loading additional channels, i.e., making the active portion greater than critical size, however, the reproduction ratio within the reactor can be brought above unity in order that a rise in density can occur. Then this excess neutron reproduction can be absorbed by neutron absorbing materials deliberately inserted into the reactor in order to hold the reproduction ratio at an average value of unity after a desired power output has been obtained, as a result of the initial rise in density, i.e., by the control rod.

Consequently, in accordance with the amount of excess reproduction ratio desired, about 1000 channels may be loaded with uranium slugs. Most of the channels not loaded with uranium may be closed by inserting plugs, preferably of graphite, in such channels in order to conserve air. Some of the channels, however, in the peripheral portions of the cube may be left open for cooling of the graphite in those portions.

One preferred form of slug construction for rod geometry is shown in Fig. 3.34. In this case, each uranium metal slug **235** is 1.1 inches in diameter and 4 inches long covered with an aluminum jacket approximately 20 mm thick in good heat conductive relation to the uranium. The slugs weight about $2\frac{1}{2}$ pounds each.

In forming the slugs **235**, the uranium portion **236** is machined to size, cleaned in trisodium phosphate and then washed in water. Aluminum or other non-fissionable metal jacket cans **237** are provided having an inside diameter some what larger than the uranium portion. This can **237** with the uranium inside is then passed through a sizing die of 1.134 inches diameter. This die, being of smaller diameter than the 1.1 inch uranium portion plus the two aluminum walls, draws the can in tight thermal contact with the uranium.

A cup-shaped cap **238** is then placed base down inside the projecting portion of the can **237** and is seam welded to the can. The projecting portion is then cut off above the seam weld **240** and the remaining projecting portion including the weld, spun over the adjacent end of the slug. Thus, each jacket completely encloses and seals the uranium preventing air from corroding the uranium and also reventing fission fragments created by nuclear fission at the surface of the uranium from entering the air stream.

The channels are loaded with uranium until the active portion is over critical size, for example, to a size where the reproduction ratio, with movable neutron absorbers removed, is about 1.005. This geometry provides, with a slug spacing of about seven inches, a volume ratio of about 47 C to 1 U, and the residual impurities, a K constant for the rod lattice of about 1.06. As the K constant of 1.06 is reduced to a reproduction ratio of 1.005 by the reduction in size of the reactor from infinity to the operating size, the reproduction ratio of 1.005 means that for every two hundred neutrons starting in each neutron generation about two hundred and one neutrons can be produced in the operating reactor over and above all losses. Under these conditions and taking onto account the fact that about one percent of the neutrons of fission are delayed in their emission for a mean time of about 5 seconds the neutron density of the reactor will double every 8 to 15 seconds. With some part of the movable neutron absorbers inserted but with the insertion of less than the amount of neutron absorbers required to make the reproduction ratio unity, the rise is slower. When the movable neutron absorbers are almost, but not entirely inserted to the critical position, a single doubling of the neutron density may take several hours. When a desired density has been reached, the reproduction ratio can be reduced to unity so that the desired density is continuously maintained by the neutron absorption in the inserted neutron absorbing material.

As in other reactors described, the neutron absorbing material may be introduced into the reactor in the form of a control rod **241** as shown diagrammatically in Fig. 3.32. This control rod extends into the graphite cube, sliding in a channel therein and is operated from outside of side shield **228** as by rack and pinion **242**. The rod is made from, or incorporates therein, an efficient neutron absorber such as cadmium or boron. Shim and safety rods **241a** and **241b** are provided.

During operation heat is released in the reactor in accordance with the neutron density therein as shown in Fig. 3.18. Most of the heat arises from the kinetic energy of the fission fragments and about 92 percent of the energy is released in the uranium. About 6 percent is released in the graphite due to neutron absorption and the slowing process therein and about 2 percent escapes from the reactor in the form of neutrons and gamma radiation. Consequently, the reactor can only be operated at a power dependent upon heat removal to the point where a stable temperature obtains. Otherwise, the reactor will accumulate heat to the point that the device may be damaged. This condition will be greatest in the center of the reactor as the heat generated is greatest there. Since aluminum melts at 658°C, stable temperatures below this value should be used although with jackets of other non-fissionable metals, such as beryllium, the stable temperature may be increased, although if the temperature should rise too high the uranium bodies might be damaged even when using beryllium jackets, as uranium of the type used in neutronic reactors melt at about 1100°C.

A stable temperature is obtained in the device by passing atmospheric air through the reactor, and in the specific example shown and described, the air is passed through the graphite channels and directly in contact with the aluminum jackets of the slugs, so that the center of the reactor is properly cooled. Under these circumstances the reactor can be operated continuously at 250 kilowatts electrical equivalent of heat by passing 32,000 cubic feet per minute through the reactor with a maximum temperature of the slugs of about 100°C, and at 500 kilowatts continuously with about 50,000 cubic feet per minute of air with a maximum metal temperature of 200°C. These heat maxima occur at the center of the reactor loaded as described, and total power output is determined by the maximum temperature permitted here. Higher temperature maxima may be used, but what these temperatures will be will depend upon the heat conductivity of the uranium, the jacket and the jacket-uranium interface, and the cooling efficiency. Such air cooled reactors have been operated continuously at 3000 kilowatts by increasing fan capacity.

To accomplish loading of the slugs **235** into the various air channels **212**, the concrete of the inlet end shield **224** is pierced with a plurality of loading apertures **245**, as shown in Figs. 3.31 and 3.35, each aperture being aligned with the axis of slug positions in the air channels **212**. Normally, during operation of the reactor, each aperture **245** is closed by a removable lead plug **246** extending through the shield **224** only.

When it is desired to load a channel with new slugs, the lead plug **246** for that channel alone is removed, and a charging tube **247** inserted, extending through the inlet and shield **224**, across the inlet chamber **225** and entering the corresponding air channel **212** as shown in Fig. 3.35. The outer end of charging tube **247** is provided with a flanged nipple **249** shaped to engage a nipple recess **250** of a plunger-operated loading mechanism indicated generally by numeral **251**. It will be noted that the charging tube is smaller than the air channel **12** and that air can pass through the channel being unloaded. The air should circulate during unloading, although it may be at reduced velocity. Slugs are forced into the air channels by plunger **251a** from loading mechanism **251**, which, being no part of the present invention, will not be described in detail.

The loading mechanism **251** is mounted on an elevator platform **256** mounted to be raised and lowered in an elevator frame **257** capable of moving along the outside of inlet end shield **224** on elevator tracks **258** (Fig. 3.33). Base **259** of the elevator frame is provided with a platform **260** projecting outwardly on the same level as the top of a supply car **261** travelling on supply car tracks **262**. Supply car **261** is used to bring a supply of slugs to the elevator for use in the loading mechanism **251**.

In the initial loading of the graphite cube **210** when uniform loading is performed, loading is started with the more central air channels until 68 slugs have been placed in the guide tube **247** and connected channel. The loading mechanism is then operated to push the slugs into the channel until the outer end of the first slug is at the outlet face **215**, leaving the outer end of the last slug about 16 inches from the inlet face **214**, leaving 16 inches of graphite for reflecting purposes.

Proceeding outwardly and preferably concentrically, additional channels are loaded, meanwhile checking the neutronic activity of the reactor as previously described. As the activity increases as the loading approaches a critical size, as previously explained, that is, the size where the reproduction ratio will be exactly unity, the approach to critical size can be predicted by extrapolation of observed neutron density values with respect to the volume of the cube loaded with uranium.

As the critical size is approached, the control rod **241** is inserted deeply into the reactor to prevent a self-sustaining chain reaction; and loading is continued until the desired maximum reproduction ratio of, for example, from 1.005 to 1.006 is attained. This ratio can be checked by removal of the control rod and measuring the time taken by the reactor to double its neutron density. From this period, the reproduction ratio can be mathematically computed.

When the desired number of channels are loaded the active core of the reactor may contain from 34 to 50 tons of uranium, and will be ready for operation. Graphite plugs for the unused air channels may be loaded in a manner similar to that described for the uranium slugs.

It will be noted that on 4 sides of the graphite cube excess graphite will be present. On the fifth side, i.e., at the inlet face, graphite will also extend 16 inches beyond the uranium. On the remaining side, i.e., the outlet face, no graphite extends beyond the uranium. Thus 5 sides of the active portion (the uranium bearing portion) are surrounded by graphite. This graphite constitutes the reflector and reduces the amount of uranium required to reach critical size. In the present instance the reduction in size is only slightly less than the optimum, as 5 sides of the active portion are surrounded with the reflecting layer.

After the reactor is loaded to give the desired shape and size of the active portion, the fan is started and the control rod is withdrawn until a rise in neutron density to a desired power output where a stable temperature in the reactor is attained. The control rod is then progressed into the reactor until a neutron balance is obtained with the reproduction ratio at unity, thus maintaining the chain reaction at the desired operating power. Small variations from the unity reproduction ratio will occur during operation, due to temperature variations of the cooling air, and to change in barometric pressure and to minor variations in air pressure delivered by the fan. However, such variations are compensated by slightly inward or outward corrective movements of the control rod, either by hand in response to indicated variations in neutron density, or automatically by direct linkage of the control rod to the output of the ionization chamber **280** and meter **281**. However, such automatic control is no part of the present invention.

During operation of the reactor the air passing through the reactor becomes radioactive due to the fact that it is subjected to intense neutron irradiation. For that reason the air passing through the operating reactor is not delivered to the atmosphere at ground level but is exhausted at a substantial distance above ground such as, for example, from the top of a 200 foot stack.

After operation of the reactor for a sufficient length of time for an amount of 94^{239} to be created sufficient for chemical separation, such as, for example, 100 days at 500 kilowatts, the reactor is shut down by inserting the control rod fully into the reactor. After about one-half hour's wait, during which all delayed neutron emission will have ceased and the more highly radioactive materials decayed sufficiently, the reactor may be unloaded.

The unloading may be accomplished in two ways, either by using the mechanism **251** to push the slugs out of the channels so that they fall by gravity out of the outlet face **215**, or by using the plunger to insert new slugs in the channels, each slug so inserted pushing an irradiated slug out of the outlet face **215**. In the first instance the graphite cube will be left empty after unloading. In the second instance the insertion of new slugs is continued until all or a predetermined part of the irradiated slugs are out of the reactor, having been replaced by fresh material. Thus, the reactor is left ready for the next run. Under ordinary circumstances the latter manner of unloading is preferred.

In either case, the slugs drop by gravity from the outlet face into outlet chamber **231**, falling on to two angularly disposed pad plates **290** positioned to intersect the falling slugs, in the bottom half of outlet chamber **231** as shown in Figs. 3.31 and 3.32. The two plates slant to a centrally disposed outlet pipe **291** extending downwardly through foundation **211** and provided with spaced valves **292** and **294**. The slugs fall by gravity into pipe **291** above valve **292**.

Outlet pipe **291** opens into a lower coffin chamber **295** that in turn connects with a tunnel **296** carrying car tracks **297** on which a coffin car **299** may be moved by means of cable **300**. Coffin car **299** supports a plurality of slug coffins **301** in position to be successively positioned beneath the lower opening of pipe **291**.

Valves **292** and **294** are operated by means of rods **302** and **304**, respectively, from behind a heavy lead shield **305**, as shown in Figs. 3.31 and 3.32. A crane **306** is used for placing coffin caps **307** on each coffin after it has been filled with irradiated slugs.

Before unloading is started, both valves **292** and **294** are closed, and the upper portion of pipe **291** is filled with water from water inlet pipe **309**. A proper water level is maintained above valve **92** by water outlet pipe **310**. The air circulation is maintained, although it may be reduced to about 25 percent of the operating value. Slugs ar then pushed out of the reactor to fall on to pad plates **290** and then roll by gravity into the water in the upper part of outlet pipe **291**.

In order that there be no material damage to the jacketing of the slugs, plates **290** are preferably padded with a soft material that does not deteriorate under neutron irradiation, and that will be able to withstand the slug impacts. A satisfactory pad has been found to be $\frac{1}{4}$ inch cotton duck on felt laid on wood backed by steel. Combinations of various synthetic elastic materials have also been found satisfactory.

After a number of slugs have been collected above valve **292** sufficient to fill a coffin **301**, unloading is stopped and valve **292** is opened, permitting the slugs and the water around them to drop through the valve and remain in the space between valves **292** and **294**. Valve **292** is then closed, the water level re-established and unloading continued. In the meantime, valve **294** is opened permitting the slugs and water to fall into one of the coffins **301**. The car is then moved to register the opening of the next coffin with the end of the outlet pipe and the first coffin is capped. The procedure is continued until all of the irradiated slugs it is desired to remove are in coffins. These coffins may then be taken to a soaking pit (not shown) to remain until the radioactivity has decayed to a point where the slugs can be submitted to chemical treatment for removal of the products formed therein by irradiation. After 100 days' operation the aging period may be of about 30 days.

Removal of the irradiated slugs under the conditions specified is performed for two reasons. Firstly, the slugs are so highly radioactive that they cannot be safely approached by personnel without adequate shielding being interposed, and, secondly, for some time after removal from the reactor this radioactivity is so intense that self-absorption of the emitted radiation causes self-heating of the slugs. By unloading during maintenance of the air stream, by dropping the slugs at once into water, and by keeping the slugs in water until the radioactivity has subsided sufficiently, melting is prevented, as the slugs are cooled as they boil the water in which they are immersed. The slugs are then stored or aged under water until ready for chemical treatment, as for example, thirty days.

In this case, it will be clear that the losses added to the reactor are mostly those due to the absorption by the aluminum jackets on the slugs or rods, as the coolant can be passed directly through the moderator. There is also a small loss due to removal of moderator material to form the air channels. However, the K reduction for the air cooled or helium cooled system is not great, about 0.005 K, for example, in the system described.

However, when a liquid coolant is to be used, pipes usually are provided to prevent the coolant from entering the moderator and the coolant itself may have a high neutron absorption characteristic to be taken into account. A neutronic reactor employing a liquid coolant will next be described.

3.11 An Illustrative Liquid-Cooled Neutronic Reactor

For powers higher than 1–3000 kilowatts, for example, liquid coolants such as water or diphenyl can be used. Pipes for the coolant are used, with the jacketed uranium slugs or rods inserted in these pipes so that the coolant will flow around the jacketed slugs or rods.

One representative structure embodying liquid cooling for high power outputs, up to 100,000 kilowatts for example is shown in Figs. 3.37, 3.38 and 3.39 and will next be only briefly described, as in many respects it is similar in design to the gas cooled reactor described previously. Specific features of this reactor are more fully described, and claimed in the application of Wigner et al. Serial No. 622,630, filed October 16, 1945.[6]

[6]@ The authors refer probably to E.P. Wigner, L.A. Ohlinger, G.J. Young, and A.M. Weinberg, *Fluid moderated reactor*, U.S. Patent No. 2,810,689, October 22, 1957.

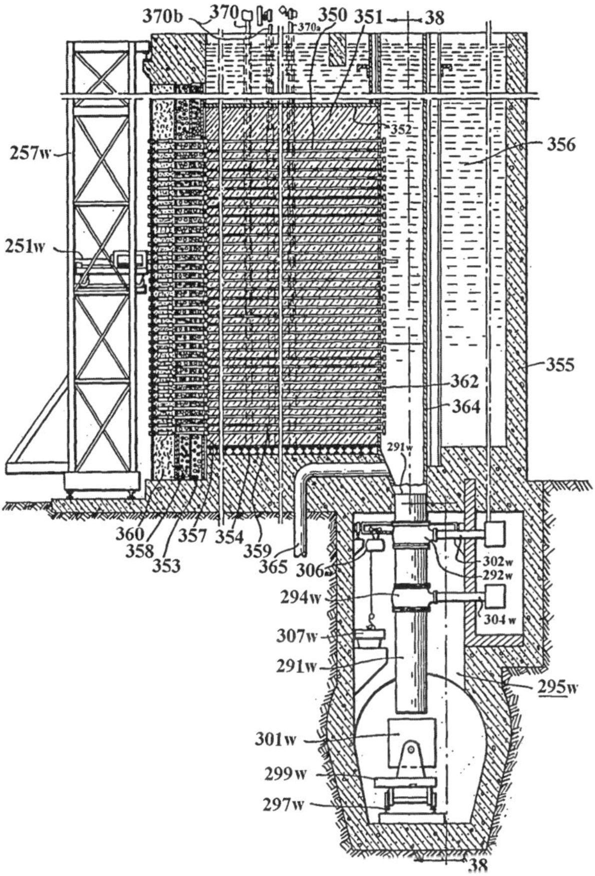

Fig. 3.37

The reactor proper **350** comprises a cylindrical structure built of graphite blocks as in the other graphite moderators described. The reactor is surrounded with a graphite reflector **351** forming an extension of the moderator and is enclosed by a fluid tight steel casing **352**, supported on I beams **354** within a concrete tank **355**, erected on foundation **353**. Tank **355** is preferably filled with water **356** to act as a shield for neutrons and gamma radiation.

The encased reactor is surrounded on all sides except one by the water **356**, and the side not surrounded, which is to be the charging face **357** of

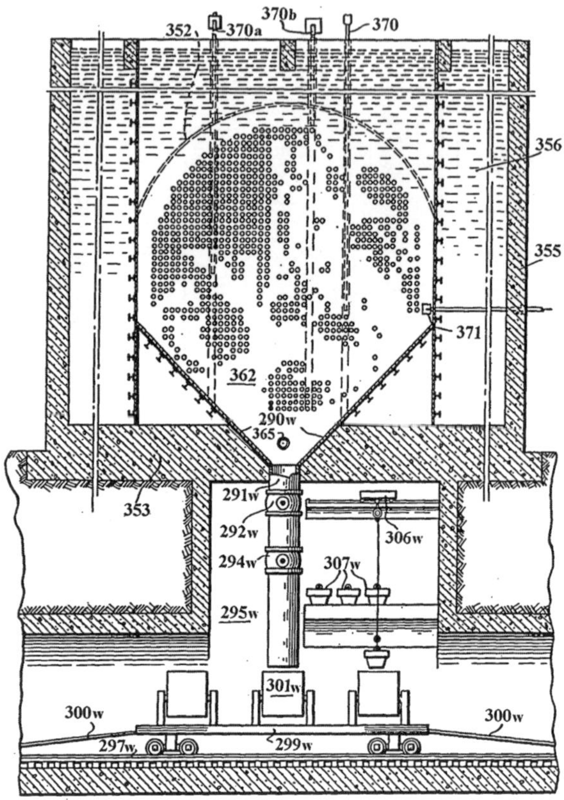

Fig. 3.38

the reactor is provided with a shield tank **358** filled, for example, with lead shot and water. Coolant tubes **359** extend through the adjacent concrete wall **360**, through shield tank **358**, through the graphite moderator block **350** to an outlet face **362** of casing **352** to empty into water **356** in tank **355**. Only a few tubes **359** are shown in Fig. 3.37 for sake of clarity of illustration. A backing wall **364** is placed in tank **355** spaced form outlet face **362**. Coolant tubes **359** are preferably of aluminum.

On the outside of tank **355** where the coolant tubes enter the reactor, the ends of coolant tubes **359** are removably capped, and are supplied with coolant under pressure from conveniently positioned manifolds. Thus

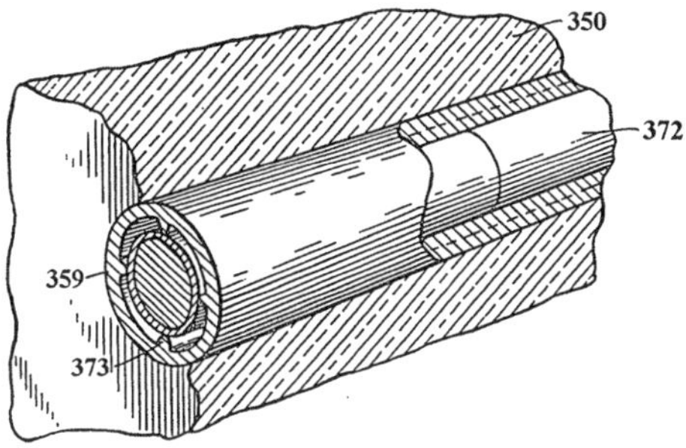

Fig. 3.39

water can be passed through tubes **359** to be discharged at outlet face **362** into tank **355**. Water, after having passed through the reactor is removed through outlet pipe **365**.

The coolant tubes **359** may then be charged with aluminum jacketed uranium slugs **372**, similar to those described in connection with the gas cooled reactor, by uncapping the tube to be loaded and pushing slugs into the tubes in end to end relationship. The reactor can then be loaded with sufficient uranium to make the reactor operative to produce high neutron densities, the heat being dissipated by the coolant circulation. This coolant may be water, for example, from a source such as a river, passed once through the reactor, and then discarded, or the water may be cooled and recirculated in a closed system. If diphenyl is used a closed system is required.

Loading and unloading are performed as in the air cooled reactor, and the same slug loading and receiving structures are used. As these have been described for the gas cooled reactor they will not be described again but the parts have been given the same numbers with a w subscript. Control is by a control rod **370** as in the other reactors described. Monitoring is by ionization chamber **371**. Shim and safety rods **370a** and **370b** are also provided, shown diagrammatically.

Referring to Fig. 3.39, which shows diagrammatically the relation of the moderator coolant pipe and rod, it will be seen that slugs **372** forming the rods are positioned in the coolant tubes **359** on projections **373** providing a uniform annulus of coolant around the slugs.

In this case, the jackets, the coolant itself and the pipes introduce parasitic losses which, for one specific example of a liquid cooled uranium-graphite reactor have been evaluated for a water cooled reactor capable of continuous operation at about 100,000 kilowatts.

For such a reactor employing uranium rods disposed in graphite in accordance with near optimum geometry conditions and utilizing uranium metal and graphite of presently obtainable purity, the value of K would be about 1.07. The value of K for the structure is determined as follows:

K for uranium rods in graphite (including residual impurities)		1.07
K reduction due to aluminum jackets and pipes	0.013	
K reduction due to coolant	0.023	
Total K reduction for cooling system	0.036	
The value of K for the structure		1.034

The principal dimensions of the reactor are as follows, using the K constant set forth above:

Axial length of active cylinder of reactor = 7 meters,
Radius of active cylinder of reactor = 4.94 meters,
Total weight of uranium metal in rods = 200 metric tons,
Weight of graphite in reactor = 850 metric tons,
Radius of uranium metal rods = 1.7 centimeters,
Thickness of aluminum jackets = 0.5 millimeters,
Thickness of aluminum pipe = 1.5 millimeters,
Thickness of liquid layer = 2.2 millimeters with water;
4 millimeters with diphenyl,
Number of rods in reactor = 1695,
Weight of aluminum in reactor = 8.7 metric tons,
Rod spacing in square array = 21.3 centimeters.

It will be noticed from the above values that the coolant annulus can be thicker with diphenyl than with water as diphenyl has a smaller danger sum for a given volume than water. This fact, together with the fact that diphenyl has a higher boiling temperature than water, makes the use of diphenyl attractive for higher powers. However, such advantages must be balanced against the requirements of a closed circulation for diphenyl, and the fact that some polymerization may take place in diphenyl, thus requiring make-up in the system to prevent the coolant from becoming too viscous for proper circulation. Each coolant has advantages for particular reactors. Liquid coolants, however, are ideal for reactors outputs up to 500,000 kilowatts.

As the total K-1 available for uranium-graphite reactors is only about 0.1 it is obvious that the amount of coolant cannot be greatly increased over the values given above, as the K constant would be so reduced as to preclude the construction of a reactor of practical size. However, when D_2O is used as a moderator, K-1 can be as high as about 0.3 when uranium rods are used. Consequently, a D_2O-uranium reactor can include a greater percentage of impurities than a uranium-graphite reactor. When D_2O is used as a moderator in a uranium rod system to be fluid cooled, the piping and jacketing procedure may be the same as for the uranium-graphite reactor, except that the coolant annulus can be increased in size. D_2O-uranium reactors can therefore be designed to operate at still higher powers, even when light water is used for cooling. However, it should be pointed out that D_2O itself can be used as a coolant, thus reducing the parasitic absorption, at least as far as the coolant alone is concerned.

It is thus apparent that by considering the coolant and the circulating elements required to be placed inside the reactor, as parasitic impurities, and then evaluating these impurities in terms of reduction in the K factor as taught herein, and then using the resultant K factor to determine the critical and operating sizes of a proposed reactor, that cooled reactors capable of operating at various desired powers can readily be designed.

3.12 Use of Different Lattices in the Same Neutronic Reactor

It will be noted that in the uranium-graphite reactor first described herein, there were two different lattice zones incorporated in the complete reactor, and that the critical size of the reactor was computed for the average K ($\overline{K}$) of the device. In this case the difference in the zones was the use of uranium in two different forms – spacing remaining substantially the same. However, reactors can also be built where there are zones different in K even though these zones are composed of wholly different moderators. For example, when a liquid or solid moderator lattice has, for some reason a low

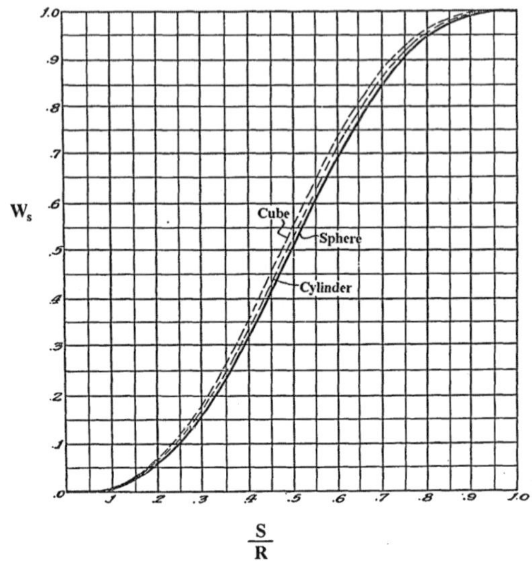

Fig. 3.40

K factor, a D_2O moderated center portion might be used to bring up the $\overline{K}$ and thus reduce the composite reactor to a practical size. Such a reactor has been disclosed in the Anderson and Brown application, filed February 2, 1944, Serial No. 520,730.[7] Furthermore a uranium H_2O lattice can be used in a reactor when a uranium D_2O lattice is used as a central zone, to give an overall K sufficiently greater than unity to enable an operating reactor to be built of practical size.

When reactors are constructed of concentric layers of materials, the average K can be calculated.

Curves are shown in Fig. 3.40 and they are down in terms of statistical weight w_s for a particular sub-side (or sub-radius) of a zone having a specific lattice therein plotted against

$$\frac{S}{R}$$

where R is the side (or radius) of the entire composite active portion of the reactor, and S is the extent of the radius (or side) of the zone incorporating a specific lattice.

[7]@ That is H.L. Anderson and H.S. Brown, *Nuclear fission chain reacting system*, U.S. Patent No. 2,990,354, June 27, 1961.

Statistical weight (w_s) may be defined as the value of a given mass of lattice weighted in accordance with its position in the reactor. Any given mass of lattice is worth more at the center of the reactor than at the edges because of the difference in neutron densities at the two positions. In general, it can be said that the effectiveness of a lattice varies in accordance with the square of the average neutron density to which it is exposed.

As the neutron density across the reactor varies, w_s represents the weight of a zone evaluated in accordance with the square of the average neutron density across that portion of the radius R through which the zone extends when the center is evaluated at unity. The curves enable the calculation of the overall or average K for a reactor consisting of concentric layers of lattices having different K factors. The overall shapes considered are (1) the cube, (2) the sphere and (3) the cylinder, with the uranium contained in the form of rods.

In using the curves, assume a cylindrical active portion of total radius R. Then if a lattice with a reproduction factor K_1 and migration length M_1 is disposed as a central cylinder of radius S_1, a second lattice of different K, say K_2, and migration length M_2, is disposed around the central cylinder up to a cylinder of radius S_2, and a third lattice of reproduction factor K_3 and migration length M_3 is disposed still further outside in a concentric layer up to the edge of the cylinder where $S_3 = R$; then the average

$$\overline{\frac{K-1}{M^2}}$$

is given by

$$\overline{\frac{K-1}{M^2}} = w\left(\frac{S_1}{R}\right)\left(\frac{K_1-1}{M_1^2}\right) + \left[w\left(\frac{S_2}{R}\right) - w\left(\frac{S_1}{R}\right)\right]\left(\frac{K_2-1}{M_2^2}\right) + \left[w\left(\frac{S_3}{R}\right) - w\left(\frac{S_2}{R}\right)\right]\left(\frac{K_3-1}{M_3^2}\right) \tag{3.15}$$

In graphite-uranium reactors of the type herein described, the migration lengths can be assumed to be the same for all the concentric zones; in that case $M^2 = M_1^2 = M_2^2 = M_3^2$ and Formula (3.15) gives the average value of $K - 1$ directly, in terms of the separate $K - 1$ for each medium. When different moderators are used the appropriate value of M^2 is inserted.

As a specific example of the use of the curves of Fig. 3.41 as applied, for example, to graphite-uranium reactors such as described herein, when the reactor is cylindrically loaded with rod geometry, if

$$\frac{S}{R} = \frac{1}{2}$$

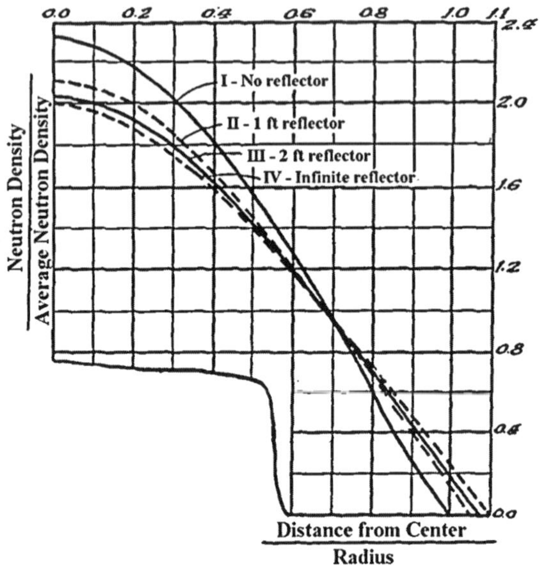

Fig. 3.41

as shown in Fig. 3.41 when $K_1 = 1.05$ and $K_2 = 1.06$ then

$$\overline{K-1} = w\left(\frac{1}{2}\right)(K_1 - 1) + \left[1 - w\left(\frac{1}{2}\right)\right](K_2 - 1). \tag{3.16}$$

Then with both migration lengths taken as being equal, the curves show

$$w\left(\frac{1}{2}\right) = 0.525 \qquad 1 - w\left(\frac{1}{2}\right) = 0.475,$$

hence

$$\overline{K} = 1.0548.$$

If $K_1 = 1.04$ and $K_2 = 1.06$, then under the same conditions $\overline{K} = 1.045$. When Formulas (3.1) and (3.2) are used for concentric cubical structures, the edge length of the cube can be used for R, with the edge lengths of the inner cubes as S_1, S_2, etc.

When the average K ($\overline{K}$) is found for the structure, this value can be used to determine what the critical size of the structure will be, as brought out in the section on critical and operating sizes.

In this manner structures can be built to proper operating sizes even though part of the structure has a K constant approaching unity. A center portion having a relatively high K constant can be used to raise the average K to a value permitting a smaller reactor than would be possible with the use only of the lattice having the lower K constant. Similarly, a central portion with a lower K constant can be used to flatten the neutron density curve across the reactor.

3.13 Critical and Operating Sizes of Neutronic Reactors

After all of the neutron losses that enter into the chain reaction have been evaluated for a specific lattice, with the exception of the loss by leakage from the exterior of the system, the size to which the system is to be built for proper operation has to be determined.

There are several ways by which critical and operating sizes are determined, and it is desirable that these sizes be found within a low margin of error so that auxiliary equipment such as shields, for example, can be constructed of proper size and not be too small to enclose the operating reactor.

One very satisfactory method of determining critical size particularly for low power reactors is to measure the value Δ in an exponential pile (referred to above) which is a structure similar in all respects but size to the structure contemplated, and then use this value to determine critical and operating size.

Δ may be evaluated by finding the relaxation distance b as heretofore set forth. In case the reactor is to be built in the form of a large spherical structure, the critical radius (R) is given by

$$R = \frac{\pi}{\sqrt{-\Delta}}. \tag{3.17}$$

In case the structure is to be a rectangular parallelepiped with sides a_1, a_2 and a_3 the critical size is given by the formula

$$-\Delta = \frac{\pi^2}{a_1^2} + \frac{\pi^2}{a_2^2} + \frac{\pi^2}{a_3^2}. \tag{3.18}$$

In case the structure is to be built up as a cylinder of height H and radius R, the critical values of these quantities may be computed from the formula

$$-\Delta = \frac{\pi^2}{H^2} + \frac{(2.405)^2}{R^2}. \tag{3.19}$$

Thus, critical size can be determined directly form the measured values of Δ without determination of a numerical value for K, for a low power chain reacting structure and with any moderator.

However, when M^2 is known, K can be determined, and it may be convenient for design purposes to refer all computations to K as a base factor, as has been done herein. Then, as

$$-\Delta = \frac{K-1}{M^2}, \tag{3.20}$$

the value of K is found when Δ and M^2 are known, as pointed out above the value of K is then used to determine critical and operating sizes for the reactor and this method is adaptable for reactors of any power.

For example, in the case of a spherical structure employing uranium bodies imbedded in graphite in the geometrics disclosed herein and without an external reflector the following formula gives the critical overall radius (R) in feet:

$$K - r = \frac{C}{R^2}, \tag{3.21}$$

when $r = 1$ by definition of critical size, and where C is a constant that varies slightly with geometry of the lattice and for normal uranium-graphite lattices may have a value close to 7.4.

For a rectangular parallelepiped structure rather than spherical, the critical size can be computed from the formula below when $r = 1$

$$K - r = C\left(\frac{1}{a^2} + \frac{1}{b^2} + \frac{1}{c^2}\right), \tag{3.22}$$

where a, b, and c are the lengths of the sides in feet.

The critical size for a cylindrical structure of uranium and graphite is given by the formula, irrespective of the shape of the uranium bodies, when $r = 1$

$$K - r = C\left(\frac{1}{H^2} + \frac{0.59}{R^2}\right), \tag{3.23}$$

where cylinder height is H feet and radius is R feet.

However, when critical size is attained, by definition, no rise in neutron density with time can be expected, as the reproduction ratio is exactly unity. It is therefore necessary to increase the size of the structure beyond the critical size, but not to the extent that the period for doubling of the neutron density is too short, as otherwise the neutron density might rise to values where the device would be damaged.

It was formerly believed the the neutron reproduction cycle would be completed in about 0.0015 seconds. Under these circumstances any size or effective size of reactor giving a neutron reproduction ratio greater than unity would have to be provided with some exceedingly quick acting control device operating within a fraction of the reproduction cycle time. However, we have found that by taking advantage of the fact that there are delayed neutrons in the cycle, we can operate and control neutronic reactors with reproduction ratios substantially exceeding unity, without encountering high and dangerous rates of neutron density rise.

We have found that there is a substantial time factor in the rise of neutron density after a reproduction ratio of unity has been exceeded due to the fact that a substantial portion, generally about one percent, of neutrons generated in a neutronic reactor are "delayed neutrons". These delayed fast neutrons may appear at any time up to several minutes after the fission has occurred. Half these neutrons are emitted within six seconds and 0.9 within 45 seconds. The mean time of delayed emission is about 5 seconds. The cycle shown in Fig. 3.1 is completed by 99 percent of the neutrons in about 0.0015 seconds, but if the reproduction ratio of the reactor is near unity, the extra 1 percent may make all the difference between an increase or a decrease in the activity. The fact that the last neutron in the cycle is held back as it were, imparts a slowness of response to the reactor that would not be present if the 100 neutrons were all emitted instantaneously.

For cases in which the reproduction ratio (r) differs from unity by (appreciably) less than 1 percent, the rise of neutron density, or more specifically the value N to which the number of neutrons has risen from an original value N_0, after a lapse of time of t seconds during and before which the pile has operated at a fixed value of r (N_0 being the number of neutrons at the beginning of t, i.e., after disappearance of transient effects due to any preceding change in r), is given by

$$N = N_0 \, e^{w\,t},$$

where

$$w = \frac{r-1}{\alpha - (r-1)} \cdot \frac{1}{T}.$$

In this formula α is the fraction of the neutrons that are delayed, i.e., $\alpha = 0.0067$, and T is the mean time of delayed emission of the delayed neutrons (5 seconds). The above formula is only approximate because it uses an average delay time.

As an example, if r is 1.001, and the system has settled down to a steady exponential rise in neutron density, then

$$w = \frac{0.001}{0.0067 - 0.001} \cdot \frac{1}{5} = \frac{1}{28.5},$$

that is $N/N_0 = 2.75$ in 28.5 seconds. Hence doubling of the neutron density occurs about every 20 seconds. The above formula thus indicates the rate of rise for relatively low values of r and shows how the reduction of the rate of the delayed neutron effect is particularly significant in the stated lower range of r values. Strictly speaking, the given equation holds only for the steady state, i.e., where r has been held constant for some time; an additional transient term must be included to obtain an accurate representation of the neutron density during the first few seconds after a sudden change of r.

If r were to be exactly 1.01, a more detailed theory shows that the neutron density would be more than triplet per second. However, if the reproduction ratio r is several percent greater than unity, so that the one percent delayed neutrons are unimportant compared with $r - 1$, the density increases at a much more rapid rate as given approximately by $r^{t/l}$ where l is 0.0015 seconds, the normal time to complete a cycle. If r were 1.02 and 1.03, the factor by which the neutron density would be multiplied per second would be 1100 and 700,000 respectively. If r were to be made 1.04, the neutron density would increase in 1.5 seconds by a factor of approximately 10^{17} over its original level.

It is thus apparent that the operating conditions must always be such that the neutron reproduction ratio does not materially exceed 1.01 as the rate of rise of neutron density could then be so fast as to be uncontrollable.

There are several ways by which such operating conditions can be met. One of the simplest is to make the actual operating size of the structure such that a reproduction ratio of 1.01 cannot be attained when all control absorbers are removed. A safe value for the maximum attainable reproduction ratio is about 1.005 at the temperature and power of operation and low power reactors can be built so that size alone limits the reproduction factor below 1.01. The size at which reproduction ratios greater than unity can be obtained, may be computed from modifications of the above formulae, when K is corrected by a poisoning factor due to operation, later to be discussed. This factor only becomes important when operating a reactor at relatively high neutron densities. For example, for active spherical structures the formula

$$K - 1.005 = \frac{C}{R^2} \tag{3.24}$$

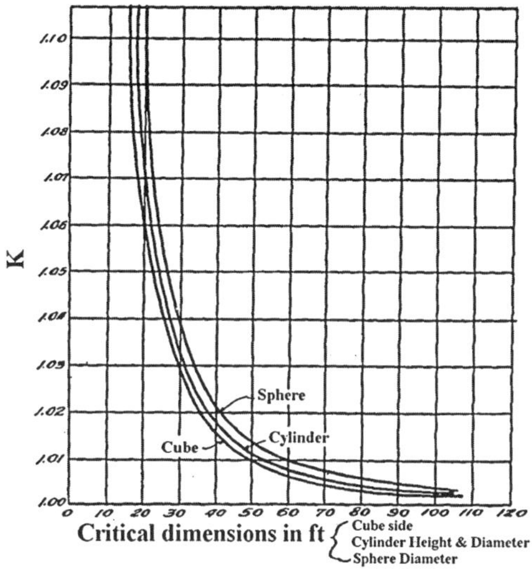

Fig. 3.42

may be used to find the operating radius R when K is known and r is 1.005. The same formula will, of course, give r for given structures for which K and R are known.

In the case of spherical structure employing uranium bodies of any shape or size imbedded in a heavy water (D_2O) moderator, the following formula gives the critical or operating overall radius, in accordance with the value of r.

$$R = \frac{56.5}{\sqrt{K - r}}, \tag{3.25}$$

where R is the radius in centimeters and r for critical size equals one.

For a parallelepiped structure rather than spherical, using a heavy water moderator, the critical and operating sizes can be computed from the formula:

$$K - r = 323\pi^2 \left(\frac{1}{a^2} + \frac{1}{b^2} + \frac{1}{c^2} \right), \tag{3.26}$$

where a, b, and c are the lengths of the sides in centimeters.

The critical size for a cylindrical structure using a heavy water moderator is given, irrespective of the shape of the uranium bodies by the formula:

$$K - r = \frac{323\pi^2}{H^2} + \frac{1868}{R^2}, \tag{3.27}$$

where H is the height in centimeters and R is the radius in centimeters, operating sizes are also determinable from these formulae by inserting the operating value of r desired.

The curves in Fig. 3.42 show approximate critical sizes for operative low power reactors or spherical, cubical and cylindrical shape utilizing graphite. When higher power is required, the sizes have to be revised upwardly to compensate for operational poisoning.

In Fig. 3.42 the vertical ordinates of the curves represent the different values of K for the critical size dimensions in feet plotted as the horizontal abscissae. The curves are labeled to show the values for the cube, the cylinder and the sphere. The cylinder, for each value, has its height equal to its diameter.

The following table 3.12 will show the approximate critical sizes for operative low power reactors utilizing D_2O, for different values of K where K is changed due to the change in geometry. In evaluating the values given in the curves and table it must be kept in mind that the critical size is that size where the chain reaction just will become self-sustaining ($r = 1$).

The use of a reflector, through the action of scattering neutrons back into the reactor, raises the density of thermal neutrons throughout the reactor and this, in turn, increases the reproduction ratio of the reactor. Thus critical size ($r = 1$) with a reflector is smaller than without a reflector. Consequently the use of a reflector permits a smaller reactor to be constructed for the same K factor.

If a reactor can be considered as being surrounded by a vacuum or by cadmium, the density of the neutrons on the outside surface of the reactor can be taken to be zero. This comes about because any neutron which passes the exterior surface of the reactor will either fly off into space, or be absorbed in the cadmium and consequently lost to the chain reaction

Table 3.12

U-metal–heavy water $K-1$	Sphere radius	Critical Sizes in Feet Cube side
0.01	18.5	32
0.02	12.3	21.5
0.05	7.7	13.4
0.1	5.45	9.45
0.2	3.84	6.65
0.3	3.14	5.45

within thc reactor.

If the reactor is immersed in air, the same assumption can be made, but it is only approximately true, because the air itself will scatter some of the neutrons which emerge back into the reactor. As another possibility, the reactor can be surrounded by some substance like graphite which will scatter most of the neutrons and absorb relatively few. In this case, neutrons emerging from the surface of the reactor will be scattered in all directions by the reflector and some of them will reenter the reactor, thereby raising the neutron density at the edge of the reactor.

The simplest discussion of the theory of a reactor reflector involves a reactor of thickness $2l$ in one direction and infinite in extent in the other two. This reduces the mathematical complexity to a problem involving a single variable. Further simplification may be obtained by assuming all the neutrons entering the reflector are thermal. Under these conditions the reactor equation may be written

$$\frac{d^2 n_i}{dx^2} + \frac{K-1}{M_i^2} n_i = 0, \tag{3.28}$$

where x is the distance from the center of the reactor measured in the direction of $2l$, n_i is the number of neutrons per cubic centimeters, K is the reproduction constant and M_i^2 is the migration area.

This equation must be solved, subject to the restriction that at the outer surfaces of the reactor, the neutron density n_i must be zero. The solution then becomes

$$n_i = A \cos \sqrt{\frac{K-1}{M_i^2}}\, x. \tag{3.29}$$

This in turn means that $2l$ must be adjusted so that

$$2l = \frac{M_i}{\sqrt{K-1}}, \tag{3.30}$$

in order for the reactor to maintain a chain reaction.

If, now, the reactor is surrounded by a reflector of thickness T with migration length M_e and neutron density n_e, we may derive a diffusion equation for the reflector. This equation will be the same as (3.28) but with K replaced by zero as there is no neutron multiplication in the reflector. Therefore

$$\frac{d^2 n_e}{dx^2} - \frac{1}{M_e^2} n_e = 0. \tag{3.31}$$

In this case, the solution of (3.28) do not go to zero at the outer surfaces of the reactor (active portion) but this condition is replaced by the requirement that n_e should be zero at the outer surfaces of the reflector, i.e., a distance $T+l$ from the center of the reactor. At the boundary of the active region, however, the solutions from (3.28) and (3.29) must give the same values of neutron density close to the boundary, and furthermore, the neutrons flowing across the boundary in one direction must be the same as those flowing in the opposite direction. Mathematically, this means that, if l' is the "new" thickness of the reactor with the reflector and τ_i and τ_e are the "mean free" paths inside the reactor and externally in the reflector respectively, then

$$n_e(l') = n_i(l')$$
$$\tau_e \frac{dn_e}{dx} = \tau_i \frac{dn_i}{dx} \text{ at } x = l'. \tag{3.32}$$

By carrying through the mathematical processes of solution the following expression will ultimately be obtained for determining l'

$$l' = \frac{M_i}{\sqrt{K-1}} \cot^{-1}\left[\sqrt{K-1}\frac{M_e\,\tau_i}{M_i\,\tau_e}\tanh\left(\frac{T}{M_e}\right)\right], \tag{3.33}$$

and thus gives a new thickness for the pile (if K is the same, etc.) which is less than that obtained formerly without a reflector.

While the discussion above has been concerned with a "sandwich" pile, the same general type of treatment may be applied to any shape as cylinders or spheres. The mathematical complexity increases rapidly, and in the case of an infinite cylinder, the final expression for l' is given by exactly the same expression in first approximation, and very little quantitative change in higher approximations.

So far it has been considered that all the neutrons escape as thermal neutrons and move with the same speed. However, in a reactor, when a thermal neutron is captured by a U^{235} nucleus, the resulting fission will generate neutrons which are moving with high speed. These fast neutrons will, in general, be slowed down in the moderator until they become thermal. However, some of these fast neutrons that are produced close to the outer surface of the reactor will be projected into the reflector. In the simplified picture just above, these neutrons were assumed absent; actually they were looked on as lost. Some of them, however, will be slowed down in the reflector material in exactly the same way as in the moderator, and may become thermal in the reflector. Some of them may then diffuse back into the reactor in the usual way.

A fast neutron will generally not go far before becoming slowed down to thermal velocities and consequently only those produced fairly close to the edge of the reactor will be slowed down in the moderator and ultimately scattered back into the reactor. For this reason the thermal neutron density near the edge of the reactor and in the reflector will be increased over what it would have been if the fast neutrons had leaked completely out of the reactor.

In Fig. 3.41, characteristic curves are shown relating to a graphite-uranium reactor, plotting the neutron density values along the radius for different points along the radius. The ordinates are

$$\frac{\text{Neutron density}}{\text{Average neutron density}},$$

and the abscissae are given as percent radius. The curves show the effect of the reflector in increasing the effective size of the reactor, a reflector of infinite thickness increasing the effective radius by about 10 percent, with a 2 ft. reflector increasing the radius about 8 percent. Accordingly, the critical radius of the active portion of the reflector may be reduced by approximately the same percentage when a reflector is used.

While the theories and allocations outlined above are only approximations, due to the fact that neutrons of all energies between that of fission and thermal energy are entering the reflector, the theories provide for sufficiently accurate design so that reactors can be, and have been built and operated, with reflectors and with proper operating characteristic as outlined herein.

A second means and method of preventing the maximum reproduction ratio from rising over about 1.01 in a neutronic reaction, during operation, is by the use of "shim" or limiting rods inserted in the reactor, when the actual size of the reactor is large enough to result in a reproduction ratio of over 1.01 under some operating conditions. The use of such rods to keep the reproduction ratio at a safe value will be described in the following section on control.

3.14 Control of the Neutronic Reaction by Variation of Neutron Losses in or from the Neutronic Reactor

Because of the fact that a reproduction ratio of unity in a neutronic reactor means by definition, that no rise in neutron density will take place, some means must be provided in a neutronic reactor whereby the neutron density can be allowed to rise to a desired value, and then act to stabilize the reaction at that value.

Again, by definition, a reproduction ratio greater than unity means that the neutron density will rise until stopped. In consequence, proper control requires the use of a means and method whereby the reproduction ratio can be controlled from unity (or less than unity) to the maximum reproduction ratio permitted.

Such control is readily feasible by controlling one or more of the losses in or from the reactor as built. Actual size can be changed, for example, especially when D_2O reactors are used, as the amount of D_2O wetting the uranium can readily be changed to provide a critical size where the reproduction ratio is unity, or a larger operating size where the reproduction ratio is greater than unity. Then when a desired neutron density has been obtained, some of the D_2O can be removed to again make the reproduction ratio unity thereby maintaining the reaction at the neutron density attained. This is called leakage factor control and is described in conjunction with the uranium D_2O reactor.

However, neutron losses due to neutron absorbers deliberately inserted into a reactor are of outstanding importance since these losses readily lend themselves to variation in any type of reactor. Cadmium and boron, for example, having high neutron absorption capabilities, are commonly used in the form of the control rods, that can be inserted in a reactor between the uranium bodies in greater or lesser amount, as set forth above in the description of the various reactors.

As pointed out in the section on critical and operating sizes of neutronic reactors, low power reactors are generally constructed of sufficient effective size to have a maximum reproduction ratio somewhat less than 1.01, which means that exterior leakage losses are not quite sufficient to reduce the reproduction ratio to unity. Under these conditions, interior neutron losses are introduced by the control rod, these rods absorbing sufficient neutrons to reduce the average reproduction ratio in the reactor to unity and below. The control rod is then made movable so that the reproduction ratio can be made well below unity when fully inserted, thus stopping the reaction; unity when partially inserted; and the maximum reproduction ratio permitted, when wholly removed from the reactor. By varying the depth of insertion of the control rod, any desired condition of the reactor can be obtained from complete shutdown to maximum rate of neutron density rise, with an intermediate position of the rod where the reproduction ratio of unity is maintained. This unity position of the rod is known as the critical position, and would always be the same except for variations in atmospheric pressure (when the reactor is open to the atmosphere): for the effect of temperature (as the neutron density rise heats the reactor), and for the effect of neutron absorbers formed in the uranium during operation, as will later be pointed out.

In a reactor of low power, using the rod type of control, the neutron density of the reactor is monitored, as by the ionization chamber and indicators previously described. This monitoring can take place in a representative portion of the system such as the outer periphery, or even exteriorly of the outer walls. From a knowledge of the neutron density distribution within the reactor (see Fig. 3.18) the neutron density in any part of the system is readily determined.

With the control rod fully inserted, the neutron density in the reactor may be about 100 times the normal natural neutron background of uranium, due to action of convergent chains in the lattice. These background neutrons provide the starting neutrons for the chains, which change to be divergent when the rod is retracted from the reactor to a point where the reproduction ratio is over unity. The neutron density will then increase, with a doubling time as determined by the effect of the delayed neutrons on the reproduction ratio, as pointed out in the preceding section on critical and operating sizes.

When a desired neutron density is reached, the rod is inserted to the position where the reproduction ratio is unity, and thereafter the attained neutron density is maintained. To reduce the neutron density the rod is inserted further into the reactor and the reaction decays. As it decays to a lower density, the rod can again be placed at the unity position to maintain the lower density. Thus the rod can be varied in position to obtain and maintain any desired neutron density with power production in the form of heat, as determined by the heat dissipation capability of the reactor.

The effect of the delayed neutrons is, of course, extremely valuable in obtaining ease and flexibility of control by the control rod. Movements of the rod only slightly outwardly from the critical position will permit a rise in neutron density for example, where minutes, hours, or even days can elapse for one doubling of the density, as desired. Such close control enables the control rod to be moved manually for example, although the rod is preferably driven by reversible electric motors, for example, under manual control of the motors by an operator watching the neutron density values.

However, with any control system there is always the possibility of accident in moving the control rod, as, for example, if motors driving the control rod should respond improperly and drive the control rod completely out of the reactor and for some reason (such as power line failure) refuse to return the rod. To take care of such an emergency, safety devices are often provided such as, for example, neutron absorbing safety rods, which are normally held completely out of the reactor while the reactor is operating, but which can be manually or automatically released at a predetermined neutron density to enter the reactor rapidly, to absorb enough neutrons to stop the reaction. Such rods are shown in Figs. 3.1, 3.21, 3.31, and 3.38, numerals **40**, **112**, **241b** and **370b** respectively. To prevent power supply failure from affecting operation of these safety rods, they are usually arranged to drop, or be pulled, by gravity into the reactor when released at a predetermined neutron density.

The use of such safety rods again brings out the importance of not permitting the reproduction ratio to greatly exceed 1.01 in an operating reactor at any time. With reactors operating at high power such as, for example, the water cooled reactor described herein, even a single doubling of the operating neutron density might be disastrous, and the safety rods must be tripped to enter into the reactor before such doubling takes place. At $r = 1.01$ the neutron density will double in about $\frac{1}{3}$ of a second. If the time of doubling became too short, the safety rods could not arrive in place before the neutron density reached undesired values. However, with minimum doubling times of several seconds, for example, the safety rods can readily operate to prevent abnormal neutron densities form being reached after the rods are tripped.

It has been pointed out above that K can change slightly because of temperature changes and because of changes in atmospheric pressure when reactors are open to the atmosphere. Such changes in K are usually small and therefore change critical and operating sizes very slightly. The resultant changes in reproduction ratio are correspondingly small and may be compensated for by relatively short movements of the control rod.

However, K may change for other reasons during operation, and in high power reactors may change by such a large amount that if not properly compensated, the reproduction ratio may fall below unity during operation. These changes in K are believed to be due to the formation of reaction by-products in the reactor, either radioactive or stable. As the production of both radioactive and stable reaction products is a function of rate of irradiation of the uranium, the effect of these by-products on neutron reproduction becomes most important in reactors operating at high neutron densities.

Several relatively long term changes in K may be expected. Isotope 94^{239} is known to produce, on the average, a fraction more neutrons per fission than U^{235}. Consequently, as U^{235} is used up by fission, and U^{238} is converted into 94^{239}, the K constant may be expected to rise. To offset this rise in K there is an accumulation of long lived or stable fission products in the uranium that are the ends of fission fragment decay chains.

In reactors operating at high neutron densities, however, radioactive elements of exceedingly high capture cross section may be formed relatively quickly in the uranium as an intermediate element in the decay chains of the fission fragments and this formation can change K during operation. One of the most important of these decay chains is believed to be the 135 fission chain starting with

$$\text{Te (short)} \rightarrow \text{I (6.6 hr.)} \rightarrow \text{Xe (9.4 hr.)} \rightarrow \text{Cs (20-30 yr.)} \rightarrow \text{barium}$$

the parenthetical times indicating half lives. The neutron absorption of tellurium, iodine, caesium and barium is relatively unimportant, but the neutron capture cross section of radioactive Xe^{135} has been measured to be about $2{,}500{,}000 \times 10^{-24}$ cm^2, many times larger than that of stable gadolinium for example, the cross section of which is about $30{,}000 \times 10^{-24}$ cm^2. Upon absorption of a neutron, Xe^{135} shifts to Xe^{136}, an element of relatively small capture cross section. The change in K corresponds in period, to the Xe^{135} appearance, and decay.

The rate of production of the Te is a function of the neutron density in which the uranium is immersed, and therefore dependent upon the power at which reactors of given type are operated. The radioactive Xe^{135} is produced with a noticeable effect on the reaction a few hours after the reaction is started and the effect is, of course, greater as the neutron density is increased and maintained. The Xe^{135} effect in high power reactors can be summarized as follows.

The reaction is started by withdrawing the control rod. The neutron density rises at a rate determined by the reproduction ratio and the effect of the delayed neutrons, until some predetermined neutron is attained. The control rod is then placed in the unity reproduction ratio position and the reaction is stabilized at the power desired. During this time radioactive iodine is formed, decaying to Xe^{135}. As more and more iodine decays, more and more Xe^{135} is formed, this Xe^{135} absorbing sufficient neutrons to reduce the reproduction ratio below unity. This absorption also converts the

Xe^{135} to Xe^{136} which has no excessive capture cross section. The neutron density drops. If no compensation were made for this drop by the rod the density might drop until background conditions prevailed, and then the reaction might automatically start up as the Xe^{135} decayed. Normally the neutron density drop is compensated for by removal of the control or equivalent rod to a new position where the reproduction ratio is again above unity. A neutron density rise occurs, bringing the density back to its former level. Again, more Xe^{135} is formed and the process is repeated until an equilibrium condition is reached where the Xe^{135} formed is transmuted by neutron absorption and by decay into isotopes of lower capture cross section, as fast as it is being formed. In the meantime, the control rod (or equivalent) has to be withdrawn by an amount thereby removing from the reactor neutron absorbers at least equal in effect to the absorption caused by the Xe^{135}.

It should also be pointed out that this Xe^{135} effect will be present when shifting from a power where the effect is stabilized, to a high power. The shift can be made and the reaction stabilized at the new power for a time, because the iodine formed from the new fissions has not had time to produce a significant amount of additional Xe^{135}. As the new Xe^{135} is formed from the decay of the newly formed iodine, the reproduction ratio will again drop and must again be increased by withdrawal of absorbers from the reactor.

However, the reduction in K due to the Xe^{135} equilibrium amount present when the neutron density is theoretically infinity in the reactor, is believed to be about 0.03 which means that to obtain a rise in neutron density to any desired density up to infinity, the rod would have to be eventually removed by an amount corresponding to an increase in the reproduction ratio of about 0.03, and somewhat less than 0.03 when finite densities are to be obtained, in accordance with the density desired the size of the reactor would have to be big enough to provide the increase. For example, in a continuously operated water cooled U-graphite reactor such as therein described, the reproduction ratio decreases due to Xe^{135} at equilibrium is about 0.0012 at 10,000 kilowatts; 0.009 at 100,000 kilowatts and would be about 0.013 and 0.020 at 200,000 kilowatts and 500,000 kilowatts respectively. As before stated reactors ordinarily are not built sufficiently large in size to provide maximum reproduction ratios of over 1.01 with all rods removed. However, if power outputs over 100,000 kilowatts are desired, the reactor must have its critical and operating sizes calculated as set forth herein using a final K constant decreased by the Xe^{135} factor for the power desired even though the amount of reduction is over 0.01. In other words a significant impurity will added during operation at high powers.

This may lead to the requirement for a reactor of such size that, if it did not acquire Xe^{135} during operation, could attain a maximum reproduction ratio of over 1.01 with all rods removed, and which, before a substantial amount of the iodine produced form fission decayed into Xe^{135}, could in consequence attain a dangerous neutron density if all rods were removed.

Such a reactor can be adequately safeguarded by the use of "shim" or limiting rods inserted in the reactor preferably to depths that will not permit a reproduction ratio of about 1.01 to be attained at any time during the operation of the reactor, even when the control rod is completely removed. Then, if a reproduction ratio of more than unity cannot be attained by outward movement of the control rod alone, due to the built-up of the Xe^{135} effect, the shim rod can be withdrawn to compensate for the Xe^{135} effect, but still be left in a position where the reproduction ratio cannot exceed 1.01, when the control rod is completely removed. Such shim rods are shown in Figs. 3.2, 3.21, 3.31 and 3.37, numerals **30**, **150**, **241a** and **370a** respectively.

It can thus be seen that compensation for the Xe^{135} effect is obtained, first by considering the Xe^{135} impurity factor for the power desired, as a reduction in K to determine a proper operating size, for a desired power, and second, by initially providing in the reactors, impurities that can be removed by amounts compensating for the Xe^{135} equilibrium amount acquired at a given power output.

Care must be taken, when shutting down a high power reactor operating with a Xe^{135} equilibrium, that sufficient neutron absorbers are inserted to prevent automatic start-up of the reaction after the Xe^{135} has decayed to the point where it does not materially affect the operation of the system. As a practical matter, shut down should include the full insertion of all control, shim and safety rods into the reactor.

Reactors operating at a few hundred watts, and operated intermittently, as when, for example, the reactors are shut down at night, are not significantly affected by Xe^{135} poisoning. Such a reactor is exemplified by the reactor first described herein. Even in the air cooled reactor described herein, operating at from 500 to a few thousand kilowatts output, the Xe^{135} effect is on the order of the temperature and pressure effects. However, in the D_2O reactor and in the water cooled reactor described herein, the effect is more pronounced and is compensated for as described.

Because of the fact that the Xe^{135} effect does not become important for several hours, reactors having an operating size too small to provide full compensation for the xenon effect at elevated neutron densities when continuously operated, nevertheless can be operated intermittently to attain such elevated densities or even higher densities for short periods until the Xe^{135} effect prevents further operation at those densities. As the Xe^{135} effect enters the reaction and stops the reaction, the density will drop. However, by waiting until the Xe^{135} decays to the point where the neutron reproduction ratio can again be made greater than unity, the reactor can again attain the desired neutron density level. Thus reactors with a small maximum reproduction ratio can be operated intermittently to attain for short periods, neutron densities far greater than could be continuously maintained.

However, the control rod can be calibrated in several ways for steady state conditions and the calibration is adequate. As the effect per inch movement of control rod is greater for the portion of the control rod nearer the center of the reactor than it is for the portion near the edge (because of the larger neutron density at the center), a unit may be chosen so that for movement of the control rod, one of said units will always have the same effect on the reproduction ratio of the reactor without regard to the actual depth of said rod in the reactor. Such a unit is based on the conventional inch, and is sometimes called a "cinch". Any movement of the control rod for the distance of one cinch has the same effect on the reproduction ratio of the reactor as a movement of the control rod one inch from the critical position.

The control rod can also be calibrated in terms of a unit known as the "inhour". One inhour is the distance the control rod must be moved from the critical or balanced position to give the reactor a period of one hour. The period of a neutronic reactor is, by definition, the time necessary for the neutron intensity to increase by a factor of "e" ($e = 2.718$).

In measuring the period of a reactor, a correction is made for any change in atmospheric pressure if the reactor is open to the atmosphere. An increase in atmospheric pressure will cause an increase in the weight of air inside the reactor. Oxygen has a small danger coefficient and, therefore, does not absorb neutrons in great quantities, but nitrogen on the other hand, has larger danger coefficient and so the great quantity of nitrogen present in the reactor has a distinct effect on the reproduction ratio r. It has been found that a change in atmospheric pressure on the reactor first described herein, is equal to 0.323 inhour for a change of one millimeter of mercury from the standard atmospheric pressure of 760 millimeters of mercury. At higher powers, a factor corresponding to the changes in operating density must be taken into account in calibration if measurements are made after the xenon effect appears.

3.15 Uses of Neutronic Reactors

In the description of the various neutronic reactors, given herein, only a little has been said regarding the uses of the reactors. As such uses are many, only a few of the most important will be mentioned here.

All the reactors described herein are primarily extremely powerful neutron and gamma ray sources. When used as neutron sources, materials to be made radioactive can be placed in or close to the periphery of the reactors, and radioactive isotopes produced, for example, in large quantities, as the materials are then exposed to the entire energy spectrum of the neutrons in or escaping from the reactor. As the leakage from the D_2O-uranium reactor is even larger than the neutron leakage from the graphite-uranium reactors, a relatively larger neutron flux can be intercepted on the exterior of this type of reactor.

One example of isotope production by exposure to reactor generated neutrons followed by transmutation, is the manufacture of U^{233} from the thorium 232, for which process thorium 233 can first be produced from thorium 232, the extent of the reaction being dependent upon the product of the slow neutron density and the time of exposure. Thorium 233 then decays to form protoactinium 233 and thence to uranium 92^{233} which is valuable as a fissionable material similar in its action to U^{235} and 94^{239}. The reaction is as follows:

$$ {}_{90}\mathrm{Th}^{232} + n \longrightarrow {}_{90}\mathrm{Th}^{233} + \text{gamma rays} $$

$$ {}_{90}\mathrm{Th}^{233} \xrightarrow[\beta^-]{\text{23.5 min.}} {}_{91}\mathrm{Pa}^{233} \xrightarrow[\beta^-]{\text{27.4 days}} {}_{92}\mathrm{U}^{233} $$

As a further example of transmutation, radioactive carbon may be produced by allowing the neutrons leaving the reactor to react with nitrogen in compound form. This reaction gives rise to carbon of mass 14 which is radioactive, and can be separated chemically from the nitrogenous compounds. Such radioactive carbon is suitable for medical and physiological uses as it may be incorporated in organic compounds and used as tracers in living organism. Well **21** can also be filled with graphite blocks and the resultant pier projected upwardly through the top. The top of this graphite pier provides a strong thermal neutron source and such a pier is known as a thermal neutron column.

It can thus be seen that the neutrons normally escaping from the reactors need not be lost, but can be put to work, and the neutrons escaping from the reactors can be utilized to produce transmutations or isotopes from elements placed in, or surrounding all parts of the active portion of the reactor. For example, the number of neutrons radiated from the external surface of a uranium-graphite reactor when operating at ten thousand kilowatts power output in the form of heat, is approximately 1.8×10^{16} neutrons per second.

In utilizing the output of reactors, internal shafts such as shaft **26** and tube **109b** play an important role. They extend to the vicinity of the center of the reactors where the highest neutron densities exist, and intense neutron bombardment of materials inserted into the bottom of these shafts will take place, particularly at high reactor powers. Furthermore, the shafts act to collimate the fast neutrons released inside of the reactors and a high density collimated beam of neutrons emerges through the external aperture, projected outwardly. Such a collimated beam, having a far greater fast neutron density than any neutron beam heretofore produced, can be utilized outside of the pile for nuclear research in all of its aspects. The number of neutrons escaping from these shafts is several times the number escaping over an area of the external surface of the reactors equal to the cross section of the shafts.

In addition, extremely high energy gamma rays are emitted during nuclear fission. These rays also escape through the shafts to the exterior of the reactors and can there be used for taking radiographs through large castings, for example, with relatively short exposures, during high power operation of the reactors. The neutrons coming from the reactors can be screened out of the gamma ray beam by the use of relatively thin sheets of materials having high neutron absorption cross sections without substantially reducing the gamma ray intensity. In addition, a bismuth filter has been found to effectively reduce the gamma rays, without substantial interference with the neutron beam. Thus shaft **26** and tube **109b** can be used either to produce a high intensity collimated neutron beam, or to produce a high energy beam of gamma rays, as desired, both for use outside the reactors.

In addition, all of the reactors described are also extremely useful in testing materials for neutron absorption and neutron production. Using one or more of the removable stringers **36a** for example, in the uncooled graphite-uranium reactor the reactor can be balanced at a given neutron density with uranium bodies of known constants in the stringer. The stringer is then withdrawn and new uranium bodies substituted of those withdrawn. The stringer is reinserted into the pile, and the neutron intensity brought to the original value. The change in position of the control rod for the balance condition, when corrected for atmospheric pressure and temperature, will at once tell whether or not the newly inserted bodies are better or worse when used as elements in the system, than those removed. From the results obtained, calculations can be made as to systems incorporating the new bodies. The effect of changes in size, impurities, coatings, and temperature on the chain reaction, can similarly be determined. In the latter case, uranium lumps can be heated and inserted to determine the effect of temperature on the reaction as measured by change in the position of the control rod. However, the stringer method of determining the effect of changes in pile construction is no part of the present invention. Similarly, materials can be tested by insertion in tube **109b** in the D_2O reactor or in the coolant channels in the other reactors.

The reactors described herein, in addition to being high power neutron generators, are capable of producing the products of the neutronic reactors in quantities related to the power at which they are operated, in that at least a portion of the uranium bodies are removable from the reactors after exposure to the reaction for varying periods of time. By proper chemical treatment, the 94^{239} and fission products can be recovered from the removed irradiated uranium bodies and thereafter utilized as desired.

With modifications, the reactors herein described can also be used as sources of power in useful form. The D_2O moderated reactors can be operated under pressure at an elevated temperature, with continuous removal of the heated D_2O for flashing into D_2O steam for operation of low pressure turbines. Enriched uranium with a light water moderator can be operated in the same manner. The gas cooled reactors, when cooled with helium under pressure, for example, can be used to produce steam by passing the heated helium through heat exchangers. Diphenyl when used as a coolant in the liquid cooled reactor can be heated above the boiling point of water and then used in heat exchangers to produce steam. With proper design, neutronic reactors can also be operated to produce steam directly in tubes passing through the reactor, utilizing the heat of vaporization for cooling the reactor, and the resultant steam for power.

This application is a division of application Serial No. 568,904, filed December 19, 1944, issued into U.S. Patent No. 2,708,656 on May 17,

1955.[8]

While the theory of the nuclear chain fission mechanism in uranium set forth herein is based on the best presently known experimental evidence, we do not wish to be bound thereby, as additional experimental data later discovered may modify the theory disclosed.

What is claimed is:

A method of operating a neutronic reactor including an active portion having a neutron reproduction ratio substantially in excess of unity in the absence of high neutron absorbing bodies, said method comprising the steps of inserting in the active portion a shim member consisting essentially of a high neutron absorbing body in an amount to reduce the neutron reproduction ratio to a value slightly higher than unity to prevent a dangerous reactivity level, controlling the reaction by moving a control member consisting essentially of a second high neutron absorbing body inwardly and outwardly in response to variations in neutron density, to maintain the neutron reproduction ratio substantially at unity, and withdrawing successive portions of the shim member to the extent necessary to enable the reactor to be controlled by movement of the control member after the neutron reproduction value has been lowered to the point where the outward movement of the control member is insufficient to maintain the neutron reproduction ratio at the desired point, and thus to maintain the range of control effected by such movement of the control member substantially constant despite diminution of neutron reproduction ratio caused by operation of the reactor, the active portion being substantially free of high neutron absorber other than the control member and the shim member.

References cited in the file of this Patent

UNITED STATES PATENTS

2,708,656 Fermi et al. .... May 17, 1955

FOREIGN PATENTS

114,150 Australia .... May 2, 1940

[8]@ The authors refer to E. Fermi and L. Szilard, *Method of operating a neutronic reactor*, U.S. Patent No. 2,708,656, May 17, 1955 (see page 217).

114,151	Australia	May 3, 1940
861,390	France	Oct. 28, 1940
233,011	Switzerland	Oct. 2, 1944
648,293	Great Britain	Jan. 3, 1951

OTHER REFERENCES

Naturwissenschaften, vol. 27 (1939), pages 402-410 (an article by Flugge).

Chapter 4

Chain Reacting System

Enrico Fermi, Santa Fe, N. Mex., and Miles C. Leverett, Oak Ridge, Tenn., assignors to the United States of America as represented by the United States Atomic Energy Commission

Application February 16, 1945, Serial No. 578,278

U.S. Patent No. 2,837,477 - Patented June 3, 1958

This Patent focuses mainly on the description of an automatic system for the control rods in a nuclear reactor (in the present case made of natural uranium and graphite) reporting, aside from several related theoretical points (already considered in previous Patents), a detailed description of it. Since the reproduction ratio of a lattice structure is reduced by the presence of neutron absorbing impurities, such impurities can be introduced in the lattice in the form of control rods, made of a material such as boron or cadmium, which will absorb large amounts of neutrons. The control procedure is based on the fact that the depth to which the control rod penetrates into the lattice will determine the value of the neutron density in the system. This relatively simple task faces the fact that the reproduction ratio of the structure can change due to changes in temperature and pressure in the system. So, a connection of the control rods with an ionization chamber, measuring neutron density, can give an automatic control of the stability of the chain reaction. Moreover, an emergency circuit for operating the safety rods is illustrated in this Patent, and, in case of failure of both the control and emergency circuits, a third automatic circuit is depicted which causes the dump of a portion of the lattice structure for interrupting the neutron density rise.

In a system of the type considered, about 92 percent of the total heat generated originates in the uranium toward the center of the lattice, about 6 percent originates in the graphite used as slowing medium, and the remaining 2 percent is generated in the structures surrounding the pile. Accordingly, the permissible power output of the reactor is limited by the rate of heat removal, so that, to prevent the accumulation of heat in the chain reaction pile, a coolant into contact with the uranium must be employed. However, the corrosive effect on uranium of the possible coolants has to be taken into account, because the presence in the system of high temperatures and intense neutron densities causes an acceleration of any normal rate of corrosion, resulting in the physical deterioration of the uranium in the system. It is essential, then, that the circulating medium be of such a character as not to destroy the uranium bodies in the system. In the present case, the cooling medium is gaseous helium circulating in the active regions of the reactor, which has the advantage of minimizing the possible corrosion of the fissile material, since it is an inert gas, and the absorption of neutrons. However, other possible choices, affecting the determination of the multiplication factor, for the coolant gas (such as air, oxigen or water vapor) are discussed as well in terms of their "danger coefficients", defined in terms of the ratio of the weight of impurity per unit mass of uranium and based on the cross section for absorption of thermal neutrons of the various elements [Fermi (1942a)].

The discussion of some methods of cooling chain reacting piles was initiated in [Fermi (1942g)], but no reference published paper exists of the material presented here.

The present invention relates to the subject of nuclear fission and more particularly to a plant wherein the heat generated as a result of the fission process can be removed at a rapid rate and preferably in such a manner that it can be utilized for the production of power. In addition, the products resulting from the fission process in the plant can readily be removed without requiring complete dismantling of the plant.

The fission process is effected in a structure commonly referred to as a chain reacting pile or reactor which may comprise a plurality of bodies containing natural uranium geometrically arranged in an efficient neutron slowing material, such as carbon in the form of graphite.

Natural uranium contains both uranium isotopes U^{235} and U^{238} in the ratio of 1 to 139. Consider the chain reaction as starting with the fast neutrons originating by nuclear fission in a uranium body. These neutrons are set free with very high energy of the order of magnitude of one million electron volts average and are therefore not in condition to be utilized efficiently to create new thermal neutron fissions in U^{235} when it is mixed with a considerable quantity of U^{238}, particularly as in the case of natural uranium. The energies of the fission-released neutrons are so high that most of the latter would tend to be absorbed by the U^{238} nuclei or lost from the system, and yet the energies are not generally high enough for production or fission by more than a small fraction of the neutrons so absorbed. For neutrons of thermal energies, however, the absorption cross section of U^{235}, to produce fission, rises a great deal more than the simple capture cross section of U^{238}; so that under the stated circumstances the fast fission neutrons, after they are created, must be slowed down to thermal energies before they are most effective to produce fresh fission by reaction with additional U^{235} atoms. If a system can be made in which neutrons are slowed down without much absorption until they reach thermal energies and then mostly enter into uranium rather than into any other element, a self-sustaining nuclear chain reaction can be obtained, even with natural uranium. Light elements, such as deuterium, beryllium, oxygen or carbon, the latter in the form of graphite, can be used as slowing agents. A special advantage of the use of the light elements mentioned for slowing down fast fission neutrons is that fewer collisions are required for slowing than is the case with heavier elements, and furthermore, the above-enumerated elements have very small neutron capture probabilities, even for thermal neutrons. Hydrogen would be most advantageous were it not for the fact that there may be a relatively high probability of neutron captured by the hydrogen nucleus. Both beryllium and deuterium are expensive and not readily available in large quantities. Carbon in the form of graphite is a relatively inexpensive, practical, and readily available agent for slowing fast neutrons to thermal energies. Recently, beryllium has been made available in sufficiently large quantities for test as to suitability for use as a neutron slowing material in a system of the type to be described. It had been found to be in every way as satisfactory as carbon.

However, in order for the premise to be fulfilled that the fast fission neutrons be slowed to thermal energies in a slowing medium with a minimum of absorption in the U^{238} isotope of the uranium, certain types of physical structure should be utilized for the most efficient reproduction of neutrons, as precautions must be taken to reduce various neutron losses and thus to conserve neutrons for the chain reaction if a self-sustaining system has to be attained.

The ratio of the number of fast neutrons produced by the fissions, to the original number of fast neutrons creating the fissions, in a system of infinite size using specific materials is called the reproduction or multiplication factor of the system and is denoted by the symbol K. This factor may be determined experimentally for any particular geometry by the process described in a copending application of E. Fermi, Serial No. 534,129, filed

May 4, 1944, now Patent No. 2,780,595.[1] If K can be made sufficiently greater than unity to create a net gain in neutrons and the system made sufficiently large so that this gain is not entirely lost by leakage from the exterior surface of the system, than a self-sustaining chain reacting system can be built to produce power by nuclear fission of natural uranium. The neutron reproduction ratio r in a system of finite size differs from K by the leakage factor, and must be sufficiently greater than unity to permit the neutron density to rise exponentially. Such a rise will continue indefinitely if not controlled at a desired density corresponding to a desired power output.

It is therefore the main object of our invention to construct a system comprising uranium and a slowing medium so that neutron losses are reduced to such an extent that a controllable self-sustaining neutron chain fission reaction is obtained therein, with resultant regulated production of neutrons, liberation of power in the form of heat and other forms, the production of radioactive fission products and new elements both radioactive and stable produced by the absorption of neutrons.

During the interchange of neutrons in a system comprising bodies of uranium of any size in a slowing medium, neutrons may be lost in four ways; that is, by absorption in the uranium metal or compound without producing fission, by absorption in the slowing down material, by absorption in impurities present in the system, and by leakage from the system. These losses will be considered in the order mentioned.

Natural uranium, particularly by reason of its U^{238} content, has an especially strong absorbing power for neutrons when they have been slowed down to moderate energies. The absorption in uranium at these energies is termed the uranium resonance absorption or capture. It is caused by the isotope U^{238} and does not result in fission but creates the isotope U^{239} which by two successive beta emissions forms the relatively stable nucleus 94^{239}. It is not to be confused with absorption or capture of neutrons by impurities, referred to later. Neutron resonance absorption in uranium may take place either on the surface of the uranium bodies, in which case the absorption is known as surface resonance absorption, or it may take place further in the interior of the uranium body, in which case the absorption is known as volume resonance absorption. It will be appreciated that this classification of resonance absorptions is merely a convenient characterization of observed phenomena, and arises, not because the neutron absorbing power of a U^{238} nucleus is any greater when the nucleus is at the surface of a body of

[1]@ See page 181.

metallic or combined uranium, but because the absorbing power of U^{238} nuclei for neutrons of certain particular energies is inherently so high that practically all neutrons that already happen to have those energies, called resonance energies as explained above, are absorbed almost immediately upon their arrival in the body of uranium metal or uranium compound, and thus in effect are absorbed at the surface of such body. Volume resonance absorption is due to the fact that some neutrons make collisions inside the uranium body and may thus arrive at resonance energies therein. After successfully reaching thermal velocities, about 40 percent of the neutrons are also subject to capture by U^{238} without fission, to produce U^{239} and eventually 94^{239}.

It is possible, by proper physical arrangement of the materials, to reduce substantially uranium resonance absorption. By the use of light elements as described above for slowing materials, a relatively large increment of energy loss is achieved in each collision and therefore fewer collisions are required to slow the neutrons to thermal energies, thus decreasing the probability of a neutron being at a resonance energy as it enters a uranium atom. During the slowing process, however, neutrons are diffusing through the slowing medium over random paths and distances so that the uranium is not only exposed to thermal neutrons but also to neutrons of energies varying between the emission energy of fission and thermal energy. Neutrons at uranium resonance energies will, if they enter uranium at these energies, be absorbed on the surface of a uranium body whatever its size, giving rise to surface absorption. Any substantial reduction of overall surface of the same amount of uranium relative to the amount of slowing material (i.e., the amount of slowing medium remaining unchanged) will reduce surface absorption, and any such reduction in surface absorption will release neutrons to enter directly into the chain reaction; i.e., will increase the number of neutrons available for further slowing and thus for reaction with U^{235} to produce fission.

For a given ratio of slowing material to uranium, surface resonance absorption losses of neutrons in the uranium can be reduced by a large factor from the losses occurring in a mixture of fine uranium particles and a slowing medium, if the uranium is aggregated into substantial masses in which the mean radius is at least 0.25 centimeters for natural uranium metal and when the mean spatial radius of the bodies is at least 0.75 centimeters for the oxide of natural uranium (UO_2). An important gain is thus made in the number of neutrons made directly available for the chain reaction. A similar gain is made when the uranium has more than the natural content of fissionable material. Consequently, where it is desired to secure a maximum K, we place the uranium in the system in the form of spaced uranium masses or bodies of substantial size, preferably either of metal, oxide, carbide, or combinations thereof. The uranium bodies can be in the form of layers, rods or cylinders, cubes or spheres, or approximate shapes, dispersed throughout the graphite, preferably in some geometric pattern. The term geometric is used to mean any pattern or arrangement wherein the uranium bodies are distributed in the graphite or other moderator with at least either a roughly uniform spacing or with a roughly systematic non-uniform spacing, and are at least roughly uniform in size and shape or are systematic in variations of size or shape to produce a volume pattern conforming to a roughly symmetrical system. If the pattern is a repeating or rather exactly regular one, a system embodying it may be conveniently described as a lattice structure. Optimum conditions are obtained with natural uranium by using a lattice of metal spheres.

The number of neutrons made directly available to the chain reaction by aggregating the uranium into separate bodies spaced through the slowing medium is a critical factor in obtaining a self-sustaining chain reaction utilizing natural uranium and graphite. The K factor of a mixture of fine uranium particles in graphite, assuming both of them to be theoretically pure, would only be about 0.785. Actual K factors as high as 1.07 have been obtained using aggregation of natural uranium in the best known geometry. With completely pure materials and uranium aggregates shaped as spheres it is possible to obtain K factors as high as 1.10 with a carbon moderator, 1.18 with a beryllium moderator and 1.3 with D_2O as a moderator.

Still higher K factors can be obtained by the use of aggregation in the case of uranium having more than the naturally occurring content of fissionable materials such as U^{233}, U^{235} or 94^{239}. Adding such fissionable materials is termed enrichment of the uranium.

It is thus clearly apparent that the aggregation of the uranium into masses separated in the slowing material is one of the most important, if not the most important factor entering into the successful construction of a self-sustaining chain reaction system utilizing relatively pure natural uranium in a slowing material in the best geometry at present known, and is also important in obtaining high K factors when enrichment of the uranium is used.

The thermal neutrons are also subject to capture by the slowing material. While carbon and beryllium have very small capture cross sections for thermal neutrons, and deuterium still smaller, an appreciable fraction of thermal neutrons (about 10 percent of the neutrons present in the system under best conditions with graphite) is lost by capture in the slowing material during diffusion therethrough. It is therefore desirable to have the neutrons reaching thermal energy promptly enter uranium.

In addition to the above-mentioned losses, which are inherently a part of the nuclear chain reaction process, impurities present in both the slowing material and the uranium add a very important neutron loss factor in the chain. The effectiveness of various elements as neutron absorbers varies tremendously. Certain elements such as boron, cadmium, samarium, gadolinium, and some others, if present even in a few parts per million, could prevent a self-sustaining chain reaction from taking place. It is highly important, therefore, to remove as far as possible all impurities capturing neutrons to the detriment of the chain reaction from both the slowing material and the uranium. If these impurities, solid, liquid, or gaseous, and in elemental or combined form, are present in too great quantity, in the uranium bodies or the slowing material or in, or by absorption from, the free spaces of the system, the self-sustaining chain reaction cannot be attained. The amounts of impurities that may be permitted in a system, vary with a number of factors, such as the specific geometry of the system, and the form in which the uranium is used – that is, whether natural or enriched, whether as metal or oxide – and also factors such as the weight ratios between the uranium and the slowing down material used – for example, whether deuterium, graphite or beryllium. Although all of these considerations influence the actual permissible amount of each impurity material, it has fortunately been found that, in general, the effect of any given impurity or impurities can be correlated directly with the weight of the impurity present and with the K factor of the system, so that knowing the K factor for a given geometry and composition, the permissible amounts of particular impurities can be readily computed without taking individual account of the specific considerations named above. Different impurities are found to affect the operation to widely different extents; for example, relatively considerable quantities of elements such as hydrogen may be present, and, as previously suggested, the uranium may be in the form of oxide, such as UO_2 or U_3O_8, or carbide, although the metal is preferred. Nitrogen may be present to some extent, and its effect on the chain reaction is such that the neutron reproduction ratio of the system may be changed by changes in atmospheric pressure. This effect may be eliminated by enclosing or evacuating the system if desired. In general, the inclusion of combined nitrogen is to be avoided.

The effect of impurities on the optimum reproduction factor K may be conveniently evaluated to a good approximation, simply by means of certain constants known as "danger coefficients" which are assigned to the various elements. These danger coefficients for the impurities are each multiplied by the percent by weight of the corresponding impurity, and the total sum of these products gives a value known as the total danger sum. This total danger sum is subtracted from the reproduction factor K as calculated for pure materials and for the specific geometry under consideration.

The danger coefficients are defined in terms of the ratio of the weight of impurity per unit mass of uranium and are based on the cross section for absorption of thermal neutrons of the various elements. These values may be obtained from physics textbooks on the subject and the danger coefficient computed by the formula

$$\frac{\sigma_i}{\sigma_u}\frac{A_u}{A_i},$$

where σ_i represents the cross section for the impurity and σ_u the cross section for the uranium, A_i the atomic weight of the impurity and A_u the atomic weight for uranium. If the impurities are in the carbon, they are computed as their percent of the weight of the uranium of the system.

Presently known values for danger coefficients for some elements are given in the following table 4.1, wherein the elements are assumed to have their natural isotopic constitution unless otherwise indicated, and are conveniently listed according to their chemical symbols.

Where an element is necessarily used in an active part of a system, it is still to be considered as an impurity; for example, in a structure where the uranium bodies consist of uranium oxides, the actual factor K would ordinarily be computed by taking that fact into account using as a base a K value computed for theoretically pure uranium.

As a specific example, if the materials of the system under consideration have 0.0001 part by weight of Co and Ag, the total danger sum in K units for such an analysis would be:

$$0.0001 \times 17 + 0.0001 \times 18 = 0.0035\ K \text{ unit}.$$

This would be a rather unimportant reduction in the reproduction factor K unless the reproduction factor for a given system, without considering any impurities, is very nearly unity. If, on the other hand, the impurities in the uranium in the previous example had been Li, Co, and Rh, the total danger sum would be:

$$0.0310 + 0.0017 + 0.0050 = 0.0377\ K \text{ unit}.$$

Table 4.1

Element	Danger Coefficient	Element	Danger Coefficient
H	10	Mo	1.0
D	0.01	Ru	2
He	0	Rh	50
Li	310	Pd	2
Be	0.04	Ag	18
B	2150	Cd	870
C	0.012	In	54.2
N	4.0	Sn	0.18
O	0.002	Sb	1.6
F	0.02	Te	1
Ne	3	I	1.6
Na	0.65	Xe	6
Mg	0.48	Cs	8.7
Al	0.30	Ba	0.30
Si	0.26	La	2.4
P	0.3	Ce	2.4
S	0.46	Pr	2.4
Cl	31	Nd	17
A	0.8	Sm	1430
K	2.1	Eu	435
Ca	0.37	Gd	6320
Sc	7	Tb	20
Ti	3.8	Dy	200
V	4	Ho	10
Cr	2	Er	40
Mn	7.5	Tm	20
Fe	1.5	Yb	10
Co	17	Lu	30
Ni	3	Hf	20
Cu	1.8	Ta	4.6
Zn	0.61	W	2.7
Ga	1	Re	18
Ge	5	Os	1.7
As	2	Ir	70
Se	6.3	Pt	2.5
Br	2.5	Au	16
Kr	6	Hg	82
Rb	0.4	Tl	0.5
Sr	0.57	Pb	0.03
Y	0.4	Bi	0.0025
Zr	0.13	Th	1.1
Cb	0.4		

This latter reduction in the reproduction factor for a given system would be serious and might well reduce the reproduction factor below unity for certain geometries so as to make it impossible to effect a self-sustaining chain reaction with natural uranium and graphite, but might still be permissible when using enriched uranium in a system having a high K factor.

The size of the system will vary, depending upon the K factor of the system, and upon other things. If the reproduction factor K is greater than unity, the number of neutrons present will increase exponentially and indefinitely, provided the structure is made sufficiently large. If, on the contrary, the structure is small, with a large surface-to-volume ratio, there will be a rate of loss of neutrons from the structure by leakage through the outer surfaces, which may overbalance the rate of neutron production inside the structure so that a chain reaction will not be self-sustaining. For each value of the reproduction factor K greater than unity, there is thus a minimum overall size of a given structure known as the critical size, above which the rate of loss of neutrons by diffusion to the walls of the structure and leakage away from the structure is less than the rate of production of neutrons within the system, thus making the chain reaction self-sustaining. The rate of diffusion of neutrons away from a large structure in which they are being created through the exterior surface thereof may be treated by mathematical analysis when the value of K and certain other constants are known, as the ratio of the exterior surface to the volume becomes less as the structure is enlarged.

In the case of a spherical structure employing uranium bodies imbedded in graphite in the geometries disclosed herein and without an external reflector the following formula gives the critical overall radius (R) in feet:

$$K - 1 = \frac{C}{R^2},$$

where C is a constant that varies slightly with geometry of the lattice and for normal graphite lattices may have a value close to 7.4.

For a rectangular parallelepiped structure rather than spherical, the critical size can be computed from the formula

$$K - 1 = C\left(\frac{1}{a^2} + \frac{1}{b^2} + \frac{1}{c^2}\right),$$

where a, b, and c are the length of the sides in feet. The critical size for a cylindrical structure is given by the formula, irrespective of the shape of the uranium bodies (cylinder height: h ft.; radius: R ft.)

$$K - 1 = C\left(\frac{1}{h^2} + \frac{0.59}{R^2}\right).$$

However, when critical size is attained, by definition no rise in neutron density can be expected. It is therefore necessary to increase the size of the structure beyond the critical size but not to the extent that the period for doubling of the neutron density is too short, as will be explained later. A desirable reproduction ratio (r) for an operating structure with all control absorbers removed and at the temperature of operation is about 1.005. The size at which this reproduction ratio can be obtained may be computed from modifications of the above formulae for critical size. For example, for spherical active structures the formula

$$K - r = \frac{C}{R^2}$$

may be used to find R when K is known and r is the reproduction ratio and is somewhat over unity. The same formula will, of course, give r for given structures for which K and R are known.

Critical size may be attained with a somewhat smaller structure by utilizing a neutron reflecting medium surrounding the surface of the active structure. For example, a 2 feet thickness of graphite having low impurity content, completely surrounding a spherical structure is effective in reducing the diameter of the uranium bearing portion by as much as 2 feet, resulting in a considerable saving of uranium or uranium compound.

The ratio of the average number of fast neutrons produced by fission of a fissionable isotope to the average number of thermal neutrons absorbed by the composition of which the fissionable isotope is a component is a constant for any particular composition commonly referred to as the "eta constant". Since fission of U^{235} produces about 2 fast neutrons per fission the eta constant for pure U^{235} would be about 2 assuming that all neutrons produced were absorbed to produce fission. However, U^{235} is usually used in conjunction with U^{238} and generally comprises but a very small portion of the composition subjected to treatment and accordingly this average constant is reduced because of absorptions by the composition by U^{238} which does not produce fission. For natural uranium an average of about 1.32 new fast neutrons will be produced by fission of the nuclei of the U^{235} isotope and this value is the "eta" constant for natural uranium. For fissionable materials other than natural uranium containing different concentrations of U^{235} and U^{238} or different fissionable isotopes such as 94^{239} the value of the eta constant will be different.

The new fast neutrons resulting from the fissions in the isotope U^{235} pass through the same cycle as just described, there being a certain portion which will produce fast fission, some which will be lost to the chain reaction and others which will be slowed down to thermal energy to be absorbed by uranium without fission.

By arranging the uranium in bodies or masses of suitable shape and sufficient size to minimize passage of neutrons having energies corresponding to or above resonance energies of U^{238} through the uranium preferably in a regular geometry resembling a crystal lattice in chemical parlance and selecting the correct volume ratio of uranium to slowing material, and, further, by suitably limiting the impurities in the uranium and the slowing material and by building the structure to a proper size, it is possible to produce in each generation more fast neutrons by fission than are originally present to start the chain, so as to perpetuate the chain reaction in the system.

As the chain reacting pile is then capable of producing more fast fission neutrons at a greater rate than the rate at which neutrons are lost from the system, there would be an exponentially rise in the neutron density, theoretically to infinity, as the pile is operated unless the rise is controlled. Removable neutron absorbers, or "impurities" in the form of control rods, can be used to control the exponential rise beyond desired limits by inserting such rods into the pile. The point at which the exponential rise is stopped is a matter of choice, and, of course, will depend upon the desired power output as well as considerations involving safety and efficiency of heat removal. It is obvious that the rate of production of heat within the pile will depend upon the operating neutron density in the system. The higher the density, the greater the production of heat in the system. Moreover the permissible power output of the reactor is limited by the rate of heat removal and therefore the permissible power output may be seriously reduced where the cooling media is correlated through improper zones of the reactor. We have found that about 92 percent of the total heat generated in a chain reacting system originates in the uranium, about 6 percent originates in the slowing medium, where graphite is used as such medium. The remaining 2 percent is generated outside the pile; that is, in the surrounding structures. Accordingly we have found that maximum output of power may be secured by passage of the coolant into contact with or closely adjacent to the uranium bodies.

Following is a table showing more specifically the type and locale of the heat generated in the pile:

SUMMARY BY TYPE

Table 4.2

	MeV/fission	Percent
Gamma radiation	23	11
Beta radiation	11	6
Kinetic Energy of fission fragments	159	79
Kinetic Energy of neutrons	7	4
	200	100

SUMMARY BY LOCALE WHERE HEAT IS GENERATED

Table 4.3

	MeV/fission	Percent
In Uranium	184	92
In Carbon	12	6
Outside Pile	4	2
	200	100

SUMMARY BY TYPE AND LOCALE

Table 4.4

	MeV per fission	Percent in U	Percent in C	Percent Outside
Kinetic Energy of fission fragments	159	100	....	....
Kinetic Energy of neutrons	7	....	90	10
Gamma radiation from fission fragments and products	5	70	25	5
Beta radiation from fission fragments and products	11	100	....	....
Nuclear affinity of neutrons (gamma radiation)	18	60	25	15

In order to control the temperature of the chain reaction and to prevent the accumulation of heat in the chain reaction pile, some suitable circulating system must be employed to convey the heat away from the pile when a large power output is desired. The design of this system within the pile proper and the type of coolant employed are critical factors which, if not properly taken into consideration, will make it impossible to design a pile capable of producing a self-sustaining chain reaction.

The problem of removing heat from a chain reacting system is complicated by various factors. The corrosive effect on uranium of most otherwise suitable circulating media is very troublesome. This factor is important primarily because of the presence in the system of high temperatures and intense neutron densities causing an acceleration of any normal rate of corrosion. One of the most serious results which may result from the corrosive action of a circulating medium on the uranium is the physical deterioration of the uranium in the system. It is essential, then, that the circulating medium be of such a character as not to destroy the uranium bodies in the system. Furthermore, many otherwise suitable cooling media absorb neutrons to such an extent that they cannot be used in the pile.

In the presently described system, helium is used as the circulating medium to remove heat from the structure. Helium is an inert gas and therefore the corrosion problem with respect to the uranium is eliminated. Furthermore, helium has practically no neutron absorption factor and is therefore very satisfactory from the point of view of its effect on the reproduction of neutrons in the chain reacting system. In other words the danger coefficient of helium is very low. The danger coefficient is defined in terms of the ratio of the weight of impurity in the chain reacting system per unit mass of uranium, and is based on the cross section for absorption of thermal neutrons. It therefore can be allowed to diffuse throughout the pile and can be circulated therein in open channels without the use of tubes which might also absorb neutrons to the detriment of the chain reaction.

The fission products resulting from the chain reaction are, at least in general, highly radioactive. The diffusion of the fission fragments within the uranium is very slow, but fission fragments originating close to the surface of the uranium can escape into the helium gas, thereby causing the gas to become radioactive. This, of course, is objectionable because it complicates the shielding problem in the circulating system outside the pile and requires steps to be taken to protect persons in the vicinity of the gas ducts from the harmful effects produced by this radioactivity. It may therefore be desirable to cover or coat the uranium surfaces with metals such as aluminum or zinc, only a thin coating being required to prevent the flying fragments of the fissions from escaping from the solid uranium. This results in a substantial reduction in the radioactivity of the circulating helium gas, thus simplifying the protection problem outside of the pile.

In addition to the usual industrial hazards, personnel must be protected from injury by gamma rays and neutrons emanating from the reactor and the circulating gas and equipment, from injury by beta rays, or by close contact with radioactive materials, and from radioactive poisoning due to inhalation of radioactive gases emanating from the reactor.

As one of the principal objects of the present invention, we provide a self-sustaining chain reacting power plant operating by virtue of nuclear fission at high power output in which temperature of the system is effectively controlled by efficient removal of heat therefrom in an effective and safe manner at high power outputs of the power plant.

Other objects and advantages of our invention will become apparent from the following description and the drawings, which illustrate a power plant operating by virtue of nuclear fission in which the heat equivalent of 100,000 kilowatts is removed by circulation therethrough of 400,000 pounds of helium per hour.

Fig. 4.1 is a diagrammatic view of the reactor and the heat extraction system;

Fig. 4.2 is a vertical sectional view through the reactor showing in elevation the graphite and uranium lattice;

Fig. 4.3 is an enlarged, fragmentary, vertical sectional view through the center portion of the reactor shown in Fig. 4.2, indicating the relationship between the uranium and graphite and further illustrating the structure at the bottom and top of the lattice;

Fig. 4.4 is an enlarged sectional view through one of the graphite uranium cartridges taken along the line **10–10** in Fig. 4.5;

Fig. 4.5 is a top plan view taken on the line **11–11** if Fig. 4.4;

Fig. 4.6 is a top plan view of an assembly of uranium plates which fit into the cartridges shown in Fig. 4.4 and is used at the center of the lattice structure where the heat is most intense;

Fig. 4.7 is an assembly of uranium plates corresponding to Fig. 4.6 but which is employed somewhat away from the center of the lattice structure where the heat generated is less intense;

Fig. 4.8 is a fragmentary, horizontal sectional view taken through the lattice on line **4–4** of Fig. 4.3 showing slightly more than one quadrant only of the reactor;

Fig. 4.9 is a perspective view illustrating one arrangement for the graphite making up the lattice structure;

Fig. 4.10 is a plan view of a graphite brick shown in Fig. 4.9;

Fig. 4.11 is a horizontal sectional view taken on the line **7–7** of Fig. 4.10;

Fig. 4.12 is a transverse sectional view taken on the line **8–8** of Fig. 4.10;

Fig. 4.13 is a vertical sectional view through three bricks illustrating the manner in which bricks are doweled together;

Fig. 4.14 is an enlarged, detailed, vertical sectional view through a portion of the reactor on line **14–14** of Fig. 4.2 showing the relationship between the lattice structure, the internal shield for the reactor, and the guide tubes which extend from the graphite through the internal shield providing a passage for charging the uranium into the lattice;

Fig. 4.15 is an enlarged, vertical sectional view through one of the discharge valves on the line **16–16** of Fig. 4.18 showing the relationship of the valve to the lattice structure;

Fig. 4.16 is a horizontal sectional view taken on the line **17–17** of Fig. 4.15;

Fig. 4.17 is a vertical sectional view through one of the valves on line **18–18** of Fig. 4.15 showing the relationship of the valve to the lattice and the lattice support;

Fig. 4.18 is a fragmentary, horizontal sectional view taken through the lower part of the reactor along line **15–15** of Fig. 4.2 showing the arrangement of the valves used in discharging the uranium from the lattice structure;

Fig. 4.19 is an enlarged, fragmentary detail sectional view showing a water cooled control rod;

Fig. 4.20 is a diagrammatic view showing the control system for the power plant, the electrical circuit being reduced to the lowest terms;

Fig. 4.21 is a longitudinal sectional view taken through one of the filters diagrammatically illustrated in Fig. 4.1;

Fig. 4.22 is a transverse sectional view taken on the line **21-21** of Fig. 4.21; and

Fig. 4.23 is an enlarged detail sectional view of the fragment marked **22** on the filter element shown in Fig. 4.21.

Referring to Fig. 4.1, the neutron chain reaction is effected in a reactor diagrammatically shown at **25**. The heat generated by virtue of the chain reaction is carried away by helium passing from the reactor through a pipe **26** into heat exchangers **27**, which may be of any well known type to provide steam for the prime mover **27a**. This steam may be used in any conventional manner to generate power as desired. The cooled helium then passes through pipe **28** into filters **29**, which remove any solid matter from the helium, and thence into a battery of water cooled compressors **31**

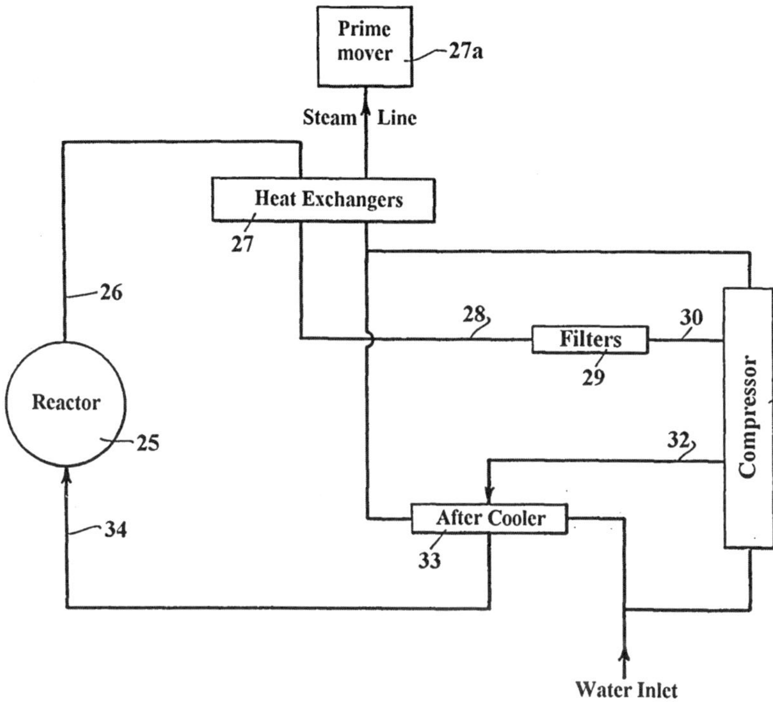

Fig. 4.1

through a pipe **30**. The compressors may be of the centrifugal or reciprocating type although the former is generally preferable. For most efficient heat removal, the helium in the reactor is maintained under pressure, and for that reason the compressors **31** serve to establish and maintain the pressure and also serve as pumps to circulate the helium. The high pressure gas leaves the compressors **31** through suitable piping **32**, and the heat resulting from the compression is removed from the gas in after-coolers **33**. From these coolers, the helium gas is returned to the reactor through piping **34**. After-coolers **33** may be used to preheat water to be turned into steam in exchangers **27**. Similarly, the jackets of the compressors **31** can be used to preheat the feed water for exchangers **27**. During operation the heat exchangers may become more or less radioactive due to the radioactivity of the helium entering the exchanger. As a consequence cleaning of the exchangers may become difficult. In order to minimize the necessity for cleaning it is found desirable to use water treated for reduction of its scale forming and corrosive properties in the heat exchangers.

The pressure of the helium gas entering the reactor is 115 pounds per square inch and the temperature is 120 degrees Fahrenheit. About 400,000 pounds of gas are circulated through the reactor shown herein per hour. The gas leaves the reactor at a pressure of 103.6 pounds per square inch and at a temperature of 800 degrees Fahrenheit.

Referring to Fig. 4.2, the reactor **25** consists of a lower chamber **35**, which serves as a gas inlet and uranium discharge chamber. Above the chamber **35** is a valve housing **36** which accommodates the uranium discharge valves presently to be described. The uranium-graphite lattice pile is contained in chamber **37**, above which is a hot gas discharge chamber **38**. A dome **39** completes the shell making up the reactor **25**. These shell segments are welded together so as to form one integral shell, and, as noted, the various segments are of spherical shape so as to give added strength to withstand the relatively high pressures existing in the shell.

Referring to Fig. 4.3, the lattice structure comprises graphite cartridges **40** surrounded by graphite bricks **41**. The cartridges contain uranium in the form of cylindrical aggregates or bodies **42** (see Figs. 4.4-4.7, inclusive). A high-grade AGOT graphite is preferably employed for the cartridges and the matrix of the lattice structure. Surrounding the lattice is dead graphite **43** (Fig. 4.3), which need not be as pure graphite as that employed in the lattice. AGX grade graphite is satisfactory for use in the dead graphite area.

The graphite cartridges **40** are arranged in vertically disposed columns extending from the bottom of the lattice to the top. The columns are disposed in parallel rows, as indicated in Fig. 4.8, so that the uranium in the graphite is arranged in a cubical lattice.

The graphite **41** may be in the form of bricks piled on top of each other, as shown in Fig. 4.9, doweled together by means of dowel rods **41a** passing through holes **41b**. As illustrated, the bricks are 22 inches long and have a cross section 11 inches by $5\frac{1}{2}$ inches. Each brick finished but before drilling weights 65.8 pounds. Each brick is drilled to provide holes **44** through which the graphite cartridges **40** will pass. As the lattice structure is being built, the graphite bricks **41** are piled up so that the holes **44** are in alignment to form a continuous vertical passage from the bottom to the top of the pile to accommodate each of the columns of graphite cartridges. Sufficient space is provided between adjacent bricks and between the cartridges and the bricks to permit expansion of the graphite.

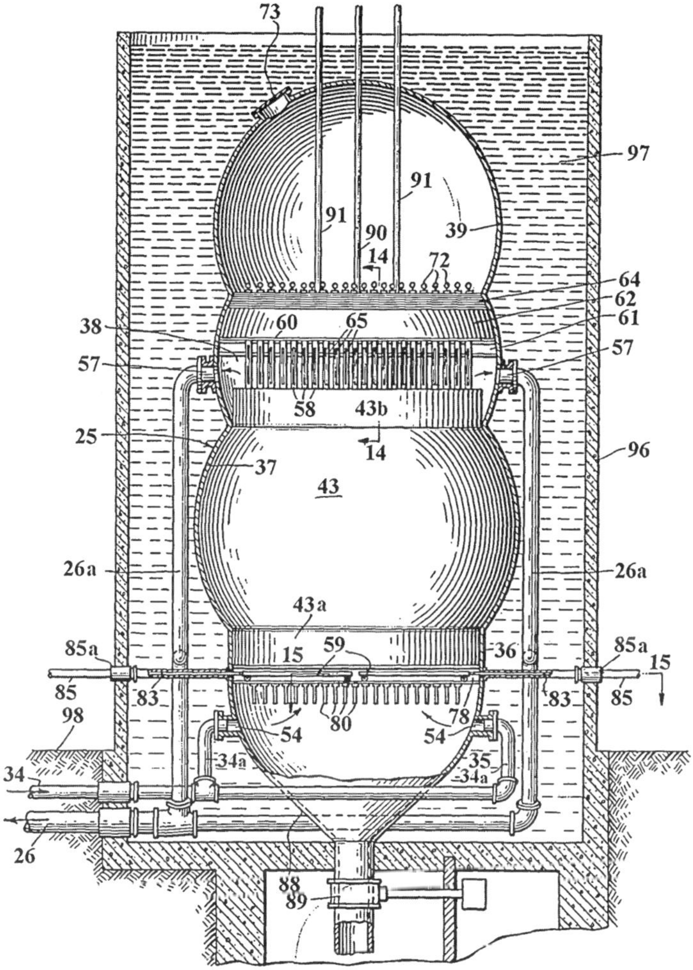

Fig. 4.2

Each of the graphite cartridges **40** (see Fig. 4.4) is provided with a longitudinal passage **45** extending throughout the entire height of the cartridge. Adjacent to the upper end of each cartridge is a cylindrical seat portion indicated at **47** provided with an annular shoulder **48** on which the uranium body **42** rests. Inasmuch as there is approximately only 50 percent free volume through the uranium bodies **42**, the velocity of helium

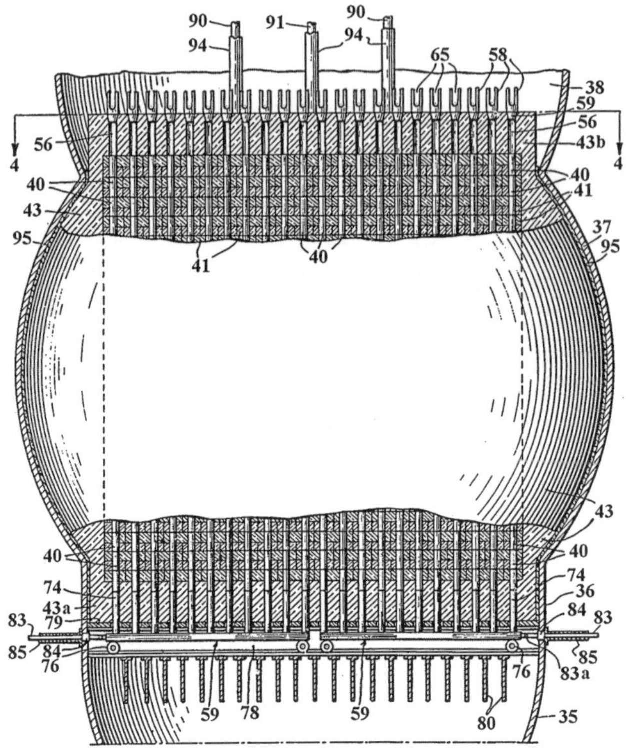

Fig. 4.3

gas passing through the uranium is considerably greater than the velocity of the gas passing through the free passage in the graphite cartridges.

Most of the heat generated as a result of the neutron chain reaction is produced in the center portion of the lattice structure and progressively less heat is generated toward the outside of the structure. Because of this, it is desirable that a greater amount of helium gas pass through the central portion of the lattice structure than through the outside portions. A satisfactory way for controlling the flow of the helium gas through the passages in the graphite cartridges is to provide a throat or constriction **46** (see Fig. 4.4) in the outer cartridges to create greater resistance to the flow of the gas through these restricted passages. In this manner, by providing the narrowest throat or constriction **46** in the outermost cartridges and further by gradually increasing the diameter of this constriction in the passages

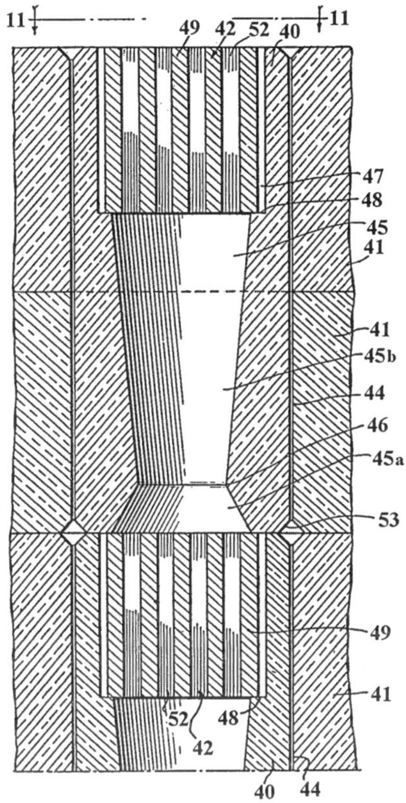

Fig. 4.4

toward the center of the lattice structure and finally providing no constriction in the center portion of the construction, it is possible to selectively control the amount of gas flow through the various locations in the lattice to most effectively remove heat from the system.

The average mass velocity of the helium passing through the ducts in the lattice is about 5.56 pounds per second per square foot, while the maximum is at the center of the lattice and is approximately 12.8 pounds per second per square foot. The average linear velocity of the gas in the duct is about 110 feet per second and the maximum at the center of the lattice is about 254 feet per second. The average mass velocity of the helium gas in the uranium elements **42** is 8.83 pounds per second per square foot, while the maximum is about 20.4 pound per second per square foot. The aver-

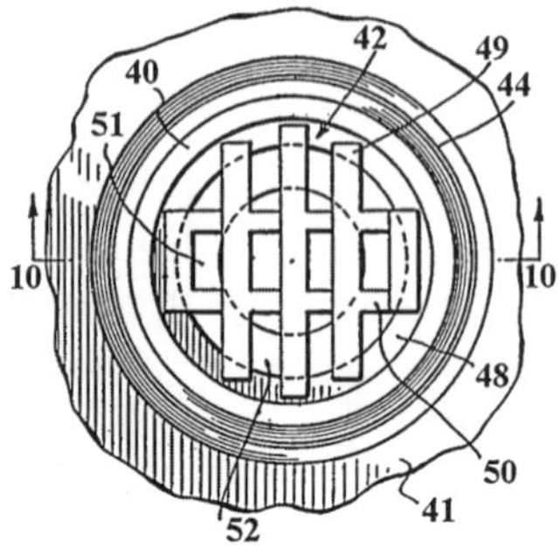

Fig. 4.5

age linear velocity of the gas in the elements is about 175 feet per second, while the maximum is 404 feet per second. On an average, about 1068 pounds of helium gas per hour pass through each column, the centermost columns conveying the most, the maximum for one column being approximately 2460 pound per hour. The heat transfer coefficient for the average column is about 163 B.t.u. per hour per square foot per degree Fahrenheit, while the coefficient for the centermost columns is about 318 B.t.u. per hour per square foot per degree Fahrenheit. 400,000 pounds of helium are circulated per hour, removing the heat equivalent of 100,000 kilowatts from the reactor. The power required for circulating the helium and running the auxiliaries of the plant is about 12,000 kilowatts, leaving a net power of around 88,000 kilowatts available to form steam for external use.

The uranium bodies **42** are each made up of a plurality of parallel plates **49** disposed vertically and spaced apart with lugs **50** serving as spacers between the adjacent plates. In this manner, a plurality of vertical passages **51** and **52** are provided between the plates. The uranium disposed

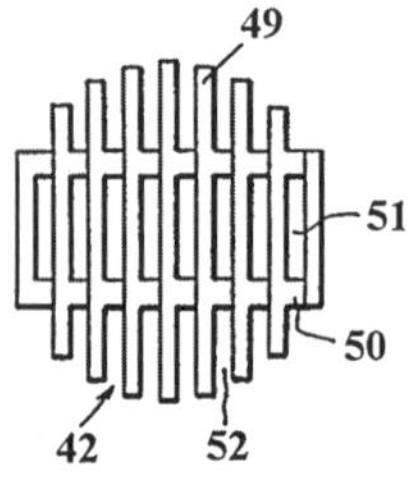

Fig. 4.6

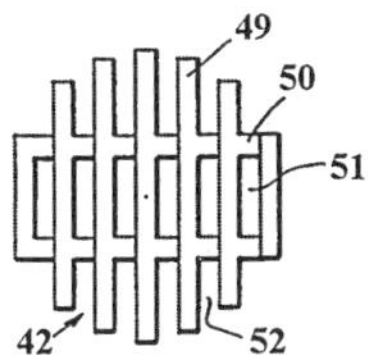

Fig. 4.7

in plate form as illustrated is thus provided with a relatively large amount of surface for cooling. The size of each uranium cylinder is about $3\frac{5}{8}$ inches by $3\frac{5}{8}$ inches, having approximately 50 percent free volume, and each weights about 13.4 pounds. This weight represents the value for uranium metal cylinders. If uranium carbide is used, the weight of each cylinder is about 14.9 pounds. The overall mass ratio of graphite to uranium in the lattice is 5.4. The overall lattice structure is in the form of a cylinder 28 feet in diameter and 26 feet high and has a two foot layer of dead graphite at the top and a layer on the sides varying in thickness from 1 foot to 4 feet. These figures represent an operative lattice, but, of course, the invention is not intended to be limited to this specific example.

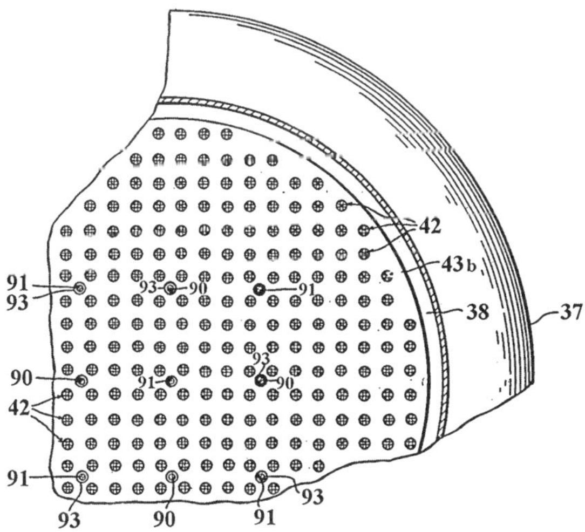

Fig. 4.8

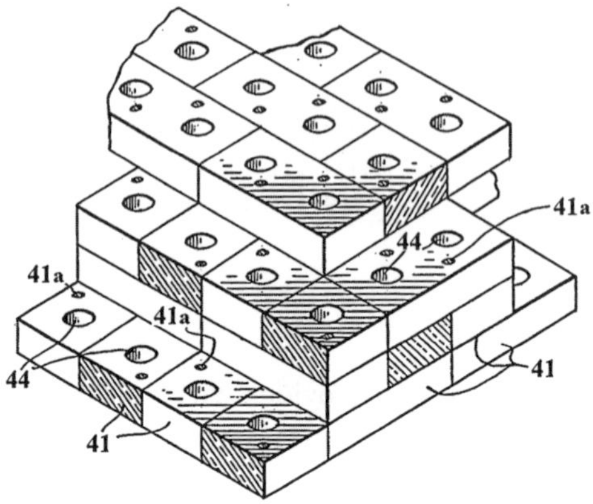

Fig. 4.9

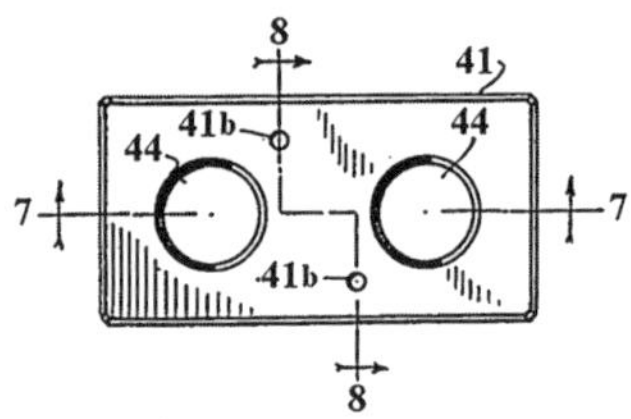

Fig. 4.10

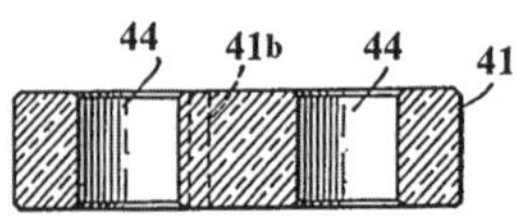

Fig. 4.11

As shown in Fig. 4.3, a space is left between the dead graphite layer **43** and the shell of the reactor and this space is filled with shredded asbestos indicated at **95** to a thickness of about one inch. The graphite will expand and contract as a result of the heat generated in the lattice and for this reason the space must be provided. The asbestos layer will compress to permit this expansion but will serve to prevent leakage of helium gas

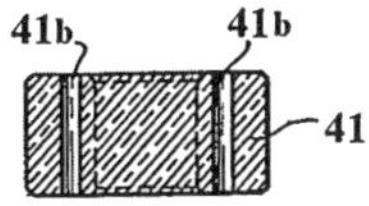

Fig. 4.12

through this space so that the gas cannot by-pass the lattice.

The inside surfaces of the reactor shell may be coated with some suitable material to prevent corrosion of the metal.

As pointed out above, most of the heat generated by the chain reaction is generated in the uranium and is produced toward the center of the lattice, and consequently in this region a greater amount of surface exposed to the circulating medium should be provided on or adjacent to the uranium and particularly in the central area of the reactor than elsewhere than in the region in the structure. This is effected, as illustrated in Fig. 4.6, passing the coolant in direct contact with the uranium bodies and by decreasing the thickness of the uranium plates **49**, thereby providing a larger number of plates with a greater number of passages between them. An intermediate condition will exist in the lattice structure in positions between the center and the top or the bottom. Thus, for these intermediate positions, uranium plates illustrated in Fig. 4.7 are employed wherein the thickness of the metal is greater than that illustrated in Fig. 4.6 but not as great as that shown in Fig. 4.5, so that the amount of cooling surface provided is somewhere between that shown in Figs. 4.5 and 4.6.

With the uranium **42** arranged in the form of vertically disposed plates **49** in the upper portion of the vertical passage **15** it is seen that a continuous passage through the graphite and the uranium is provided throughout the entire length of each of the cartridges **40**. As shown in Fig. 4.4, each of the cartridges is chamfered, as indicated at **53**, so as to facilitate the insertion

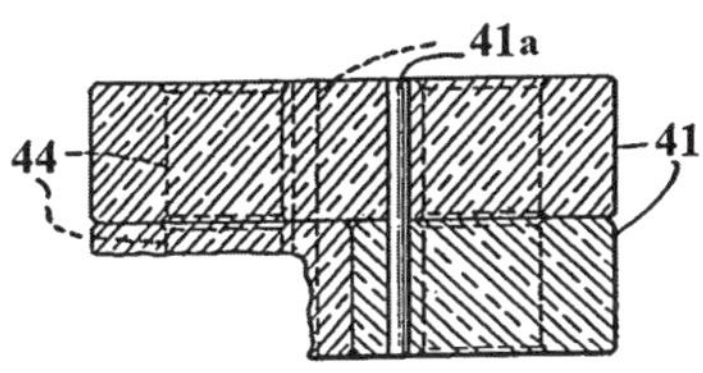

Fig. 4.13

of the cartridges into the openings **44** in the bricks.

The lattice structure filled to capacity usually contains about 12,000 uranium elements.

Referring again to Fig. 4.2, the helium gas enters the reactor from the pipe **34** (Fig. 4.1) through intake nozzles **54**, passes upwardly in the direction of the arrows through vertical passages in the lower layer **43a** of dead graphite (Fig. 4.3) and then continues upwardly through the passages **45** in the graphite cartridges **40** (Figs. 4.3 and 4.4) throughout the entire height of the lattice, and finally through passages **56** in the upper layer **43b** of the dead graphite into discharge chamber **38**. The hot gas leaves the reactor from the discharge chamber **38** through discharge nozzles **57**. From the reactor, as previously explained, the helium gas passes through the cooling circuit shown in Fig. 4.1.

At the top of the graphite **43** over each column of graphite cartridges **40** is a guide pipe **58** (Figs. 4.2, 4.3, and 4.14) socketed at its lower end in the graphite at **59**.

The top of the gas discharge chamber **38** is bounded by a steel floor **60** (Fig. 4.14) supported on a plurality of I-beams **61**, which, in turn, are carried by the steel shell. Above the floor **60** is an internal neutron and gamma ray shield, generally indicated at **62**, consisting of a 3 foot layer of graphite **63** and a layer of steel plates **64**. The lowermost steel plate **64a** is one inch thick, and the next plate **64b** is a one inch steel plate with about 2 percent to 10 percent boron in the metal. The next 14 plates are of mild steel or cast iron, and each plate is about one inch thick.

The guide tubes **58** extend upwardly through the hot gas chamber **38** and through the graphite layer **63** and steel layer **64** making up the internal shield **62**, and terminate at the top of the steel layer **64**. Slots **65** are provided in the guide tubes **58** throughout substantially the height of the hot gas chamber **38** through which helium gas passing up through the lattice is discharged. At the top of the guide tubes **58** is a steel and graphite plug **66** consisting of a steel layer **66a** corresponding in thickness to the thickness of the steel layer **64** of the internal shield and a graphite layer **66b** corresponding in thickness to the graphite layer **63** of the internal shield. A threaded rod **67** passes through the longitudinal center of the graphite portion **66b** of the plug and is threaded into the steel portion **66a**, as indicated at **68**. A suitable nut **69** is threaded onto the lower end of the rod **67** and a washer **70** is disposed between the nut **69** and the graphite **66b** engaging the lower face of the graphite portion **66b** of the plug. When the rod is drawn up tight into the steel portion **66a** of the

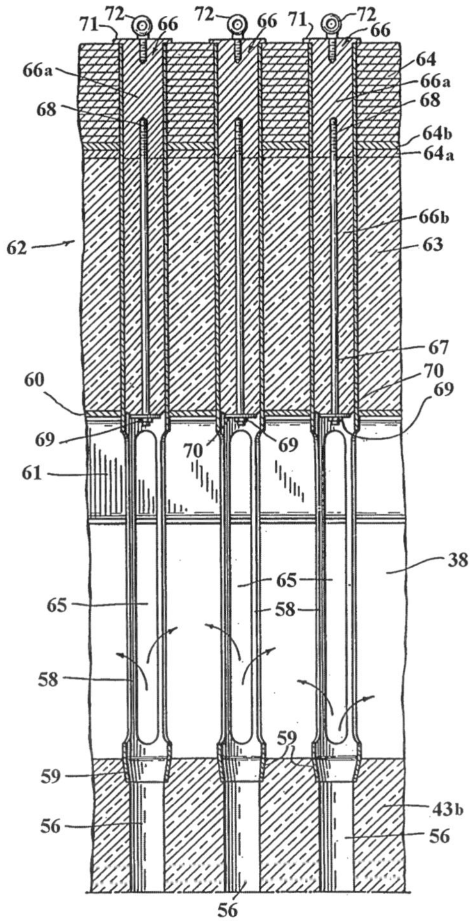

Fig. 4.14

plug, the assembly becomes rigid. The plug is provided with a flange **71** which rests on the upper surface of the internal radiation shield **62**, and a screw eye **72** is threaded into the top of the plug, providing means for grasping the plug for withdrawal and insertion into the shield.

The dome **39** encloses a chamber above the steel and graphite plug **62**. A covered manhole opening **73** is provided in the dome **39** for admitting persons in and out of this chamber for loading uranium into the lattice. As previously explained, the uranium is mounted in the graphite cartridges

40 (Fig. 4.4) and the loaded cartridges are lowered through the guide tubes **58** into the vertical passages **56** in the upper layer **43b** of the dead graphite, and finally down into the lattice structure. In order to prevent establishment of a substantial concentration of radioactive material in the dome it is found desirable to establish a helium pressure within the chamber somewhat higher than that in adjoining chambers thereby preventing or at least minimizing diffusion of radioactive gases into the dome.

At the bottom of the lattice, each of the columns of graphite cartridges **40** is supported on a dead graphite cylinder **74** (see Fig. 4.3) of AGX grade graphite through which is a vertical passage for conveying the helium gas into the passages **45** through the graphite cartridges **40**. The graphite cylinders, in turn, are supported on dump valves generally indicated in Fig. 4.2 at **59**. There is a plurality of such valves for discharging the uranium from the lattice.

During the operation of the present device, the transuranic element 94 is produced, together with radioactive fission products. After long periods of operation, the fission products may so poison the materials in the device by neutron absorption as to lower the reproduction ratio of the system. In order to perpetuate the chain reaction, it is essential that the value of the reproduction ratio remain above unity. Thus it may be desirable to remove the fission products from the lattice from time to time. This is done by removing the uranium from the lattice and replacing it with fresh material. The radioactive fission products and element 94 can then be separated from the uranium by extraction methods. The radioactive fission products are useful in medicine and as radiation sources, and element 94, being fissionable in a manner similar to U^{235}, may be used to enrich natural uranium to increase its efficiency in chain reacting systems, for example. The separation processes form no part of the present invention, so that no purpose will be served in describing herein the details thereof.

Dump valves **59** are especially useful in case of extreme emergency to prevent the reactor from being destroyed in case of failure of the control and safety rods to effectively limit the chain reaction to safe values. All or part of the uranium bodies can be dumped rapidly into chamber **35**, destroying the geometrical arrangement of the uranium bodies in the graphite and thus preventing the maintenance of the chain reaction. Such procedure is only resorted to after failure of the control and safety rods to act. The dumping may be manual and/or automatic upon rise of the neutron density in the system to a dangerous level, as will be brought out later.

Each dump valve **59** in effect is a flat car mounted on wheels **76**, oper-

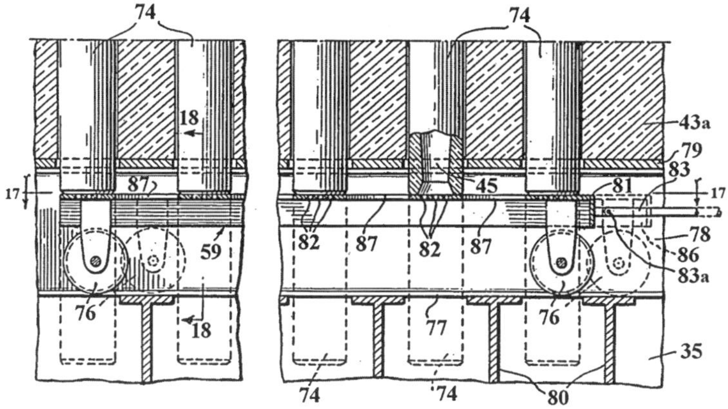

Fig. 4.15

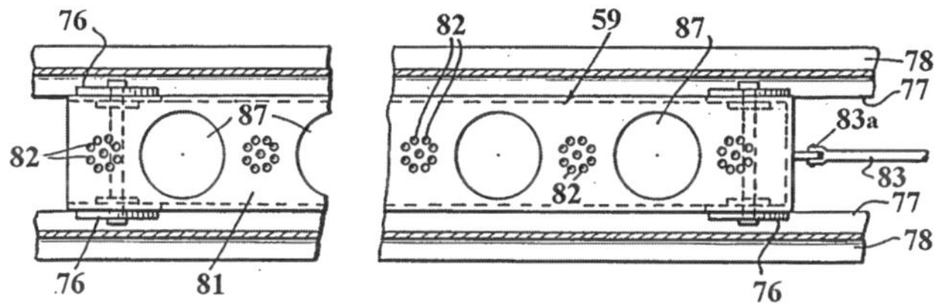

Fig. 4.16

ating on the bases **77** of adjoining I-beams **78**, the bases serving as rails. The I-beams **78** are spaced apart, as best shown in Fig. 4.17, and on their upper flanges support a floor **79** on which is mounted the dead graphite **43a** (see Fig. 4.3). The I-beams **78**, in turn, are supported on T-beams **80**, which are carried by the reactor shell.

Referring to Figs. 4.15-4.17, inclusive, each dead graphite cylinder **74** supporting a column of uranium cartridges **40** rests on the top **81** of the dump valves or cars **59** directly over a plurality of perforations **82** through the top **81** of the cars **59**, the perforations **82** being in alignment with the vertical passages through the dead graphite cylinders **74** and the passages **45** through the column of graphite cartridges **40**. Thus, with the dump valve or car **59** disposed in the solid line position shown in Fig. 4.15, helium gas passes upwardly from the inlet chamber **35** between the T-beams **80** and I-beams **78** through the perforations **82** in the top **81** of

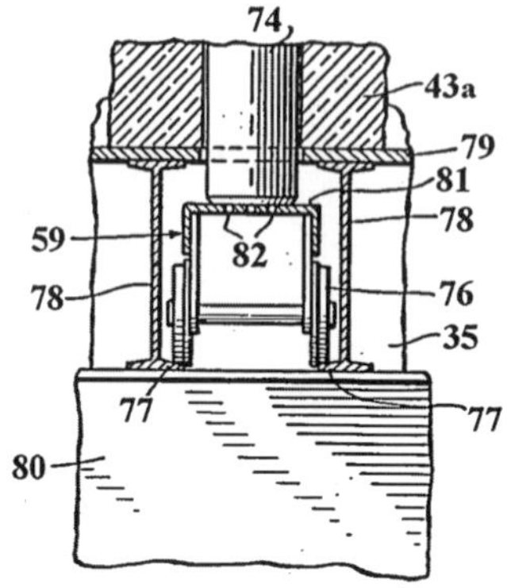

Fig. 4.17

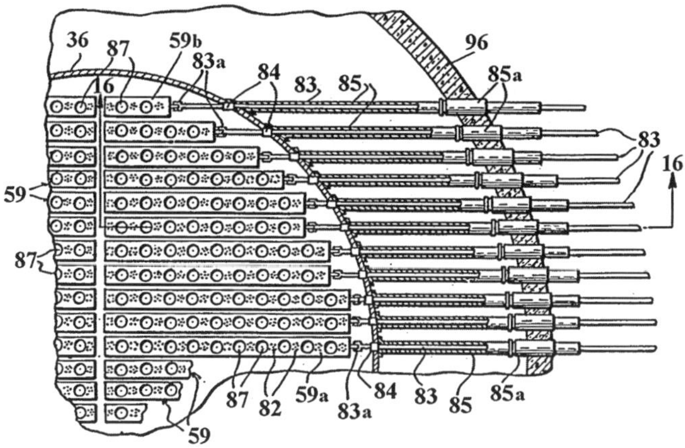

Fig. 4.18

the valve cars, and then through the vertical passages leading through the lattice structure.

The dump valves or cars **59** vary in length, depending upon their position in the lattice, as shown in Fig. 4.18, the centermost car **59a** being adapted to support **10** columns of cartridges, whereas the outermost car, shown at **59b**, supports only two columns.

As illustrated in Fig. 4.18, the cross section of the reactor, insofar as the dump valves are concerned, is divided into four quadrants, each quadrant being provided with 11 dump valves or cars and being served entirely independently of the other quadrants of the reactor. One quadrant only is completely shown in Fig. 4.18, though each of the other three quadrants in an exact duplicate of the one illustrated, except that in the left-hand half of the reactor shown in Fig. 4.18, the dump valves or cars operate in the opposite direction from that shown for the right-hand half of the reactor.

Extending horizontally from the front of each of the dump valves or cars **59** is a dump rod **83** suitably fastened to the car as shown at **83a** in Figs. 4.15 and 4.16. The dump rods **83** project through the shell of the reactor, as shown in Fig. 4.18, and extend horizontally to a valve control position (not shown) outside the rector shell. Stuffing boxes, generally indicated at **84** in Fig. 4.18, are provided around each of the dump rods **83** where the rod passes through the shell of the reactor to prevent escape of gases from inside the reactor shell. The dump rods **83** are slidable in guide tubes **85**. Stuffing boxes **85a** surround the guide tubes **85** where they pass through the concrete wall **96**. The guide tubes may be provided with suitable means (not shown) for absorbing or obstructing the path of neutron and gamma radiation therethrough as will be understood by those skilled in the art.

When dumping or discharging the uranium from the lattice structure for purposes of treating the uranium, all of the columns of uranium loaded graphite cartridges **40** supported on a single dump valve car are dumped or discharged simultaneously. In other words, referring to Fig. 4.18, all 10 columns of uranium loaded graphite cartridges supported on the dump valve car **59a** are discharged simultaneously. This is done by moving the car to the dotted line position **86** (see Fig. 4.15) wherein large circular openings **87**, shown in Figs. 4.15 and 4.16, become disposed directly under the dead graphite cylinders **74**, thereby permitting the dead graphite cylinders **74** and the column of uranium loaded graphite cartridges **40** supported thereon to discharge through the openings **87** in the dump valve car. The graphite and uranium thus discharged passes down into a downwardly extending chute **88** shown in Fig. 4.2, and accumulates at a gate valve **89**. Upon openings of this gate valve **89**, the uranium and graphite thus discharged can be passed into a suitable car or container.

A self-maintaining chain reaction, once started, must be controlled, for otherwise the neutron density may increase so rapidly that the reaction will reach violent proportion. The rate of heat generation in the lattice may exceed the rate of heat removal by the heat extraction system so that the temperature in the reactor will rise beyond a safety limit, even to the point of causing the uranium to melt with resulting breakdown of the lattice structure.

Referring to Figs. 4.2, 4.3, and 4.8, four control rods **90** and five safety rods **91** pass vertically into the lattice through vertical openings **93** in the graphite **41** between the cylinders of uranium **42**. These rods extend about two-thirds of the vertical distance through the lattice and serve to absorb neutrons so as to stabilize or stop the chain reaction.

The safety rods **91** are movable vertically in the lattice but are to be disposed either in their lowermost position in the lattice or in their outermost position, the latter position being assumed when the chain reaction is in effect. The safety rods are lowered into the lattice only in the event of emergency or when the operation of the reactor is to be entirely stopped for any reason.

The control rods **90** also are movable vertically in the lattice but their positions will vary, their function being to stabilize the chain reaction and, with the help of the helium gas, to maintain a constant temperature in the lattice of about 800°F as the maximum.

All of the rods **90** and **91** may be controlled automatically, as will be explained later, or manually. Though nine rods in all are shown more or less rods may be used if desired.

The absorption of the neutrons by the control rods **90** is accompanied by considerable heat. Referring to Fig. 4.19, each control rod **90** may be in the form of a hollow tube closed at its lower end and surrounding an inner hollow tube **90a**. The outer and inner tubes are spaced apart to form an annular passage **90b**. Cooling water may thus be circulated through the control rod **90** entering the rod through a flexible tube **90c** which conveys the water to the inside passage through the inner tube **90a**, the water flowing downwardly through this tube to the bottom of the outer tube and then upwardly through the outer passage **90b** finally leaving the rod through a second flexible tube **90d**. The water may then be conveyed externally of the pile through a suitable cooling circuit (not shown) and then may be returned to the inlet tube **90c**.

Extending above the lattice and surrounding each of the rods **90** and **91** is a guide pipe **94** (Fig. 4.3) to guide the movement of its corresponding rod. These pipes also may be provided with suitable means (not shown) for preventing escape of radiations from the reactor.

Referring to Figs. 4.2 and 4.18, a concrete shield **96** completely surrounds the reactor and extends vertically throughout the entire height of the reactor, terminating approximately eight feet above the top of the reactor dome **39**. The concrete shield is cylindrical in cross section and is

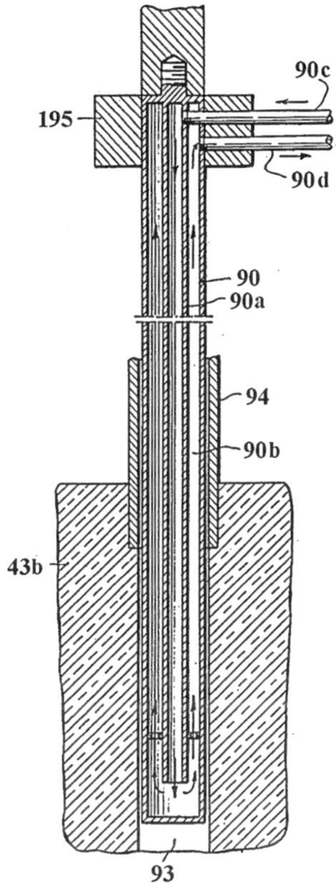

Fig. 4.19

spaced a minimum distance of ten feet from the nearest point on the reactor shell. This concrete shield **96** is filled with water, shown at **97** in Fig. 4.2.

Extending downwardly from the discharge nozzles **57** are discharge pipes **26a** which are enclosed within the concrete shield **96** and are disposed as closely as possible to the reactor shell so that there is a substantial thickness of water between the discharge pipes **26a** and the concrete wall **96**. These discharge pipes extend down to the bottom of the concrete shield to a position below the level of the ground indicated at **98**, and are connected to the main discharge header **26**, which likewise is disposed below the ground level.

Similarly, the gas intake header **34** (Fig. 4.2) is disposed below the level of the ground and passes through the concrete wall **96** of the shield entering the reactor through branch headers **34a**, which are completely submerged in the water shield.

The construction of the lattice is commenced with the excavation for the foundations for the reactor shell proper and for the concrete structure or shield containing the water in which the reactor is immersed. The reactor shell itself, with its underground connections, dump valves, and graphite supporting beams is then built up to the level of the upper or internal radiation shield **62**, and any necessary elevators and temporary super-structure required is erected. The graphite bricks **41**, which have previously been machined and bored, are then laid, doweling them together and taking particular care to keep the brick surfaces perfectly true and clean. When all the graphite bricks are laid, a temporary platform is laid over the top of the graphite bricks, the platform being provided with a hole directly over the location of each of the graphite cartridges. With a special reaming tool, each hole through the graphite is properly gauged throughout its length to detect the presence of any shoulders or projections, and such projections and shoulders are removed with the reaming tool. Following this operation, the temporary platform is removed and the beams **61** for supporting the upper or internal graphite shield **62** are placed into position. The upper shield **62** is then built up of AGX or AGR graphite to its final thickness of about three feet. The beams for supporting the steel plates are then laid and the 16 one-inch plates making up the steel portion **64** of the shield are laid on top of the beams. The guide pipes **58** are then inserted in place and suitable termocouples (not shown) for measurements of temperature within the reactor adjacent the point at which the cooling fluid is removed, installed on them. The steel and graphite plugs **66** are then inserted and the control and safety rod guides installed. The piping for the helium gas is then installed and fitted to the reactor shell and then the concrete shield **96** is poured. The mechanism for operating the control and safety rods are finally installed in their proper place, and the entire unit is then in condition for operation except for the loading of the uranium.

From within the dome **39** of the reactor shell the graphite cartridges **40** loaded with the uranium plates **42** are lowered into the openings in the graphite. Several cartridges arranged in a column may be lowered into place at a time.

During the loading of the uranium, both the control rods and safety rods **90** and **91** respectively are disposed in their innermost position in the lattice. As the charging of the uranium approaches the quantity necessary to produce an operating lattice, the loading operation will be suspended long enough for trial runs to be made. During the trial runs, the safety rods are moved to their outermost positions and the control rods are moved to various positions. Readings of the neutron density are made by the use of indium foils inserted in suitable testing slots in the lattice. The indium foil is inserted in the lattice and left there for a predetermined period of time, during which the foil is subjected to neutron bombardment. Then by suitable Geiger-Muller counters, the beta radiations from the indium foil radioactivity created by the neutron bombardment are counted. When it is found by this experimental means that the quantity of uranium in the lattice will support a chain reaction at the highest temperatures contemplated, the loading of the uranium is discontinued. All empty portions of the columns are then filled with dead graphite so as to complete the lattice structure. The ratio of graphite to carbon and the size of the uranium plates may be determined in accordance with the principles discussed in copending application of Fermi and Szilard, Serial No. 568,904, filed December 19, 1944, now Patent No. 2,708,656.[2]

The power plant above described is ideally adapted for automatic control to maintain the neutron density within the reactor substantially constant at a predetermined level, and thus give a substantially constant power output. Due to the fact that large masses of materials are utilized in the reacting portion of the structure, there is a temperature lag therein. Consequently, it is convenient to monitor and control the reaction by means of ionization chambers which will measure the neutron density at the periphery of the lattice portion of the structure. As the rate of neutron diffusion out of a chain reacting system is always proportional to the rate of generation of neutrons within the structure, the ionization chambers can readily placed at the periphery of the pile or lattice, and in fact are preferably so positioned in order that they be not subjected to the extremely high neutron densities existing within the lattice.

Before proceeding to a description of one type of control system that may be utilized, it is desirable to point out the manner in which the control rods operate to regulate the neutron density. In any self-sustaining chain reacting structure adapted to produce power, the neutron multiplication ratio of the system must be greater than unity. For any value over unity, by an amount sufficient to overcome impurity losses in the system, the chain reaction becomes self-sustaining and the neutron density without control will increase exponentially in point of time, until the device is destroyed.

[2]@ See page 217.

For proper control, the system must be held in balance by maintaining the chain reaction at some point where the production of new neutrons is balanced with the neutrons initiating the chain. Under these conditions, the reacting portion of the structure will continue to maintain the neutron density therein which obtained when the system was balanced.

However, in order to enable the reactor to reach a desired neutron density, the system must be permitted, for a period of time, to rise in neutron density until the desired density is reached. After the desired density has been reached, it is necessary thereafter to hold the system in balance.

Inasmuch as the reproduction ratio of the lattice structure is reduced by the presence of neutron absorbing impurities, such impurities can be introduced in the lattice in the form of the control rod which can be of a material such as boron or cadmium, which will absorb large amounts of neutrons. The depth to which this control rod penetrates into the lattice will determine the reproduction ratio of the lattice and a range can be obtained between a condition providing a neutron reproduction ratio which is greater than unity and a condition at which no chain reaction can be maintained. The exponential rise of neutron density can be made relatively fast or relatively slow in accordance with whether the multiplication ratio is permitted to be much greater than one, or only slightly greater than one. There is a small percentage of delayed neutrons emitted in the fission process. These delayed neutrons cause the neutron density to rise in an appreciable time rather than almost instantaneously. The time for doubling the neutron density increases as the multiplication ratio approaches unity, and by adjustment of the control rods any desired rate of rise can be obtained up to the maximum corresponding to the reproduction ratio characteristic of the given structure when all control and safety rods are removed.

The broad method of control preferred is to withdraw all safety and control rods from the structure to a point where there is an exponential, and preferably low, rise in neutron density within the structure. When a desired neutron density has been reached, the control rod is then returned into the pile to a point where the reaction is balanced. This balance is then maintained to maintain a constant power output in the reactor. The maintenance of the balance point with the control rod would be relatively simple were it not for the fact that changes in temperature change the reproduction ratio of the structure slightly, and in any chain reacting structure where there is any variation of pressure of the circulating medium for example, or in atmospheric pressure in case of a structure exposed to the atmosphere, the reproduction ratio of the system will slightly change. It is desirable, therefore, that the control rods be so manipulated that they maintain a constant neutron density within the system. Such a method of control may be accomplished by automatic connection of the control rods with an ionization chamber measuring neutron density, positioned within the reactor close to the lattice.

Furthermore, due to the exponential rise of neutron density within the reacting structure when the multiplication ratio is greater than unity, all possible precautions must be taken to prevent a continued exponential rise in neutron density in case of failure of the control rod to return to the balance position. It is for this reason that safety rods are provided. In case the neutron density has risen to a very large figure before the safety rods operate, it might then be impossible for the safety rods to absorb a sufficient number of neutrons to reduce a dangerously high neutron density to safe limits in a sufficiently short period of time. Under these circumstances, there is no alternative but to dump the uranium bodies and thus destroy the lattice arrangement by which the self-sustaining chain reaction is made possible.

While there are many means by which the control rods, the safety rods, and the dumping can be operated, it is believed that by the illustration and description of one simplified circuit, other and fully equivalent circuits will be made apparent to those skilled in the art.

Referring therefore to Fig. 4.20, which shows diagrammatically and reduced to lowest terms one form of control circuit that may be used for regulating the output of the power plant hereinbefore described, and referring first to control circuit **A**, a control ionization chamber **120** is placed within the reacting structure adjacent to the periphery of the lattice and filled with boron fluoride. A central electrode **121** is provided within the chamber **120** and connected to wire **122** leading outside of the reactor to a movable contact **123** on a resistor **125**. Resistor **125** is connected across a relay coil **126**. One side of relay coil **126** is connected to battery **127**, the other end of which is connected to shield **129** around wire **122**. Shield **129** is grounded, as is chamber **120**. Alpha ray ionization due to neutron reaction with the boron within chamber **120** is proportional to the neutron density. Usual auxiliaries such as current amplifiers, supplemented relays etc. may be employed if desired in order to increase the sensitivity of the device. Thus the current in resistor **125** is varied in accordance with neutron density reaching the ionization chamber. Relay coil **126** operates a relay armature **130**, which is sprig biased by spring **131** to contact one

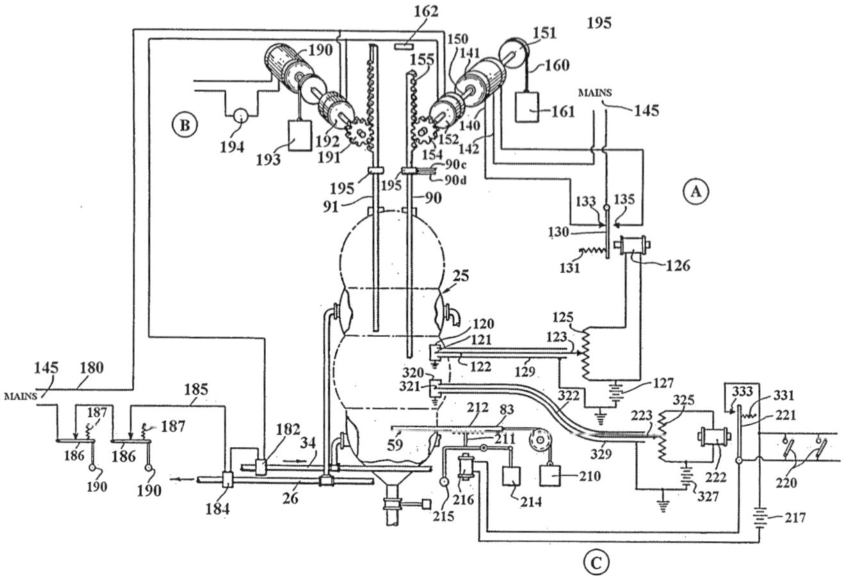

Fig. 4.20

motor contact **133**, and is urged by current in coil **126** to contact a second motor contact **135**. Contacts **133** and **135** connect to the outside of split winding **140** of motor **141**, the center connection **142** of which is connected through power mains **145** to armature **130**. Motor **141** operates shaft **150**, having on one side thereof a pulley **151**, the other end thereof being connected through a magnetic clutch **152** to a control rod gear **154**. Control rod gear **154** meshes with a rack **155** on a control rod **90**. Pulley **151** has a cable **160** wound thereon connected to a counter-weight **161** so that the weight of the control rod is substantially balanced by counter-weight **161**, thus permitting motor **141** to run easily in either direction. As several control rods are to be utilized to control the pile, it is preferable that each rod be controlled by a separate ionization chamber and that the chambers be distributed around the periphery of the lattice.

Having described a circuit for controlling the position of a control rod, we shall now describe its operation, considering the safety rods withdrawn. Slider **123** on resistor **125**, having previously been calibrated in terms of neutron density, is moved to the density position at which it is desired the reactor to operate, taking into account the difference in neutron density at the center of the lattice and at the periphery thereof during operation. This difference is a constant ratio at various operating densities. The reactor having at rest a neutron density much lower than the desired density at which relay coil **126** will receive enough current to operate armature **130**, very little ionization takes place in ionization chamber **120**, thus causing armature **130** to rest against contact **133**. Motor **141** is thus energized to withdraw the control rod from the reactor to a point as determined by a limit stop **162**, where the multiplication ratio of the reactor is just sufficiently greater than unity to permit an exponential rise in neutron density with the reactor. The motor **141** will stall at stop **162**, and should be of a type permitting stalling for the required time. The reaction at this position of the control rod becomes self-sustaining and the neutron density rises. In consequence, the ionization within chamber **120** rises. As the ionization in chamber **120** increases, more and more current passes through relay coil **126** until the desired value has been reached. Relay coil **126** then operates to cause armature **130** to connect with contact **135**, thus reversing the motor **141** to drive the control rod into the reactor to a point where the neutron density starts to decay. The control rod **90** will thereafter hunt between a point above the balance position where the neutron density rises and a point below the balance position where the neutron density decays, providing and average neutron density within the reactor as determined by the setting of slider **123** on resistor **125**. As the mass of the reactor causes any temperature change to lag behind any neutron density change, the temperature of the reactor is maintained substantially constant. If desired, any of the well known anti-hunting circuits may be utilized, as will be apparent to those skilled in the art.

The main purpose of the control circuit **A** is to regulate the control rods to substantially balance the neutron density to produce any desired average temperature within the reactor.

Due to the fact that it might be possible for the control system as described to fail and thereby leave the control rods in a position where the neutron density would continuous to rise indefinitely, both the safety rods **91** and the control rods **90** are preferably to be operated so as to enter the pile immediately upon any failure of the control rod system. One such emergency circuit for operating the rods is illustrated in circuit **B**. The circuit for each of the safety rods is the same.

The circuit **B** comprises a connection to main power line **145**, one side of which leads through wire **180** to the magnetic clutch **152** on the control rod drive. Wire **180**, after passing through clutch **152**, passes in series through a pressure switch **182** in the helium input pipe **34** and then in a series with a thermostat **184** in the helium output pipe **26**. Return wire **185** from the thermostat **184** then passes through a series of emergency switches **186**, each of which being held closed by springs **187**, and opened by manual operating handles **190**. Thus, magnetic clutch **152** on the control rod motor shaft is in series with the power mains, the helium output temperature thermostat **184**, the helium input pressure switch **182**, and the emergency switches **186**. The opening of any one of the switches will thus de-energize clutch **152** to remove the urge of the counter-weight from the control rods. The control rods will fall by gravity into the reactor. Furthermore, any failure of power in supply line **145** will also de-energize clutch **152** and permit the control rod to fall into the pile.

Each safety rod **91** may be raised out of the pile by safety rod motor **190** driving safety rod gear **191** through safety rod magnetic clutch **192**, the safety rod being provided with a balancing couter weight **193** similar to that used on the control rod. Motor **190** is hand controlled by switch **194**. Safety rod magnetic clutch **192** is connected in parallel with control rod magnetic clutch **152**. Thus, upon any failure of power in the mains or the opening of any of the switches **182**, **184**, or **186**, all of the rods will be relieved of their counter-weights and will fall into the pile by gravity. Stops **195** may be provided on the safety rods, and any suitable braking action used to reduce the impact shock on the reactor.

However, in an abundance of caution, it is arranged that in case of failure of both the control and emergency circuit just described to stop the chain reaction, a portion of the lattice structure may be dumped by the use of circuit **C**. Dump valves **59** can be used for this purpose by connecting one or more of the dump valve rods **83** to a dump counter-weight **210** urging the dump valve to open position. The dump valve is maintained closed by a dump valve latch **211** held in engagement with rack **212** on the dump valve rod **83** by latch counter-weight **214**. Latch **211** may be withdrawn manually by means of handle **215**, or electrically by means of coil **216**, which is energized by a battery **217**, which preferably is separate and apart from power line **145**. The circuit for coil **216** may be closed in two ways, either by hand switches **220** or by relay armature **221** moved by a relay coil **222** controlled by an emergency circuit identical in all respects with the ionization chamber control circuit previously described for circuit **A**. The individual portions of this circuit will therefore be numbered with numbers plus **200** corresponding to those used in circuit **A**, previously described.

The emergency ionization circuit is set by means of slider **223** to respond to a neutron density approaching the danger limit of the structure. Thus, in case the neutron density approaches the danger limit, the relay operates and portions of the lattice are automatically dumped. In case of failure of the automatic circuit to dump, the hand switches can be utilized; and, in case of failure of the hand switches, the handle **215** may be operated to dump the system. Obviously, such dumping is resorted to only in extreme emergency.

To place the system in operation, it is first tested under 150 pounds air pressure and examined for leaks, using the soap and water method. Each section of the structure that can be blocked off is separately evacuated, then filled with helium at 115 pounds pressure. This helium is then evacuated and the system filled with new helium with all radiation shields in place and with the control and safety rods completely inserted into the reactor. The circulation of water is started through the compressors, the after-coolers, and the heat exchangers. The compressors are then started, to circulate the helium. The control rods are left in the reactor, and the safety rods withdrawn and held. The control circuit **A** is then set for a neutron density corresponding to about a 70°F rise in temperature in the reactor. After checking all parts of the system for proper operation, the temperature of the outlet gas is then stepped up at the rate of about 70°F a day until the desired temperature of 800°F is attained in the outlet gas. The control circuit **A** is then set to keep the neutron density sufficiently constant to maintain the desired gas output temperature and the safety circuits **B** and **C** set to operate to shut down the system in case of any substantial rise in neutron density or output gas temperature, failure of the circulating system or suspension of power supply from the mains. In accordance with good engineering practice, sufficient standby auxiliaries should be provided for substitution in the system in case of failure of individual units, such as, for example, a compressor. A standby generator may also, if desired, be provided for immediate switch-over in case of power main failure. Such standby equipment, however, does not obviate the necessity for adequate safety and emergency control.

In addition to the usual industrial hazards during operation of the power plant, operating personnel must be protected from injury by gamma rays and neutrons generated in the pile reactor and from radiation from the circulating gas and equipment, from beta rays, from close contact with radioactive materials, and from radioactive poisoning due to inhalation of radioactive gases.

The problem of radiation from the pile is met by building the concrete shield, of one foot minimum thickness, around the reactor shell, and filling the intervening space with water which covers the top of the shell to a depth of 8 feet. The minimum thickness of the water layer between the steel shell and the concrete shield is 10 feet. The gas discharge pipes pass downward through this water as near to the shell as practical, so that no extra shielding for them is required until they emerge from the concrete shield somewhat below ground level. These precautions will reduce radiation from the pile structure to 0.10 Roentgen per 8 hr day at the point of closest possible approach.

Walkways may be built around the concrete shield at the level of the dump valves, and at the level of the housings for the ionization chambers. A stairway and elevator may also be installed, with platforms at each outlet through the radiation shield. Roughly 2200 ft^3/day of an explosive mixture of hydrogen and oxygen will be evolved from the water surrounding the shell. While this gas production is the basis for a "no-smoking" rule near the top of the system and for a requirement for spark-proof electrical equipment, natural ventilation makes other precautions unnecessary.

Many of the problems in connection with protection of personnel arise from the fact that the helium gas will carry some of the fission products with it, emitting both gamma rays and neutrons as it flows. The fraction of the fission products formed which escapes into the gas is, however, less than 2×10^{-4}. Local deposition of the fission products in, for example, the heat exchangers or the filters will intensify the radiation problem there, but will alleviate it in other locations, such as the compressors.

For protection against radiation from fission products which are in the circulating gas, all lines and equipment are either placed in concrete lined trenches and covered with a sufficient quantity of dirt to provide adequate protection, or housed in thick walled concrete structures. All valves are constructed with an extended bonnet and long stem, as shown in Fig. 4.18, which permits the stuffing boxes to be carried through the radiation shield for tightening. Stuffing boxes on all large valves may be vented in order to permit the helium that leaks through the stuffing boxes to be collected and returned to the system. Motor driven valves are preferably located at key points in the circulating system so that parts of the system may be blocked off quickly in the event of a line break or serious leak in the system. The valve control system is arranged so that any valve may be operated either from a point near the valve itself of from a main instrument house remote from the system. On any piece of equipment which may be cut out of the system, double valves are installed, one motor driven and one hand

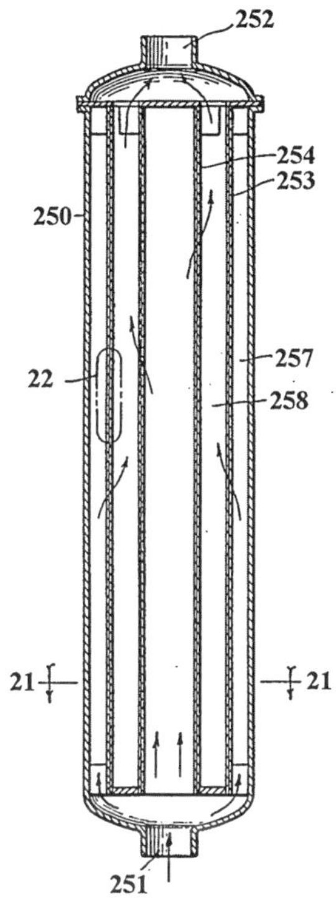

Fig. 4.21

operated.

The primary purpose of the gas cleaners or filters is to remove some of the solid fission products from the circulating gas. A less important function is to remove graphite dust from the system. A filter which will remove fine smokes is effective in taking out at least a substantial portion of the solid fission products. A filter element of a special rock wool, of fiber

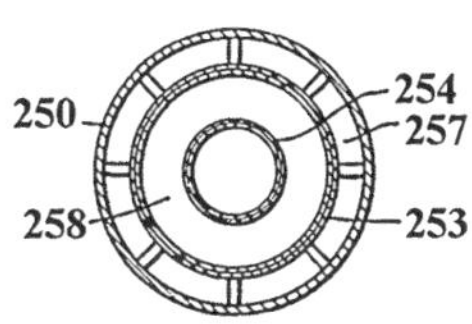

Fig. 4.22

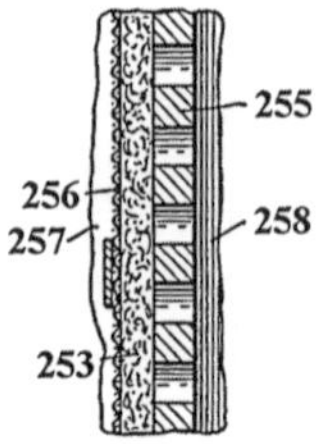

Fig. 4.23

size of the order of 2 to 3 microns, and containing about one pound per square foot of the rock wool, is utilized in the example described.

Accordingly, the filters **29**, shown in Fig. 4.1, in the circulating system, have been designed on the basis of the smoke filters. Each filter has a cylindrical shell **250** (see Figs. 4.21 to 4.23) about 6 feet in diameter and 24 feet long, with 24 inch inlet and outlet nozzles **251** and **252** respectively. The filter consists of two concentric cylinders **253** and **254** of rock wool, supported on a perforated steel plate **255** and held in place with suitable screen **256**. The gas flows from the outside **257** of the outer filter and in the inside of the inner filter into the annular space **258** between the two. Gas velocity through the filters is about 0.75 foot per second, and the pressure drop is less than one pound per square inch. Radiation protection around the filters of 8 feet of well packed earth is sufficient and convenient.

Each compressor may be housed in a concrete buildings with walls 4 feet thick, carrying a 3 foot thick roof made up of precast slabs which can be removed. A crane may be erected on the roof and may be used for erection of the units and for moving heavy parts for repair. Each compressor house may be individually ventilated with a fan, and each compressor may have its own instrument board. The compressor itself may become radioactive in time, but with ten operating compressors and two standbys, for example, it is evident that in a systematic rotation of compressors each machine will be available for examination and cleaning at least once in 30 days. Special precautions to detect compressor leakage may be taken, for example, by using recording chambers in the vent of each stall.

The inhalation hazard for personnel operating the structure can be considered on the basis that the maximum helium leakage which might occur from the system is 1 percent of the volume contained per day, and on the basis that the maximum safe radiation permissible in the lungs is that corresponding to 0.1 Roentgen per day over the whole body.

All ordinary safety precautions can be taken to remove this inhalation danger, and several means can be used dependent on locale of the plant. Among these are (1) to use an odorant such as, for example, ammonia in the helium gas in order to indicate small helium leaks, (2) the use of a powerful fan to blow any escaping helium away from valve packing and other equipment on which work is being done, (3) the use of oxygen masks to preventing breathing escaping helium in locations where the use of a fan is impractical, (4) careful check, under competent supervision, of the radiation to which men are exposed, and relieving any men who have received more than a safe dose of radiation in any one day, (5) periodic medical examination of all personnel.

While the invention has been described with particular reference to a graphite type of moderator, it is to be understood that it is not limited to this moderator and that other neutron slowing materials which do not have an excessive neutron absorbing property may be utilized. For example, heavy water (D_2O) or beryllium may be used very effectively for this purpose.

In accordance with a further modification, it is also possible to change the form of the bodies of the uranium bodies which are dispersed in the moderator. For example, the uranium bodies may be in the form of cylinders, spheres or cubes dispersed in a moderator similar to the plates herein described. In such a case some modification of the volume ratio of uranium to moderator may be required. Since the establishment of the proper ratio has been described in other copending applications, for example, an application of Fermi, Serial No. 534,129, filed May 4, 1944,[3] and does not form an essential feature of the present invention it is believed unnecessary to discuss herein the principles involved.

In accordance with a further modification coolants other than helium may be suitable; for example, air, oxygen, or water vapor. The problem in selecting a proper coolant is dependent upon the tendency of the coolant to absorb neutrons. Thus if an excessive quantity of neutrons are absorbed by the coolant, the number of neutrons remaining for absorption by uranium and fission of U^{235} will be insufficient to permit establishment and maintenance of a self-sustaining chain reaction. Consequently it is desirable to utilize a cooling agent which has but a slight tendency to absorb neutrons

[3]@ That is E. Fermi, *Test exponential file*, U.S. Patent No. 2,780,595, February 5, 1957 (see page 181).

and in any event to avoid establishment of a quantity of cooling agents in the reactor of such magnitude that the amount of neutrons absorbed will prevent maintenance of the chain reaction. Coolant gases having a danger coefficient below about 2.5 are found preferable although gases having a danger coefficient as high as 15 are sometimes used.

Within limits the approximate effect of any quantity of coolant may be computed by computing the danger sum of the total quantity of coolant in the reactor at any given time. The danger sum of this coolant may be computed in the manner previously described for computing the danger sum of impurities or neutron absorbers in the reactor simply by determining the weight of coolant in the reactor per unit weight of uranium and the danger coefficient for each impurity or absorber. In no case should the danger sum of the coolant in the reactor be so great as to reduce the reproduction factor K below unity. The permissible danger sum will be determined, therefore, by the magnitude of K of the reactor without the coolant. For example, with a carbon moderator, maximum K obtainable is about 1.1 and in no case can the danger sum of the coolant exceed 0.1 or the chain reaction will no longer be self-sustaining. Preferably in such case the danger sum of the coolant should be maintained below about 0.05. On the other hand, with a moderator which has less tendency to absorb neutrons such as D_2O, a maximum K obtainable is approximately 1.3 and in such a case, the danger sum of the coolant should certainly be less than 0.3. Likewise when natural uranium is used in a beryllium moderator, the danger sum should be less than 0.18. Where other neutron absorbers such as impurities or controls are present in the reactor the permissible quantity of coolant is decreased by the danger sum of such absorbers. Moreover most reactors have a substantial leakage factor usually not less than 0.01 K units and thus permissible coolant must be diminished by this factor also. The maximum permissible danger sum may be increased where enriched materials containing concentrations of U^{235} or 94^{239} greater than the concentration of U^{235} in natural uranium but in all events the volume of coolant in the reactor should be correlated in accordance with the reproduction factor and/or reproduction ratio to prevent these ratios from decreasing below unity and preferably to permit establishment of a reactor in which the reproduction ratio r is not less than about 1 but not more than about 1.01.

While the invention is particularly concerned with reactors in which natural uranium is used as the source of fissionable isotope, other compositions consisting of or containing fissionable isotopes which upon fission yield neutrons may be utilized. For example, enriched compositions containing U^{235}, U^{233}, or 94^{239} dispersed in U^{238} or Th^{232} may be used for this purpose. Compositions containing 5 or 10 percent or even more of the fissionable isotope in such cases may be used and under such conditions a greater latitude in the amount of coolant which may be permitted in the reactor is possible.

Where compositions comprising abnormal amounts of U^{235} or uranium or therein containing U^{233} or 94^{239} or similar isotopes are used in the chain reaction some change in the rate of neutron production and consequently some change in the permissible latitude in the amount of impurities or leakage may be encountered. For example as the reactor herein described proceeds in its operation, a quantity of 94 is generated which being fissionable takes part in the reaction. Moreover the fission products produced being impurities tend to reduce the K factor but this effect is counteracted to a substantial degree by the neutron output of 94^{239} upon fission. After a substantial period of operation of the reactor the K factor may be increased substantially due to the fact that somewhat more neutrons are evolved upon fission of 94^{239} than are obtained upon fission of U^{235}.

At all events to maintain a self-sustaining chain reaction, $Z - (A + B + C + L)$ must not be less than about one, where:

Z = number of neutrons produced by fission per neutron consumed in fission of fissioning isotope or isotopes;
A = number of neutrons absorbed by the non-fissioning isotope for example U^{238}, Th^{232} producing fissioning isotope per neutron consumed in fission of fissioning isotopes;
B = the number of neutrons absorbed by the moderator per neutron consumed in fission;
C = number of neutrons absorbed by other absorbers impurities or control rods per neutron consumed in fission; and
L = number of neutrons lost in leakage per neutron consumed in fission.

Z may be ascertained by bombardment of a pure isotope with neutrons of the energies developed in a reactor.

A, B and C may be computed to approximate values by computing the danger sum basing the computation upon the ratio of neutron capture cross section and concentration of the moderator, non-fissioning isotope or other absorber to that of the fissioning isotope or isotopes in the manner previously described or may be determined experimentally and L may be determined experimentally. Preferably the value $Z-(A+B+C+L)$ should not exceed unity by more than the fraction of neutrons evolved which are delayed neutrons. In the case of U^{235} this fraction is about 0.01.

Since a coolant is a neutron absorber the amount of coolant within a reactor must be controlled so as to prevent the value C from reaching a value such that the difference between Z and the sum of A, B, C, and L becomes less than one. Preferably this value $A+B+C+L$ should be such that the value $Z-(A+B+C+L)$ is not less than about one nor more than about 1.01.

In this disclosure the helium gas has been shown as being under pressure. Cooling can be effected with the gas at or near atmospheric pressure as shown, for example in a copending case of Enrico Fermi and Leo Szilard, Serial No. 596,465, filed May 29, 1945.[4]

During the operation of the neutronic reactor, particularly at high neutron densities, radioactive elements of exceedingly high capture cross section may be formed in the uranium as an intermediate element in the decay chains of fission fragments and this formation will lower the value of the reproduction factor for the system. Radioactive xenon^{135} is an example of such an intermediate element, this production having a half life of about 9 hours and being formed mostly from radioactive iodine which has an half life of about 6.6 hours and decay to barium. There should be sufficient excess in the available reproduction ratio of the reactor so that, in the event the reproduction factor is reduced as a result of the formation of an intermediate decay element having a high capture cross section for neutrons, the control rods may be withdrawn sufficiently to maintain the operating reproduction ratio at a value of unity while maintaining the power output at the desired level. It might be desirable to initially construct the reactor sufficiently over-sized to supply this excess reproduction ratio when needed and in this event removable impurities for example in the form of additional shim or control rods may be initially placed in the reactor and kept there at all times until the reproduction ratio commences to fall as the result of the formation of these intermediate decay elements.

While the theory of the nuclear chain fission mechanism in uranium set forth herein is based on the best presently known experimental evidence, we do not wish to be bound thereby, as additional experimental data later discovered may modify the theory disclosed. Any such modification of theory, however, will in no way affect the results to be obtained in the practice of the invention herein described and claimed.

[4]@ That is E. Fermi, *Air cooled neutronic reactor*, U.S. Patent No. 2,836,554, May 27, 1958 (see page 409).

We claim:

1. A device for effecting a self-sustaining neutron chain reaction comprising uranium bodies geometrically arranged in graphite, a shell surrounding the graphite, the shell being spaced slightly from the graphite to provide an expansion space for the graphite, a quantity of helium in the shell and circulated under super atmospheric pressure through the graphite and in heat exchange relationship with the uranium, and a resilient packing in the space between the graphite and the shell to prevent passage of helium therethrough.

2. A device for effecting a self-sustaining neutron chain reaction comprising uranium bodies geometrically arranged in graphite, a shell surrounding the graphite, the shell being spaced slightly from the graphite to provide an expansion space for the graphite, a quantity of helium in the shell and circulated under super atmospheric pressure through the graphite and in heat exchange relationship with the uranium, and asbestos fibre filling the space between the graphite and the shell to prevent passage of helium therethrough.

References cited in the file of this Patent

UNITED STATES PATENTS

2,708,656 Fermi et al. .... May 17, 1955

FOREIGN PATENTS

114,150 Australia .... May 2, 1940
114,151 Australia .... May 3, 1940
861,390 France .... Oct. 28, 1940
233,011 Switzerland .... Oct. 2, 1944
648,293 Great Britain .... Jan. 3, 1951
(Corresponding to French Patent 861,390)

OTHER REFERENCES

Power, July 1940, p. 58.
Smyth: "Atomic energy for Military Purposes", August 1945.

Goodman: "The Science and Engineering of Nuclear Power", vol. 1, p. 275, Addison-Wesley Press, Inc. (1947).

Kelly et al.: Phys. Rev. 73, 1135-9 (1948).

Chapter 5

Neutronic Reactor I

Enrico Fermi, Santa Fe, N. Mex., assignor to the United States of America as represented by the United States Atomic Energy Commission

Application May 12, 1945, Serial No. 593,510

U.S. Patent No. 2,931,762 - Patented Apr. 5, 1960

This Patent is an "evolution" of the previous Patents (especially Patent No. 2,708,656).

Particular attention is paid here to the problem of removing heat from a chain reacting device. The system proposed (and carried into effect) is to cool the moderator (but not directly the uranium) with a liquid, for example water or diphenyl, circulating in tubes of aluminium, lead or copper. This method has several advantages. First of all, a medium circulating in a pile, in direct contact with the uranium, produces the physical deterioration of the uranium, due to the fact that high temperatures and intense neutron densities cause an acceleration of the normal rates of corrosion. Furthermore, many coolants absorb neutrons to such an extent that they cannot be used in large quantities in a pile. Another problem which can be avoided, or at least minimized, with the system proposed here, is the contamination of the coolant by radioactive fission products.

Apart from the presentation of the novel kind of reactor mentioned above, some details about the construction and operation of the system, including interesting tricks, are reported. Several new physical data are presented, for example on the neutron density inside the active portion of the reactor during its construction and on the temperature at various places inside the reactor.

The main subject of this Patent does not appear in any other published papers.

My invention relates to the general subject of nuclear fission and particularly to the establishment of self-sustaining neutron chain reactions, compositions of matter and methods of producing such compositions suitable for use in creating a self-sustaining chain reaction by nuclear fission of uranium by slow neutrons in a neutronic reactor.

Certain isotopes, U^{235}, U^{233}, 94^{239} or other isotopes of element 94 can be split of fissioned by bombardment with thermal neutrons, i.e., neutrons in or close to thermal equilibrium with the surrounding medium, and the neutrons producing fission give rise to new fission neutrons in sufficiently large numbers to overcome the various neutron losses in the system. Since the result of the fission of the uranium nucleus is the production of two lighter elements with great kinetic energy, plus approximately 2 neutrons for each fission along with beta and gamma radiation, a large amount of power can be made available.

Most of the neutrons arising from the fission process are set free with the very high energy of above one million electron volts average and are therefore not in condition to be utilized most efficiently to create new thermal fissions in U^{235} or similar isotope. The fast fission neutrons, after they are created, are slowed down to thermal energies before they produce fresh fission by bombardment of additional U^{235} atoms. When the system is made in which neutrons are slowed down without much absorption until they reach thermal energies and then mostly enter into uranium rather than into any other element, a nuclear chain reaction is obtained even with natural uranium. Light elements, such as deuterium, beryllium, or carbon, the latter in the form of graphite, can be used as slowing agents, or moderators. Carbon in the form of graphite, for example, is a relatively inexpensive, practical and readily available moderator for slowing fast neutrons to thermal energies.

During the chain reaction heat is developed and in order that the reaction may be conducted at a rapid rate it is necessary to remove this heat from the reactor. In accordance with the present invention it has been found that the reactor may be effectively cooled by passing a coolant through the moderator through channels which are spaced from the uranium bodies disposed in the moderator. By cooling in this manner, problems which arise due to corrosion of the uranium or contamination of the coolant when the coolant is passed in contact with or in heat exchange relationship with the uranium, are substantially minimized or eliminated.

Certain principles respecting neutron reactors and their application to the present invention are discussed below.

The ratio of the fast neutrons produced in one generation by the fissions to the original number of fast neutrons in a system of infinite size using specific materials is called the reproduction or multiplication factor of the system, and is denoted by the symbol K. If K can be made sufficiently greater than unity to create a net gain in neutrons in the system and the system made sufficiently large so that this gain is not entirely lost by leakage from the exterior surface of the system, then a self-sustaining chain reacting system of finite and practical size can be built to produce power by nuclear fission of natural uranium. The neutron reproduction ratio (r) in a system of finite size differs from K by the external leakage factor, and this must be sufficiently greater than unity to permit the neutron density to rise exponentially. Such rise will continue indefinitely if not controlled at a desired density corresponding to a desired power output. Such control is feasible and will be later described.

It is therefore important to so construct a system comprising uranium and a slowing medium in which neutron losses are reduced to such extent that a controllable self-sustaining neutron chain fission reaction is obtained therein, with resultant regulated production of neutrons, liberation of power in the form of heat, the production of radioactive fission products and new elements both radioactive and stable produced by the absorption of neutrons in uranium.

During the interchange of neutrons in a system comprising bodies of uranium of any size disposed in a neutron moderator, neutrons may be lost to the chain reaction in four ways:

(1) By absorption or capture in the uranium content of the bodies without producing fission,
(2) By absorption or capture in the moderator material itself,
(3) By leakage out of the system through the periphery thereof, and
(4) By absorption or capture by the impurities present in both the uranium bodies and the moderator.

These losses will be considered in the order named, as any one of these losses, if too high, can prevent a self-sustaining chain reaction from being attained.

Natural uranium, particularly by reason of its U^{238} content, has an especially strong absorbing power for neutrons when they have been slowed down to moderated so-called resonance energies. The absorption in uranium at these energies is termed the uranium resonance absorption or capture. The neutrons are captured by the isotope U^{238}. This capture does not result in fission but leads to the creation of the relatively stable nucleus 94^{239}. Neutron resonance absorption in uranium may take place either on the surface of the uranium bodies, in which case the absorption is known as surface resonance absorption, or it may take place further in the interior of the uranium body, in which case the absorption is known as volume resonance absorption. Volume resonance absorption is due to the fact that some neutrons make collisions inside the uranium body and may thus arrive at resonance energies therein. After successfully reaching thermal velocities, a portion of the neutrons are subject to capture by U^{238} without fission leading to the production of 94^{239}.

When fission occurs by neutron capture, the following reaction takes place:

$$_{92}U^{235} + n \rightarrow A + B + \text{about } 2n \quad \text{(average)}$$

A = "Light" fission fragment, e.g., Br, Kr, Rb, Sr, Y, Zr, Cb, Mo, 43, Ru, Rh. Atomic mass 83-99 inclusive. Atomic number, 34-45 inclusive.

B = "Heavy" fission fragment, e.g., Sb, Te, I, Xe, Cs, Ba, La, Ce, Pr, Nd. Atomic mass 127-141 inclusive. Atomic number, 51-60 inclusive.

The fission elements are unstable and radioactive with half-lives varying in length in accordance with the element formed.

When a neutron is absorbed by U^{238} without fission taking place the following reaction occurs:[1]

$$92^{238} + n \rightarrow 92^{239} + [\text{6 MeV of } \gamma \text{ rays, not necessarily all of one frequency}]$$

$$92^{239} \xrightarrow{\text{23 min.}} 93^{239} + \beta^- \; [\text{1 MeV } \beta^-, \text{ no } \gamma \text{ rays}]$$

$$93^{239} \xrightarrow{\text{2.3 days}} 94^{239} + \beta^- \begin{bmatrix} \text{600 keV upper } \beta^- \text{ energy limit. Also} \\ 2\ \gamma \text{ rays, 400 keV and 270 keV, about} \\ \text{1/2 converted to electrons} \end{bmatrix}$$

It is possible by proper physical arrangement of the materials to reduce substantially the uranium resonance absorption. By the use of light elements as described above for moderators a relatively large increment in energy loss is achieved in each collision and therefore, fewer collisions are required to slow the neutrons to thermal energies, thus decreasing the probability of a neutron being at a resonance energy as it enters a uranium

[1]@ In the third reaction, the original version of this Patent reports incorrectly "upper γ" instead of "upper β^-".

atom. During the moderation, however, neutrons are diffusing through the slowing medium over random paths and distances so that the uranium is not only exposed to thermal neutrons, but also to neutrons of energies varying between the energy acquired by reason of fission and thermal energy. Neutrons at uranium resonance energies will, if they enter uranium at these energies, be absorbed on the surface of a uranium body, whatever its size, giving rise to surface absorption. Any substantial reduction of overall surface of the same amount of uranium will reduce surface absorption, and any such reduction in surface absorption will release neutrons to enter directly into the chain reaction.

For a given ratio of slowing material to uranium, surface resonance absorption losses of neutrons in the uranium can be substantially reduced by a large factor when the uranium is aggregated into substantial masses in which the mean spatial diameter is at least about 0.5 centimeter for natural uranium metal and somewhat larger when the bodies are of a uranium compound as hereinafter more fully discussed. For example with UO_2 the minimum radius is 0.75 centimeters and with other uranium compounds a similar variation from metallic uranium may be observed. The degree of this variation is dependent upon the density of the uranium compound, its bulk density, and the absorption coefficient of other elements therein for neutrons. In any event the uranium may be placed in the system in the form of geometrically spaced uranium masses or bodies of substantial size, preferably either of metal, oxide, carbide, or combinations thereof. The uranium bodies can be in the form of layers, rods, or cylinders, cubes or spheres, or approximate shapes, dispersed throughout the graphite. Optimum conditions are obtained with natural uranium by using metal spheres.

The K factor of a mixture of fine uranium particles in graphite assuming both of them to be theoretically pure, would only be about 0.785. Actual K factors as high as 1.09 have been obtained using aggregation of natural uranium in the best known geometry in graphite, and with as pure materials as is presently possible to obtain.

Assuming theoretically pure graphite, and theoretically pure natural uranium metal of the highest presently obtainable densities of about 18 g/cm^3, the maximum possible K factor theoretically obtainable is about 1.1 when the uranium is aggregated with optimum geometry. Still higher K factors can be obtained in uranium having more than the naturally occurring content of fissionable elements or by the use of different moderators in place of carbon. Adding such fissionable material is termed enrichment of the uranium.

Neutrons are also subject to capture by the moderator. While carbon and beryllium have very small capture cross sections for thermal neutrons, and deuterium still smaller, an appreciable fraction of thermal neutrons, about 10 percent of the neutrons present in the system under best conditions when graphite is used, is lost by capture in the graphite atoms during diffusion therethrough. It is therefore desirable to have the neutrons reaching thermal energy promptly enter uranium.

When the system has a K factor greater than unity, the uranium will support a chain reaction providing an exponential rise in neutron density, if the overall size of the system is sufficiently large to overcome the loss of neutrons escaping from the system. Thus the overall size is important.

The overall size of the system will vary, depending upon the value of the K factor of the system. If the reproduction factor K is greater than unity, the number of neutrons present will increase exponentially and indefinitely, provided the structure is made sufficiently large. If, on the contrary, the structure is small, with a large surface-to-volume ratio, there can be a rate of loss of neutrons from the structure by leakage through the outer surfaces, which may overbalance the rate of neutron reproduction inside the structure so that a chain reaction will not be self-sustaining. For each value for the reproduction factor K greater than unity, there is thus a minimum overall size of the lattice structure known as the critical size wherein the reproduction ratio is unity. The rate of diffusion of neutrons away from a large structure in which they are being created through the exterior surface thereof may be treated by mathematical analysis when the value of K is known.

In the case of a spherical structure employing uranium bodies of any shape or size imbedded in a carbon slowing material, the following formula gives the approximate overall radius for critical or operating size:

$$K - r = \frac{C}{R^2},$$

where R is the radius in feet, r is the reproduction ratio which for an operative reactor cannot be below unity and for critical size is unity, and C is a constant which is dependent on the detail of the lattice. For the aforementioned carbon lattice, the value of C is of the order of 7 to 8 whereas with heavy water as the moderator the value of C is about 3.

For a parallelepiped structure rather than spherical, the critical or operating size can be computed from the formula:

$$K - r = C\left(\frac{1}{a^2} + \frac{1}{b^2} + \frac{1}{c^2}\right),$$

where a, b, and c are the lengths of the sides in feet. The critical or operating size for a cylindrical structure is given by the formula:

$$K - r = C\left(\frac{1}{h^2} + \frac{0.59}{R^2}\right),$$

where h is the height and R is the radius, both in feet irrespective of the shape of the uranium bodies. When operating sizes are to be found the value of r desired (for example $r = 1.005$) is to be used in the computation.

However, even when the first three above-mentioned losses are reduced to a practical minimum, no self-sustaining chain reaction can be obtained in any system unless impurities in the materials used for the reaction are reduced to such an extent that the loss by parasitic capture by such impurities will not prevent the reaction from becoming self-sustaining. Impurities present in both the uranium and the moderator consequently constitute a very important neutron loss factor in the chain. The effectiveness of various elements as neutron absorbers varies tremendously.

Certain elements such as boron, cadmium, samarium, gadolinium, and some others, for example, if present even in a few parts per million, could very likely prevent a self-sustaining chain reaction from taking place. It is highly important, therefore, to remove as far as possible all impurities capturing neutrons to the detriment of the chain reaction from both the slowing material and the uranium. If these impurities are present in too great quantity, the self-sustaining chain reaction cannot be attained. The permissible amounts of impurities will vary for each specific geometry, depending upon such considerations as the form in which the uranium is used – that is, whether natural or enriched, whether as metal or oxide. The type of slowing down material used – for example, whether deuterium, graphite, or beryllium – also influences the effect of impurities, as do the weight ratios between the uranium and the slowing down material. Elements such as oxygen may be present, and, as previously suggested, the uranium may be in the form of oxide, such as UO_2 or U_3O_8, a carbide, or fluoride, but the metal is preferred since it is more dense and more free from neutron absorbers. Nitrogen may be present in the reactor in fairly large amounts, and its effect on the chain reaction is such that the neutron reproduction ratio of the system may be changed by changes in atmospheric pressure. This effect may be eliminated by excluding nitrogen from the system, or by sealing the system from the effect of changes of atmospheric pressure.

The effect of impurities on the optimum reproduction factor K may be conveniently evaluated by means of certain constants known as "danger coefficients" which are assigned to the various elements. These danger coefficients for the impurities are each multiplied by the percent by weight of the corresponding impurity, with respect to the weight of uranium in the system, and the total sum of these coefficients gives a value known as the total danger sum. This total danger sum is subtracted from the reproduction factor K as calculated for pure materials and for the specific geometry and materials under consideration.

The danger coefficients are defined in terms of the ratio of the weight of impurity per unit mass of uranium and are based on the cross section for absorption of thermal neutrons of the various elements. These values may be obtained from physics textbooks on the subject and the danger coefficient computed by the formula[2]

$$\frac{\sigma_i}{\sigma_u}\frac{A_u}{A_i},$$

wherein σ_i represents the cross section for the impurity and σ_u the cross section for the uranium, A_i the atomic weight of the impurity and A_u the atomic weight for uranium. Regardless whether the impurities are in the carbon or in the uranium, they are computed as their percent by the weight of the uranium in the system.

Danger coefficients for some elements are given in the following table 5.1, wherein the elements are listed in order of their atomic number.

The sum of the danger coefficients of the impurities in any given composition entering into a reactor as multiplied by the percent by weight of the uranium in the reactor is known as total danger sum of the composition. This figure is a dimensionless constant like K and accordingly can be directly subtracted from K. It will be noted that the danger coefficients given are related to the neutron absorption value of unity for uranium.

As a specific example, if the materials of a system under consideration have 0.01 percent by weight of each of the elements Co and Ag with respect to the weight of the uranium in the system, the total danger sum in K units for such an analysis would be:

$$0.0001 \times 18 + 0.0001 \times 17 = 0.0035.$$

This figure can then be subtracted from the K calculated for theoretically pure materials to give the actual K factor for the geometry and materials used. This would be a rather unimportant reduction in the reproduction factor K unless the reproduction factor K for a given geometry and materials without considering impurities, is very nearly unity. If, on the other

[2]@ In the original text the symbols T_i, T_u appear instead of σ_i, σ_u.

Table 5.1

Element	Danger Coefficient	Element	Danger Coefficient
H	10	Mo	1.0
D	0.01	Ru	∼ 2
He	0	Rh	50
Li	310	Pd	∼ 2
Be	0.04	Ag	18
B	2150	Cd	870
C	0.012	In	54.2
N	4.0	Sn	0.18
O	0.002	Sb	1.6
F	0.02	Te	1
Ne	<3	I	1.6
Na	0.65	Xe	<6
Mg	0.48	Cs	8.7
Al	0.30	Ba	0.30
Si	0.26	La	<2.4
P	0.3	Ce	<2.4
S	0.46	Pr	<2.4
Cl	31	Nd	∼ 17
A	∼ 0.8	Sm	∼ 1430
K	2.1	Eu	435
Ca	0.37	Gd	∼ 6320
Sc	<7	Tb	∼ 20
Ti	3.8	Dy	∼ 200
V	4	Ho	∼ 10
Cr	2	Er	∼ 40
Mn	7.5	Tm	∼ 20
Fe	1.5	Yb	∼ 10
Co	17	Lu	∼ 30
Ni	3	Hf	∼ 20
Cu	1.8	Ta	4.6
Zn	0.61	W	2.7
Ga	∼ 1	Re	∼ 18
Ge	(<5)	Os	<1.7
As	2	Ir	∼ 70
Se	6.3	Pt	∼ 2.5
Br	2.5	Au	16
Kr	<6	Hg	82
Rb	∼ 0.4	Tl	0.5
Sr	0.57	Pb	0.03
Y	0.4	Bi	0.0025
Zr	∼ 0.13	Th	1.1
Cb	<0.4		

hand, the impurities in the uranium are Li, Co, and Rh in the same percentage, the total danger sum would be:

$$0.0310 + 0.0017 + 0.0050 = 0.0377 \text{ reduction in } K \text{ due to impurities.}$$

The maximum possible K factors for neutronic reaction systems when optimum geometry is used, and where the materials used are assumed to be theoretically pure, have been calculated approximately as follows:

Table 5.2

Materials	K for Pure Materials
U metal-graphite moderator	1.1
U oxide-graphite moderator	1.07
U metal-beryllium moderator	1.18
U metal-beryllium oxide moderator	1.09
U metal-heavy water moderator	1.3

It can readily be seen from the above tabulation that the total danger sum for impurities in both the uranium and moderator must be less than 0.3 in order that the K factor remains equal to or greater than unity with a moderator such as a deuterium moderator, 0.18 for a beryllium moderator, and 0.1 for a graphite moderator.

The permissible reduction in K due to impurities will vary in accordance with the type of reactor to be built. In the first place, the maximum K factors listed above are, as pointed out above, present only in systems of infinite size, whereas a practical system, in order to be capable of construction, must be within the bounds of space, weight and economy. The following table 5.3 will show the critical sizes required for operative reactors utilizing graphite, and deuterium in the form of heavy water, for different values of K. In evaluating the table it must be kept in mind that the critical size is that size where the chain reaction just will become self-sustaining. In any practical system the ratio of neutrons produced by fission to the neutrons absorbed to produce fission in each generation must be slightly greater than unity to permit the neutron density to rise to a desired power output. This reproduction ratio in a system of finite size is generally preferred to be around 1.005 but may be lower. Consequently the reactors as built would have to be slightly larger in size than the critical sizes given below, in order to be operative without neutron reflectors. With neutron reflectors somewhat smaller reactors may be operated.

Table 5.3

	Approximate critical Sizes in Feet	
	Sphere radius	Cube side
U Metal-graphite, $K - 1$:		
0.01	27.2	47
0.02	19.4	33.6
0.03	15.7	27.2
0.05	12.1	21.0
0.1	8.6	14.9
U Metal-heavy water, $K - 1$:		
0.01	17	30
0.02	12	21
0.05	8	13
0.1	5.5	9.5
0.2	4	7
0.3	3	5.5

During the neutron chain reaction heat is evolved. About 92 percent of total heat generated in a chain reacting system originates in the uranium, about 6 percent originates in the moderator, where graphite is used. The remaining 2 percent is generated outside the pile, that is, in the surrounding structures. Heat also is conducted into the moderator from the uranium bodies.

Tables 5.4-5.6 show more specifically the type and locale of the heat generated in a uranium-carbon system.

In order to control the temperature of the chain reaction and to prevent the accumulation of heat in the chain reacting system, a suitable circulating system can be employed to convey the heat away from the pile, when a large power output is desired. The design of this system within the pile proper, and the type of coolant employed are critical factors which, if not properly taken into consideration, will make it impossible to design a cooled pile

SUMMARY BY TYPE

Table 5.4

	MeV/fission	Percent
Gamma radiation	23	11
Beta radiation	11	6
Kinetic energy of fission fragments	159	79
Kinetic energy of neutrons	7	4
	200	100

SUMMARY BY LOCALE WHERE HEAT IS GENERATED

Table 5.5

	MeV/fission	Percent
In uranium	184	92
In carbon	12	6
Outside pile	4	2
	200	100

capable of producing a self-sustaining chain reaction.

The problem of removing heat from a chain reacting system by means of a coolant is complicated by various factors. The corrosive effect on uranium of most otherwise suitable circulating media is very troublesome. This factor is important primarily because of the presence in the system of high temperatures and intense neutron densities causing an acceleration of normal rates of corrosion. One of the most serious consequences which may result from the corrosive action of a circulating medium on the uranium is the physical deterioration of the uranium in the system. It is essential then that the circulating system be of such a character as not to destroy the uranium bodies in the system. Furthermore, many otherwise suitable cooling media absorb neutrons to such an extent that they cannot be used in large quantities in the pile, and only certain metals can be used for tubing to conduct the coolant through the pile.

In the present system, a liquid, such as water or diphenyl, or other coolant, is used as the circulating medium to remove heat from the structure, conducted through tubing of aluminum, lead or copper, for example. The water is circulated through a controlled number of such tubes placed in

SUMMARY BY TYPE AND LOCALE

Table 5.6

	MeV per fission	Percent in U	Percent in C	Percent Outside
Kinetic energy of fission fragments	159	100	....	....
Kinetic energy of neutrons	7	....	90	10
Gamma radiation from fission fragments and products	5	70	25	5
Beta radiation from fission fragments and products	11	100	....	....
Nuclear affinity of neutrons (gamma radiation)	18	60	25	15

the moderator of the system and spaced from the uranium. Therefore, the corrosion problem with respect to the uranium and furthermore contamination of the coolant with radioactive fission products from the uranium is avoided or at least minimized. The heat is transferred from the uranium to the coolant by conduction through the moderator, despite the fact that most of the heat is generated in the uranium, cooling the moderator permits effective operation of the reactor at relatively high power outputs for example from 5,000 to 10,000 kilowatts, continuously and without stopping the chain reaction. At such powers moderator cooling presents the further advantage of simplicity, and permits loading and unloading of the reactor without interference with the cooling system.

It is, therefore, an object of my invention to provide a means and method of cooling a moderator of a chain reacting pile at a point or points spaced from the fissionable component disposed in the moderator.

It is still another object of my invention to dissipate sufficient heat during the operation of a chain reacting system to permit continuous operation of the system at high neutron densities.

A further object of my invention is to provide a means and method of circulating a neutron absorbing cooling fluid through a chain reacting system, under conditions such that a chain reaction therein may be maintained.

The above and other objects, features and advantages of my invention will become apparent to those skilled in the art when taken in connection with the following description and the accompanying drawings wherein:

Fig. 5.1 is a schematic perspective view of the exterior of a chain reacting system;

Fig. 5.2 is a horizontal cross section on line **2–2** in Fig. 5.1 with the active portion of the reactor shown in plan;

Fig. 5.3 is a fragmentary horizontal section with the safety rod in plan, and showing the safety rod and slot;

Fig. 5.4 is a fragmentary vertical section on line **4–4** in Fig. 5.3;

Fig. 5.5 is a vertical section on line **5–5** in Fig. 5.2 with the headers and part of the concrete walls removed and with a portion broken away;

Fig. 5.6 is a view partly in elevation and partly in section as indicated by the line **6–6** in Fig. 5.2;

Fig. 5.7 is a top view, broken away, of a live graphite block;

Fig. 5.8 is an irregular vertical section substantially on line **8–8** in Fig. 5.7;

Fig. 5.9 is an elevation, broken away, of a dead graphite block;

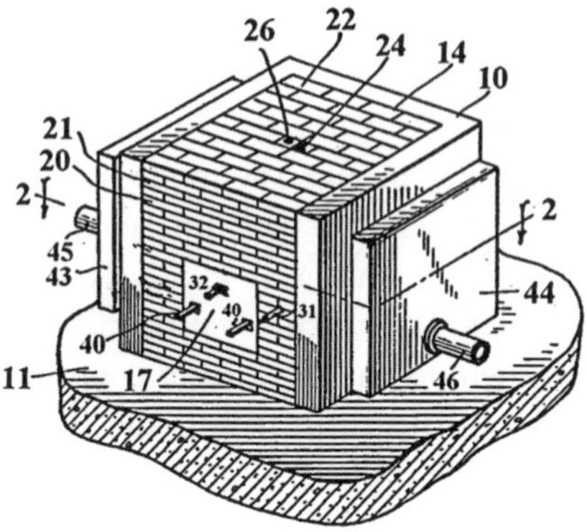

Fig. 5.1

Fig. 5.10 is a diagram showing a neutron density measuring device and circuit;

Fig. 5.11 is a chart or diagram showing neutron density distribution across a chain reacting system; and

Fig. 5.12 is a vertical section of a cooling tube in graphite.

As a self-sustaining chain reaction is not known to be possible with natural uranium and graphite without aggregation of the uranium into bodies of substantial size, I prefer to illustrate my invention by description of a moderator-cooled, self-sustaining chain reacting system, in which the active portion is cubical, and in which highly purified uranium metal cylindrical lumps are utilized, imbedded in high quality graphite. The effect of impurities in the graphite is substantially the same as in the uranium since the neutrons diffuse freely in both materials.

Referring first to Figs. 5.1 and 5.2 illustrating the completed structure, side walls **10** are erected on a heavy foundation **11**, both preferably of poured concrete about 5 feet thick, leaving a vault space **14** inside walls **10** in which is built up, as will be explained later, an active portion **15** (Fig. 5.2) surrounded by a graphite reflector **16** which essentially encases the reactor itself. The reflector is built with a graphite projection **17** which extends to a plane flush with the outer surface of front wall **20** also 5 feet thick, formed of concrete bricks **21**. The top of the structure is closed with covering bricks **22** also of concrete and 4 feet thick. The top covering is pierced by an aperture **24** which constitutes the mouth of a well **25**

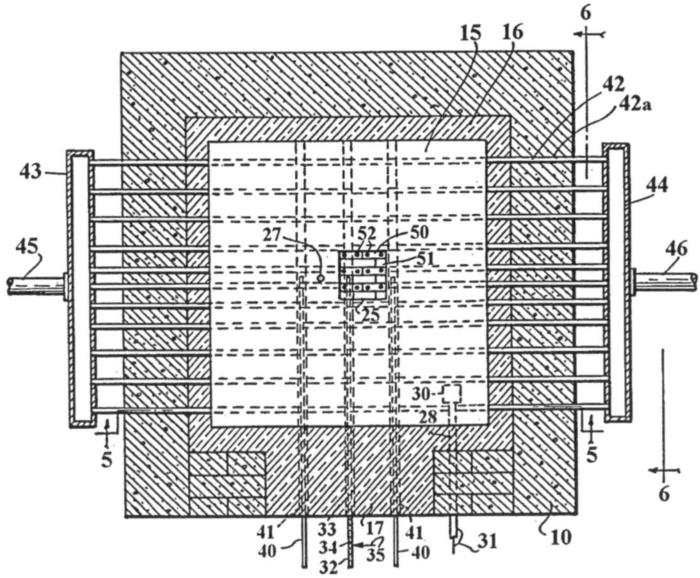

Fig. 5.2

extending to at least the peripheral layer of uranium bodies in the active portion **15**. A second aperture **26** is the exterior opening of a shaft **27** extending to the center of the structure.

Also entering the active portion **15** is a slot **28** at the end of which is positioned an ionization chamber **30** connected to the exterior by wire line **31**.

Extending from the outside of projection **17** into the center of active portion **15** is a control rod **32** sliding in an aluminum-lined slot **33** in the graphite. A scale **34** is provided on control rod **32** cooperating with a suitable index diagrammatically illustrated at **35** to indicate the depth of penetration of the control rod **32** as shown in Fig. 5.2. Also extending into the active portion **15** through projection **17** are two spaced safety rods **40** operating in slots **41**. Safety rods **40** and control rod **32** are preferably flat sheets of cadmium or cadmium clad steel, as shown in Figs. 5.3 and 5.4, and are arranged to move freely in and out of slots **33** and **41**, respectively.

Water tubes **42** pass horizontally through holes **42a** in the concrete side walls and pass through the active portion **15**. On each side of the system the tubes enter headers **43** and **44** having connecting pipes **45** and **46**,

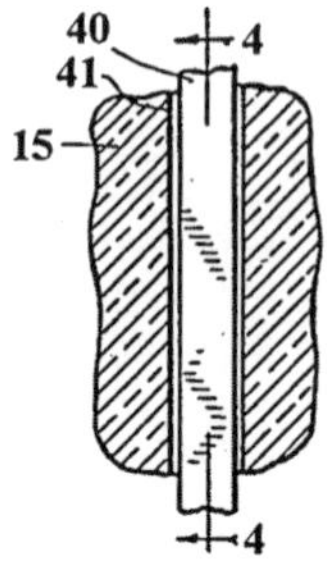

Fig. 5.3

respectively, for coolant connection. A total of 97 tubes are used, disposed in a special pattern, as will later be described.

This completes the gross structure of the device. The active portion **15** will next be described.

The active self-sustaining chain reacting portion **15** to be built up within vault **14** is designed to so reduce neutron losses as to make a self-sustaining chain reaction possible, as has been previously outlined, using specific types and sizes of uranium metal masses, embedded in graphite, all of best obtainable purity, the uranium being spaced with a specific geometry.

In the system illustrated, the graphite is in the form of blocks or bricks having dimensions of 4″ × 4″ × 16″, these blocks having been planned by woodworking machinery to have smooth rectangular sides. The danger sum of impurities in the graphite is below about 0.02. Certain of the blocks **50** are drilled with two holes spaced 8″ (20 cm) center to center to receive the uranium bodies which are placed therein as shown in Figs. 5.5 to 5.8,

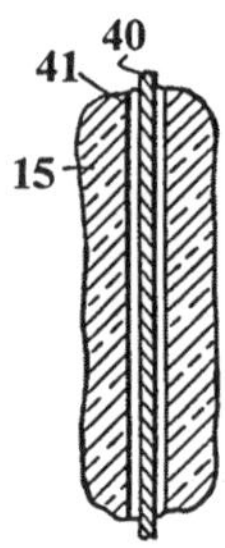

Fig. 5.4

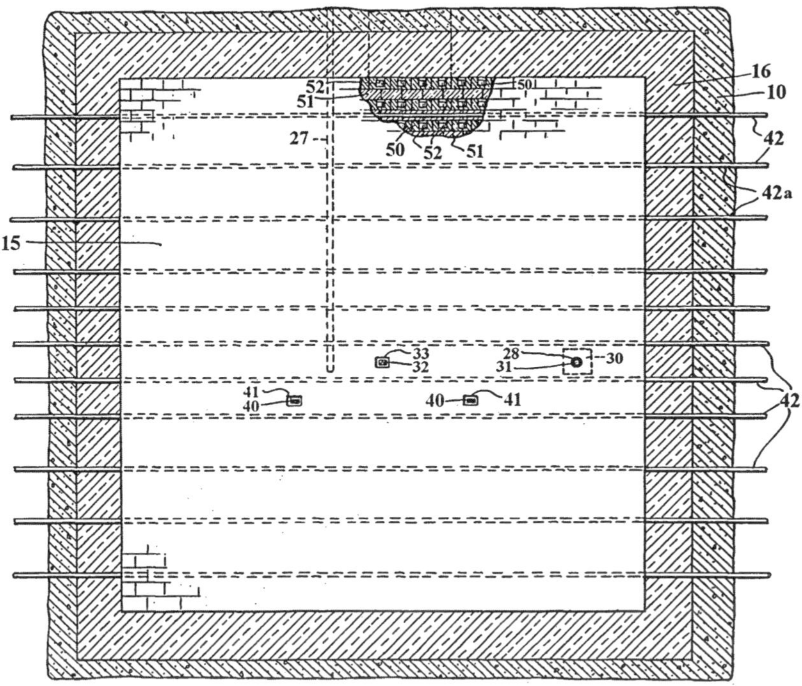

Fig. 5.5

inclusive. These graphite blocks **50** are termed live graphite. Other blocks **51** as shown in Fig. 5.9, contain no uranium and may be termed dead graphite. The uranium bodies are cast uranium metal cylinders **52** having a density of 18 g per cubic centimeter and a danger sum for impurities less than 0.01 each cylinder being about $2\frac{1}{4}''$ in diameter, and $2\frac{1}{4}''$ in height, weighting about 6 pounds each.

The active portion **15** of the structure, built inside the vault, has about 30 layers of live graphite, alternating with layers of dead graphite to form a cube having side dimensions of approximately 6 meters. A total amount of graphite of approximately 240 tons is used in the active portion. Approximately 80 tons of exceptionally pure cast uranium metal bodies are used for the cylinders. The total number of uranium bodies is about 27,000. The volume ratio is about 35 carbon to 1 uranium.

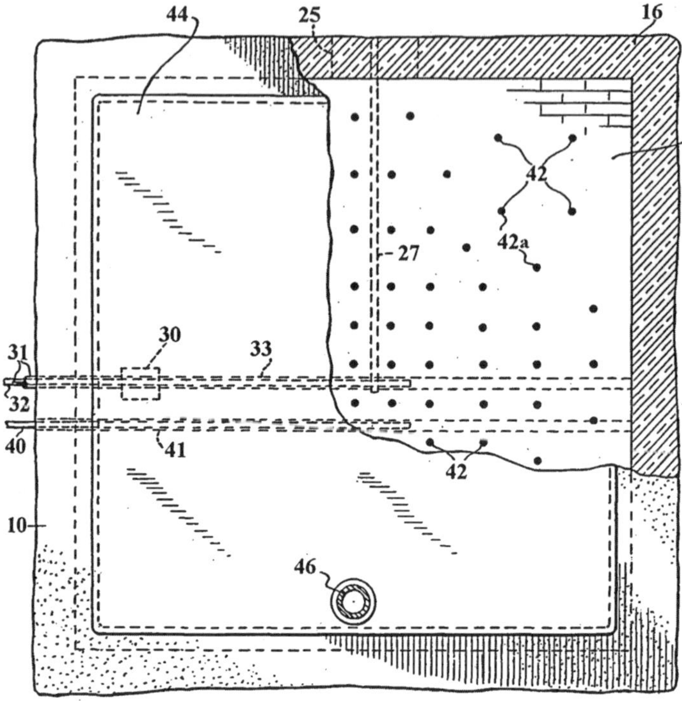

Fig. 5.6

The spacing of the uranium elements in the graphite is such that each unit cell comprising one uranium body and surrounding graphite has a side size of 20 centimeters. The arrangement of the bodies in the graphite is similar to that of a cubic crystal lattice. The K factor for the metal bodies **52** and graphite alone is about 1.07, without including the tubes **42**. This K factor is sufficiently high so that a system of practical size can be built to have a reproduction ratio greater than unity even when

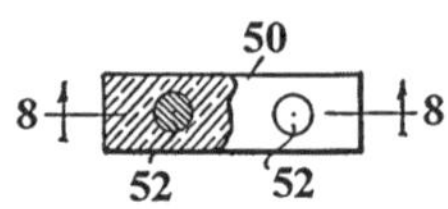

Fig. 5.7

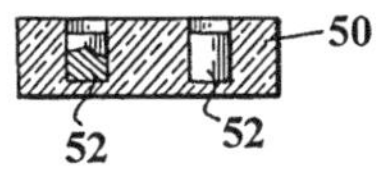

Fig. 5.8

impurities in the form of 97 tubes having a wall thickness of from 1.5 to 2 millimeters and an internal diameter of 1.5 centimeters filled with a coolant such as water are included in the system. The tubes and the contained water only reduce the K factor to about 1.06, this K being still sufficiently high to permit the side dimension of 6 meters when built in the form of a cube. In such a reactor it will be noted that the amount of coolant is apportioned so that the reproduction ratio (r) is at least unity after losses for leakage, pipes, and other absorbers has been taken into consideration. Since for control purposes it is undesirable to permit the reproduction ratio from exceeding about 1.01 failure of coolant flow must be avoided since the removal of coolant from the system might increase the reproduction factor above this figure. Thus the amount of coolant in the reactor must be carefully apportioned so that r does not exceed about 1.01. Lead absorbs slightly more neutrons than aluminum, but still can be used without greatly enlarging the system. Copper tubes and water introduce a K loss of about 3 percent and while copper can be used, the reduction in K requires that the system be considerably increased in size. Aluminum tubes are, therefore, preferred to permit the smallest size of structure.

The actual construction of the pile is started by placing on the foundation of vault **14** several layers of dead graphite blocks to start reflector **16**, after which the active portion **15** may be started by piling up alternate layers of dead graphite and live graphite, filling the edges of both live and dead layers out to the concrete side walls with dead graphite to continue the reflector **16**. The blocks are closely packed without cement to eliminate air spaces as far as possible. Because of the block type of construction, the system is termed a pile. As the stacking continues, the tubes **42** are installed. This is done as shown in Fig. 5.12 by splitting a row of dead

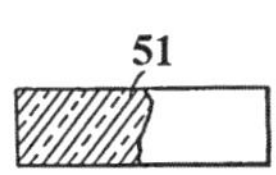

Fig. 5.9

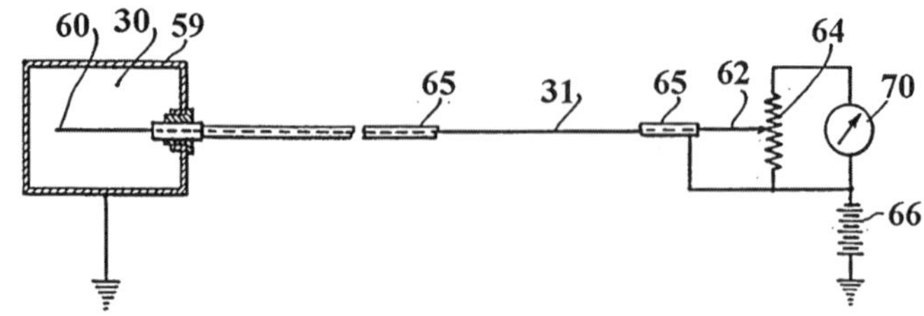

Fig. 5.10

graphite blocks longitudinally in halves **53**, routing out channels for the tubes, laying the tubes in the lower halves and then covering with the upper halves. The clearance between graphite and tube should be as small as possible, from 0.002″ to 0.005″.

When slightly less than half the required number of layers have been

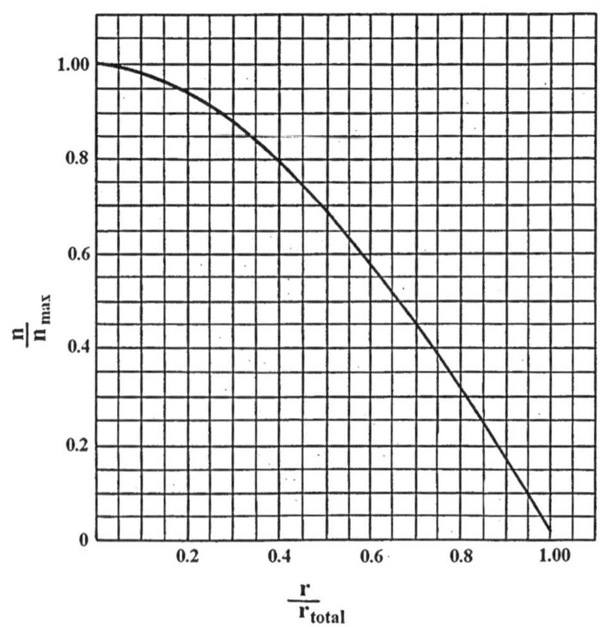

Fig. 5.11

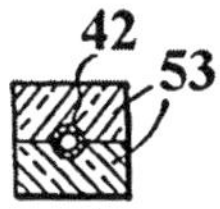

Fig. 5.12

piled in the vault, the aluminum-lined slots **41** are provided in a dead carbon layer through which the safety rods slide, and the safety rods **40** installed. After a few more layers are added to the pile, the aluminum-lined control rod slot **33** is built in, the control rod **32** inserted, and the ionization chamber **30** installed. Wire line **31** is led to the outside and connected to a monitoring circuit, as shown in Fig. 5.10.

Ionization chamber **30** may comprise a sealed metal casing **59** containing approximately 18 liters of boron fluoride at one atmosphere pressure into which projects a central electrode **60**. The central electrode is connected through central wire **31** to a contact **62** adjustable across resistor **64**. One end of resistor **64** is connected to an outer wire shield **65** and to a battery **66** of about 450 volts potential, the other end of which is grounded, as is casing **59**. The other end of resistor **64** is connected through galvanometer **70** to the wire shield side of battery **66**. Neutron absorption by the boron in the chamber causes alpha ray ionization in the chamber, the amount thereof being measured by galvanometer **70**. Only the ionization chamber need be in the pile, with the galvanometer positioned up to 300 feet away, and as the neutron density at any given point in the reactor bears a substantially constant relationship with the neutron density at any other point in the reactor the chamber may be in the outer layers of the active portion. The galvanometer deflection is only approximately linear with neutron density, but is reproducible and readily calibrated in terms of neutron density, power or heat, as desired.

Construction is continued, layer by layer including the **42**, with the control rod and safety rods fully inserted into the pile. Matching blocks bored with a 3″ hole between the uranium bodies in the live blocks, and similar holes in proper position in the dead blocks are aligned as layers are added to provide shaft **27**.

Preferably, at least from the halfway point, the natural neutron density in the pile is monitored as layers are added with water placed in such tubes **42** as are already installed. By plotting the change in neutron density value within the pile as layers are added thereto, with the control rod and safety rods withdrawn, a prediction can be made in advance as to when the chain reaction in the structure will become self-sustaining. The personnel building the pile can thus be warned that the critical size is being approached.

With the safety and control rods fully inserted into the structure, the active portion **15** and the reflector **16** are completed to the final size desired, with additional dead carbon blocks being positioned over the top thereof to complete the reflector **16**, leaving well **25** and the shaft **27** open. Usually the reactor is constructed somewhat oversized and sufficient neutron absorbers inserted as fixed limiting or shim rods which limit the ultimate neutron reproduction ratio so that it cannot exceed a maximum of 1.01 and usually about 1.005. Such rods or a portion may be removed later if the reproduction factor should change due to poisoning by fission products or for other cause. When the active portion is complete the top and front of the vault are closed, except for the outward continuation of well **25** and shaft **27** by means of the cement blocks. The headers **43** and **44** are connected to tubes **42**. A structure is thus obtained having an active portion **15** therein, surrounded by a carbon reflector **16**, and enclosed in concrete walls on all sides except for the carbon projection **17** utilized to carry the control and safety rods to the exterior of the vault, and provided with means for circulating fluid through the moderator of the active portion.

The structure has been shown as surrounded with a layer **16** of graphite, which will reflect back into the active portion some of the neutrons which might otherwise diffuse out of the system. As neutrons diffuse through this layer in random paths, a small percentage of the neutrons entering the layer are turned around and re-enter the active portion **15**. To the extent that the neutrons are returned to the active portion **15**, the graphite layer acts as a reflector. This material is known as a scattering material and preferably has a low neutron absorption characteristic. Some elements as lead, deuterium oxide or bismuth may also be used. The thickness of the graphite reflector **16** may be on the order of one to three feet or more, depending upon the size of the active portion and the degree of scattering required. By using this peripheral layer of scattering material, the overall size of the active portion **15** can be slightly smaller than in a case where no scattering layer is employed, as neutron losses to the exterior are reduced. Other arrangements and combinations may be used to produce a satisfactory reflector, but for present purposes, since the reflector per se forms no part of this invention, the foregoing specific example will suffice.

The concrete walls serve as the main shield to prevent gamma radiation escaping outside the structure, as carbon being a light material absorbs gamma rays only to a relatively small extent. The water in the concrete also serves to absorb escaping neutrons.

In Fig. 5.11 a curve is shown wherein the ratio of neutron density to maximum neutron density at the center of the active portion is plotted against the ratio of the partial radius to the total radius of the active portion. It will be noted that the graph is a cosine curve and that there is a flux representing the exterior loss, extending beyond the total radius. While this loss in a carbon-uranium pile represents about 4 percent of the neutrons generated, the surface density at any point is small with respect to the maximum density. Since the number of fissions in the active portion is proportional to neutron density and the heat is due to the fissions, the diagram clearly indicates where the heat is released. Consequently, the heat distribution curve from the center to the side of the active portion approximates the neutron density distribution curve.

Accordingly, I have disposed the tubes **42**, as indicated in Fig. 5.6, with a greater concentration near the center of the active portion **15** where the heat development is larger. In the central portion I dispose tubes **42** in a square pattern of 40 centimeters side, and distribute the remaining tubes in the outlying portions in such a way as to substantially equalize the heat dissipated in each tube. For a total energy production of 10,000 kilowatts the heat dissipated by each tube is about 25,000 calories per second. The heat dissipated per unit length of tubing is also greater near the center of the active portion than at the ends. Near the center of the pile, the heat dissipation amounts to about 65 calories per second per centimeter.

Allowing for a temperature rise of 25°C of the cooling water, the flow in each pipe is 1000 cubic centimeters per second. The pressure drop for tubes of 1.5 centimeters internal diameter is 1.2 atmosphere within the pile. The water may be circulated by gravity or by pumps between headers **43** and **44**. The water may be cooled and recirculated or used only once and disposed of.

While the most efficient cooling of a pile such as described herein can be obtained by directly cooling the uranium, where the largest amount of heat is released, the graphite will be cooled directly and the uranium will be cooled by conduction with the illustrated cooling system, to such an extent that the pile can operate at from 5,000 to 10,000 kilowatts continuously without melting the uranium. The following approximate temperatures given at various places for tubes in the central portion of the pile are tabulated for 1.5 centimeters internal diameter aluminum tubes, with an average clearance of 0.002 inch between the tubes and the graphite, and between the uranium and the graphite, when 1000 cc/s of water at an entering temperature of 40°C is passed through the tubes.

Table 5.7

Position	10,000 kw	5,000 kw
	°C	°C
Entering water	40	40
Internal surface of tube	62	51
External surface of tube	82	61
Graphite near pipe	312	176
Bulk of graphite	637	339
Graphite near uranium	725	383
Surface of uranium	775	408
Center of lump	861	451

As the melting point of the uranium used in the pile is about 1100°C it can be seen from the above tabulation that the pile can be safely operated at 10,000 kilowatts, when clearances are kept below 0.002 inch. If it should be desirable to increase clearances or operate further away from the melting temperature of uranium, then operation at lower heat output is preferable. While piles only conductively cooled in air can operate at 5,000 to 10,000 kilowatts, they can do so for short periods only since the heat evolved tends to overheat the reactor within a very short time. By cooling the reactor it may be operated at high neutron densities continuously and thus the rate of production of the products of reaction may be increased.

After the structure has been completed, it is ready for operation. Full circulation of the coolant is started, and the neutron absorbing safety rods **40** are withdrawn from the active portion **15**. The neutron absorbing control rod **32** is then slowly retracted until the galvanometer **70** indicates that the neutron density is rising. If a slowly rise is desired, the rod is retracted only enough to indicate a rise. If a faster rise is desired, the rod is retracted further to increase the reproduction ratio in the system.

When any desired neutron density is reached, the control rod is pushed back into the pile until a point is found where the neutron density remains constant. At this point the system is balanced with a neutron reproduction ratio of unity. No special source of neutrons is needed in the structure, as the natural neutrons always present and constantly diffusing through the pile are sufficient to start the reaction.

To again increase the neutron density, the control rod is moved outwardly until the new desired neutron density is attained at a desired rate, and then the control rod is moved into the point where the system is again balanced, thus stabilizing the reaction at the new neutron density. If it is desired to decrease the neutron density, the control rod is pushed into the pile until the reproduction ratio is less than unity and the neutron density decays to the desired value. Then the control rod is retracted to the position at which balance is re-established. In this manner, any desired neutron density within the heat dissipating limits of the structure may be obtained and maintained. The position of the control rod in the system, therefore, does not govern the neutron density in the pile, but only the rate of change of the density.

The pile herein described has a maximum possible reproduction ratio r of about 1.005, causing a doubling of the neutron density every 8 seconds with the control and safety rods completely removed. This slow rise renders the pile always easy to control and safe to operate. The low reproduction ratio in the system obtained from a geometry giving an overall K factor of 1.06 is due to the fact that the system is built close to critical size and that large quantities of neutrons are deliberately permitted to leak from well **25** and shaft **27** for use outside the pile.

The unity reproduction ratio position of the control rod within the pile for maintaining any desired neutron density would always be the same were it not for the fact that the temperature within the pile changes to some extent and influences the neutron losses in the materials and also for the fact that the pile is exposed to varying atmospheric pressure. Changes in the nitrogen content of the pile accompany changes in atmospheric pressure, and cause a variation in the K factor as nitrogen is a neutron absorbing impurity. The unity reproduction ratio point on the control rod, therefore, changes slightly in accordance with the temperature at which the pile is being operated and with the atmospheric pressure. Changes due to ambient temperature of the air and the cooling water are relatively small and the balanced position of the rod can readily be calibrated for temperature effect. Changes due to atmospheric pressure are also small and can be compensated for.

If at any time it is desired to stop the reaction, the control rod is inserted deep within the pile until the reproduction ratio becomes less than the value necessary to maintain the reaction with this size of structure. The neutron density then decays to that of the natural neutrons. The safety rods are also inserted when the pile is left unattended. If desired, the safety rods may be pulled into the system by weights tripped by an emergency latch (not shown) in case of failure of the control to control the reaction for any reason, or upon failure of coolant circulation. It is also possible to control the position of the control rod **32** by means of automatic connection to the monitoring circuit, although such automatic control forms no part of the present invention and is not required in the system described.

The reactor as described has many uses. Radioactive fission products and new elements, both radioactive and stable, are produced by fission and absorption of neutrons in the uranium. Large quantities of neutrons and penetrating gamma rays are produced and can be utilized.

The radioactive fission products and the 94^{239} produced by normal operation of the pile can be recovered after a predetermined exposure of the uranium bodies in the pile to high neutron densities by removing the uranium bodies, dissolving them in a suitable solution, and extracting the desired elements. This separation does not constitute any part of the present invention and, therefore, no details are herein given. Radioactive fission products are valuable as gamma radiation sources and for use as biological tracers in medicine, while 94^{239} is useful as a fissionable material to increase K factors when added to natural uranium, or when used alone, in chain reacting systems.

Well **25**, extending through the concrete and carbon shield and exposing the outer layer of uranium bodies, permits the entire energy spectrum of neutrons in the pile to reach the interior of the well. Therefore, well **25** can be packed with any material desired and the scattering effect of the material determined, for example. Again, isotopes of, or transmutation to various elements can be produced in large quantities by packing these elements into the well and exposing them to the action of the neutrons escaping from the periphery of the reactor. One example of isotope production, followed by transmutation, is the manufacture of U^{233} from thorium 232, for which process thorium 233 can first be produced from thorium 232 in well **25** in accordance with the slow neutron density in the well and time of exposure, the extent of the reaction being dependent on the product of those factors. Thorium 233 then decays to form protoactinium 233 and thence to uranium 92^{233} which is valuable as a fissionable material similar in its action to U^{235} and 94^{239}. The reaction is as follows:

$$ {}_{90}\mathrm{Th}^{232} + n \longrightarrow {}_{90}\mathrm{Th}^{233} + \text{gamma rays} $$

$$ {}_{90}\mathrm{Th}^{233} \xrightarrow[\beta^-]{23.5\text{ min.}} {}_{91}\mathrm{Pa}^{233} \xrightarrow[\beta^-]{27.4\text{ days}} {}_{92}\mathrm{U}^{233} $$

As a further example of transmutation, radioactive carbon may be produced by allowing the neutrons leaving the pile to react with nitrogen in compound form in well **25**. This reaction gives rise to carbon of mass **14** which is radioactive, and can be separated chemically from the nitrogenous compounds. Such radioactive carbon is suitable for medical and physiological uses as it may be incorporated in organic compounds and used as tracers in living organisms.

It can thus be seen that the neutrons normally escaping from the pile need not be lost, but can be put to work. Continuous operation at from 5,000 to 10,000 kilowatts accelerates the effect of the neutrons by permitting continuous bombardment at high neutron densities.

In utilizing the output of the pile, shaft **27** also plays an important role. It extends to the center of the pile where the highest neutron density exists, and intense neutron bombardment of materials inserted into the bottom of this shaft will take place, particularly at high pile powers. Furthermore, the shaft acts to collimate the fast neutrons released at the center of the pile and a high density collimated beam of neutrons emerges through the aperture, projected upwardly. This collimated beam, having a far greater neutron density than any produced by other means, can be utilized outside of the pile for nuclear research in all of its aspects. The number of neutrons escaping from shaft **27** is about four or five times the number escaping over an area of the external surface of the pile equal to the cross section of shaft **27**.

In addition, extremely high energy gamma rays are emitted during nuclear fission. These rays also escape through shaft **27** to the exterior of the pile and can there be used for taking radiographs through large castings, for example, with relatively short exposures during high power operation of the pile. The neutrons coming from the pile can be screened out of the gamma ray beam by the use of relatively thin sheets of materials having high neutron absorption cross sections without substantially reducing the gamma ray intensity. In addition, a bismuth filter has been found to effectively reduce the gamma rays without substantial interference with the neutron beam. Thus shaft **27** can be used either to produce a high intensity collimated neutron beam, or to produce a high energy beam of gamma rays, as desired both for use outside the pile. When using the pile in this manner, a heavy sheathing of lead is preferably positioned over the top of the pile and apertured for well **25** and shaft **27**. The personnel working on the top of the pile will then be adequately protected from the effects of gamma rays generated by the pile and escaping around the periphery thereof.

While I have described my invention as applied to a structure in which the moderator only is cooled by circulating a liquid therethrough, it is distinctly to be understood that in certain structures both the moderator and the uranium can be directly cooled, although the direct cooling of the uranium is no part of the present invention. Consequently, I do not desire to limit my invention to a chain reacting system wherein only the moderator is directly cooled but desire to apply the invention described and claimed herein to any chain reaction system whether or not otherwise cooled. It is also to be understood that my invention is equally applicable to other types of reactors including those having liquid moderators such as deuterium oxide or other reactors constructed in accordance with the general principles described in the present application or copending application of Enrico Fermi and Leo Szilard, Serial No. 568,904, filed December 19, 1944, now Patent No. 2,708,656, dated May 17, 1955.[3] For example uranium containing greater than usual concentrations of U^{235} may be used. Alternatively the invention may be applied to other fissionable compositions such as U^{233}, 94^{239} etc. or to combinations of these fissionable isotopes with U^{238} or Th^{232} or similar isotope which yields a fissionable isotope by neutron absorption.

While the theory of the nuclear chain fission mechanism in uranium set forth herein is based on the best presently known experimental evidence, I do not wish to be bound thereby, as additional experimental data later discovered may modify the theory disclosed. Any such modification of theory, however, will in no way affect the results to be obtained in the practice of the invention herein described and claimed.

What is claimed is:

1. A neutronic reactor comprising an active portion consisting of blocks of graphite arranged in layers to form a cube six meters on a side, about 27,000 bodies of natural uranium having a total weight of about 80 tons disposed in holes in alternate layers of graphite blocks, the spacing of the uranium elements in the graphite being such that each unit cell comprising one uranium body and surrounding graphite has a side size of 20 centimeters, 97 horizontal aluminum tubes containing flowing water disposed in the layers of graphite blocks which do not contain uranium, and means for controlling the reactor.

[3]See page 217.

2. The reactor according to claim 1 wherein the tubes are more closely spaced at the center of the reactor than toward the outside of the reactor to equalize the removal of heat.

References cited in the file of this Patent

UNITED STATES PATENTS

400,160	Babbitt	Mar. 26, 1889
1,769,994	Hendryx	July 8, 1930
2,006,649	Modine	July 2, 1935
2,127,193	Toulmin	Aug. 16, 1938
2,206,634	Fermi et al.	June 2, 1940
2,708,656	Fermi et al.	May 17, 1955

FOREIGN PATENTS

114,150	Australia	May 2, 1940
114,151	Australia	May 3, 1940
233,011	Switzerland	Oct. 2, 1944

OTHER REFERENCES

Naturwissenschaften, vol. 27 (1937), pages 402-410.

Smyth: "Atomic energy for Military Purposes", pages 103, 104, August 1945. Copy may be purchased from the Supt. of Documents, Washington 25, D.C.

Kelly et al.: "Physical Review", 73, 1135-9 (1948).

Chapter 6

Air Cooled Neutronic Reactor

Enrico Fermi, Santa Fe, N. Mex., and Leo Szilard, Chicago, Ill., assignors to the United States of America as represented by the United States Atomic Energy Commission

Application May 29, 1945, Serial No. 596,465

U.S. Patent No. 2,836,554 - Patented May 27, 1958

The specific reactor considered in this Patent is an uranium-graphite one cooled by air, circulating within the porous graphite, and with control and safety rods made of cadmium or boron. The air serving as coolant passes only once through the reactor, so that it is not too much enriched in radioactive ^{41}Ar*. Furthermore, it is exhausted at a substantial distance above ground (from the top of a stack), in order that the radioactive argon in the cooling air is sufficiently dispersed in and diluted by fresh atmospheric air before reaching any person on the ground.*

Since one of the main object of this Patent is to produce plutonium, some constructional details aimed at removing plutonium for the reactor, when a certain concentration of it is achieved, are illustrated. In particular, the mechanism for the loading or unloading of the uranium slugs is made of iron or lead in order to shield it from the radioactive bars in case they are loaded. It is also interesting to note with the authors that even after the uranium slugs have been extracted, they are so exceedingly radioactive that the produced heat would melt themselves if not immersed in water.

The production of plutonium was considered by Fermi in some previously issued reports (see, for example, [Fermi (1962)] on pages 391 and 411), but what discussed here in so great detail (including the basic air cooling) is present in no other published paper.

The present invention relates to a neutronic reactor which is capable of numerous uses but is particularly adapted to use for the production of the transuranic element 94^{239} and/or radioactive fission products by neutrons released during a self-sustaining nuclear chain reaction through fission of uranium with slow neutrons. More particularly, our invention relates to the removal of the heat of the neutronic reaction to such an extent that the reaction may be conducted at a more rapid rate and the production of element 94^{239} and/or fission products may be accelerated. Natural uranium may be used in the reaction and contains the isotopes 92^{238} and 92^{235} in the ratio of approximately 139 to 1. Hereinafter in the specification and the claims the term uranium is to be understood as referring to uranium and its chemical compositions of normal isotopic content or equivalent compositions, unless otherwise indicated by the context.

When it became known that the isotope U^{235} in natural uranium could be split or fissioned by bombardment with thermal neutrons, i.e., neutrons at or near thermal equilibrium with the surrounding medium, many predictions were made as to the possibility of obtaining a self-sustaining chain reacting system operating at high neutron densities. In such a system, the fission neutrons produced give rise to new fission neutrons in sufficiently large numbers to overcome the neutron losses in the system. Since the result of the fission of the uranium nucleus is the production of two lighter elements with great kinetic energy, plus approximately 2 fast neutrons on the average for each fission along with beta and gamma radiation, a large amount of power can be made available in a self-sustaining system.

It has been found that most of the neutrons arising from the fission process are set free with the very high energy of the order of one million electron volts and are therefore not immediately in condition to be utilized efficiently to create new thermal neutron fissions in U^{235} when it is mixed with a considerable quantity of U^{238}, as is the case with natural uranium. The energies of the fission-released neutrons are so high that most of the latter would tend to be absorbed by the U^{238} nuclei, and yet the energies are not generally high enough for production or fission by more than a small fraction of the neutrons so absorbed. For neutrons of thermal energies, however, the absorption cross section of U^{235}, to produce fission, rises a great deal more than the simple capture cross section of U^{238}; so that under the stated circumstances the fast fission neutrons, after they are created, must be slowed down to thermal energies before they are most effective to produce fresh fission by reaction with additional U^{235} atoms. When a system larger than critical size is made in which neutrons are slowed down without much absorption until they reach thermal energies and then mostly enter into uranium rather than into any other element, a self-sustaining nuclear chain reaction is obtained, even with natural uranium. Light elements, such as deuterium, beryllium, oxygen or carbon, the latter in the form of graphite, can be used as slowing agents or neutron moderators. A special advantage of the use of the light elements mentioned for slowing down fast fission neutrons is that fewer collisions are required for slowing than is the case with heavier elements, and furthermore, the above-enumerated elements have very small neutron capture probabilities, even for thermal neutrons. Hydrogen would be most advantageous were it not for the fact that there may be a relatively high probability of neutron capture by the hydrogen nucleus. Carbon in the form of graphite is a relatively inexpensive, practical, and readily available agent for slowing fast neutrons to thermal energies. Recently, beryllium has been made available in sufficiently large quantities for test as to suitability for use as a neutron slowing material in a system of the type to be described. It has been found to be in every way as satisfactory as carbon. Deuterium compounds such as deuterium oxide while more expensive are even more satisfactory.

However, in order for the premise to be fulfilled that the fast fission neutrons be slowed to thermal energies in a slowing medium without too large an absorption in the U^{238} isotope of the uranium, certain types of physical structure are utilized for the most efficient reproduction of neutrons, as precautions must be taken to reduce various neutron losses and thus to conserve neutrons for the chain reaction if a self-sustaining system is to be attained.

An initial number of fast neutrons in the system by going through the process of absorption and fission reproduces in the next generation a number of neutrons generally different from the initial number. The ratio of the number produced after one generation to the initial number for a system of infinite size is called the reproduction or multiplication factor of the system and is denoted by the symbol K. If K can be made sufficiently greater than unity to create a net gain in neutrons for the system of infinite size and the system made smaller but still sufficiently large so that this gain is not entirely lost by leakage from the exterior surface of the system, than a self-sustaining chain reacting system can be built to produce power (in the form of heat) by nuclear fission of natural uranium. The neutron reproduction ratio in a system of finite size differs from K by the leakage factor (neutrons lost from the system through leakage), and must be sufficiently greater than unity to permit the neutron density to rise exponentially. Such a rise will continue indefinitely if not controlled at a desired density corresponding to a desired power output.

During the interchange of neutrons in a system comprising bodies of uranium of any size in a slowing medium, neutrons may be lost in four ways, by absorption in the uranium metal or compound without producing fission, by absorption in the slowing down material, by absorption in impurities present in the system, and by leakage from the system. These losses will be considered in the order mentioned.

Natural uranium, particularly by reason of its U^{238} content, has an especially strong absorbing power for neutrons when they have been slowed down to moderate energies. The absorption in uranium at these energies is termed the uranium resonance absorption or capture. It is caused by the isotope U^{238} and does not result in fission but creates the isotope U^{239} which by two successive beta emissions forms the relatively stable nucleus 94^{239}. It is not to be confused with absorption or capture of neutrons by impurities referred to later. Neutron resonance absorption in uranium may take place either on the surface of the uranium bodies, in which case the absorption is known as surface resonance absorption, or it may take place further in the interior of the uranium body, in which case the absorption is known as volume resonance absorption. It will be appreciated that this classification of resonance absorptions is merely a convenient characterization of observed phenomena, and arises, not because the neutron absorbing power of a U^{238} nucleus is any greater when the nucleus is at the surface of a body of metallic, or combined uranium, but because the absorbing power of U^{238} nuclei for neutrons of certain particular energies is inherently so high that practically all neutrons that already happen to have those energies, called resonance energies as explained above, are absorbed almost immediately upon their arrival in the body of uranium metal or uranium compound, and thus in effect are absorbed at the surface of such body. Volume resonance absorption is due to the fact that some neutrons make collisions inside the uranium body and may thus arrive at resonance energies therein. After successfully reaching thermal velocities, about 40 percent of the neutrons are also subject to capture by U^{238} without fission, to produce U^{239} and eventually 94^{239}.

It is possible, by proper physical arrangement of the materials, to reduce substantially uranium resonance absorption. By the use of light elements as described above for neutron moderators, a relatively large increment of energy loss is achieved in each collision and therefore fewer collisions are required to slow the neutron to thermal energies, thus decreasing the probability of a neutron being at a resonance energy as it enters a uranium atom. During the slowing process, however, neutrons are diffusing through the slowing medium over random paths and distances so that the uranium is not only exposed to thermal neutrons but also to neutrons of energies varying between the emission energy of fission and thermal energy. Neutrons at uranium resonance energies will, if they enter uranium at these energies, be absorbed on the surface of a uranium body whatever its size, giving rise to surface absorption. Any substantial reduction of overall surface of the same amount of uranium relative to the amount of moderator (i.e. the amount of moderator remaining unchanged) will reduce surface absorption, and any such reduction in surface absorption will release neutrons to enter directly into the chain reaction, i.e., will increase the number of neutrons available for further slowing, and thus for reaction with U^{235} to produce fission.

For a given ratio of moderator to uranium, surface resonance absorption losses of neutrons in the uranium can be reduced by a large factor from the losses occurring in a mixture of fine uranium particles and a slowing medium, when the uranium is aggregated into substantial masses in which the mean spatial radius is at least 0.25 centimeters for natural uranium metal and when the mean spatial radius of the bodies is at least 0.75 centimeters for the oxide of natural uranium (UO_2). An important gain is thus made in the number of neutrons made directly available for the chain reaction. A similar gain is made when the uranium has more than the natural content of fissionable material. Consequently, the uranium is placed in the system in the form of spaced uranium masses or bodies of substantial size, preferably either of metal, oxide, carbide, or combinations thereof. The uranium bodies can be in the form of layers, rods or cylinders, cubes or spheres, or approximate shapes, dispersed throughout the graphite, preferably in some geometric pattern. The term geometric is used to mean any pattern or arrangement wherein the uranium bodies are distributed in the graphite or other moderator with at least either a roughly uniform spacing or with a roughly systematic non-uniform spacing, and are at least roughly uniform in size and shape or are systematic in variations of size or shape to produce a volume pattern conforming to a roughly symmetrical system. If the pattern is a repeating or rather exactly regular one, a system embodying it may be conveniently described as a lattice structure. Optimum conditions are obtained with natural uranium by using a lattice of metal spheres.

The number of neutrons made directly available to the chain reaction by aggregating the uranium into separate bodies spaced through the slowing medium is a critical factor in obtaining a self-sustaining chain reaction utilizing natural uranium and graphite. The K factor of a mixture of fine uranium particles in graphite, assuming both of them to be theoretically pure, would only be about 0.785. Actual K factors as high as 1.07 have been obtained using aggregation of natural uranium in the best known geometry, and with as pure materials as it is presently possible to obtain.

Assuming theoretically pure carbon and theoretically pure natural uranium metal, both of the highest obtainable densities, the maximum possible K factor theoretically obtainable is about 1.1 when the uranium is aggregated with optimum geometry. Moreover when beryllium is used as the moderator, a K factor as high as 1.18 is obtainable and if D_2O is used a K factor of about 1.3 may be secured with pure materials. Still higher K factors can be obtained by the use of aggregation in the case of uranium having more than the naturally occurring content of fissionable materials such as U^{233}, U^{235} or 94^{239}. Adding such fissionable material is termed enrichment of the uranium.

It is thus clearly apparent that the aggregation of the uranium into masses separated in the slowing material is one of the most important, if not the most important factor entering into the successful construction of a self-sustaining chain reacting system utilizing relatively pure natural uranium in a slowing material in the best geometry at present known, and is also important in obtaining high K factors when enrichment of the uranium is used.

The thermal neutrons are also subject to capture by the slowing material. While carbon and beryllium have very small capture cross sections for thermal neutrons, and deuterium still smaller, an appreciable fraction of thermal neutrons (about 10 percent of the neutrons present in the system under best conditions with graphite) is lost by capture in the slowing material during diffusion therethrough. It is therefore desirable to have the neutrons reaching thermal energy promptly enter uranium.

In addition to the above-mentioned losses, which are inherently a part of the nuclear chain reaction process, impurities present in both the slowing material and the uranium add a very important neutron loss factor in the chain. The effectiveness of various elements as neutron absorbers varies tremendously. Certain elements such as boron, cadmium, samarium, gadolinium, and some others, if present even in a few parts per million, could prevent a self-sustaining chain reaction from taking place. It is highly important, therefore, to remove as far as possible all impurities capturing neutrons to the detriment of the chain reaction from both the slowing material and the uranium. If these impurities, solid, liquid, or gaseous, and in elemental or combined form, are present in too great quantity, in the uranium bodies or the slowing material or in, or by absorption from, the free spaces of the system, the self-sustaining chain reaction cannot be attained. The amounts of impurities that may be permitted in a system, vary with a number of factors, such as the specific geometry of the system, and the form in which the uranium is used – that is, whether natural or enriched, whether as metal or oxide – and also factors such as the weight ratios between the uranium and the slowing down material, and the type of slowing down or moderating material used – for example, whether deuterium, graphite or beryllium. Although all of these considerations influence the actual permissible amount of each impurity material, it has fortunately been found that, in general, the effect of any given impurity or impurities can be correlated directly with the weight of the impurity present and with the K factor of the system, so that knowing the K factor for a given geometry and composition, the permissible amounts of particular impurities can be readily computed without taking individual account of the specific considerations named above. Different impurities are found to affect the operation to widely different extents; for example, relatively considerable quantities of elements such as hydrogen may be present, and, as previously suggested, the uranium may be in the form of oxide, such as UO_2 or U_3O_8, or carbide, although the metal is preferred. Nitrogen may be present to some extent, and its effect on the chain reaction is such that the neutron reproduction ratio of the system may be changed by changes in atmospheric pressure. This effect may be eliminated by enclosing or evacuating the system if desired. In general, the inclusion of combined nitrogen is to be avoided.

The effect of impurities on the optimum reproduction factor K may be conveniently evaluated to a good approximation, simply by means of certain constants known as "danger coefficients" which are assigned to the various elements. These danger coefficients for the impurities are each multiplied by the percent by weight of the corresponding impurity, and the total sum of these products gives a value known as the total danger sum. This total danger sum is subtracted from the reproduction factor K as calculated for pure materials and for the specific geometry under consideration.

The danger coefficients are defined in terms of the ratio of the weight of impurity per unit mass of uranium and are based on the cross section for absorption of thermal neutrons of the various elements. These values may be obtained from physics textbooks on the subject and the danger coefficient computed by the formula

$$\frac{\sigma_i}{\sigma_u} \cdot \frac{A_u}{A_i},$$

wherein σ_i represents the cross section for the impurity and σ_u the cross section for the uranium, A_i the atomic weight of the impurity and A_u the atomic weight for uranium. If the impurities are in the carbon, they are computed as their percent of the weight of the uranium of the system.

Presently known values for danger coefficients for some elements are given in the following table 6.1, wherein the elements are assumed to have their natural isotopic constitution unless otherwise indicated, and are conveniently listed according to their chemical symbols:

Table 6.1

Element	Danger Coefficient	Element	Danger Coefficient
He	0	Cu	1.8
Li	310	Zn	0.61
B	2150	Ga	~ 1
N	4.0	As	2
F	0.02	Se	6.3
Na	0.65	Br	2.5
Mg	0.48	Rh	50
Al	0.30	Ag	18
Si	0.26	Cd	870
P	0.3	In	54.2
S	0.46	Sn	0.18
Cl	31	Sb	1.6
K	2.1	I	1.6
Ca	0.37	Ba	0.30
Ti	3.8	Sm	~ 1430
V	4	Eu	435
Cr	2	Gd	~ 6320
Mn	7.5	Pb	0.03
Fe	1.5	Bi	0.0025
Co	17	Th	1.1
Ni	3		

Where an element is necessarily used in an active part of a system, it is still to be considered as an impurity; for example, in a structure where the uranium bodies consist of uranium oxide, the actual factor K would ordinarily be computed by taking that fact into account using as a base K a value computed for theoretically pure uranium.

As a specific example, if the materials of the system under consideration have 0.0001 part by weight of Co and Ag, the total danger sum in K units for such an analysis would be:

$$0.0001 \times 17 + 0.0001 \times 18 = 0.0035\ K \text{ units.}$$

This would be a rather unimportant reduction in the reproduction factor K unless the reproduction factor for a given system, without considering any impurities, is very nearly unity. If, on the other hand, the impurities in the uranium in the previous example had been Li, Co, and Rh, the total danger sum would be:

$$0.0310 + 0.0017 + 0.0050 = 0.0377\ K \text{ units.}$$

This latter reduction in the reproduction factor for a given system would be serious and might well reduce the reproduction factor below unity for certain geometries so as to make it impossible to effect a self-sustaining chain reaction with natural uranium and graphite, but might still be permissible when using enriched uranium in a system having a high K factor.

This strong absorbing action of some elements renders a self-sustaining chain reacting system capable of control. By introducing neutron absorbing elements in the form of rods or sheets into the interior of the system, for instance in the slowing material between the uranium masses, the neutron reproduction ratio of the system can be changed in accordance with the amount of absorbing material exposed to the neutrons in the system. A sufficient mass of the absorbing material can readily be inserted into the system to reduce the reproduction ratio of the system to less than unity and thus stop the reaction. Consequently, it is another object of our invention to provide a means and method of controlling the chain reaction in a self-sustaining system.

When the uranium and the slowing material are of such purity and the uranium is so aggregated that fewer neutrons are parisitically absorbed than are gained by fission, the uranium will support a chain reaction producing an exponential rise in neutron density if the overall size of the system is sufficiently large to overcome the loss of neutrons escaping from the system. Thus the overall size is important.

The size of the system will vary, depending upon the K factor of the system, and upon other things. If the reproduction factor K is greater than unity, the number of neutrons present will increase exponentially and indefinitely, provided the structure is made sufficiently large. If, on the contrary, the structure is small, with a large surface-to-volume ratio, there will be a rate of loss of neutrons from the structure by leakage through the outer surfaces, which may overbalance the rate of neutron production inside the structure so that a chain reaction will not be self-sustaining. For each value of the reproduction factor K greater than unity, there is thus a minimum overall size of a given structure known as the critical size, above which the rate of loss of neutrons by diffusion to the walls of the structure and leakage away from the structure is less than the rate of production of neutrons within the system, thus making the chain reaction self-sustaining. The rate of diffusion of neutrons away from a large structure in which they are being created through the exterior surface thereof may be treated by mathematical analysis when the value of K and certain other constants are known, as the ratio of the exterior surface to the volume becomes less as the structure is enlarged.

In the case of a spherical structure employing uranium bodies imbedded in graphite in the geometries disclosed herein and without an external reflector the following formula gives the critical overall radius (R) in feet:

$$K - 1 = \frac{C}{R^2},$$

where C is a constant that varies slightly with geometry of the lattice and for normal graphite lattices may have a value close to 7.4.

For a rectangular parallelepiped structure rather than spherical, the critical size can be computed from the formula

$$K - 1 = C\left(\frac{1}{a^2} + \frac{1}{b^2} + \frac{1}{c^2}\right),$$

where a, b, and c are the lengths of the side in feet. The critical size for a cylindrical structure is given by the formula, irrespective of the shape of the uranium bodies

$$K - 1 = C\left(\frac{1}{h^2} + \frac{0.59}{R^2}\right)$$

(Cylinder height h ft Radius R ft).

However, when critical size is attained, by definition no rise in neutron density can be expected. It is therefore necessary to increase the size of the structure beyond the critical size but not to the extent that the period for doubling of the neutron density is too short, as will be explained later. A desirable reproduction ratio for an operating structure with all control absorbers removed and at the temperature of operation is about 1.005. The operating size at which any given reproduction ratio can be obtained may be computed from modifications of the above formulae for critical size. For example, for spherical active structures the formula

$$K - (\text{reproduction ratio}) = \frac{C}{R^2}$$

may be used to find R when K is known and the reproduction ratio is somewhat over unity. The same formula will, of course, give the reproduction ratio for given structures for which K and R are known.

Critical size may be attained with a somewhat smaller structure by utilizing a neutron reflecting medium surrounding the surface of the active structure. For example, a 2 feet thickness of graphite having low impurity content, completely surrounding a spherical structure is effective in reducing the diameter of the uranium bearing portion by as much as 2 ft, resulting in a considerable saving in uranium or uranium compound.

The neutronic chain reaction referred to can be made self-sustaining in a device known as a neutronic reactor wherein uranium bodies are dispersed in an efficient neutron slowing medium or moderator, when the reactor is made to be just above a critical size where the rate of neutron generation inside the reactor is slightly greater than the rate of neutron loss from the exterior of the reactor. Under these conditions, a self-sustaining nuclear chain reaction can be obtained within the reactor having any neutron density desired, up to infinity. However, to prevent destruction of the reactor, the heat of the reaction must be controlled, and then removed by an amount providing a stable temperature in the reactor at some predetermined and controlled operating level. As the greater the number of fissions, the greater the number of neutron are present to produce 92^{239} converting to 94^{239} by successive beta decay, the production of 94^{239} is accelerated by operating the reactor at high neutron density levels.

A stable temperature in a neutronic reactor composed entirely of moderator and fissionable material such as, for example, graphite and uranium metal, can only be attained at a relatively low power output as the heat generated can be dissipated only by conduction out of the reactor. Higher power outputs with greater production of 94^{239} require additional heat removal.

It is the principal object of our invention to increase the removal of heat from a neutronic reactor during operation in order that the production of 94^{239} and/or other products of the neutronic reaction can be accelerated.

It is another object of our invention to conduct a gaseous coolant such as air through a neutronic reactor in such a manner that heat is removed from the reactor without interfering with the operation of the reactor.

As the 94^{239} and the various fission products are formed directly in and become a part of the uranium bodies, it is desirable to be able to remove the irradiated uranium bodies from the reactor when a desired concentration of 94^{239} is under formation or has been formed therein, in order that chemical separation be most readily accomplished.

It is therefore still another object of our invention to provide a neutronic reactor in which the uranium is readily removed and replaced.

In accordance with the present invention it has been found that heat may be removed from a neutronic reactor most effectively by passing the coolant in contact with or closely adjacent to uranium bodies. Thus it has been ascertained that most of the heat developed in the reactor is developed in the uranium and therefore that to secure effective cooling the coolant should be passed sufficiently close to uranium to ensure rapid heat removed therefrom.

For the sake of simplicity the invention has been described with particular reference to the use of air as a coolant although it is to be understood that other coolants may be used without departing from the spirit of the invention.

While coolant gas such as air, in many respects is not as efficient a cooling fluid as a liquid, such as water for example, it has many advantages over liquids in that liquids require that piping be inserted in the reactor. These pipes, as well as the liquids that would be circulated through them, absorb neutrons. For example, the pipes and water carried by the pipes in a water cooled reactor absorb sufficient neutrons to increase the overall parasitic neutron absorption by an appreciable amount as will be apparent from the table of danger coefficients given above.

On the other hand, air requires no pipes and can be circulated through channels cut directly in the solid moderator. Graphite, for example, is porous and when piled in blocks to form the reactor framework contains about 25 percent air. The provision of the air channels, which in effect merely removes some of the moderator, has very little effect on the chain reaction when the proper volume ratio of moderator to uranium is used, and most of the effect of the air channels is to introduce slightly more air in an air cooled neutronic reactor than would be present in an uncooled reactor. Nitrogen does absorb neutrons, but as the air in the channels is only under slight pressure while passing through the reactor, the net effect of the channels and the air is small.

It is therefore, another object of our invention to provide a means and method for cooling a neutronic reactor with a minimum of interference with the chain reaction.

When air is passed through a neutronic reactor operating at high neutron densities, components of the air itself, such as argon, for example, become radioactive. It is highly desirable, therefore, particularly for the biological safety of operating personnel, that air having any radioactivity associated therewith, be disposed of in a safe manner.

A still further object of our invention is to provide a means and method of safely disposing of air having radioactive components therein.

Broadly stated, our invention includes a neutronic reactor comprising a mass of neutron moderator having uranium containing bodies therein, the moderator being pierced by channels through which air may be circulated to remove the heat of reaction to the extent that a stable temperature can be attained in the device at high neutron densities.

Other objects and advantages of our invention may be more clearly understood by reference to the following description and the attached drawings which illustrate, as an example, one form our invention may take. This example is not to be taken as limiting as other forms within the scope of the appended claims will be readily apparent to those skilled in the art. Requisite information and criteria for operativeness of neutronic reactors are set forth in the copending application of the present inventors filed December 19, 1944, Serial No. 568,904, now Patent No. 2,708,656, dated May 17, 1955.[1]

In the drawings:

Fig. 6.1 is a longitudinal view partly in section and partly in elevation of an air cooled neutronic reactor system illustrating our invention.

Fig. 6.2 is a cross sectional view, partly in elevation, taken as indicated by the line **2–2** in Fig. 6.1.

Fig. 6.3 is a plan view of the system shown in Figs. 6.1 and 6.2.

Fig. 6.4 is a longitudinal sectional view partly in elevation of a jacketed slug.

Fig. 6.5 is a longitudinal sectional view, partly in elevation, of a horizontal channel during a loading and unloading operation.

Fig. 6.6 is a cross sectional view taken as indicated by the line **6–6** in Fig. 6.5.

Fig. 6.7 is a longitudinal sectional view, partly in elevation, of one form of loading device.

Fig. 6.8 is a view partly in section and partly in elevation taken as indicated by the line **8–8** in Fig. 6.7.

Fig. 6.9 is a top plan view of the loading device shown in Figs. 6.7 and 6.8.

Fig. 6.10 is a side view of the loading device shown in Figs. 6.7, 6.8 and

[1]@ See page 217.

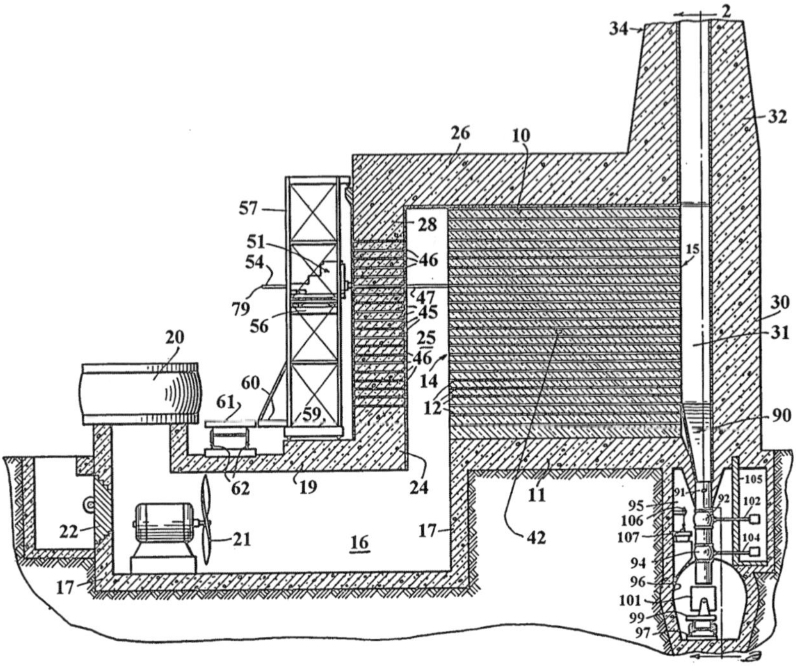

Fig. 6.1

6.9.

Fig. 6.11 is a view partly in section and partly in elevation of a friction drive used in the loading device.

Fig. 6.12 is a diagram or chart showing K factors for various radii of uranium metal rods.

Referring to the drawings, we have chosen to illustrate our invention by reference to a graphite-uranium reactor, sometimes known as a pile.

Such a rector broadly comprises a mass of graphite blocks closely piled or stacked into a cube **10** as shown in Figs. 6.1 and 6.2. This graphite cube may be, for example, 24 or 25 feet on a side and rest on a concrete foundation **11**. The graphite cube **10** is pierced with horizontal air channels **12**, of square cross section, with one of the diagonals vertical. The channels may be readily made by grooving adjacent blocks. The channels are 1.75 inches on a side and extend completely through the reactor, from an inlet face **14** to an outlet face **15**. About 1500 channels may be provided, and as will be later brought out, any unused channel can be plugged. Only a

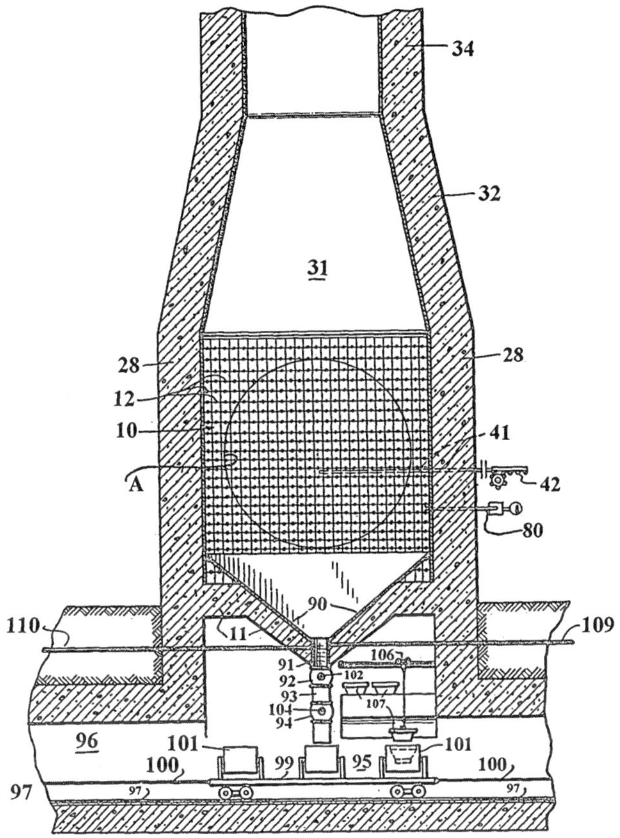

Fig. 6.2

few of the channels are shown in the drawings for sake of clarity.

Adjacent the inlet face **14** of the cube, the foundation is continued downwardly to form the floor of an inlet air duct **16** extending outwardly. The inlet air duct **16** is completed by concrete side walls **17** and top **19**.

At some distance away from the graphite cube **10** the inlet duct is turned upwardly to terminate in an air filter **20**, relatively close to the surface of the ground. A fan or blower **21**, here illustrated as electrically driven, is installed on the floor of the inlet duct just below the air filter, access to the fan being conveniently obtained through duct door **22**, behind the fan.

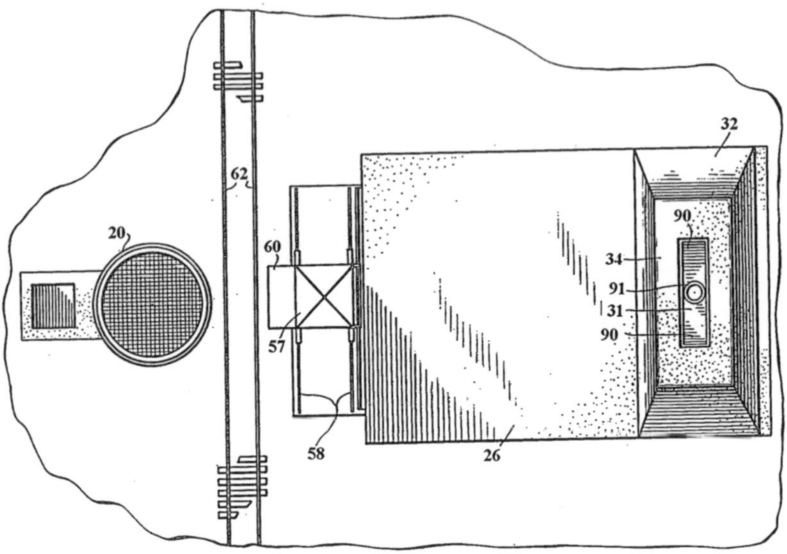

Fig. 6.3

The concrete top **19** of the inlet air duct is continued upwardly as an inlet end shield **24**, positioned parallel to but spaced away from inlet face **14** of the cube **10** to form an inlet chamber **25** communicating with air channels **12**.

Above the inlet chamber **25** and the cube **10** the concrete is continued horizontally to form a top shield **26**, and side shields **28** are built up from the foundation **11** to enclose cube **10**. Shields **26** and **28** closely approach the top and side faces of the cube, to minimize air flow around the outside of the cube. A small amount of air circulation, however, may be desirable over the top and side faces to cool these faces.

At the outlet face **15**, an outlet end shield **30** of concrete is provided. End shield **30** is parallel to and spaced from the outlet face **15** of the graphite cube to form an outlet chamber **31** communicating above with the base **32** of a stack **34** projecting upwardly and formed as a continuation of the concrete top side and outlet end shields. Thus the cube **10** is completely enclosed by concrete shields or shields of similar inorganic material containing water of crystallization or other neutron absorbing material, with a duct system operating by virtue of pressure provided by fan **21** to conduct air from close to ground level through channels **12** into the stack and then into the atmosphere well above ground level at the top of the stack. The concrete shields may be from five to ten feet in accordance with the maximum desired operating power of the reactor and serve as shields to reduce escape of neutrons and gamma radiation.

As a neutronic reaction will take place when uranium bodies are properly spaced in a moderator mass of a certain finite size, the above described device can be made chain reacting by placing uranium bodies in the horizontal channels in such a manner and in such an amount that a neutron reproduction ratio of slightly over unity is obtained, exclusive of all neutron losses within the reactor and from the exterior of the reactor. This reproduction ratio may be defined as the ratio of the number of neutrons gained by fission to the total number of neutrons lost by absorption in the uranium, absorption in the moderator, absorption by impurities in the reactor and by leakage from the reactor for a reactor of finite size.

Using the graphite mass as the moderator to slow fast neutrons to energies where they again are able to create fission in 92^{235}, the device will have a reproduction ratio of unity when approximately 700 of the channels **11** in the graphite cube are each loaded with 68 aluminum jacketed uranium slugs **35** lying end to end, with a channel spacing of 7 inches measured center to center, and with the loaded channels roughly forming a cylinder as indicated by line **A** in Fig. 6.2. Both graphite and uranium should be of highest possible purity.

However, more than a unity reproduction ratio is required, as when the reproduction ratio is exactly unity no rise in neutron density will occur. Under such conditions the device will not develop high neutron densities or power in the form of heat. By loading additional channels, i.e., making the active portion greater than critical size, however, the reproduction ratio within the reactor can be brought above unity in order that a rise in density can occur. Then this excess neutron reproduction can be absorbed by neutron absorbing materials deliberately inserted into the reactor in order to hold the reproduction ratio at an average value of unity after a desired power output has been obtained, as a result of the initial rise in density.

Consequently, in accordance with the amount of excess reproduction ratio desired, about 1000 channels may be loaded with uranium slugs. Most of the channels not loaded with uranium may be closed by inserting plugs, preferably of graphite, in such channels in order to conserve air. Some of the channels, however, in the peripheral portions of the cube may be left

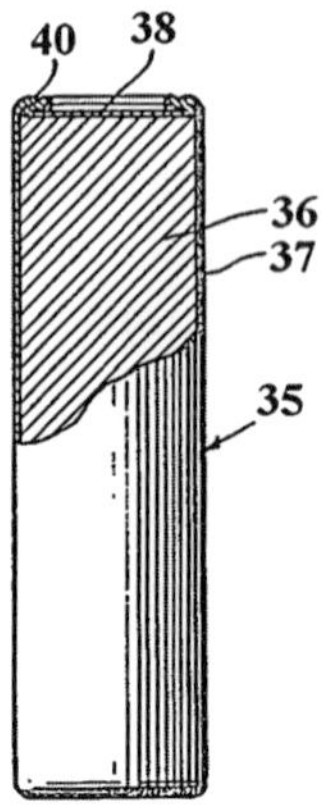

Fig. 6.4

open for cooling of the graphite in those portions.

One preferred form of slug construction is shown in Fig. 6.4. Each uranium metal slug is 1.1 inches in diameter and 4 inches long covered with an aluminum jacket approximately 20 mm thick in good heat conductive relation to the uranium. The slugs weight about $2\frac{1}{2}$ pounds each.

In forming the slugs **35**, the uranium portion **36** is machined to size, cleaned in trisodium phosphate and then washed in water. Aluminum, zinc or other non fissionable metal of low neutron absorption jacket cans **37** are provided having an inside diameter somewhat larger than the uranium portion. This can **37**, open at one end only, is slipped over the uranium after being cleaned in benzine and hot water. The can **37** with the uranium inside is then passed through a sizing die of 1.134 inches diameter. This die, being of smaller diameter than the 1.1 inch uranium portion plus the 20 millimeter walls, draws the can in tight thermal contact with the uranium.

A cup-shaped cap **38** is then placed base down inside the projecting portion of the can **37** and is seam welded to the can. The projecting portion is then cut off above the seam weld **40** and the remaining projecting portion including the weld, spun over the adjacent end of the slug. Thus each jacket completely encloses and seals the uranium, preventing air from corroding the uranium and, as will be pointed out later, also preventing or restraining fission fragments created by nuclear fission at the surface of the uranium from entering the air stream.

The channels are loaded with uranium until the reproduction ratio, with neutron absorbers removed and the coolant flowing through the reactor, is about 1.005 to 1.006. This means that for every two hundred neutrons starting in each neutron generation about two hundred and one neutrons are produced in the reactor over and above all losses. Under these conditions and taking into account the fact that about one percent of the neutrons of fission are delayed in their emission for a mean time of about 5 seconds, the neutron density of the reactor will double every 8 to 15 seconds. With some part of the neutron absorbers inserted but with the insertion of less than the amount of neutron absorbers required to make the reproduction ratio unity, the rise is slower. When the neutron absorbers are almost but not entirely inserted the doubling of the neutron density may take several hours. Then when a desired density has been reached, the reproduction ratio can be reduced to unity so that the desired density is continuously maintained by the introduction of neutron absorbing material into the reactor.

The neutron absorbing material is introduced into the reactor by means of a control rod **41** as shown in Fig. 6.2. This control rod extends into the graphite cube, sliding in a channel therein and is operated from outside of side shield **28** as by rack and pinion **42**. The rod is made from, or incorporates therein, an efficient neutron absorber, such as for example, cadmium or boron. A sheet of cadmium riveted to a steel strip forms a satisfactory control rod. As the depth of insertion of the rod determines the amount of neutron absorbing material inside the reactor, the critical position of the rod is where the rate of neutron absorption by the rod balances the reproduction ratio at unity. Thus, by moving the rod outwardly from the critical position the neutron density in the reactor will rise. Moving the rod inwardly from the critical position causes the reproduction ratio to fall below unity, and the reaction stops. Thus the reaction is always under control, and as the rise in neutron density is exceptionally slow as the rod approaches the critical position, manual control is possible. Other and similar rods (not shown) may be provided, if desired, for rapid progression into the reactor to stop the reaction in case of failure of the control rod to stop the rise in neutron density for any reason. Such rods are termed safety rods.

During operation heat is released in the reactor in accordance with the neutron density therein. Most of the heat arises from the kinetic energy of the fission fragments and about 92 percent of the energy is released in the uranium. About 6 percent is released in the graphite due to neutron absorption therein and about 2 percent escapes from the reactor in the form of neutrons and gamma radiation. Consequently, the reactor can only be

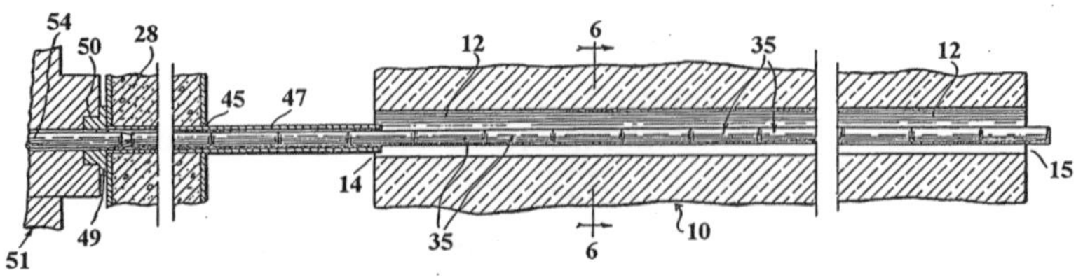

Fig. 6.5

operated at a power dependent upon heat removal to the point where a stable temperature obtains. Otherwise, the reactor will accumulate heat to the point that the device may be damaged. Since aluminum melts at 658°C, stable temperatures below this value should be used although with jackets of other non-fissionable metals, such as beryllium, the stable temperature may be increased, although if the temperature should rise too high the uranium bodies might be damaged even when using beryllium jackets as uranium of the type used in neutronic reactors melts at about 1100°C.

A stable temperature is obtained in the device of the present invention by passing atmospheric air through the reactor, and in the specific example shown and described, the air is passed through the graphite channels and directly in contact with the aluminum jackets of the slugs. Under these circumstances the reactor can be operated continuously at 250 kilowatts electrical equivalent of heat by passing 32,000 cubic feet per minute through the reactor with a maximum temperature rise of the slugs to about 100°C, and at 500 kilowatts continuously with about 50,000 cubic feet per minute of

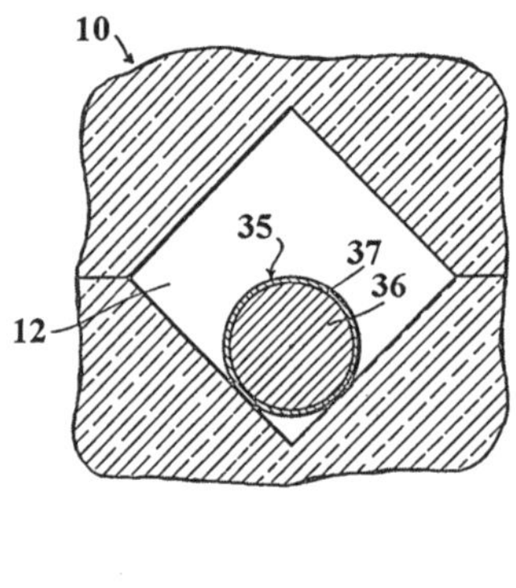

Fig. 6.6

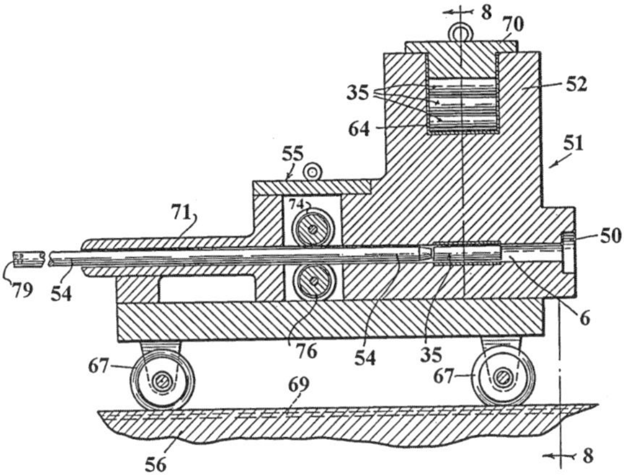

Fig. 6.7

air with a maximum metal temperature of 200°C. The output of the reactor can be stabilized at still higher powers by passage of a greater volume of air therethrough if desired.

Having discussed generally the operation of the reactor, and the temperature stabilization thereof by air cooling at elevated powers and neutron densities, we will now describe one means and method by which the reactor can be loaded and unloaded, in order that the neutron irradiated uranium can be removed for further processing such as the recovery of 94^{239} formed in the uranium, and fresh uranium inserted for subsequent operation of the reactor.

To accomplish loading of the slugs **35** into the various air channels **12**, the concrete of the inlet end shield **24** is pierced with a plurality of loading apertures **45**, as shown in Figs. 6.1 and 6.5, each aperture being aligned with the axis of slug positions in the air channels **12**. Normally, during operation of the reactor, each aperture **45** is closed by a removable lead plug **46** extending through the shield **28** only.

When it is desired to lead a channel with new slugs, the lead plug **46** for that channel alone is removed, and a charging tube **47** inserted, extending through the inlet end shield **24**, across the inlet chamber **25** and entering the corresponding air channel **12** as shown in Fig. 6.5. The outer end of charging tube **47** is provided with a flanged nipple **49** shaped to engage a nipple recess **50** of a loading mechanism indicated generally by numeral **51**. It will be noted that the charging tube is smaller than the air channel **12**

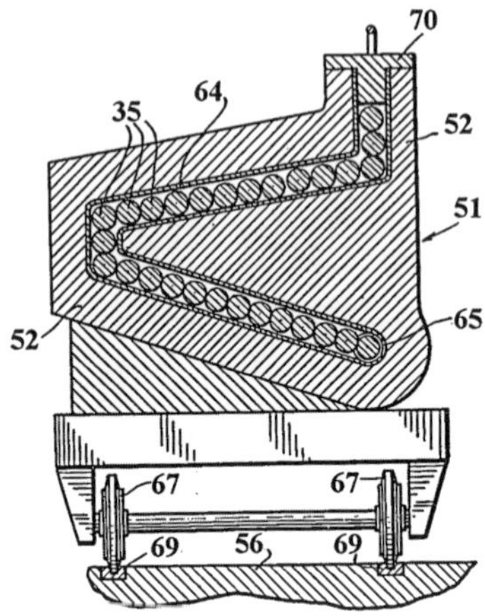

Fig. 6.8

and that air can pass through the channel being unloaded. The air should circulate during unloading, although it may be at reduced velocity.

Loading mechanism **51** comprises a loading magazine **52**, a loading plunger **54**, and a plunger drive **55**, as shown in Figs. 6.7 to 6.10 inclusive.

The loading mechanism **51** is mounted on an elevator platform **56** mounted to be raised and lowered in an elevator frame **57** capable of moving along the outside of inlet end shield **24** on elevator tracks **58**. Base **59** of the elevator frame is provided with a platform **60** projecting outwardly

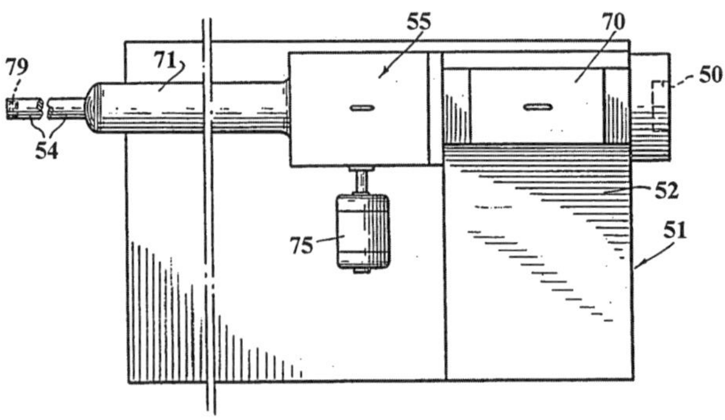

Fig. 6.9

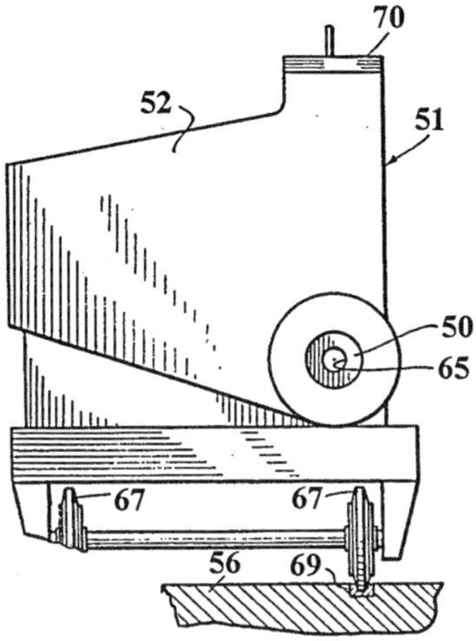

Fig. 6.10

on the same level as the top of a supply car **61** travelling on supply car tracks **62**. Supply car **61** is used to bring a supply of slugs to the elevator for use in the loading mechanism **51**.

The slugs **35** when received at the elevator, are loaded into an inclined loading channel **64** in the loading magazine, in side by side relationship and feed by gravity to the bottom of the roof. The bottom of loading channel **64** is a part of plunger bore **65** extending through the loading magazine ending in the nipple recess **50** cooperating with flanged nipple **49** on charging tube **47** so that the plunger bore **65** and the loading bore in guide tube **47** are in concentric alignment. To provide engagement and disengagement of nipple recess **50** and nipple **49**, the entire loading mechanism is movable with respect to elevator platform **56** on wheels **67** running in guides **69** on the elevator platform.

It will be noted that loading magazine **52** is massive. In some instances it may be desirable to load slugs already partially irradiated and in consequence radioactive. The thick walls of the magazine then act as a shield for the radioactive slugs, and in this case a heavy cap **70** may close the upper opening of the loading channel **64**. Iron or lead may be used for the body of the magazine. In addition, the use of the thick metal in the magazine, particularly around the plunger bore **65**, reduces radiation that might pass through the interconnected loading aperture **45** and guide tube **47** either

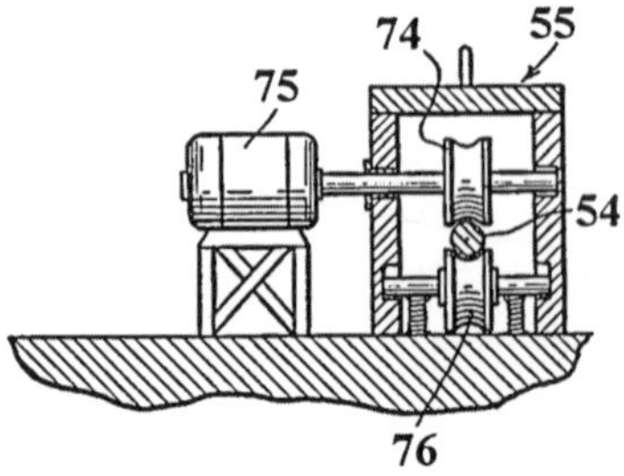

Fig. 6.11

from radioactive slugs therein or from the irradiated slugs in the reactor when charging tube **47** is empty.

The slugs are fed from magazine **52** by a reciprocating motion of plunger **54** operating in plunger bore **65**. Plunger **54** may be of iron to act as a shield when inserted into charging tube **47** and is supported outwardly by plunger bearing **71** on the opposite side of a plunger drive **55**.

Plunger drive **55** in simplified form may be a friction wheel **74** driven by motor **75** as shown in Figs. 6.7, 6.9 and 6.11 opposed by an idler wheel **76** pressed against plunger **55** by spring **77**. Motor **75** is reversible and under control of the loading operator. Plunger **54** is sectional, having a threaded end **79** capable of making connection with additional plunger sections. Sufficient sections are provided to insert the plunger entirely through an air channel **12** when required.

In the initial loading of the graphite cube **10**, loading is started with the more central air channels until 68 slugs have been placed in the guide tube **47** and connected channel. Plunger **54** is then operated to push the slugs into the channel until the outer end of the first slug is at the outlet face **15**. The plunger is then withdrawn, leaving the outer end of the last slug about 16 inches from the inlet face **14**, for purposes explained later.

Proceeding outwardly and preferably concentrically, additional channels are loaded, meanwhile checking the neutronic activity of the reactor. As the activity increases as the loading approaches a critical size, that is, the size where the reproduction ratio will be exactly unity, the approach to critical size can be predicted by extrapolation of observed neutron density values with respect to the volume of the cube loaded with uranium. The neutron density values can be obtained from an ionization chamber **80** (Fig. 6.2), for example, using any well-known indicating circuit, or by measuring the radioactivity of indium foils, for example, induced by neutron irradiation when inserted into the reactor.

As the critical size is approached, the control rod **41** is inserted deeply into the reactor to prevent a self-sustaining chain reaction; and loading is continued until the desired maximum reproduction ratio of, for example, from 1.005 to 1.006 is attained. This ratio can be checked by removal of the control rod and measuring the time taken by the reactor to double its neutron density. From this period, the reproduction ratio can be mathematically computed.

When the desired number of channels are loaded the active core of the reactor may contain from 34 to 50 tons of uranium, and will be ready for operation. Graphite plugs for the unused channels may be loaded in a manner similar to that described for the uranium slugs.

It will be noted that on 4 sides of the graphite cube excess graphite will be present. On the fifth side, i.e., at the inlet face, graphite will also extend 16 inches beyond the uranium. On the remaining side, i.e., the outlet face, no graphite extends beyond the uranium. Thus, 5 sides of the active portion (the uranium bearing portion) are surrounded by graphite. This graphite constitutes a reflector and reduces the amount of uranium required to reach critical size.

As stated above, graphite and other neutron moderators can be used around the active portions of a neutronic reactor, as such materials scatter neutrons passing through then and thereby change the direction of neutrons. Statistically, due to the scattering action, a large part of the neutrons that otherwise would be lost to the active portion are returned to it, thus reducing exterior loss and thereby reducing critical size. In the present instance the reduction in size is slightly less than the optimum, as only 5 sides of the active portion are surrounded with the reflecting layer.

It will be noted that more air channels are originally provided in the graphite cube than are required for the disposition of the uranium containing slugs. The manner in which the uranium is disposed in the graphite is known as the geometry of the system, and this geometry may be varied to suit conditions, for example, the system as described may be said to have rod geometry with cylindrical loading, as the end to end relationship of the slugs forms, in effect, a long rod of uranium. Further, the use of full length rods in all channels concentrically arranged around the center of the active portion, gives that portion a generically cylindrical shape.

In Fig. 6.12, contour lines have been plotted for lattices of theoretically pure uranium metal rods of a density of approximately 18 g/cm^3 embedded in theoretically pure graphite.

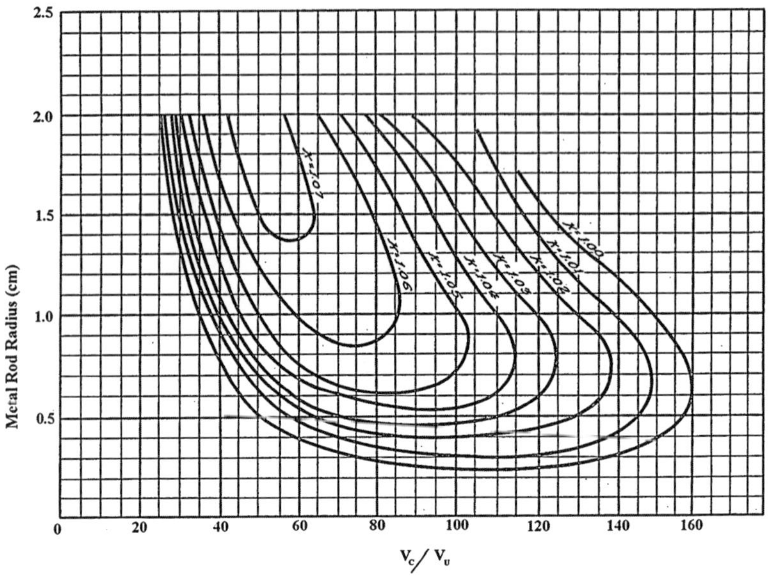

Fig. 6.12

Along the ordinates of the graph are plotted the radii of the rods. Along the abscissae are plotted volume ratios of graphite to uranium. These values give the volume ratio of carbon to uranium for the unit cell that is repeated to form the lattice and for the structure, as any particular lattice geometry is a mere repetition of like cells.

The curves of Fig. 6.12 have been drawn from a number of observed values of K, and, in addition, from many computed values. The computed values are based on the fact that K is proportional to the product of three factors: $p \times f \times \epsilon$, where p is the probability of a fast fission neutron escaping resonance capture and becoming a thermal neutron; f is the fraction of the thermal neutrons absorbed by uranium (both by simple capture and to produce fission) rather than by the graphite; and ϵ is a small factor by which the number of neutrons is increased because of additional neutrons due to fission produced by the fast fission neutrons before leaving the lump of uranium. Each of these factors may be computed separately by methods known to physicists, using experimentally determined constants appropriate to these phenomena. The proportionality factor required to obtain K from the product of these three factors has been determined from the measured values of K in certain special cases and also checked by independent experiments. The K contour lines for natural uranium metal rods or cylinders are presented in Fig. 6.12. The values set forth on the graphs are only accurate within a reasonable margin of error due to the necessity of extrapolation and interpolation of the results obtained by more accurate and actual measurements of lattice arrangements, such as, for example, the results obtained by the measurements of lattice structures too small to support a self-sustaining chain reaction, as set forth in Enrico Fermi's application, Serial No. 534,129, filed May 4, 1944, now Patent No. 2,780,595, dated February 5, 1957.[2] However, if structures are built well within the limits of the curves shown, using materials of a purity presently obtainable, and to the proper critical sizes, a self-sustaining chain reacting system will result.

The highest values for the reproduction factor K are obtained where both the neutron resonance absorption in the uranium and the neutron absorption in the graphite have minimum values. As the uranium bodies are enlarged, for any given volume ratio, surface resonance absorption will decrease but carbon absorption will increase. The increase in carbon absorption in this case is due to the fact that the larger the uranium bodies are, for any given volume ratio, the grater will be the distance between the bodies. The neutrons reduced to thermal energies will then have to diffuse in the carbon over a longer path, thus increasing the probability of capture by carbon, before entering uranium. Likewise, as the size of the uranium bodies is decreased, uranium resonance absorption increases, and in fact, increases faster than the decrease of carbon absorption. Consequently, with all other factors remaining constant, as the size of the uranium bodies and the ratio of the volume of graphite to uranium departs from the optimum, the reproduction factor will decrease in value, as shown in Fig. 6.12.

It can be seen from the curves in Fig. 6.12 that if the radii of the uranium rods are less than 0.25 cm the value of the reproduction factor K is less than unity for all volume ratios of graphite to uranium, so that for rod sizes of natural uranium metal less than this value, it would be impossible to build a self-sustaining chain reacting system irrespective of the overall size of the structure. For rods of greater radius than 0.25 cm, it is possible to obtain values for the reproduction factor K greater than unity, providing the ratio between graphite volume and the volume of the uranium is within certain limits shown on the graph. The innermost contour shown

[2]@ See page 181.

in Fig. 6.12 represents a value for the reproduction factor K of 1.07. At approximately the center of this contour a higher reproduction factor K, for example, about 1.08 would be found. This highest value is for optimum conditions with theoretically pure uranium rods of about 1.7 cm in radius, and a volume ratio of about 52 carbon to 1 uranium.

Lump geometry may also be used in the reactor, and is accomplished by separating the individual slugs by graphite rods, for example 4 inches long and of about the same diameter. In this case additional changes will be loaded to bring the amount of uranium almost to the same value as that used in the rod geometry, although the lump geometry is slightly more efficient than rod geometry.

Furthermore, a spherical shape is somewhat more efficient for the active portion than a cylindrical shape, and can be approached with either rod or lump geometry by shortening the extent of uranium loading in the channels as the peripheral concentric layers are added, until an approximate sphere is formed. A suitable compromise is the use of one or more concentric rings of outer channels around a cylindrically loaded central portion, with only half the number of slugs in the outer channels and disposed with the ends of the slug row equally distant form the ends of the fully loaded rows. Other arrangements will be apparent to those skilled in the art.

In the presently described device using 1.1 inch diameter slugs inserted in end-to-end relation to form 22 feet rods spaced 7 inches center to center, the curves indicate a K factor of about 1.06, and a volume ratio of about 47 carbon to 1 uranium. The active portion is 22 feet long and about 21 feet in diameter (circular cross section) for operating conditions, i.e., it is above critical size.

It will be noted that the formulae for critical size and operating size are given herein for reactors with no reflector. It will be further noted that the curves in Fig. 6.12 are given for theoretically pure graphite and uranium.

As pointed out previously the presence of impurities or other neutron absorbers in the reactor may reduce the K factor of the reactor substantially. However, use of a reflector substantially reduces loss of neutrons due to leakage and consequently such reduction permits use of uranium of lower purity.

After the reactor is loaded the fan is started and the control rod is withdrawn until a rise in neutron density to a desired power output where a stable temperature in the reactor is attained. The control rod is then progressed into the reactor until a neutron balance is obtained with the reproduction ratio at unity, thus maintaining the chain reaction at the desired operating power. Small variations from the unity reproduction ratio will occur during operation, due to temperature variations of the cooling air, and to change in barometric pressure and to minor variations in air pressure delivered by the fan. However, such variations are compensated by slight inward or outward corrective movements of the control rod, either by hand in response to indicated variations in neutron density, or automatically by direct linkage of the control rod to the output of the ionization chamber **80**. However, such automatic control is no part of the present invention.

During operation of the reactor, the air passing through the reactor becomes radioactive due to the fact that it is subjected to intense neutron irradiation. Investigation has proved that the only significant radioactivity present in the air after having passed through the operating reactor is that of Ar^{41}, having a 110 minute half life. At 500 kilowatt power, however, this activity is present in the exhaust air to the point that it would be biologically dangerous to operating personnel unless highly diluted during its radioactive decay. For that reason the air passing through the operating reactor is not delivered to the atmosphere at ground level but is exhausted at a substantial distance above ground, such as for example from the top of a 200 feet stack, with the result that when and if any of the radioactive Ar^{41} reaches ground level it is so dispersed in and diluted by fresh atmospheric air that less than 0.1 Roentgen per day will be received by any persons on the ground, either close to or away from the stack. Thus, the air is only passed once through the reactor and does not acquire excessive radioactivity.

The sole presence of the above noted type of radioactivity, however, is predicated on the use of the jackets capable of restraining escape of fission products sealed around the uranium bodies. Aluminum is preferred for the jackets, as aluminum has a relatively low neutron capture probability and, consequently, can be used in substantial amounts in the reactor without absorbing or capturing sufficient neutrons to prevent a self-sustaining chain reaction from occurring. Aluminum also corrodes very slowly in hot air. Other metals such as zinc, stainless steel or lead may be used.

The jackets have two functions, both of which reduce radioactivity in the cooling air. The first is to prevent oxidation of the uranium. While considerable oxide could be tolerated in the reactor itself if the uranium were to be used in unprotected condition, some of the oxide particles would be picked up by and exhausted in the cooling air. As these particles would be highly radioactive and relatively heavy, the proper dispersal thereof would be a difficult problem.

In addition, if operation should be accomplished in the reactor with bare uranium, fission fragments from nuclear fissions occurring on the surface of the uranium would also be projected into the air stream and would be carried out by the air stream. These fragments are exceptionally radioactive and could not safely be dispersed into the atmosphere. When jackets are used, these fragments are, however, stopped by the jackets and cannot enter the air stream.

Thus, the jackets prevent corrosion of the uranium and prevent fission fragments and corrosion products of uranium from entering the air stream. As fission fragments will pass through an extremely small hole, one method of monitoring the reactor for jackets failure, such as for example a weld crack, is to monitor the radioactivity of the stack gas. If the stack gas shows any substantially radioactivity other than that of Ar^{41} then it is clear that a jacket failure has occurred. Such monitoring of the stack gas is normally a routine procedure of an operating air cooled reactor, but forms no part of the present invention.

After operation of the reactor for a sufficient length of time for an amount of 94^{239} to be created sufficient for chemical separation, such as for example 100 days at 500 kilowatts, the reactor is shut down by inserting the control rod fully into the reactor. After about one-half hour's wait, during which all delayed neutron emission will have ceased and the more violent radioactivity subsided, the reactor may be unloaded.

The unloading may be accomplished in two ways, either by using the plunger to push the slugs out of the channels so that they fall by gravity out of the outlet face **15**, or by using the plunger to insert new slugs in the channels, each slug so inserted pushing an irradiated slug out of the outlet face **15**. In the first instance the graphite cube will be left empty after unloading. In the second instance the insertion of new slugs is continued until all or a predetermined part of the irradiated slugs are out of the reactor, having been replaced by fresh material. Thus, the reactor is left ready for the next run. Under ordinary circumstances the latter manner of unloading is preferred.

In either case the slugs drop by gravity from the outlet face into outlet chamber **31**, falling on to two angularly disposed pad plates **90** positioned to intercept the falling slugs, in the bottom half of outlet chamber **31** as shown in Figs. 6.1 and 6.2. The two plates slant to a centrally disposed outlet pipe **91** extending downwardly through foundation **11** and provided with spaced valves **92** and **94**. The slugs fall by gravity into pipe **91** above valve **92**.

Outlet pipe **91** opens into a lower coffin chamber **95** that in turn connects with a tunnel **96** carrying car tracks **97** on which a coffin car **99** may be moved by means of cable **100**. Coffin car **99** supports a plurality of slug coffins **101** in position to be successively positioned beneath the lower opening of pipe **91**.

Valves **92** and **94** are operated by means of rods **102** and **104**, respectively, from behind a heavy lead shield **105**, as shown in Fig. 6.1. A crane **106** is used for placing coffin caps **107** on each coffin after it has been filled with irradiated slugs.

Before unloading is started, both valves **92** and **94** are closed, and the upper portion of pipe **91** is filled with water from water inlet pipe **109**. A proper water level is maintained above valve **92** by water outlet pipe **110**. The air circulation is maintained, although it may be reduced to about 25 percent of the operating value. Slugs are then pushed out of the reactor to fall on to pad plates **90** and then roll by gravity into the water in the upper part of outlet pipe **91**.

In order that there be no material damage to the jacketing of the slugs, plates **90** are preferably padded with soft material that does not deteriorate under neutron irradiation, and that will be able to withstand the slug impact. A satisfactory pad has been found to be $\frac{1}{4}$ inch cotton duck on felt laid on wood backed by steel. Combinations of various synthetic elastic materials have also been found satisfactory.

After a number of slugs have been collected above valve **92** sufficient to fill a coffin **101**, unloading is stopped and valve **92** is opened, permitting the slugs and the water to drop through the valve and remain in the space between valves **92** and **94**. Valve **92** is then closed, the water level reestablished and unloading continued. In the meantime, valve **94** is opened permitting the slugs and water to fall into one of the coffins **101**. The car is then moved to register the opening of the next coffin with the end of the outlet pipe and the first coffin is capped. The procedure is continued until all of the irradiated slugs are in coffins. These coffins may then be taken to a soaking pit (not shown) to remain until the radioactivity has decayed to a point where the slugs can be submitted to chemical removal of the products formed therein by irradiation. After 100 days' operation the aging period may be about 30 days.

Removal of the irradiated slugs under the conditions specified is performed for two reasons. Firstly, the slugs are so highly radioactive that they cannot be safely approached by personnel without adequate shielding being interposed, and, secondly, for some time after removal from the reactor this radioactivity is so intense that self absorption of the emitted radiation causes self heating of the slugs sufficient to melt the slugs if not cooled in some manner. By unloading during maintenance of the air stream, by dropping the slugs at once into water, and by keeping the slugs in water until the more violent radioactivity has subsided, melting is prevented, as the slugs are cooled as they boil the water in which they are immersed. The slugs are then stored or aged under water until ready for chemical treatment, as for example, for thirty days.

It will thus be seen that we have provided a means and method of removing the heat of reaction from a neutronic reactor to the extent that the reactor may be continuously operated at neutron densities well above those safely obtainable in an uncooled reactor. The increased neutron densities lead to more rapid production of 94^{239} and other products of neutron irradiation, and by means of our invention, the irradiated uranium containing these products can readily be removed from the reactor and fresh uranium inserted. In addition, the radioactive gases in the cooling air are safely disposed of.

It is also to be noted that reactor operation at 250 to 500 kilowatts heat equivalent has been mentioned. These powers, however, are in no way maxima as the operating powers are dependent solely on the air supply available and the permissible maximum slug temperatures. Ordinarily slug temperatures of from 100°C to slightly over 400°C are permissible without special treatment of the uranium bodies before jacketing. In case, however, it is desired to operate the reactor with a maximum slug temperature above about 430°C, then precautions should be taken to remove occluded hydrogen from the uranium to less than 0.0002 percent in order to prevent swelling of the packets by release of this hydrogen. At temperatures below 430°C no swelling occurs from the release of occluded hydrogen, as uranium hydride is formed, preventing rise of internal pressure. Above 430°C, however, internal pressure from released hydrogen may, with unprocessed uranium, swell the jackets. The occluded hydrogen, however, is easily removed from the uranium bodies by heating them to a temperature of from 580°C to 600°C under continuous evacuation by a vacuum pump until equilibrium is reached. The uranium bodies are then cooled 10 to 20 hours, preferably in an argon atmosphere, and then jacketed. When processed in this manner no swelling will occur at any temperature. Thus, if desired, the reactor can be operated at powers of from 1000 kilowatts to 5000 kilowatts, when the proper amount of air is supplied, and the hottest slugs permitted to rise in temperature to from 400°C to 500°C.

While the invention has been described with particular reference to a graphite type of moderator, it is to be understood that it is not limited to this moderator and that other neutron slowing materials which do not have an excessive neutron absorbing property may be utilized. For example, heavy water (D_2O) may be used very effectively for this purpose and in such a case suitable channels may be provided in the liquid D_2O for passage of the coolant therethrough. Moreover, beryllium is also found suitable.

In accordance with a further modification coolants other than air may be suitable, for example, helium or oxygen or fluorine are suitable for this purpose. In addition other coolants such as water, deuterium oxide, etc. may be used in liquid or vapor state. The problem in selecting a proper coolant is dependent upon the tendency of the coolant to absorb neutrons. Thus if an excessive quantity of neutrons are absorbed by the coolant, the number of neutrons remaining for absorption by uranium and fission of U^{235} will be insufficient to permit establishment and maintenance of a self-sustaining chain reaction. Consequently, it is desirable to utilize a cooling agent which has but a slight tendency to absorb neutrons and in any event it is desirable to avoid establishment of a quantity of cooling agents in the reactor of such magnitude that the amount of neutrons absorbed will prevent maintenance of the chain reaction. Where gaseous coolants are used it is preferred to use those having a danger coefficient below about 15 preferably below 2.5.

Where a coolant gas is expensive as is the case with helium, suitable means, not shown, may be provided for collecting and recirculating the cooling gases and the entire reactor assemblage may be enclosed in a gas tight shell to minimize or prevent substantial loss of the coolant gas.

Within limits the approximate effect of any quantity of coolant may be computed by computing the total quantity of coolant in the reactor at any given time. Where piping or conduit is used in the channels or other portion of the reactor as is the case when a liquid coolant is used or where a liquid moderator is used, the danger sum of the piping also must be considered. The danger sum of the circulating system including coolant and piping if used may be computed in the manner previously described for computing the danger sum of impurities or neutron absorbers in the reactor simply by determining the weight of coolant and piping in the reactor per unit weight of uranium and the danger coefficient for each impurity or absorber. In no case should the danger sum of the coolant and piping in the reactor be so great as to reduce the reproduction factor K below unity. The permissible danger sum will be determined therefore by the magnitude of K of the reactor without the coolant and piping.

For example, with natural uranium and a carbon moderator, the maximum K obtainable is about 1.1 and in no case can the danger sum of the coolant and conduit therefore exceed 0.1. Preferably in such case the danger sum of the coolant and conduit should be maintained below about 0.05. On the other hand, with a moderator which has less tendency to absorb neutrons, such as D_2O, the maximum K obtainable is approximately 1.3 and in such a case, the danger sum of the coolant and piping or conduit should certainly be less than 0.3, likewise when natural uranium is used in a beryllium moderator, the danger sum should be less than 0.18. Where other neutron absorbers such as impurities, controls, etc. are present, the permissible quantity of coolant is decreased by the danger sum of such absorbers. Moreover most reactors have a substantial leakage factor usually not less than 0.01 K units and thus permissible coolant must be diminished by the factor. The maximum permissible danger sum may be increased where enriched materials containing concentrations of U^{235} or 94^{239} greater than the concentration of U^{235} in natural uranium but in all events the volume of coolant in the reactor should be correlated in accordance with the reproduction factor and/or reproduction ratio to prevent these ratios from decreasing below unity or at least to secure and average reproduction ratio of unity.

In general it is found to be hazardous to permit the reproduction ratio from becoming excessively high since in such a case the rate of increase of neutron reproduction is so rapid that control of the reaction may be impossible. It has been found that while most of the neutrons are released almost instantly upon fission a small portion thereof are delayed and are released from one-half to 60 seconds or more after neutron bombardment. In the case of U^{235} approximately one percent of the neutrons evolved are delayed neutrons. Control of a reactor may be effected without difficulty by maintaining the reproduction ratio at a value of at least about one but not over one by more than a fraction equal to the fraction of neutrons evolved which are delayed neutrons. Thus in the case of U^{235} and similar products the reproduction ratio should not exceed 1.01 and the amount of coolant introduced should be such as to maintain an average reproduction ratio of about one and in any case at least about one and no more than about 1.01. For ease of control it is preferred that this ratio should not exceed 1.005.

While the invention is particularly concerned with reactors in which natural uranium is used as the source of fissionable isotope, other compositions consisting of or containing fissionable isotopes which upon fission yield neutrons may be utilized. For example, enriched compositions containing U^{235}, U^{233}, or 94^{239} dispersed in U^{238} or Th^{232} may be used for this purpose. Compositions containing 5, 10 percent or even more of the fissionable isotope in such cases may be used and under such conditions a greater magnitude in the amount of coolant which may be permitted in the reactor is possible.

Where compositions comprising abnormal amounts of U^{235} or 94^{239} or similar isotopes are used in the chain reaction some change in the rate of neutron production and consequently some change in the permissible magnitude in the amount of impurities or leakage may be encountered. For example, as the reactor herein described proceeds in its operation a quantity of 94 is generated which, being fissionable, takes part in the reaction. Moreover the fission products produced being impurities tend to reduce the K factor but this effect is counteracted to a substantial degree by the neutron output of 94^{239} upon fission. After a substantial period of operation of the reactor the K factor may be increased substantially due to the fact that somewhat more neutrons are evolved upon fission of 94^{239} than are obtained upon fission of U^{235}.

At all events to maintain a self-sustaining chain reaction $Z - (A + B + C + L)$ must be not less than about one, where:

Z = number of neutrons produced by fission per neutron.

A = fraction of neutrons absorbed other than by fission by any of the isotopes of uranium or thorium. Such absorption may lead to the formation of new bodies fissionable with slow neutrons.

B = the number of neutrons absorbed by the moderator per neutron consumed in fission.

C = number of neutrons absorbed by other absorbers (impurities or control rods per neutron consumed in fission) and

L = number of neutrons lost in leakage per neutron consumed in fission.

Z may be ascertained by bombardment of a pure isotope with neutrons of the energies developed in a reactor.

A, B, and C may be computed to approximate values by computing the danger sums basing the computations upon the ratio of neutron capture cross section and concentration of the moderator, non-fissioning isotope or other absorber to that of the fissioning isotope or isotopes in the manner previously described, or may be determined experimentally and L may be determined experimentally.

Preferably the value $Z - (A + B + C + L)$ should not exceed unity by more than the fraction of neutrons evolved which are delayed neutrons; in the case of U^{233} this value is about 0.01.

Since a coolant is a neutron absorber, the amount of coolant within a reactor must be controlled so as to prevent the value C from reaching a value such that the difference between Z and the sum of A, B, C and L becomes less than one. Preferably this value $A+B+C+L$ should be such that the value $Z-(A+B+C+L)$ is not less than about one or more than about 1.01.

During the continued operation of a reactor in which a graphite moderator is used it has been found that as the operation proceeds the K factor of the reactor increases. This is due to the fact that upon neutron bombardment certain impurities in the graphite, such as boron, which are high neutron absorbers, are converted to an isotope which has less tendency to absorb neutrons. Thus continued neutron bombardment of a moderator such as graphite containing an isotope, such as B^{10}, which is a high neutron absorber reduces the content of this isotope converting it to another isotope of a lower neutron capture cross section such as B^{11}. Broadly speaking this process is applicable to the removal of any isotope of high neutron capture cross section present in low concentration, for example of the order of one or several parts per million, and generally less than 0.1 percent which, upon neutron absorption, is converted to an isotope of lower neutron capture cross section. Such isotopes may be removed from any material such as carbon, D_2O or beryllium which itself has a relatively lower neutron capture cross section.

During the operation of the neutronic reactor particularly at high neutron densities radioactive elements of exceedingly high capture cross section may be formed in the uranium as an intermediate element in the decay chains of fission fragments and this formation will lower the value of the reproduction factor for the system. Radioactive Xe^{135} is an example of such an intermediate element, this product having a half life of about 9 hours and being formed mostly from radioactive iodine which has a half life of about 6.6 hours and decays to barium. There should be sufficient excess in the reproduction ratio of the reactor so that in the event the reproduction factor is reduced as a result of the formation of an intermediate decay element having a high capture cross section for neutrons the control rods may be withdrawn sufficiently to maintain the reproduction ratio at a value of unity while maintaining the power output at the desired level. It might be desirable to initially construct the reactor sufficiently oversized to supply this excess reproduction ratio when needed and in this event removable impurities for example in the form of additional shim or control rods may be initially placed in the reactor and kept there at all times until the reproduction ratio commences to fall as the result of the formation of these intermediate decay elements.

While the theory of the nuclear chain fission mechanism in uranium set forth herein is based on the best presently known experimental evidence, we do not wish to be bound thereby, as additional experimental data later discovered may modify the theory disclosed. Any such modification of theory, however, will in no way affect the results to be obtained in the practice of the invention herein described and claimed.

We claim:

1. A neutronic reactor comprising a cubical mass of graphite approximately 25 feet in each dimension, horizontal air channels of approximately 1.75 inches square cross section extending between opposite faces of said mass in a repetitive square pattern with approximately 7 inches between centers of adjacent channels each of said channels having one of its corners extending downward, 68 cylindrical uranium slugs approximately 1.1 inches in diameter and 4 inches long lying end-to-end centrally of each of the central 700 to 1000 channels in the downwardly extending corner thereof, at least one movable body of cadmium extending inwardly from one of the outer faces, means for selectively positioning said body of cadmium, means for forcing air through said channels from one of said faces to the other of said faces, and a shield of concrete from five to ten feet in thickness surrounding said mass.

2. A neutronically reactive composite assembly suitable for the construction of a neutronic reactor active portion consisting in neutronic essentials of a mass of graphite having parallel air filled chambers of approximately 1.75 inch square cross section extending therethrough in a repetitive square pattern with approximately 7 inches between centers of adjacent channels, each of said channels having one of its corners extending down ward, and cylindrical uranium slugs approximately 1.1 inches in diameter in the downwardly extending corner of each channel.

3. A neutronic reactor comprising a cubical mass of graphite approximately 25 feet in each dimension, horizontal air channels of approximately 1.75 inches square cross section extending between opposite sides of said mass in a repetitive square pattern with approximately 7 inches between centers of adjacent channels, each of said channels having one of its corners extending downward, 68 cylindrical uranium slugs approximately 1.1 inches in diameter and 4 inches long lying end-to-end centrally of each

of the central 700 to 1000 channels in the downwardly extending corner thereof, at least one movable body of cadmium extending inwardly from one of the outer faces, means for selectively positioning said body of cadmium, a shield around said mass, the wall of said shield being spaced away from each of said faces to form separate chambers connected by said channels, said shield having an air duct connecting one of said chambers to the atmosphere near ground level and a second air duct connecting the outer member to the atmosphere substantially above ground level, and a fan in one of said ducts to create an air flow through said channels and ducts in a direction exhausting said air through said second duct.

References cited in the file of this Patent

UNITED STATES PATENTS

2,127,193	Toulmin	Aug. 16, 1938
2,161,985	Szilard	June 13, 1939
2,708,656	Fermi et al.	May 17, 1955

FOREIGN PATENTS

440,023	Great Britain	Dec. 12, 1935
114,150	Australia	May 2, 1940
861,390	France	Oct. 28, 1940
233,011	Switzerland	Oct. 2, 1944

OTHER REFERENCES

Naturwissenschaften, vol. 27, 1937, pages 402-410.

Power, July 1940, pp. 56-59.

Handbook of Chemistry and Physics, 26th ed., p. 2462, Chemical Rubber Pub. Co. (1942).

Rutherford: "Radioactive Transformations", pp. 204-207, Charles Scribner's Sons (1906).

Smyth: "Atomic energy for Military Purposes", pp. 103, 104, Aug. 1945.

Kelly et al.: Phys. Rev. 73, 1135-9 (1948).

Chapter 7

Testing Material in a Neutronic Reactor

Enrico Fermi, Santa Fe, N. Mex., and Herbert L. Anderson, Hartford, Conn., assignors to the United States of America as represented by the United States Atomic Energy Commission

Application August 28, 1945, Serial No. 613,156

U.S. Patent No. 2,768,134 - Patented Oct. 23, 1956

The main object of this Patent is to give a suitable method for determining neutron absorption by different materials, when they are irradiated with the neutrons coming from a nuclear pile. Such a reactor has deliberately a low reproduction factor, due to the apertures made in the system, from which neutrons are lost (the use of a coolant, in particular, is not provided).

The test is carried out by means of a comparison of the effects produced by the testing material with respect to those of the standard active material in the reactor. If the equilibrium position of the control rod with the test material in the reactor is further out of the reactor than it was with the standard lump in the reactor, then the test material absorbs more neutrons than the standard metal did. The opposite conclusion is, instead, reached if the control rod must be pushed further into the reactor to achieve equilibrium with the test material into the system.

Several details about the calibration of the control rod with different units are given. In particular, two kind of units are defined here. The first one is the "inhour", that is the distance that the control rod must be moved from the critical position to give the reactor a period of one hour, where the period is the time necessary for the neutron intensity to increase by a factor of e. *The second one, the "cinch", has the characteristic that a movement of the control rod for one cinch has the same effect on the reproduction ratio as a movement of the control rod one inch from the critical position. Together with the definition of these units, a discussion of the corrective effects due to a pressure change is given.*

The main subject of this Patent does not appear in any other published paper.

Our invention relates to the general subject of nuclear fission and more particularly to a means and method for testing materials by means of a self-sustaining nuclear chain reaction system. Such a chain reaction system may be created by the nuclear fission of uranium by thermal neutrons, utilizing natural uranium having a U^{235} isotope content of as low as the natural ratio of approximately 1/139 of U^{238} or an enriched uranium having a higher U^{235} content.

It is known that the isotope U^{235} in natural uranium can be split or fissioned by bombardment with thermal neutrons, i.e., neutrons at or near thermal equilibrium with the surrounding medium, and it has been found possible to obtain a self-sustaining chain reacting system operating at high neutron densities. In such a system, the fission neutrons produced give rise to new fission neutrons in sufficiently large numbers to overcome the neutron losses in the system. Since the result of the fission of the uranium nucleus is the production of two lighter elements with great kinetic energy, plus approximately 2 fast neutrons on the average for each fission along with beta and gamma radiation, a large amount of power can be made available in such a self-sustaining system.

However, most of the neutrons arising from the fission process are set free with the very high energy of above one million electron volts average and are therefore not in condition to be utilized efficiently to create new thermal neutron fissions in U^{235} when it is mixed with a considerable quantity of U^{238}, particularly as in the case of natural uranium. The energies of the fission-released neutrons are so high that most of the latter would tend to be absorbed by the U^{238} nuclei, and yet the energies are not generally high enough for production of fission by more than a small fraction of the neutrons so absorbed. For neutrons of thermal energies, however, the absorption cross-section of U^{235}, to produce fission, rises a great deal more than the simple capture cross-section of U^{238}; so that under the stated circumstances the fast fission neutrons, after they are created, must be slowed down to thermal energies before they are most effective to produce fresh fission by reaction with additional U^{235} atoms. In certain systems in which neutrons are slowed down without much absorption until they reach thermal energies and then mostly enter into uranium rather than into any other element, a self-sustaining nuclear chain reaction can be obtained, even with natural uranium. Light elements, such as deuterium, beryllium, oxygen or carbon, the latter in the form of graphite, can be used as slowing agents. A special advantage of the use of the light elements mentioned for slowing down fast fission neutrons is that fewer collisions are required for slowing than is the case with heavier elements, and furthermore, the above-enumerated elements have very small neutron capture probabilities, even for thermal neutrons. Carbon in the form of graphite is a relatively inexpensive, practical, and readily available agent for slowing fast neutrons to thermal energies and our invention will be described as embodied in a graphite and uranium system, although it is equally applicable to other systems.

However, in order for the premise to be fulfilled that the fast fission neutrons be slowed to thermal energies in a slowing medium without too large an absorption in the U^{238} isotope of the uranium, certain types of physical structures should be utilized for the most efficient reproduction of neutrons, as precautions must be taken to reduce various neutron losses and thus to conserve neutrons for the chain reaction if a self-sustaining system is to be attained.

The ratio of the number of fast neutrons produced by the fissions, to the original number of fast neutrons creating the fissions, in a system of infinite size using specific materials is called the reproduction or multiplication factor of the system and is denoted by the symbol K. If K can be made sufficiently greater than unity to create a net gain in neutrons and the system made sufficiently large so that this gain is not entirely lost by leakage from the exterior surface of the system, then a self-sustaining chain reacting system can be built to produce power by nuclear fission of natural uranium. The neutron reproduction ratio r in a system of finite size differs from K by the leakage factor, and must be sufficiently greater than unity to permit the neutron density to rise exponentially. Such a rise will continue indefinitely if not controlled at a desired density corresponding to a desired power output.

During the interchange of neutrons in a system comprising bodies of uranium of any size in a slowing medium, neutrons may be lost in four ways: By absorption in the uranium metal or compound without producing fission, by absorption in the slowing down material, by absorption in impurities present in the system, and by leakage from the system. These losses will be considered in the order mentioned.

Natural uranium, particularly by reason of its U^{238} content, has an especially strong absorbing power for neutrons when they have been slowed down to moderate energies. The absorption in uranium at these energies is termed the uranium resonance absorption or capture. It is caused by the isotope U^{238} and does not result in fission but creates the isotope U^{239} which by two successive beta emissions forms the relatively stable nucleus 94^{239}. It is not to be confused with absorption or capture of neutrons by impurities, referred to later. Neutron resonance absorption in uranium may take place either on the surface of the uranium bodies, in which case the absorption is known as surface resonance absorption, or it may take place further in the interior of the uranium body, in which case the absorption is known as volume resonance absorption. It will be appreciated that this classification of resonance absorption is merely a convenient characterization of observed phenomena, and arises not because the neutron absorbing power of a U^{238} nucleus is any greater when the nucleus is at the surface of a body of metallic, or combined uranium, but because the absorbing power of U^{238} nuclei for neutrons of certain particular energies is inherently so high that practically all neutrons that already happen to have these energies, called resonance energies as explained above, are absorbed almost immediately upon their arrival in the body of uranium metal or uranium compound, and thus in effect are absorbed at the surface of such body. Volume resonance absorption is due to the fact that some neutrons make collisions inside the uranium body and may thus arrive at resonance energies therein. After successfully reaching thermal velocities, about 40 percent of the neutrons are also subject to capture by U^{238} without fission, to produce U^{239} and eventually 94^{239}.

It is possible, by proper physical arrangement of the materials, to reduce substantially uranium resonance absorption. By the use of light elements as described above for slowing materials, a relatively large increment of energy loss is achieved in each collision and therefore fewer collisions are required to slow the neutrons to thermal energies, thus decreasing the probability of a neutron being at a resonance energy as it enters a uranium atom. During the slowing process, however, neutrons are diffusing through the slowing medium over random paths and distances so that the uranium is not only exposed to thermal neutrons but also to neutrons of energies varying between the emission energy of fission and thermal energy. Neutrons at uranium resonance energies will, if they enter uranium at these energies, be absorbed on the surface of a uranium body whatever its size, giving rise to surface absorption. Any substantial reduction of overall surface of the same amount of uranium relative to the amount of slowing material (i.e., the amount of slowing medium remaining unchanged) will reduce surface absorption, and any such reduction in surface absorption will release neutrons to enter directly into the chain reaction, i.e., will increase the number of neutrons available for further slowing and thus for reaction with U^{235} to produce fission.

For a given ratio of slowing material to uranium, surface resonance absorption losses of neutrons in the uranium can be reduced by a large factor from the losses occurring in a mixture of fine uranium particles and a slowing medium, if the uranium is aggregated into substantial masses in which the mean spatial diameter is at least 1 centimeter for natural uranium metal and when the mean spatial diameter of the bodies is at least 2.4 centimeters for the oxide of natural uranium (UO_2). An important gain is thus made in the number of neutrons made directly available for the chain reaction. A similar gain is made when the uranium has more than the natural content of fissionable material. Consequently, we place the uranium in the system in the form of spaced uranium masses or bodies of substantial size, preferably either of metal, oxide, carbide, or combinations thereof. The uranium bodies can be in the form of layers, rods or cylinders, cubes or spheres, or approximate shapes, dispersed throughout the graphite, preferably in some geometric pattern. The term geometric is used to mean any pattern or arrangement wherein the uranium bodies are distributed in the graphite or other moderator with at least either a roughly uniform spacing or with a roughly systematic non-uniform spacing, and are at least roughly uniform in size and shape or are systematic in variations of size or shape to produce a volume pattern conforming to a roughly symmetrical system. The structure of uranium and slowing material is hereinafter called a neutronic reactor. It is sometimes known as a "pile".

The number of neutrons made directly available to the chain reaction by aggregating the uranium into separate bodies spaced through the slowing medium is a critical factor in obtaining a self-sustaining chain reaction utilizing natural uranium and graphite. The reproduction factor K of a mixture of fine uranium particles in graphite, assuming both of them to be theoretically pure, would only be about 0.785. Actual K factors as high as 1.07 have been obtained using aggregation of natural uranium metal in the best known geometry, and with as pure materials as it is presently possible to obtain.

Assuming theoretically pure carbon and theoretically pure natural uranium metal, both of the highest obtainable densities, the maximum possible K factor theoretically obtainable is about 1.1 when the uranium is aggregated with optimum geometry. Still higher K factors can be obtained by the use of aggregation in the case of uranium having more than the naturally occurring content of fissionable elements. Adding such fissionable material is termed enrichment of the uranium.

The thermal neutrons are also subject to capture by the slowing material. While carbon has a very small capture cross-section for thermal neutrons, and deuterium still smaller, an appreciable fraction of thermal neutrons (about 10 percent of the neutrons present in the system under best conditions with graphite) is lost by capture in the slowing material during diffusion therethrough. It is, therefore, desirable to have the neutrons reaching thermal energy promptly enter uranium.

In addition to the above-mentioned losses, which are inherently a part of the nuclear chain reaction process, impurities present in both the slowing material and the uranium add a very important neutron loss factor in the chain. The effectiveness of various elements as neutron absorbers varies tremendously. Certain elements such as boron, cadmium, samarium, gadolinium, and some others, if present even in a few parts per million, could very likely prevent a self-sustaining chain reaction from taking place. It is highly important, therefore, to remove as far as possible all impurities capturing neutrons to the detriment of the chain reaction from both the slowing material and the uranium. If these impurities, solids, liquid, or gaseous, and in elemental or combined form, are present in too great quantity, in the uranium bodies or the slowing material or in, or by absorption from, the free spaces of the system, the self-sustaining chain reaction cannot be attained.

It is therefore an object of this invention to provide a method and means of testing materials in order to determine the neutron absorption of said materials.

Another object of this invention is to provide a quick, simple method whereby material may be introduced into an operating nuclear chain reaction system and the effect of said material on the nuclear chain reaction of the system may be determined.

The amounts of impurities that may be permitted in a system, vary with a number of factors, such as the specific geometry of the system, and the form in which the uranium is used – that is, whether natural or enriched, whether as metal or oxide – and also factors such as the weight ratios between the uranium and the slowing down material, and the type of slowing down or moderating material used – for example, whether deuterium, graphite or beryllium. Although all of these considerations influence the actual permissible amount of each impurity material, by the use of our method and apparatus, the effect of any given impurity or impurities can be correlated directly with the weight of the impurity present and with the K factor of the system, so that knowing the K factor for a given geometry and composition, the permissible amounts of particular impurities can be readily computed without taking individual account of the specific considerations named above.

The effect of impurities on the optimum reproduction factor K has in the past been approximated by means of certain constants known as "danger" coefficients which are assigned to the various elements. These danger coefficients for the impurities are each multiplied by the percent by weight of the corresponding impurity, and the total sum of these products gives a value known as the total danger sum. This total danger sum is subtracted from the reproduction factor K as calculated for pure materials and for the specific geometry under consideration. In order to use this method, the kind and amount of each impurity must be determined. This is difficult because even tracer amounts of certain impurities have a considerable effect on the reproduction factor K. However, by the use of our method the total reduction in K factor for a specific material containing impurities may be ascertained quickly and accurately without the necessity for analyzing the material either qualitatively or quantitatively.

The strong absorbing action of some elements renders a self-sustaining chain reacting system capable of control. By introducing neutron absorbing elements in the form of rods or sheets into the interior of the system, for instance in the slowing material between the uranium masses, the neutron reproduction ratio of the system can be changed in accordance with the amount of absorbing material exposed to the neutrons in the system. A sufficient mass of an absorbing material can readily be inserted into the system to reduce the reproduction ratio of the system to less than unity and thus stop the reaction.

For instance, in one system, which will be further described hereinafter, the reproduction ratio is controlled by inserting a long, thin cadmium rod called a control rod a chosen distance into a transverse slot near the center of the structure. Such a rod may be calibrated in convenient units and we have discovered that by inserting a sample of material to be tested into the chain reacting system and ascertaining the amount of the cadmium rod that must be inserted into the system to bring said system to a desired reproduction ratio, the effect of the sample on the reproduction factor of the system may be determined.

It is thus another object of our invention to provide a method and means of determining the effect of a change in the materials or the structure of a chain reacting system on the reproduction factor K of said system.

When the uranium and the slowing material are of such purity and the uranium is so aggregated that fewer neutrons are parasitically absorbed than are gained by fission, the uranium will support a chain reaction producing an exponential rise in neutron density if the overall size of the system is sufficiently large to overcome the loss of neutrons escaping from the system. Thus the overall size is important.

The size of the system will vary, depending upon the K factor of the system, and upon other things. If the reproduction factor K is greater than unity, the number of neutrons present will increase exponentially and indefinitely except as affected by a temperature effect, provided the structure is made sufficiently large. If, on the contrary, the structure is small, with a large surface-to-volume ratio, there will be a rate of loss of neutrons from the structure by leakage through the outer surfaces, which may overbalance the rate of neutron production inside the structure so that a chain reaction will not be self-sustaining. For each value of the reproduction factor K greater than unity, there is thus a minimum overall size of a given structure known as the critical size, above which the rate of loss of neutrons by diffusion to the walls of the structure and leakage away from the structure is less than the rate of production of neutrons within the system, thus making the chain reaction self-sustaining. The rate of diffusion of neutrons away from a large structure in which they are being created through the exterior surface thereof may be treated by mathematical analysis when the value of K and certain other constants are known, as the ratio of the exterior surface to the volume becomes less as the structure is enlarged.

In the case of a spherical structure employing uranium bodies imbedded in graphite in the geometries disclosed herein and without an external reflector the following formula gives the critical overall radius (R) in feet:

$$K - 1 = \frac{C}{R^2},$$

where C is a constant that varies slightly with geometry of the lattice and for normal graphite lattices may have a value of the order of 7 to 8.

For a rectangular parallelepiped structure rather than spherical, the critical size can be computed from the formula

$$K - 1 = C\left(\frac{1}{a^2} + \frac{1}{b^2} + \frac{1}{c^2}\right),$$

where a, b, and c are the lengths of the sides in feet. The critical size for a cylindrical structure is given by the formula, irrespective of the shape of the uranium bodies, in which h=cylinder height in feet, and R=radius in feet:

$$K - 1 = C\left(\frac{1}{h^2} + \frac{0.59}{R^2}\right).$$

However, when merely critical size is attained, by definition no rise in neutron density can be expected. It is therefore necessary to increase the size of the structure beyond the critical size but not to the extent that the period for doubling of the neutron density is too short, as will be explained later. A desirable reproduction ratio (r) for an operating structure with all control absorbers removed and at the temperature of operation is about 1.005. The size at which this reproduction ratio can be obtained may be computed from modifications of the above formulae for critical size. For example, for spherical active structures the formula

$$K - r = \frac{C}{R^2}$$

may be used to find R when K is known and r is somewhat over unity. The same formula will, of course, give r for given structures for which K and R are known.

As stated above the period for doubling of intensity must not be too short. If it were too short the reactor would be incapable of control. Control by means of a neutron absorber such as a cadmium rod is feasible only because of the fact that when fission takes place all of the neutrons emitted do not leave the U^{235} immediately. Instead from about 0.05 percent to 1 percent of the neutrons are emitted up to several minutes after fission takes place. Half of these "delayed" neutrons are emitted in six seconds and 90 percent within 45 seconds. The mean time of delayed emission is about 5 seconds. The fact that these percentage of neutrons is delayed, imparts a slowness of response to the reactor that would not be present if all the neutrons were emitted instantaneously. Therefore, for ease of control the reproduction ratio r with all control absorbers removed is kept about 1.005; the excess of 0.005 being within the percentage of delayed neutrons.

Critical size may be attained with a somewhat smaller structure by utilizing a neutron reflecting medium surrounding the surface of the active structure. For example, a two feet thickness of graphite having low impurity content, completely surrounding a spherical structure is effective in reducing the diameter of the uranium bearing portion by as much as 2 feet, resulting in a considerable saving of uranium or uranium compound.

For further consideration of the theory and characteristics of neutronic reactor systems, reference is made to the copending application of Enrico Fermi and Leo Szilard, Serial No. 568,904, filed December 19, 1944, now Patent No. 2,708,656 issued May 17, 1955.[1]

[1]@ See page 217.

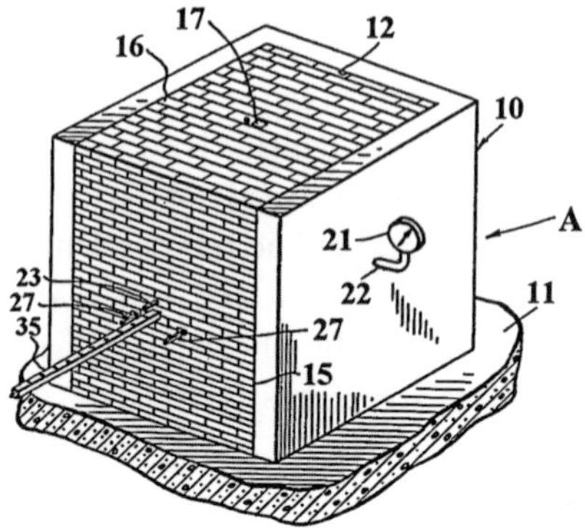

Fig. 7.1

We will further describe one embodiment of our invention as used with a carbon moderated chain reacting system using uranium oxide and uranium metal for the fissionable material. The objects and advantages of our invention will become apparent from the following description read in conjunction with the drawings, in which:

Fig. 7.1 is a perspective view of a neutronic reactor completely enclosed in a shield of concrete blocks;

Fig. 7.2 is an enlarged elevational view partially in cross-section of the front face of the reactor with the shield removed;

Fig. 7.3 is an enlarged side elevational view partially in cross-section of the neutronic reactor with the shield removed;

Fig. 7.4 is a plan view of the top of the reactor with the shield removed;

Fig. 7.5 is an enlarged plan view partially in cross-section of one of the uranium bearing graphite blocks of which the active portion of the reactor is constructed;

Fig. 7.6 is a side elevational view partially in cross-section taken on line **6–6** of Fig. 7.5;

Fig. 7.7 is a side elevational view partially in cross-section of one of the solid graphite blocks;

Fig. 7.8 is an enlarged longitudinal sectional view of the removable graphite stringer in which material to be tested is placed; and

Fig. 7.9 is a cross-sectional view taken on line **9–9** of Fig. 7.8.

A self-sustaining chain reaction in not possible in graphite with natural uranium without aggregation. Such a chain reaction is possible with ag-

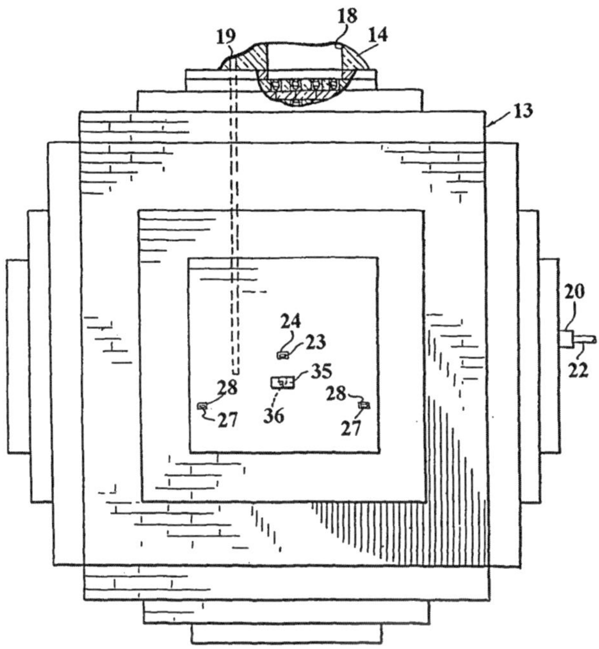

Fig. 7.2

gregation and we will describe a conductively cooled self-sustaining chain reacting system, in which both natural uranium metal and natural uranium oxide (UO_2) bodies are utilized, imbedded in high quality graphite and in which the active portion is roughly spherical.

Referring first to Fig. 7.1 illustrating the completed structure, side walls **10** are erected on a heavy foundation **11**, both preferably of poured concrete about 5 feet thick, leaving a vault space **12** inside walls **10** in which is built up, as will be explained later, an active portion **13** (Fig. 7.2) surrounded by a graphite reflector **14**. The sides of the structure are closed by a front wall **15**, and a top **16** both 5 feet thick and formed of concrete bricks. The top covering is pierced by an aperture **17** leading to a well **18** extending to at least the peripheral layer of uranium bodies in the active portion **13**. A second aperture is the exterior opening of a shaft **19** extending to the center of the structure. The well **18** and shaft **19** are used as neutron passageways when bodies outside the reactor are to be exposed to radiation.

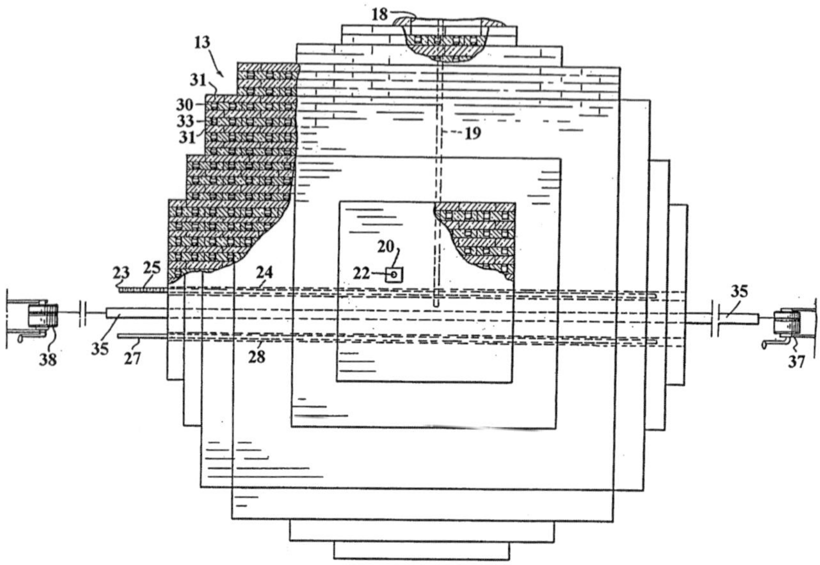

Fig. 7.3

Entering the side wall **10** opposite the central portion of active portion **13** is a slot at the end of which is positioned an ionization chamber **20** connected to a galvanometer **21** by wire line **22**.

Extending from the outside of the structure into the center of active portion **13** is a control rod **23** sliding in slot **24** in the graphite (Fig. 7.2). A scale **25** is provided on control rod **23** to indicate the depth of penetration of the control rod **23** into the active portion **13**. Also extending into the active portion **13** through the front wall **15** are two spaced safety rods **27** operating in slots **28**. Safety rods **27** and control rod **23** are preferably thin flat rods of cadmium, backed with aluminum for strength, and are arranged to move freely in and out of slots **28** and **24** respectively. This completes the overall structure of the device. The active portion **13** will next be described.

The self-sustaining chain reacting unit **13** to be built up within vault **12** is designed to so reduce neutron losses as to make a self-sustained chain reaction possible, as has been previously outlined, using specific types and sizes of uranium masses, both of metal and oxide in graphite, all of best obtainable purity, and spaced with a specific geometry. An all metal struc-

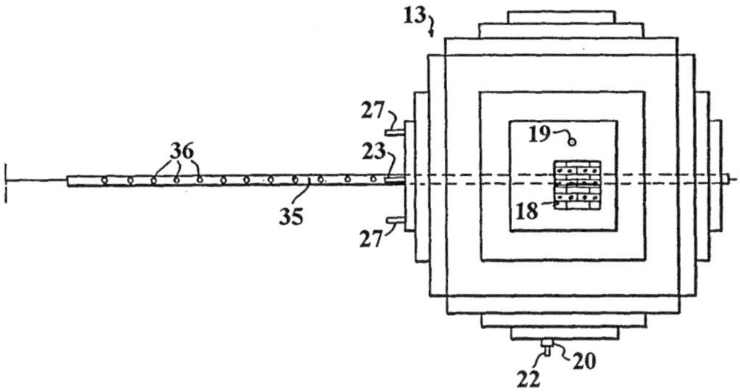

Fig. 7.4

ture could be built, i.e., wherein all of the uranium bodies are of metallic uranium, but the combination of metal and oxide in the present example is utilized for economy, uranium metal being at present more expensive to produce than uranium oxide.

The graphite is in the form of blocks or bricks having dimensions of $4\frac{1}{8}'' \times 4\frac{1}{8}'' \times 16\frac{1}{2}''$, these blocks having been planned by woodworking machinery to have smooth rectangular sides. Certain of the blocks **30** are drilled with two holes spaced $8\frac{1}{4}''$ center to center to receive the uranium bodies which are placed therein as shown in Figs. 7.5 and 7.6. These graphite blocks **30** are termed live graphite. Other blocks **31**, as shown in Fig. 7.7, contain no uranium and may be termed dead graphite. The uranium bodies are in two forms, one form being cast uranium metal cylinders **32** having a density of about 18 g/cm^3, each cylinder being $2\frac{1}{4}''$ in diameter, and $2\frac{1}{4}''$ in height, weighting about 6 pounds each. The other uranium bodies **33** are in the form of pseudospheres of uranium oxide, UO_2, $3\frac{1}{4}''$ in diameter and slightly less than $3\frac{1}{4}''$ in height, similar to the metal cylinders with the exception of a beveling on top and bottom to approximate spherical contours weighting about 4.7 pounds each. The oxide in the

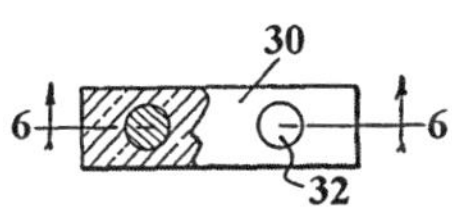

Fig. 7.5

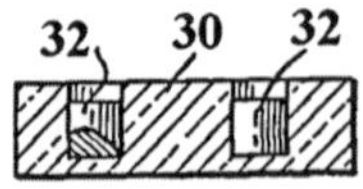

Fig. 7.6

pseudospheres is compressed to a density of about 6.1 g/cm^3 by hydraulic pressure.

The active portion **13** of the structure, which is built inside the vault, has 75 layers of live graphite, alternating with layers of dead graphite to approximate a sphere having a diameter of approximately 24 feet. A total amount of graphite of approximately 290 tons is used in the active portion. Approximately 6 tons of cast uranium metal bodies are used for the cylinders. This metal, being more effective in the reaction, is placed at the center of the active structure **13** with the oxide pseudospheres concentrically arranged around the metal bearing center as shown in cut-away portions in Fig. 7.3. Approximately 40 tons of compressed oxide bodies are used, and the total number of uranium bodies is about 20,000, of which 2,000 are metal. The volume ratio of the metal containing section is about 60 carbon to 1 uranium and of the oxide containing section approximately 25 carbon to 1 uranium considered as oxide in the described compressed form.

The spacing of the uranium elements in the graphite is such that each unit cell comprising one uranium body and surrounding graphite has a side size of $8\frac{1}{4}''$ and a volume of 0.324 cubic feet. The arrangement of the bodies in the graphite is that of a simple cubic lattice of the uranium bodies. The K factor for the metal bodies **32** and graphite is about 1.07, and for the oxide bodies **33** and graphite about 1.039, danger coefficients for both the graphite and the uranium included. The average K for the entire structure is about 1.05. Such a structure will sustain a chain reaction when assembled roughly in the shape of a sphere approximately 24 feet in average diameter and when surrounded by a suitable neutron reflector of dead graphite.

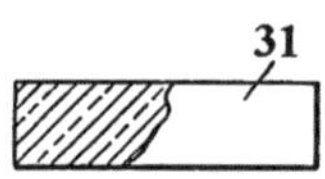

Fig. 7.7

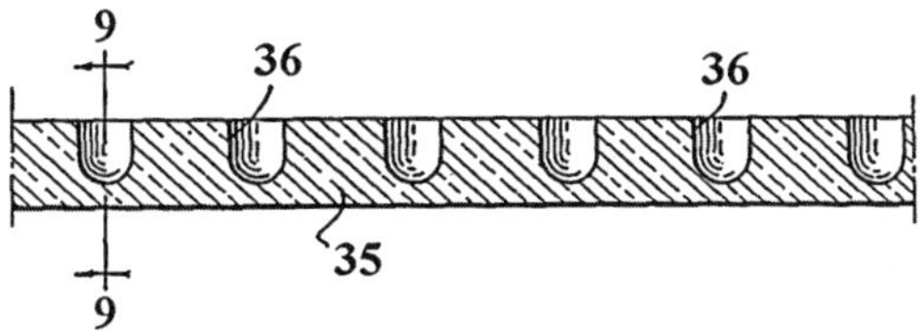

Fig. 7.8

The actual construction of the neutronic reactor is started by placing on the foundation of vault **12**, several layers of dead graphite blocks to start reflector **14**, after which the active portion may be started by piling up alternate layers of dead graphite and live graphite, filling the edges of both live and dead layers out to the concrete side walls with dead graphite to continue the reflector **14**. The blocks are closely packed, without cement, to eliminate air spaces as far as possible.

When slightly less than half the required number of layers have been piled in the vault, the slots **28** are provided in a dead carbon layer through which the safety rods slide, and the safety rods **27** installed. After a few more layers are added to the pile, the control slot **24** is built in, the control rod **23** inserted, and the ionization chamber **20** installed on the outer boundary of the active structure. Wire line **22** is led to the outside and connected by a monitoring circuit (not shown) to galvanometer **21**.

Ionization chamber **20** comprises a sealed metal casing containing approximately 18 l of boron trifluoride at one atmosphere pressure into which projects a central electrode (not shown). The ionization chamber **20** is connected through a suitable circuit to galvanometer **21**. Neutron absorption by the boron in the chamber releases alpha particles by nuclear reaction and causes alpha ray ionization in the chamber, the amount thereof being measured by galvanometer **21**. The neutron density varies from the center of the reactor to the periphery, being highest in the center. However, to

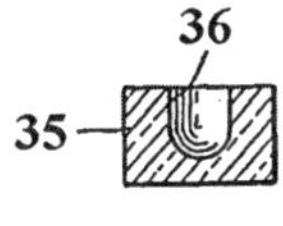

Fig. 7.9

avoid poisoning of the pile by the material of the chamber it is placed at the periphery of the reactor. The density at the periphery is proportional to the density at other points in the reactor. The galvanometer deflection is only approximately linear with neutron density, but is reproducible and readily calibrated in terms of neutron density or power, if desired.

Construction is continued, layer by layer, with the control rod and safety rods fully inserted into the reactor. Along lines passing close to the center of the reactor, one or more removable stringers **35** of live carbon are provided so that one or more complete rows of uranium bodies can be removed from the reactor as may be desired for test purposes, as will later be described.

Stringer **35** is composed of one long block or several blocks of graphite secured together. It is slightly smaller in cross-section than carbon blocks **30** so that it may be easily moved in and out of the reactor. Depressions or cells **36** are formed in the stringer **35** for receiving material to be tested. The stringer may be conveniently made twice as long as the depth of the reactor so that one-half of said stringer is out of the reactor ready for loading or unloading while the other half is in said reactor. Winches **37** and **38** are used for pulling said stringer in and out of the reactor.

With the safety rods and control rod fully inserted into the structure, the active portion **13** and the reflector **14** are completed to the final size desired. The top **16** and front **15** are then closed with cement blocks. A structure is thus obtained having an active portion **13**, surrounded by a carbon reflector **14** and enclosed in concrete walls on all sides except for the slots **24** and **28** through which the control and safety rods slide to the interior of the structure and slots for well **18** and shaft **19**.

The graphite reflector **14** is used to reflect back into the active portion some of the neutrons that might otherwise diffuse out of the system. By using such a reflector the overall size of the reactor may be somewhat smaller than in a reactor where no reflector is employed because the neutron losses to the exterior are reduced.

The concrete walls serve as the main shield to prevent the escape of gamma radiation outside the structure as carbon, being a light material, absorbs gamma rays only to a small extent.

After the structure of the neutronic reactor has been completed, it is ready for operation, utilizing control by the control rod **23** described above. As has been explained, cadmium has a high danger coefficient and is a very good neutron absorber and therefore the greater the quantity of rod **23** in the reactor, the more neutrons that are absorbed. Even small increments of movement of the control rod into the reactor have an effect on the reproduction ratio r of the reactor. To commence operation safety rods **27** are withdrawn entirely from the active portion **13** and the control rod **23** is then slowly withdrawn to a point where galvanometer **21** indicates that the neutron density is rising. If a faster rise is desired, the rod is retracted further to increase the reproduction ratio r of the system.

When any desired neutron density is reached, the control rod is pushed back into the reactor until a point is found where the neutron density remains constant. At this point the system is balanced, with a neutron reproduction ratio r of unity. No special source of neutrons is needed in the structure, as the natural neutrons always present and constantly diffusing through the reactor are sufficient to start the reaction.

To again increase the neutron density, the control rod is moved outwardly in such manner and to such extent that the neutron density rises at a desired rate and attains the new desired value, and then the control rod is moved into the point where the system is again balanced, thus stabilizing the reaction at the new neutron density. If it is desired to decrease the neutron density, the control rod is pushed into the reactor until the reproduction ratio is less than unity and the neutron density decays to the desired value. Then the control rod is retracted to the position at which balance is reestablished. In this manner, any desired neutron density within the heat dissipating limits of the structure may be obtained and maintained. The position of the control rod in the system therefore does not govern the neutron density in the reactor, but only the rate of change of neutron density.

The reactor herein described has a maximum possible reproduction ratio r of 1.0042, causing a doubling of the neutron density about every 8 seconds with the control and safety rods completely removed. This slow rise renders the system always easy to control and safe to operate. The low reproduction ratio in the system obtained from a geometry giving an overall K factor of 1.054 is due to the fact that the structure is close to the critical size and to the fact that large quantities of neutrons are deliberately permitted to leak from well **18** and shaft **19** for use outside the reactor.

The unity reproduction ratio position or "balanced" position of the control rod within the reactor for maintaining any desired neutron density would always be the same were it not for the fact that the temperature within the reactor changes to some extent and influences the neutron losses in the materials and also for the fact that in any reactor exposed to atmospheric pressure, changes in the nitrogen content of the reactor accompanying changes in the atmospheric pressure, change the r ratio, as nitrogen is an absorbing impurity. The unity reproduction ratio point on the control rod therefore changes slightly in accordance with the temperature of the reactor and with the atmospheric pressure. Changes due to temperature are relatively small and the balanced position of the rod can readily be calibrated for temperature effect.

If at any time it is desired to stop the reaction, the control rod is inserted deep within the reactor until the reproduction ratio becomes less than unity, i.e., less than the value necessary to maintain the reaction with this size of structure. The neutron density then decays to that of the natural neutrons multiplied by a factor of about 100 which is the inherent multiplication when the control and safety rods are inserted. If desired, the safety rods may be pulled into the system by weights tripped by an emergency latch in case of failure of the control rod to control the reaction for any reason. It is also possible to control the position of the control rod **23** by means of automatic connection to the monitoring circuit, although such automatic control forms no part of the present invention and is not required in the system described.

Having described the general structure and operation of a neutronic reactor, we will now describe how the reactor may be adapted for testing materials in accordance with our invention.

A "standard" batch of uranium cylinders is placed in one-half of the depressions **36** of "double length" removable stringer **35**, and this loaded half of the stringer is moved into the reactor by means of winches **37** and **38**. The reactor is then balanced by means of control rod **23** at any convenient neutron density.

While the reactor is being balanced, the second half of the stringer **35**, which is outside of the reactor, is loaded with the metal to be tested. The weights of standard and test metal must be equal or a correction must be made for the inequality. The stringer **35** is then moved so that the standard metal is outside of the reactor and test metal is inside and the reactor is again balanced. If the control rod is further out of the reactor for equilibrium with the test metal in the reactor than it was with the standard metal in the reactor, then the test metal is parasitically absorbing more neutrons than did the standard metal. Therefore, said test metal is not as good for reactor construction as is the standard metal. However, if the control rod must be pushed further into the reactor to achieve equilibrium with the test metal in the reactor, then the test metal is parasitically absorbing less neutrons than did the standard metal. Corrections must be made for any change in barometric pressure or reactor temperature between readings of the control rod. The double length stringer **35** makes it possible to substitute material without stopping the neutronic reaction.

In order to determine quantitatively what the effect of the test metal is on the reproduction factor K, we have calibrated the control rod **23** in several different ways.

As the effect per inch movement of control rod **23** is greater for the portion of the control rod nearer the center of the reactor than it is for the portion near the edge, a unit may be chosen so that for movement of the control rod **23** one of said units will always have the same effect on the reproduction ratio of the reactor without regard to the depth of said rod in the reactor. Such a unit is the conventional inch sometimes called a "cinch". Any movement of the control rod the distance of one cinch has the same effect on the reproduction ratio of the reactor as a movement of the control rod one inch from the balanced position. In order to calibrate the rod in cinches, the reactor is balanced so that the neutron density is constant. The rod is then withdrawn one inch and the rate of rise of neutron density is measured by timing the interval necessary for the neutron density to double as shown n the galvanometer **21**. The rod **23** is then withdrawn until the rate of rise is twice the amount just measured, and this position is marked "2 cinches". The same procedure is followed until the entire rod is calibrated in terms of cinches. If the atmospheric pressure or reactor temperature should change during the calibration, a suitable correction must be made for said change. A movement of the control rod of the described reactor approximately 2.3 cinches has been found to equal a 1 percent change in the excess reproduction ratio $(r - 1)$.

We have also calibrated the control rod in terms of a unit known as the "inhour". One inhour is the distance that the control rod must be moved from the critical or balanced position to give the reactor a period of one hour. The period of a neutronic reactor is by definition the time necessary for the neutron intensity to increase by a factor of "e" (e = 2.718).

In order to calibrate the control rod **23** in inhours, it is necessary to determine the critical point of said control rod at which the reactor is balanced and then measure the period of various positions of the control rod. The period can be measured by taking a reading of neutron density on galvanometer **21** and then measuring the time necessary for this density to be multiplied by a factor of 2.718.

The relation between inhours and the period (T) of the reactor may be expressed approximately by:

inhr. at balanced condition - inhr. =

$$\frac{64}{T}+\frac{245}{T+3.57}+\frac{688}{T+10.1}+\frac{1938}{T+34.5}+\frac{665}{T+83}.$$

This formula was derived from experiments on the described reactor. By substituting in said formula for the periods, and solving for inhours, the control rod **23** may be marked in inhours.

In measuring the period of the reactor, a correction must be made for any change in atmospheric pressure. As described above, the reactor is open to the atmosphere. Therefore, an increase in atmospheric pressure will cause an increase in the weight of air inside the reactor. Oxygen has a small danger coefficient and, therefore, does not absorb neutrons in great quantities, but nitrogen on the other hand has a larger danger coefficient and so the great quantity of nitrogen present in the reactor has a distinct effect on the reproduction ratio r. By experiment it has been found that a change in atmospheric pressure on the described reactor is equal to 0.323 inhour for a change of one millimeter of mercury for the standard atmospheric pressure of 760 mm of mercury.

A rise in temperature of the reactor has an effect on the reproduction ratio that has been determined by experiment to reduce said ratio 0.814 inhour per degree centigrade for the described reactor.

Also by experiment a change in the average weight of metal in each cell **36** from a standard weight of 1900 g has an effect for which a correction must be made. This correction may be expressed in inhours for this reactor by

$$0.00535\ (1900 - w).$$

It may be explained that the procedure of the invention is greatly facilitated by calibration of the control rod or other control element or device of the reactor, in units significant of the reactivity, such as inhours (abbreviated "ih."). A very direct measure of the reactivity of a neutronic reactor is the period, i.e., the time for the neutron density to change by the factor "e" as explained hereinabove, and although the excess reproduction ratio Δr (which is $r - 1$, or the amount by which the ratio departs from unity when the reactor is unbalanced) bears a generally inverse relation to the period, it has been found more convenient in numerous testing operations to adopt a special unit that is proportional to Δr, and equal to the reciprocal of the period for small values of such reciprocal. A suitable unit of that character is the inhour, which as stated hereinabove is the amount of reactivity that gives the reactor a period of one hour.

In marking or calibrating the control rod in inhours, an appropriate zero point may be arbitrarily or otherwise adopted, for instance, so that the zero inhour point represents the position of the control rod completely inserted into the reactor. Determinations of the period can be made for a multiplicity of rod positions between the position of normal balance (actually $r = 1$) and the position of complete withdrawal, and by substituting values of the period (T) in seconds in the formula given above relative inhour values may be obtained for the selected positions of the rod. It will be noted that the formula gives the number of inhours by which the control rod departs form balanced condition and the actual markings for a rod are each equal to the calculated value of the formula (in inhours) for the point to be marked augmented by the inhour value of the position of normal balance. Although other methods may be followed, this procedure leads to the useful convention of considering an inward displacement of the control rod as representing a positive change in the number of inhours and withdrawal of the rod as representing a negative change. The readings of inhours for the balance point of the control rod represents the reactivity of the reactor; i.e., corresponds to the reproduction factor K: for example, if absorbing impurities other than the control rod are inserted in the reactor, its reactivity is reduced and the rod must be pulled out further in order to maintain the actual reproduction ratio at unity. The decrease in K introduced by the absorption of the impurity may then be determined from the difference in the inhour readings of the control rod taken at balanced conditions both before and after the insertion of the impurity. The relation between change in number of inhours and change in position of the control rod may be called the "sensitivity" of the rod in inhours per centimeter and for the usual type of control rod differs in numerical value for different positions of the rod.

To obtain accurate markings for positions of the rod further inside the reactor than the normal critical or balance point, i.e., portions of the scale representing greater reactivity such as might be obtained by replacing portions of the reactor structure with superior materials tending to increase the value of K, and to determine the inhour reading for the position of normal balance, various procedures may be adopted as will now be apparent to those skilled in the art. For instance, if the operating neutron density at critical position is made high enough to permit significant readings of decay, one method is to make determinations of the time of decay of neutron intensity, as for decay by a factor of (e), corresponding to a multiplicity of rod positions where the actual reproduction ratio r is less than unity, and then to substitute these determinations, as negative values of the period, in the formula for inhours.

A further way of making or checking the calibration of positions on both sides of the normal balance or critical point, involves measuring the neutron intensity at various localities along the control rod slot while the rod itself is held at the critical position. The neutron intensity may be measured with copper foils, as by determining the radioactivity, or "activity", induced in copper foils of a standard size and character. Taking the positions of the foils in the slot as corresponding, for example, to points on the control rod scale that would be indicated if the inner end of the rod were disposed successively at the foil positions, it is found that at least for a considerable range on either side of the balance point, the square of the copper foil activity is proportional, in first approximation, to the sensitivity of the control rod in inhours per centimeter. The proportionality constant can be determined by integrating this differential curve (i.e., the stated relation) from the zero position to a position at which a period determination was made and for which the number of inhours is thus known; in actual instances, such calculations made for a plurality of different period-determined positions have been in good agreement. Using the proportionality constant, a multiplicity of foil activity measurements can then be employed to extend or check the calibration of the control rod.

As explained above, in assigning values of inhours to various positions of the control rod, and in subsequently taking readings of control rod position in inhours, correction for pressure effect (i.e., decrease of K or r with rise of pressure) must be made in the use of a reactor that is not sealed from the atmosphere. Such corrections for the reactor shown can be conveniently computed for readings of pressure that are translated to an appropriate standard temperature; for example, the atmospheric pressure readings of a suitable barometer can be corrected to 0°C, and then the control rod readings or markings corrected to a selected standard pressure, e.g., 740 mm, as employed in the specific example and table set forth hereinbelow. As above stated, the pressure effect in such that a rise of pressure of one mm of mercury produces such decrease of reproduction ratio as to require outward displacement of the control rod by 0.323 inhour to rebalance the reactor after such rise. Although the temperature effect is of corresponding magnitude per degree C as stated hereinabove, circumstances of operation of the reactor may be such that its temperature changes are inconsequential during the period in which calibration is being effected, or during any series of testing operations.

For the specific reactor described, it has been found that the relation between the reproduction factor K and the scale of inhours as marked on or for the control rod is such that one inhour corresponds to a change in K for the reactor, or in the reproduction ratio r (since the reproduction ratio may be considered as equal to K diminished by a quantity representative of neutron leakage loss), of 3.04×10^{-5}. However, for most purposes of the tests made in accordance with the present invention, as where other structure or materials are compared in a removable stringer with a standard structure and material, it is desirable to ascertain the change in K for the type of cell under test, or in other words, to ascertain how the K of a reactor would change if its entire structure were modified in the manner of the removable stringer, as compared with a reactor embodying throughout the material and structure utilized as a standard of comparison in the stringer. It will be understood that the reproduction factor is a constant that equally characterizes a unit cell comprising a uranium body and its surrounding portion of the moderator, or a group of such cells, or any reactor constituting a regular repetition of such unit cells.

It has advantageously been found that the changes of control rod position necessary to rebalance the reactor with various samples under test may be directly calibrated in change of K for cells of the character tested, i.e., in the amount by which the reproduction factor of a reactor would be increased or decreased by appropriately utilizing, throughout the entire device, the material of the sample under test, and each such calibration may be employed for a variety of different tests. A simple example will serve to illustrate one manner of making a calibration of this sort. In a number of tests it was desired to investigate the characteristics of various uranium metal bodies. Sixteen cells, i.e., sixteen pockets **36** of the stringer **35** were used for the tests, and in the preliminary calibration alternate readings of control rod balance position were taken, with the 16 pockets first occupied by uranium bodies of a standard character, and then occupied by pressed bodies of uranium oxide (UO_2), it being understood that in such comparisons it is preferable to use oxide bodies providing thermal neutron utilization equivalent to that of the metal bodies, in the reactor cells constituted when the stringer is in place. With appropriate pressure correction, the critical positions were found to differ by 6.0 inhours, the corrected position for UO_2 being 124.4 inhours and for the U metal 130.4 inhours. By independent tests of so-called "exponential piles" in accordance with the procedure disclosed in the copending application of Enrico Fermi, Serial No. 534,129, filed May 4, 1944,[2] for Nuclear Chain Reacting Systems, it was

[2]@ The authors refer to E. Fermi, Test exponential pile, U.S. Patent No. 2,780,595, February 5, 1957 (see page 181).

determined that in a lattice arrangement of the same type as this reactor the difference in K between metal bodies and oxide bodies of the precise size, weight and character used in the stringer was 0.03. Dividing this value by the figure of 6.0 given above, the change of K per inhour is found to be 0.005, which may also be written as 0.5% change of K per inhour, since the actual values of K usually contemplated in these tests are close to unity. In other words, according to this calibration, when the sixteen test pockets of the stringer are occupied by bodies to be compared with bodies previously in such pockets, a control rod position change of one inhour represents a change of 0.5% in the factor K for the cells completed by the test pockets, or in the reproduction factor for an entire reactor embodying the material tested in the stringer pockets.

Where the metal bodies under test in the sixteen pockets differ in weight from standard metal lumps, it is desirable to eliminate from the results the difference in reactivity of the reactor due to such difference in weight, and correction for weight is made as indicated hereinabove. Corresponding correction can be effected where the bodies to be compared with uranium metal are not metallic but are, for instance, uranium oxide or carbide; in such case, calculation should first be made, as will now be readily understood by persons skilled in the art, to determine an equivalent weight of metal (i.e., the amount of metallic uranium that would afford the same thermal neutron utilization in a cell, of graphite and metal, of similar size), which is then corrected. Where oxide bodies, for example, are to be compared with bodies of other batches of the same oxide of uranium, weight correction can be simplified (as in the case of metal comparisons) by plotting a weight correction curve from inhour (critical position) readings for a variety of weights of one batch of the oxide in the pockets.

Reverting to the calibration procedure whereby differences in balance position of the control rod are converted to changes in K, it will now be appreciated that changes in inhours may be readily correlated with other properties or characteristics to be investigated. For instance, where stringers of graphite or other material are to be tested for neutron absorption, preliminary comparison may be made between materials of known absorption properties and a relation thus obtained between change of inhours and change of, say, atomic absorption cross section in square centimeters. Likewise, other calibration methods can be used for determining change of K per inhour, as by observing the control rod balance positions for a set of standard metal lumps first without and then with measured quantities of known neutron absorbing materials, adjacent the lumps, and then calculating the difference in K (i.e., K for the test cells of the stringer) represented by the known difference in neutron loss and finally correlating such difference in K with the observed difference in inhours.

With relationships and calibrations of the above described sort ascertained, it is possible to determine accurately the effect on the reproduction factor, or other significant characteristics, of any new batch of material that it is desirable to test. By way of specific example, the following table 7.1 gives the result of tests on three different batches of material, i.e., metallic uranium of natural isotopic content, identified in the first column of the table as lots numbered 4, 5, and 6 respectively. The test were made with the reactor specifically described herein, by substituting bodies of the tested material, in each instance, in the sixteen pockets of the stringer. The control rod positions are those of balance, i.e., for an actual reproduction ratio of unity. The next to last column gives the amount (in inhours) and direction of the departure of control rod position from that for the standard metal bodies in the stringer pockets. The last column sets forth the corresponding increase or decrease of reproduction factor that would be characteristic of each lot under test (if used in a reactor), expressed in percent of K as determined by the above calibration for the reactor described herein, viz. $\Delta K\% = 0.5 \times \Delta ih$. Note that the standard material is returned to the reactor after each three batches of metal are tested, so that any change in control rod position is noted. If there is a change between the control rod position of one test of the standard batch and the control rod position of a second test of the standard batch, the average of the two readings is taken as standard. Note also that no correction is made for temperature because the temperature varies so slightly between testing the standard and the sample.

By the use of our invention practically any material used in a neutronic reactor may be tested. Thus by the use of a removable graphite stringer similar to stringer **35** but without the metal carrying cells **36**, new supplies of graphite may be tested against standard graphite. Also by removing a larger section from the reactor containing several rows of live graphite and substituting various arrangements of graphite and uranium, new types of geometry may be tested. It is also possible to coat the uranium masses with various materials and determine the effect of such coatings on K. From these measurements the effect of coating each uranium mass may be discovered.

Many other uses and changes in our method and apparatus will be apparent to those skilled in the art. We do not wish to be limited to the

Table 7.1

Lot No.	Av. Wt. per cell, g	Press., 0° C	Control Rod Position, cm	Control Rod Position, inhrs	Correction for Press., inhrs	Correction for Weight, inhrs	Corrected Control Position, inhrs	S ih.	S $(r-1)$ percent
Std	1.987	736.6	436.0	127.9	−1.1	−0.5	126.3	....	....
4	1.976	736.6	436.0	127.9	−1.1	−0.4	126.4	+0.1	+0.05
5	1.809	736.6	433.05	126.5	−1.1	+0.5	125.9	−0.4	−0.20
6	1.834	736.6	433.00	126.5	−1.1	+0.4	125.8	−0.5	−0.25
Std	1.937	736.6	435.95	127.9	−1.1	−0.5	126.3	....	....

embodiments of our invention that are described since these are merely illustrative. The scope of the invention is limited only by the attached claim.

We claim:

A method of producing an indication of the relative neutronic purity of a sample of fissionable material with respect to a standard body of fissionable material by utilizing a graphite moderated neutronic reactor having a graphite stringer movable therethrough and in length approximately twice the depth of the reactor, which consists in loading standard bodies of fissionable material in depressions in a first half of the movable stringer outside the reactor; moving the stringer to position the standard bodies in the reactor; balancing the reactor at a predetermined neutron level by positioning a control rod therein; said control rod having a scale to show the depth of penetration of the rod in the reactor; loading the sample bodies of fissionable material in depressions in the second half of the movable stringer outside the reactor; again moving the stringer to position the sample bodies in the reactor while simultaneously removing the first half of the stringer containing the standard bodies outside the reactor; and again balancing the reactor at the said neutron level by again positioning the control rod therein, whereby the difference in the depth of penetration of the control rod as shown by the scale thereon when the standard and sample bodies respectively are in the reactor indicates the relative neutronic purity of the sample with respect to the standard.

References cited in the file of this Patent

UNITED STATES PATENTS

2,288,718 Kallman et al. .... July 7, 1942

FOREIGN PATENTS

114,150 Australia .... May 2, 1940
114,151 Australia .... May 3, 1940
233,011 Switzerland .... Oct. 2, 1944

OTHER REFERENCES

Anderson et al.: Physical Review 72, pages 16-23 (1947). Report of Work Done in 1943-1944.

Chapter 8

Neutron Velocity Selector

Enrico Fermi, Santa Fe, N. Mex., assignor to the United States of America as represented by the United States Atomic Energy Commission

Application September 18, 1945, Serial No. 617,121

U.S. Patent No. 2,524,379 - Patented Oct. 3, 1950

This Patent presents a detailed description of the construction and operation of a velocity selector for neutrons with velocities up to 6000÷7000 *m/s. This apparatus employs a rotating shutter designed in such a way that neutrons are passed during a portion of each rotation of the shutter, the shutter blocking all neutron radiation at other times.*

The selector is built up with alternate laminations of a material with high neutron capture cross section (such as, for example, cadmium, boron or gadolinium), and parallel laminations of a material with low capture probability (such as, for example, aluminium, magnesium or beryllium). This is required in order to provide a path through the shutter to the neutrons, which then pass into a ionization chamber.

The timing mechanism, adopted to activate or deactivate the neutron detection, and measuring means at given times following each opening or closing of the shutter, is electronic (not mechanic), controlled by a photocell unit.

The reference published article for the main topic of the present Patent is [Fermi (1947)].

The present invention relates to neutron velocity selector apparatus and particularly to apparatus of this type which utilizes a rotating shutter.

In various physical investigations it is necessary to measure the numbers and effects of neutrons at various velocities. The usual apparatus for accomplishing this includes a beam source of neutrons, neutron detection or measuring means, and a device known generally as a neutron velocity selector which is located intermediate the beam source and the detection or measuring means. The neutron velocity selector includes a shutter means for periodically interrupting the neutron beam and means for preventing the activation of the neutron detection and measuring means except for short periods of time following each opening of the shutter, velocity selection of neutrons being thereby affected. The best known type of neutron velocity selector employs a rotating shutter which is so designed that neutrons are passed during a portion or portions of each rotation of the shutter, the shutter effectively blocking all neutron radiation at other times. With this apparatus, there is used a mechanical timing mechanism operable to activate the neutron detection or measuring means at the desired time following each opening of the shutter, and reasonably satisfactory results have been obtained with this arrangement.

There are, however, serious inherent limitations in mechanical timing devices of this character and in the other known prior apparatus, and these limitations act to so restrict the range of operation and the accuracy of the devices that there is a great existing need in the art for an improved neutron velocity selector. Accordingly, a principal object of the present invention is to produce a novel improved neutron velocity selector, and particularly a velocity selector which shall be simple in design and easy to operate, and which shall be free of the undesirable limitations and deficiencies of the known prior structures.

As will hereinafter appear, this object is accomplished by the provision of a neutron velocity selector apparatus which utilizes a novel improved rotating shutter of special design and which employs an electronic, photoelectrically controlled means, for effecting the activation and the deactivation of the associated neutron detection and measuring means at predetermined times following the opening and the closing of the shutter. The particular shutter involved and the combination of that shutter with an electronic timing means, while producing a simple, substantially completely automatic instrument, greatly increases the fundamental accuracy of the selector apparatus and at the same time greatly extends its useful range of operation.

The various novel features and structural combinations which contribute to the successful operation of the apparatus of the invention will be made more apparent in the following description and the accompanying drawings of one preferred embodiment thereof. In the drawings,

Fig. 8.1 is an end elevation of the rotatable shutter unit and the driving motor thereof of a velocity selector constructed in accordance with the teachings of the present invention;

Fig. 8.2 is a plan view of the rotating shutter shown in end elevation in Fig. 8.1;

Fig. 8.3 is a side elevation view partially in section of the shutter unit of Figs. 8.1 and 8.2;

Fig. 8.4 is a sectional view on the line **4-4** of Fig. 8.3;

Fig. 8.5 is an enlarged fragmentary cross-sectional view showing certain of the details of construction of the shutter unit; and

Fig. 8.6 is a diagrammatic view illustrating a neutron velocity selector apparatus which includes a rotating shutter unit and a photoelectric timing means in accordance with the present invention.

The features of the rotating shutter unit are illustrated in Figs. 8.1 to 8.5. The active portion of this unit comprises a fabricated, cylindrical member **11** having a laminated central portion **13** which consists of a plurality of superposed, alternate laminations **15** and **17** of sheet aluminum and thin cadmium sheet or foil, respectively. In Figs. 8.3, 8.4 and 8.6 the solid lines between cross-hatched spaces represent the cadmium sheets **17**, since the

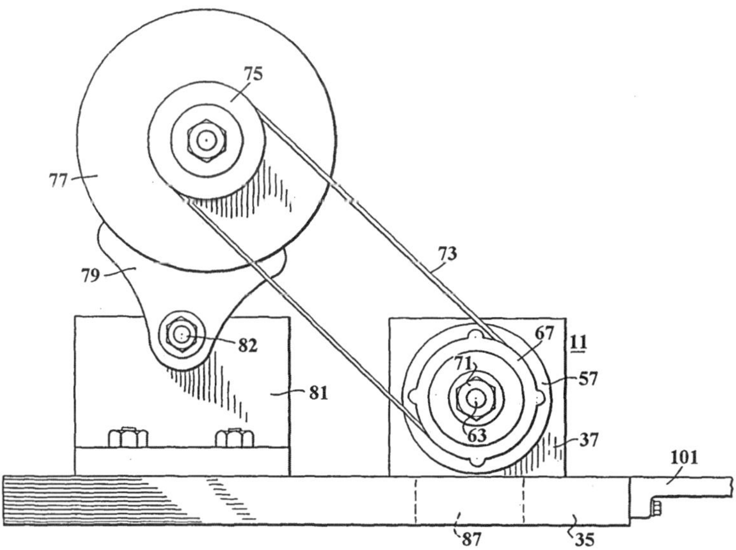

Fig. 8.1

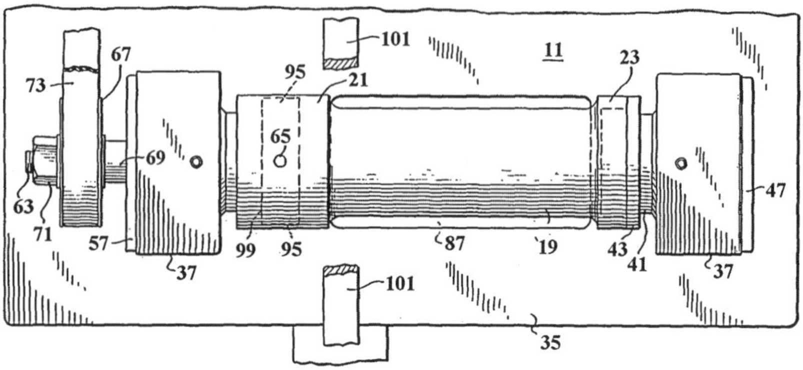

Fig. 8.2

thickness thereof is only a small fraction of the thickness of the aluminum sheets **15**, as is pointed out below. The laminations **15** and **17** are held in proper relation within a thin walled, cylindrical, tubular support **19** which is of aluminum or other material having low neutron capture capabilities and which has a relatively heavy, solid hub **21** at one end thereof and a hollow hub **23** at the other end. The laminations **15** and **17** which comprise the central portion **13** of the shutter unit are held centrally of the tubular support **19** by four, suitably shaped, arcuate, spacer strips **25**, preferably of metal, which extend for the full length of the laminations and which are held apart by elongated radially positioned shield strips **27** of cadmium or

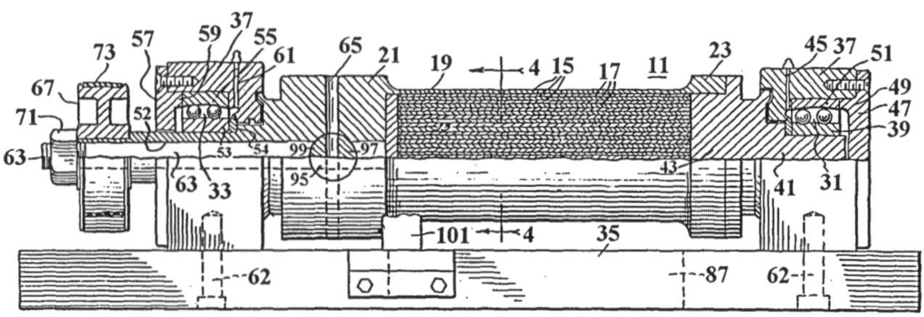

Fig. 8.3

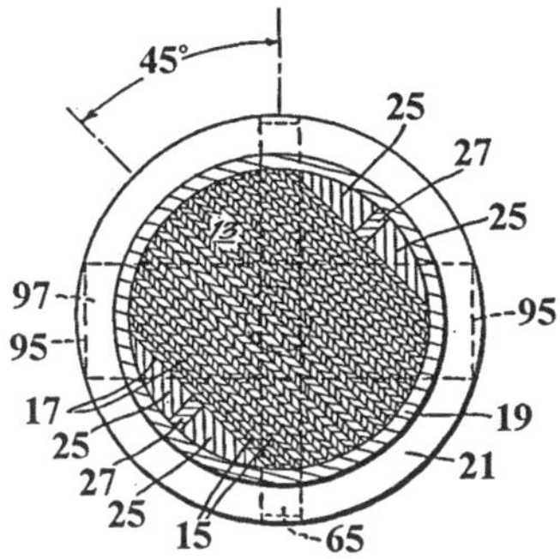

Fig. 8.4

other material having high neutron capture capabilities. To obtain a balanced unit which can be rotated at high speed without excessive vibration, the laminations **15** and **17** are symmetrically disposed within the hollow, cylindrical support **19**, and the surface planes thereof extend parallel to the longitudinal axis of the support, as illustrated.

In a particularly satisfactory embodiment of the invention, the aluminum laminations **15** were approximately $4\frac{3}{4}''$ long with a maximum width along the diameter of the support **19** of about 2 inches. The laminations **15** were cut from a $\frac{1}{32}$ inch aluminum sheet, and the cadmium sheet or foil laminations **17** which were interposed between the aluminum sheets **15** had a thickness of the order of 0.004 to 0.008 inch.

The aluminum laminations provide a relatively low loss path through

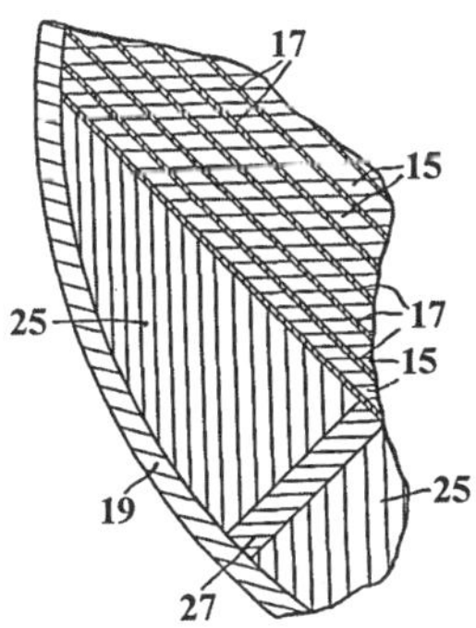

Fig. 8.5

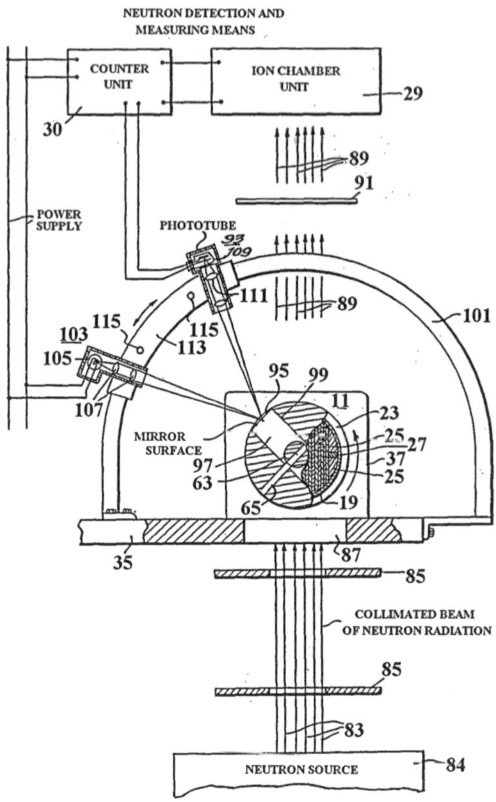

Fig. 8.6

the shutter in the shutter position where the laminations align with the direction of movement of the neutrons. The thickness of cadmium is sufficient to cause the shutter to block all neutrons incident thereon, except those of extremely high velocity, when the shutter is turned more than about $1\frac{1}{2}$ degrees from the position where the laminations align with the direction of movement of the neutrons. The cadmium shield strips **27** in this embodiment of the invention were $\frac{1}{16}$ of an inch thick. The sole function of the shield strips **27** is to provide shielding at the side portions of the shutter which prevents neutron transmission in those areas when the shutter is in operation.

While the use of alternate cadmium and aluminum laminations is preferred in the shutter construction, it will be understood that other materials can be used. The important requirement is that there shall be alternate laminations of material which has high neutron capture capabilities, for example, cadmium, boron, or gadolinium, and parallel laminations of a material having low capture capabilities, such as aluminum, magnesium, or beryllium, to provide a neutron path through the shutter.

In order to utilize the apparatus successfully without greatly extending the distance from the shutter to the neutron detection and measuring means, which may include an ion chamber unit and a counter unit as illustrated generally at **29** and **30** in Fig. 8.6, it is desirable to rotate the shutter at high speeds, those of the order of 10,000 to 15,000 R.P.M.[1] being frequently encountered. To facilitate the convenient obtaining of speeds of this order, the active portion **11** of the shutter is supported for free rotation about the longitudinal axis thereof by means of a pair of precision bearings **31** and **33** which are in turn supported upon a rigid base **35** by means of suitable bearing blocks **37**. The inner race **39** of the bearing **31** at the hollow hub end of the shutter tube **19** engages a suitable stub shaft member **41** having an enlarged end **43** which is press fitted into the hollow hub **23** of the support tube **19** for the shutter laminations **15** and **17**. The inner end portion **43** of this stub shaft **41** serves the further function of holding the shutter laminations in place. Lubricant is admitted to the bearing **31** through a suitable passageway **45**, and the outer end of the bearing assembly is closed by means of a plate member **47** having an internal rib **49** for holding the outer race **51** of the bearing **31** properly seated against the bearing block **37**.

The bearing construction at the other end of the shutter is substantially identical to the structure just described, except that in this instance the shaft **52** which engages the inner race **53** of the bearing **33** is hollow and is formed integrally with the hub **21** at the end of the shutter tube **19**. The inner race **53** abuts against an oil thrower **54** which in turn abuts against a suitable shoulder provided on the bearing shaft **52**, and the outer race **55** is held in position in the bearing block **37** by an outer cap member **57** having an internal rib **59** similar to the member **49** previously described. Lubricant is admitted through a suitable passageway **61**.

The bearing blocks **37** are both rigidly bolted to the base **35** by suitable bolt members **62** as indicated in Fig. 8.3. If desired, dowel pins (not shown) may be used to assure accurate positioning of the bearing blocks **37** relative

[1]@ That is, revolution per minute.

to the base **35**. Motive power is applied to the shutter mechanism by means of a short shaft **63** which is press fitted into the hollow shaft **52** of the shutter tube **19** and is keyed thereto by a pin **65**. The axis of the shaft **63** is concentric with the axis of the shutter.

Externally of the bearing support and end closure at this end of the shutter there is a belt pulley **67** which is supported upon and keyed to the shaft **63** by a suitable spacer and keying member **69** and a nut **71**, which engages a threaded portion on the outer end of the shaft **63**. The pulley **67** is mechanically connected by a belt **73** to the drive pulley **75** of a suitable variable speed motor **77** which is adjustably supported upon the base **35** by means of a bracket **79** affixed to the motor frame, an angle iron **81** affixed to the base **35**, and a hinge bolt connection **82** which permits convenient adjustment of the position of the motor (Fig. 8.1). It is intended that pulleys of various diameters may be applied to the drive shaft **63** at the drive end of the shutter and to the drive shaft of the motor **77**. The adjustable support provided by the connection **82** facilitates the changing of pulleys and the obtaining of correct belt tension after such change has been made.

During the operation of the device, as has been previously pointed out, it is contemplated that the rotatable shutter shall be placed in the path of a beam of neutron radiation, most conveniently a collimated beam, and that the beam, after having been periodically interrupted as the shutter rotates, shall impinge upon an ion chamber or neutron detecting means. Such an arrangement is illustrated diagrammatically in Fig. 8.6, where the arrows **83** represent a collimated beam of neutron radiation which originates in a suitable source, such as a neutronic reactor, indicated generally at **84**, and which passes through suitably orificed shield or collimating plates **85** arranged to limit the dimensions of the beam to a width not greater than that of the laminated central portion **13** of the shutter. The beam of neutron radiation after passing through the shield plates **85**, and a suitable opening **87** in the base **35**, reaches the central portion **13** of the shutter and during those periods when the laminations **15** and **17** are in line with direction of neutron movement in the neutron beam, neutron radiation, as indicated by the arrows **89**, will pass through the shutter and reach the ion chamber **29**. Material under test may be located at a sample position illustrated at **91**. The ion chamber **29** is connected to a suitable indicating apparatus such as the counter mechanism indicated at **30**.

In order for the apparatus to be susceptible of convenient use, it is necessary that means be provided for timing the operation of the indicator or counter. The only result of the operation of the shutter is to periodically interrupt the neutron beam in such manner that the neutron radiation of the beam is transmitted through the shutter only during periods of time roughly equal to $\frac{1}{240}$ of the period of time required for each revolution of the shutter, these openings of the shutter to transmit neutron radiation occurring at exactly the same point during each half revolution thereof. Having available an interrupted neutron beam of this character, it becomes apparent that if some means be available for activating the neutron detection or measuring means at a predetermined time following each opening of the shutter, the activation persisting for a very short interval only, it will be possible to select and study the effects of neutron radiation of various velocities. In the prior known arrangements, attempts have been made to accomplish this selection by timing means which utilizes mechanical expedients, but as previously indicated, these mechanical arrangements are of limited capability and utility.

The apparatus of the present invention utilizes an electronic, photoelectrically controlled, timing mechanism, which, although of simple design, is capable of extremely accurate operation over a much greater range of velocity than has heretofore been considered possible. This timing means includes a photocell control unit **93** which is adapted to be energized to effect activation of the neutron detection and measuring means in response to a beam of light reflected into the photocell by one of a pair of diametrically spaced mirrors **95** which are carried by the rotating shutter. The mirrors **95** are provided by polishing a mirror surface at either end of a pin **97** which is supported in the solid hub **21** at the drive end of the shutter tube **19**. The pin **97** and the hole **99** within which it fits are accurately ground so that the flat mirror surfaces **95** are perpendicular to the longitudinal axis of the pin **97** and so that the longitudinal axis of the pin intersects the axis of rotation of the shutter and is accurately positioned relative to the parallel planes of the laminations **15** and **17**. While almost any relative positioning of the mirror pin **97** and the laminations **15** and **17** may be used, it has been found convenient to displace the axis of the mirror pin 45 degrees from the planes of the laminations, and such an arrangement is illustrated in the drawings. The axis of the mirror pin **97** is preferably at right angles of the axis of rotation of the shutter, and the mirror pin may be conveniently held in place by the taper pin **65** which locks the shaft **63** in position. It will be noted that the inner end of the shaft **63** also passes through the mirror pin **97**, but this is of no constructional importance.

To permit convenient adjustment of the timing interval and thereby further facilitate the use of the apparatus, an arcuate support **101** is provided for the photocell control unit **93** and the beam light source **103** used in conjunction therewith. The arcuate support **101** is bolted to the base **35** and is concentric to the axis of rotation of the shutter unit, and is preferably located in a plane which is parallel to the plane of revolution of the longitudinal axis of the mirror pin **97**, as is illustrated in the drawings. The beam light source **103** is of unitary construction and includes a source of light **105** and suitable collimating lenses **107**. The photocell unit **93** includes a phototube **109** and a similar set of collimating lenses **111**. A common support **113** which slidably engages the main arcuate support **101** is provided for both the photocell and the light source units. The optical constants of the beam source **103** and the photocell unit **93** are such that a spot of light will be concentrated on the mirror surface **95** at one end of the mirror pin **97** and will be reflected into the phototube **109** at each instant when the axis of the mirror pin **97** is midway between the beam source **103** and the phototube unit **93**, i.e., once during each half revolution of the shutter. Further, since the main arcuate support **101** for the light beam source and photocell assemblage is concentric with the axis of rotation of the shutter, and parallel to the plane of rotation of the axis of the mirror pin **97**, this condition will pertain for any position which the light beam source and photocell assemblage may occupy along the arcuate support **101**.

The phototube **109** is connected to the counter unit **30** so as to effect a triggering activation of that unit during each of the very short periods when the phototube is illuminated. As is well known in the art, the operation of an electronic circuit of this character involves a time interval of the order of only a few microseconds, and it will be apparent that the arrangement permits extremely accurate timing. Further, since the light beam source and photocell assemblage may be moved a full 180° relative to the neutron admitting position of the shutter, it is possible to vary the timing over the complete operative range of the device.

In the position shown in Fig. 8.6, the light beam source and photocell combination is in position to trigger and activate the neutron detection and measuring means at the instant the shutter opens. This is the maximum time interval which is practical in operation of the device since the only neutrons which will affect the ion chamber in this instance are those which were released during the last opening of the shutter, and obviously, these will be neutrons of the slowest type. If the position of the light beam source and photocell assemblage is moved in either direction from the position illustrated in Fig. 8.6, the triggering will take place so as to decrease the time interval and to indicate or measure the effect of neutrons which reach the ion chamber in a lesser interval of time. Knowing the speed of rotation of the shutter, the distance from the shutter to the ion chamber or other detection unit, and the relative angular position of the light beam source and photocell assemblage, it is possible to determine the velocity of the neutrons whose effect is being observed to a very high degree of accuracy. In an apparatus having the shutter dimensions previously described and operating at reasonably obtainable speeds of rotation with a separation of the shutter and ion chamber of 2 to 10 meters, for example, it is possible to accurately separate and measure neutrons at velocities up to 6,000 or 7,000 meters per second, which is far beyond the range of any of the known selector mechanism using mechanical timing means for activating and deactivating the neutron detection and measuring means.

In the foregoing there has been described a new and improved neutron velocity selector apparatus which possesses important advantages over the prior known devices. The particular features of the invention which are believed to be new are expressly pointed out in the appended claims.

What is claimed is:

1. In apparatus of the class described, means for alternately blocking and permitting to pass a neutron stream, said means comprising a plurality of superposed, alternate laminations of sheet material having high neutron capture capabilities and sheet material having low neutron capture capabilities, means supporting said laminations for rotation about an axis which is parallel to the surface planes of said laminations, means for detecting neutrons permitted to pass by said first means, and means for counting detected neutrons of predetermined energy during predetermined periods of time, said counting means including reflecting means rotating with said first means.

2. In apparatus of the class described, shutter means for alternatively blocking and permitting to pass a neutron stream, said means comprising a fabricated cylindrical member having a laminated central portion which comprises a plurality of superposed, alternate laminations of sheet material having high neutron capture capabilities and sheet material having low neutron capture capabilities, said laminations being so supported that the surface planes thereof extend parallel to the axis of said cylindrical member, means supporting said cylindrical member for rotation about the axis

thereof, means for detecting neutrons permitted to pass by said shutter means, means for counting detected neutrons, and means for selectively operating said neutron counting means including a light source, reflecting means rotating with said cylindrical member, and a reflected light receiving means.

3. Neutron velocity selector apparatus comprising a rotatable shutter which includes a plurality of superposed, alternate laminations of material having high neutron capture capabilities and material having low neutron capture capabilities, said shutter being movable to a closed, neutron blocking position and to an open, neutron transmitting position during each revolution thereof, a mirror rotatable with said shutter, a source of light, and a photocell positioned to receive light reflected from said source by said mirror at a predetermined time following each opening of said shutter.

4. Neutron velocity selector apparatus comprising a rotatable shutter which includes a fabricated, cylindrical member having a laminated central portion consisting of a plurality of superposed, alternate laminations of sheet material having high neutron capture capabilities and sheet material having low neutron capture capabilities, said laminations being so supported that the surface planes thereof extend parallel to the longitudinal axis of said cylindrical member, said shutter being movable to a closed, neutron blocking position and to an open, neutron transmitting position during each revolution thereof, a mirror rotatable with said shutter, a source of light, and a photocell positioned to receive light reflected from said source by said mirror at a predetermined time following each opening of said shutter.

5. Neutron velocity selector apparatus comprising a rotatable shutter which includes a fabricated, cylindrical member having a laminated central portion consisting of a plurality of superposed, alternate laminations of sheet material having high neutron capture capabilities and sheet material having low neutron capture capabilities, said laminations being so supported that the surface planes thereof extend parallel to the longitudinal axis of said cylindrical member, said shutter being movable to a closed, neutron blocking position and to an open, neutron transmitting position during each revolution thereof, a pair of angularly spaced mirrors rotatable with said shutter, a source of light, and a photocell positioned to receive light reflected from said source by one of said mirror at a predetermined time following each opening of said shutter.

6. Neutron velocity selector apparatus comprising a rotatable shutter which includes a hollow, cylindrical support member of a material having

low neutron capture capabilities and a plurality of superposed, alternate laminations of sheet material having high neutron capture capabilities and sheet material having low neutron capture capabilities, symmetrically disposed within said support member with the surface planes thereof extending parallel to the axis of said member, said shutter being movable to a closed, neutron blocking position and to an open, neutron transmitting position during each half revolution thereof, a pair of angularly spaced mirrors rotatable with said shutter, a source of light and a photocell positioned to receive light reflected from said source by one of said mirror at a predetermined time following each opening of said shutter.

7. In combination, a beam source of neutrons, normally inactive, neutron detection apparatus including an ion chamber which is positioned to receive neutron radiation from said beam source, neutron velocity selector means located intermediate said ion chamber and said beam source, said velocity selector means comprising a rotatable shutter which includes a cylindrical member having a laminated central portion consisting of a plurality of superposed, alternate laminations of sheet material having high neutron capture capabilities and sheet material having low neutron capture capabilities, said laminations being so supported that the surface planes thereof extend parallel to the longitudinal axis of said cylindrical member, said shutter being movable to a closed, neutron blocking position and to an open, neutron transmitting position during each revolution thereof, and means actuable to activate said neutron detection means for a predetermined short period of time following each opening of said shutter, said actuating means including a mirror rotatable with said shutter, a source of light, and a photocell positioned to receive light reflected from said light source by said mirror at a predetermined time following each opening of said shutter.

8. In combination, a beam source of neutrons, normally inactive, neutron detection apparatus including an ion chamber which is positioned to receive neutron radiation from said beam source, neutron velocity selector means located intermediate said ion chamber and said beam source, said velocity selector means comprising a rotatable shutter which includes a cylindrical member having a laminated central portion consisting of a plurality of superposed, alternate laminations of sheet material having high neutron capture capabilities and sheet material having low neutron capture capabilities, said laminations being so supported that the surface planes thereof extend parallel to the longitudinal axis of said cylindrical member, said shutter being movable to a closed, neutron blocking position and to

an open, neutron transmitting position during each half revolution thereof, and means actuable to activate said neutron detection means for a predetermined short period of time following each opening of said shutter, said activating means including a pair of angularly spaced mirrors rotatable with said shutter, a source of light, and a photocell positioned to receive light reflected from said light source by one of said mirrors at a predetermined time following each opening of said shutter.

9. In combination, a beam source of neutrons, normally inactive, neutron detection apparatus including an ion chamber which is positioned to receive neutron radiation from said beam source, neutron velocity selector means located intermediate said ion chamber and said beam source, said velocity selector means comprising a rotatable shutter which includes a hollow, cylindrical support member of a material having low neutron capture capabilities, and a plurality of superposed, alternate laminations of sheet material having high neutron capture capabilities and sheet material having low neutron capture capabilities, symmetrically disposed within said support member with the surface planes thereof extending parallel to the axis of said member, said shutter being movable to a closed, neutron blocking position and to an open, neutron transmitting position during each half revolution thereof, and means actuable to activate said neutron detection means for a predetermined short period of time following each opening of said shutter, said actuating means including a pair of angularly spaced mirrors rotatable with said shutter, a source of light, and a photocell positioned to receive light reflected from said source by one of said mirrors at a predetermined time following each opening of said shutter.

10. A shutter for use in neutron velocity selector apparatus comprising a plurality of superposed, alternate laminations of sheet material having high neutron capture capabilities and sheet material having low neutron capture capabilities, and means supporting said laminations for rotation about an axis which is parallel to the surface planes of said laminations.

11. A shutter for use in neutron velocity selector apparatus comprising a fabricated, cylindrical member having a laminated central portion which comprises a plurality of superposed, alternate laminations of sheet material having high neutron capture capabilities and sheet material having low neutron capture capabilities, said laminations being so supported that the surface planes thereof extend parallel to the axis of said cylindrical member, and means supporting said cylindrical member for rotation about the axis thereof.

12. A shutter for use in neutron velocity selector apparatus compris-

ing a hollow, cylindrical member of a material having low neutron capture capabilities, a plurality of superposed, alternate laminations of sheet material having high neutron capture capabilities and sheet material having low neutron capture capabilities, symmetrically disposed within said hollow cylindrical member with the surface planes of said laminations extending parallel to the axis of said member, and means supporting said cylindrical member for rotation about the axis thereof.

13. A shutter for use in neutron velocity selector apparatus comprising a plurality of superposed, alternate laminations of sheet aluminum and sheet cadmium, and means supporting said laminations for rotation about an axis which is parallel to the surface planes of said laminations.

14. A shutter for use in neutron velocity selector apparatus comprising a fabricated, cylindrical member having a laminated central portion which comprises a plurality of superposed, alternate laminations of sheet aluminum and sheet cadmium, said laminations being so supported that the surface planes thereof extend parallel to the axis of said cylindrical member, and means supporting said cylindrical member for rotation about the axis thereof.

15. A shutter for use in neutron velocity selector apparatus comprising a hollow, cylindrical member of a material having low neutron capture capabilities, a plurality of superposed, alternate laminations of sheet aluminum and sheet cadmium, symmetrically disposed within said hollow cylindrical member with the surface planes of said laminations extending parallel to the axis of said member, and means supporting said cylindrical member for rotation about the axis thereof.

16. In combination, a beam source of neutrons, means for alternately blocking the beam and permitting the beam to pass mounted in the path of said beam, said means comprising a plurality of superimposed alternate laminations of sheet material having high neutron capture capabilities and sheet material having low neutron capture capabilities, means supporting said laminations for rotation about an axis which is parallel to the surface planes of said laminations, normally inactive means for detecting neutrons permitted to pass by said means for blocking neutrons, and means to actuate said neutron detecting means for at least one predetermined short interval of time following the passing of neutrons through the means for blocking neutrons, said means including at least one mirror rotatable with the laminations, a source of light, and at least one photocell positioned to receive light reflected from said light source by a mirror at a predetermined time following each passage of neutrons through the blocking means.

17. The method of measuring the intensity of neutrons of a given velocity in a beam of neutrons comprising placing a number of spaced parallel sheets of neutron absorbing material in the path of a beam of neutrons with the surfaces of the sheets parallel to the path of the neutrons, turning the sheets whereby the beam of neutrons is interrupted, and measuring the intensity of those neutrons impinging on a given point at a distance from the sheets during an interval of time taken after the lapse of a period of time from the time that the sheets obstructed the beam.

18. The method of measuring the intensity of a velocity component in a neutron beam comprising: placing a number of spaced parallel sheets of neutron absorbing material in the path of the neutron beam, continuously rotating said sheets about a central axis which is parallel to the surface of the sheets, and measuring the intensity of those neutrons impinging on a given point at a distance from the neutron absorbing sheets during an interval of time taken after the lapse of a period of time from the time that the sheets obstructed the path of the beam.

19. The method of measuring the intensity of a velocity component in a neutron beam comprising: placing a number of spaced parallel sheets of neutron absorbing material in the path of a beam of neutrons, continuously rotating said sheets about a central axis which is parallel to the surface of the sheets, and measuring the intensity of those neutrons impinging on a given point at a distance from the neutron absorbing sheets during an interval of time, said interval commencing after the lapse of a period of time after the parallel sheets obstruct the neutron beam equal to the time required for the slowest neutron of those neutrons having a velocity greater than the velocity component desired to traverse the distance from the sheets to the given point, and terminating just before the period of time elapses that is required for the fastest neutron of those having a velocity less than the velocity component desired to be measured to traverse the aforedefined distance.

Enrico Fermi

REFERENCES CITED

The following references are of record in the file of this Patent:

UNITED STATES PATENTS

2,275,748 Fearon .... Mar. 10, 1942

Chapter 9

Neutronic Reactor II

Enrico Fermi, Santa Fe, N. Mex., and Leo Szilard, Chicago, Ill., assignors to the United States of America as represented by the United States Atomic Energy Commission

Application October 11, 1945, Serial No. 621,838

U.S. Patent No. 2,807,581 - Patented Sept. 24, 1957

This Patent gives a detailed description of a variant of the reactor presented in Patent No. 2,708,656 (see page 217) by the same authors; it makes use of uranium arranged in plates, instead of spheres or rods. Such a different geometry is particularly efficient when a liquid moderator (for example heavy water) is used; in this case the moderator itself serves also as a coolant. In the Patent, however, the use of solid moderators (like graphite or beryllium) is discussed as well.

The adoption of the given geometry leads to greater neutron losses in the reactor (due to resonant capture in uranium), but they are compensated by the mentioned use of a liquid moderator/coolant.

The main subject of this Patent does not appear in any other published paper.

The present invention relates to the general subject of nuclear fission and particularly to the establishment of self-sustaining neutron chain fission reactions in systems embodying uranium having a natural isotopic content.

This application is a continuation in part of our co-pending application Serial Number 568,904 filed December 19, 1944, now maturated into U. S. Patent 2,708,656, issued on May 17, 1955.[1]

When it became known that the isotope U^{235} in natural uranium could be split or fissioned by bombardment with thermal neutrons, i.e., neutrons

[1]@ See page 217.

at or near thermal equilibrium with the surrounding medium, many predictions were made as to the possibility of obtaining a self-sustaining chain reacting system operating at high neutron density. In such a system, the fission neutrons produced give rise to new fission neutrons in sufficiently large numbers to overcome the neutron losses in the system. Since the result of the fission of the uranium nucleus is the production of two lighter elements with great kinetic energy, plus approximately 2 fast neutrons on the average for each fission along with beta and gamma radiation, a large amount of power could be made available if a self-sustaining system could be built.

In order to attain such a self-sustaining chain reaction in a system of practical size, the ratio of the number of neutrons produced in one generation by the fissions, to the original number of neutrons initiating the fissions, must be known to be greater than unity after all neutron losses are deducted, and this ratio is, of course, dependent upon the values of the pertinent constants.

In the co-pending application of Enrico Fermi, Serial Number 534,129, filed May 4, 1944, and entitled "Nuclear Chain Reacting Systems", now Patent No. 2,780,595, dated February 5, 1957,[2] there is described and claimed a means and method of determining the neutron reproduction ratio for any type of uranium containing structure, directly as a result of a simple measurement which can be performed with precision. Accurate values for all of the pertinent nuclear constants need not be known.

In a self-sustaining chain reaction of uranium with slow neutrons, as presently understood, 92^{238} is converted by neutron capture to the isotope 92^{239}. The latter is converted by beta decay to 93^{239} and then to 94^{239}, also by beta decay. Other isotopes of 93 and 94 may be formed in small quantities. By slow or thermal neutron capture, 92^{235} on the other hand, can undergo nuclear fission to release energy appearing as heat, gamma and beta radiation together with the formation of fission fragments appearing as radioactive isotopes of elements of lower mass numbers, and with the release of secondary neutrons.

The secondary neutrons thus produced by the fissioning of the 92^{235} nuclei have a high average energy, and must be slowed down to thermal energies in order to be in condition to cause slow neutron fission in other 92^{235} nuclei. This slowing down, or moderation of neutron energy, is accomplished by passing the neutrons through a material where the neutrons

[2]@ This corresponds to the Patent in Chapter 2, though the title is different.

are slowed by collision. Such a material is known as a moderator. While some of the secondary neutrons are absorbed by the uranium isotope 92^{238} leading to the production of element 94, and by other materials such as the moderator, enough neutrons can remain to sustain the chain reaction, when proper conditions are maintained.

Under these proper conditions, the chain reaction will supply not only the neutrons necessary for maintaining the neutronic reaction, but also will supply the neutrons for capture by the isotope 92^{238} leading to the production of 94, and excess neutrons for use as desired.

As 94 is a transuranic element, it can be separated from the unconverted uranium by chemical methods, and as it is fissionable by slow neutrons in a manner similar to the isotope 92^{235}, it is valuable, for example, for enriching natural uranium for use in other chain reacting systems of smaller overall size. The fission fragments are also valuable as sources of radioactivity.

An initial number of fast neutrons in the system by going through the process of absorption and fission produce in the next generation a number of neutrons generally different from the initial number. The ratio of the number produced after one generation to the initial number for a system of infinite size is called the reproduction or multiplication factor of the system and is denoted by the symbol K. For any finite system, some neutrons will escape from the periphery of the system. Consequently a system of finite size may be said to have a K constant, even though the value thereof would only exist if the system as built were extended to infinity without change of geometry or materials. Thus, when K is referred to herein as a constant of a system of practical size, it always refers to what would exist in the same type of system of infinite size. If K can be made sufficiently greater than unity to indicate a net gain in neutrons in the theoretical system of infinite size, and an actual system is built to be sufficiently large so that this gain is not entirely lost by leakage from the exterior surface of the system, then a self-sustaining chain reacting system of finite and practical size can be built to produce power and related by-products by nuclear fission of natural uranium. The neutron reproduction ratio in a system of finite size, therefore, differs from K by the external leakage factor, and by a factor due to the neutron absorption by localized neutron absorbers, and the reproduction ratio must still be sufficiently greater than unity to permit the neutron density to rise exponentially with time in the system as built.

Progressive empirical enlargement of any proposed system for which the factor K is not accurately known in a attempt to attain the overall size of a structure of finite size above which the rate of loss of neutrons by diffusion through the periphery of the structure is less than the rate of production of neutrons in the system leads only to an expensive gamble with no assurance of success. The fact that K is greater than unity and the fact that the critical size is within practical limits must be known rather accurately in advance, as otherwise a proposed structure having a K factor less than unity, or even a K factor greater than but close to unity, might not sustain a chain reaction even if all of the uranium in the world were included.

The earliest attempts to predict a structure capable of sustaining a chain reaction, using natural uranium, involved the use of fine uranium particles such as uranium oxide powder, dispersed in hydrogen in combined form as the slowing agent. However, these attempts were not successful, and analysis of experiments made has indicated that the neutron losses in such a system when natural uranium is used, will prevent a chain reaction from being sustained, irrespective of the size of the system, due to neutron absorption at resonance, in the U^{238} content of the uranium.

However, we have pointed out in our above cited application, it is possible by proper physical arrangement of the materials to reduce substantially uranium resonance absorption. By the use of light elements for moderators, fewer collisions are required to slow the neutrons to thermal energies with large increments of energy loss per collision, thus decreasing the probability of a neutron being at a resonance energy as it encounters a uranium atom. During the moderation, however, neutrons are moving through the slowing medium over random paths and distances so that the uranium is not only exposed to thermal neutrons but also to neutrons of energies varying between the energy of fission and thermal energy. Neutrons at uranium resonance energies will, if they enter uranium at these energies, be absorbed on the surface of a uranium body whatever its size, giving rise to surface absorption. Any substantial reduction of overall surface of the same amount of uranium will reduce surface absorption, and any such reduction in surface absorption will release neutrons to enter directly into the chain reaction.

For a given ratio of moderator to uranium, surface resonance absorption losses of neutrons in the uranium can be substantially reduced by a large factor when the uranium is aggregated into spaced substantial masses in the moderator. The uranium may be placed in the system in the form of geometrically spaced uranium masses or bodies of substantial size, preferably either of metal, oxide, carbide, or combinations thereof. The term geometric is used to mean any pattern or arrangement wherein the uranium bodies are distributed in the moderator with at least a roughly uniform spacing and are roughly uniform in size and shape, or are systematic in variations of size, shape or spacing to produce a volume pattern conforming to a generally symmetric system. If the pattern is a repeating or rather exactly regular one, the structure may be conveniently described as a lattice.

The resonance losses in uranium constitute one of the critical factors in the total losses permissible in a neutronic reactor. Proper sizes and shapes of the uranium bodies and volume ratios of uranium to moderator must be fairly accurately known in order that optimum geometry be approached, or if the used of near-optimum geometry is not desirable, then the permissible ranges of departure from the optimum should be determined, so that a reproduction ratio greater than unity can be maintained in a reactor of practical size.

Neutrons are also subject to capture by the moderator. While carbon and beryllium have very small capture cross sections for thermal neutrons and deuterium still smaller, a fraction of the thermal neutrons present in the system under best conditions is lost by capture in the moderator during diffusion therethrough. It is therefore desirable to have the neutrons reaching thermal energy enter uranium as promptly as possible. This may be taken care of by using optimum or near-optimum geometry where the resonance absorption is substantially equal to absorption in the moderator.

Moderators differ in their ability to slow down neutrons and in their capacity to absorb neutrons. An important criterion of the ability to slow down neutrons is what is known as the scattering cross section of the nucleus. The ability to absorb or capture neutrons is expressed by what is known as the capture cross section of the nucleus. The ratios of absorption cross section to scattering cross section for various moderators are approximately as follows:

Light water (H_2O)	0.00478
Diphenyl	0.00453
Beryllium	0.00127
Graphite	0.000726
Heavy water (D_2O)	0.00017

The choice of moderators, therefore, will depend on many considerations, as will be apparent from further discussions herein.

The patent application cited above has set forth that K factors greater than unity can be obtained by aggregating the uranium in the form of spheres, rods and layers, and has shown the structure, K factor and volume ratio ranges for uranium spheres and rods in various moderators. The present application deals solely with aggregation of the uranium into layers, sometimes called plates or slabs, in various moderators, and K factors and volume ratios for operative neutronic reactors embodying such construction.

While plate or slab geometry is not as efficient in reducing resonance absorption as sphere or rod geometry, due to the fact that the aggregation of the uranium into plates does not reduce the exposed surface as much as the other geometries mentioned, reactors embodying plate geometry have several advantages, especially when the plates are used in liquid moderators. The plates are thin in section for practical K factors, and in consequence can be easily cooled by circulation of the moderator. Furthermore, less jacketing material need be used for protection from the effects of moderator corrosion or to prevent fission product contamination of the moderator or coolant than in the other forms, and in consequence K reduction due to neutron absorption in jacket material is less. Fewer units need be used and handling problems are thus reduced. Therefore, in spite of a somewhat reduced efficiency, plate geometry has a definite place in reactor design.

It is, therefore, an object of the present invention to provide plate geometry neutronic reactors operative to sustain a chain fission reaction with natural uranium disposed in a moderator.

It is a further object of the present invention to set forth the ranges within which operative reactors can be constructed of alternate layers of natural uranium and a neutron moderator.

The objects and advantages of the present invention will be more readily understood from the following description read by reference to the drawings, which show two illustrative forms the present invention may take, as follows:

Fig. 9.1 is a diagrammatic vertical sectional view of the basic structure of a neutronic reactor incorporating spaced uranium plates immersed in a heavy water moderator;

Fig. 9.2 is a cross-sectional view taken as indicated by the line **2-2** in Fig. 9.1;

Fig. 9.3 is a cross-sectional view of a fragment of a uranium plate in a protective jacket;

Fig. 9.4 is a longitudinal sectional view taken as indicated by line **4-4** of Fig. 9.3;

Fig. 9.5 is a diagrammatic side view, partly in central section, of the basic structure of a neutronic reactor embodying uranium plates disposed in a solid moderator;

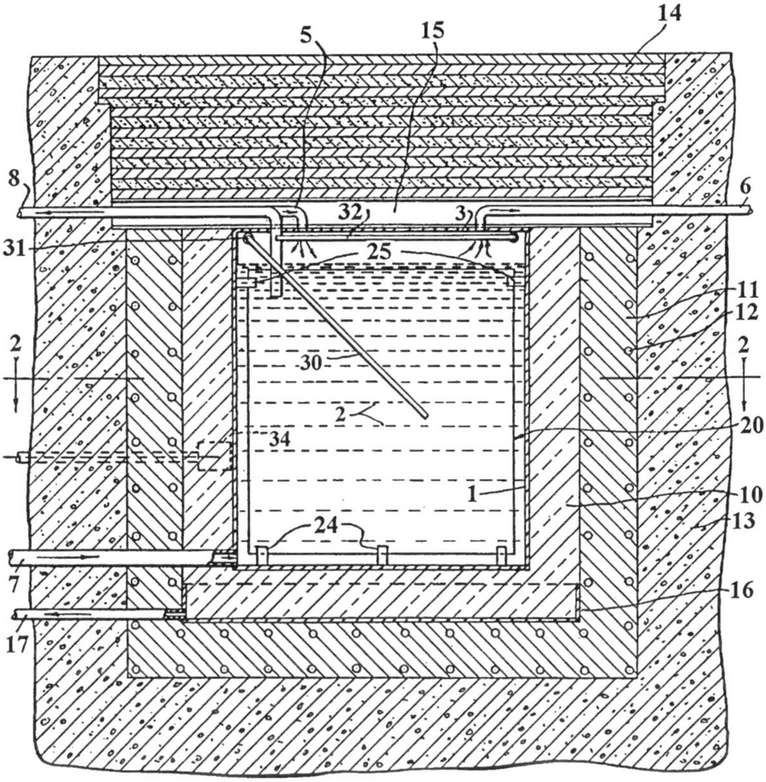

Fig. 9.1

Fig. 9.6 is a graph or diagram showing K factors for uranium plates in terms of volume ratios of D_2O moderator to uranium and half thickness of uranium plate;

Fig. 9.7 is a graph or diagram for uranium plates in terms of volume ratios of Be metal moderator to uranium and half thickness of uranium plate; and

Fig. 9.8 is a graph or diagram for uranium plates in terms of volume ratios of graphite moderator to uranium and half thickness of uranium plate.

Referring first to Figs. 9.1 to 9.4 inclusive, in Fig. 9.1 a substantially cubic reactor tank **1** is shown preferably of aluminum or thin stainless steel containing a quantity of heavy water as indicated by liquid lines **2**. The

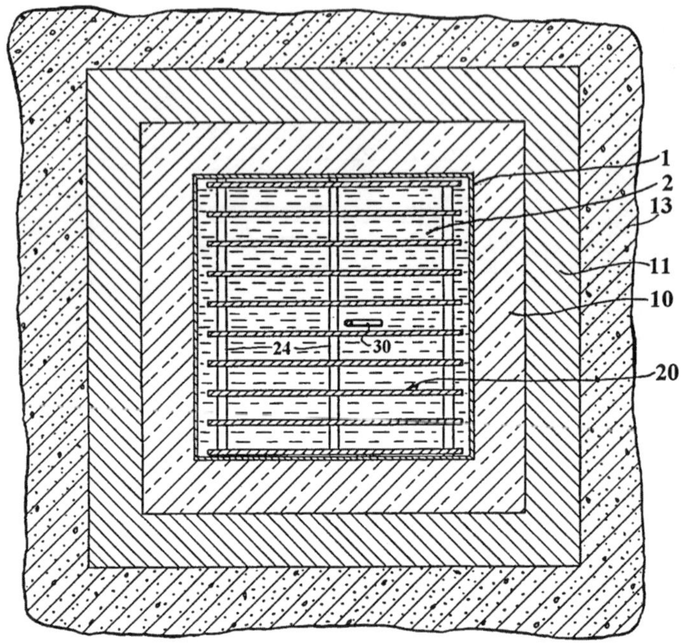

Fig. 9.2

top of the tank **1** is closed by a cover plate **3** forming a space above the heavy water. A helium inlet pipe **5** and a helium outlet pipe **6** enter this space through the cover plate **3** of tank **1**. In order that the heavy water may be circulated, a heavy water inlet pipe **7** is provided at the bottom of the tank and a heavy water outlet pipe **8** passes through the cover plate **3** of the tank and extends downwardly into the heavy water.

The tank **1** is surrounded on five sides by a graphite reflecting layer **10** preferably built up from machined graphite blocks. This graphite layer is in turn surrounded on five sides by a shielding layer **11** formed from bricks of a cadmium lead alloy and cooled by water pipes **12**. This shield is in turn surrounded by thick concrete walls **13**, and at the top of the reactor is a stepped back opening to be filled during operation with shielding material **14**, supported on beams **15**, preferably removable, to complete the shielding around the sixth side of the reactor. In order to conserve heavy water in case of a leak in tank **1**, a bottom cup-shaped lining **16** is provided between

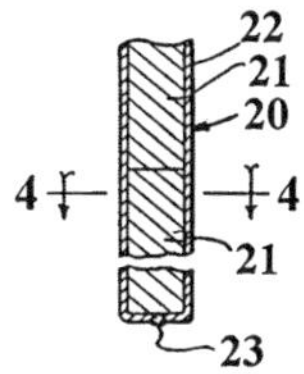

Fig. 9.3

the graphite reflecting layer **10** and the cadmium shield **11**, the lining **16** being drained through drainage pipe **17**. Before shield **14** closes the top of the reacting system, fissionable material is inserted in the heavy water consisting in this case of spaced plates or slabs **20** formed from natural uranium. As shown particularly in Figs. 9.1 and 9.2, the plates or slabs **20**, formed of natural uranium, separate and isolate the heavy water or other moderator mass into distinct and individual layers, the width and length of each plate being substantially equal to the width and length of the adjoining moderator layers, each plate **20** having an interrupted surface area substantially equal to the cross-sectional area of the active portion of the reactor. A plurality of plates **20** is used to provide definite volume ratios of uranium to heavy water, as will be brought out later. Only a few plates are shown in Fig. 9.2 for clarity of illustration.

As it is at present impractical to roll or otherwise form complete sheets of uranium metal several centimeters thick, it is preferred to form each plate from a plurality of strips of uranium 10 to 20 centimeters wide and the full height of the plate, such strips being readily fabricated by rolling extruded uranium rods, for example, into flat strips. These strips **21** are shown in Figs. 9.3 and 9.4, and are positioned in edge to edge relationship to form a plate of the width desired and the composite plate thus made is enclosed in a thin aluminum jacket **22** comprising aluminum plates rolled

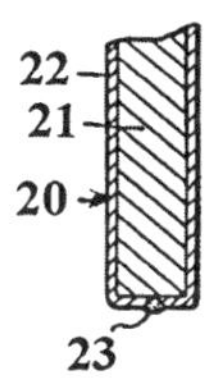

Fig. 9.4

in firm contact with the uranium and secured by turning the edges thereof and welding the aluminum sheets around the entire edge of the plate as indicated by weld **23**.

To properly space the uranium plates **20** inside the reactor, spacer bars **24** are provided on the bottom of the reactor notched to receive the plates, and similar spacer bars **25** are positioned near the top of the uranium plates, so that the plates **20** are tied together with the assembly standing erect with the plates in parallel relationship inside the reactor. When the proper volume ratio of uranium to moderator is used with the plates of definite thicknesses and extent, a chain reaction will take place at elevated neutron densities. To remove the heat of the reaction from the reactor, it is preferred that the heavy water be circulated outside the reactor, cooled in heat exchangers not shown, and returned to the reactor.

The reactor is controlled first by the use of a control rod **30** (Fig. 9.1) which is movable from an upper corner bearing **31** by a shaft extending outwardly through the shields so that a greater or less extent of the rod will enter the heavy water between plates in the reactor. For safety reasons, a second rod **32**, which may be termed a safety rod, is normally held horizontally above the level of the heavy water to be released at will to swing into the heavy water between the plates in case a predetermined neutron density is exceeded. The neutron density in the reactor can be monitored in several ways, such as, for example, monitoring the temperature of the outgoing heavy water, or more directly by measuring the neutron density just outside the tank **1** by means of an ionization chamber **34** connected to an indicating device in view of the operator of the reactor. Such ionization chamber will also permit the operator to start and stop the reactor and to reach a desired neutron density level by varying the amount of absorbing material in control rod **30** within the reactor. The size of the reactor is preferably such as to provide a neutron reproduction ratio of unity with the control rod material approximately half-way inserted into the reactor. Under these conditions, insertion of more of the rod into the reactor will stop the reaction while the removal of some of the rod material will permit a reproduction ratio over unity to be obtained so that the neutron density can rise to a desired level and then be held there by moving the control rod back to its position where the reproduction ratio is again unity. Thereafter, small movements of the control rod will permit the chosen neutron density value to be maintained.

Slab or plate geometry for the disposition of uranium in a moderator can also be used with solid moderators, such as, for example, graphite,

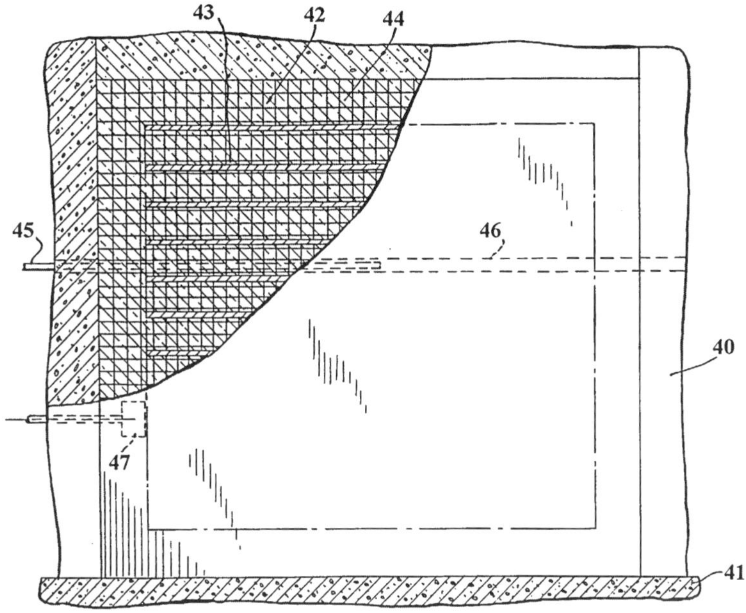

Fig. 9.5

or graphite and beryllium metal of high purity. In this case, a simple nuclear reactor can be made, for example, as shown diagrammatically in Fig. 9.5, in which construction the uranium plates need not be jacketed. In Fig. 9.5, the reactor comprises a heavy concrete shield wall **40** mounted on a concrete base **41** enclosing a cube **42** of solid moderating material preferably formed from machined bricks of graphite or beryllium metal to form a closely packed shield of moderating material. In this moderator cube **42** layers of uranium **43** are horizontally positioned by the bricks and spaced vertically to provide the desired geometry. The uranium layers, however, do not extend entirely to the surface of the moderator cube but terminate on all sides and above and below short of these surfaces to provide a reflecting layer **44** completely enclosing the uranium-bearing portion. In this case, control is by a horizontally movable control rod extendable into and out of the reactor through a slot **46** in the bricks, and the neutron density is monitored by an ionization chamber **47** positioned in the reflector. Control of the reaction in this case is as before by inserting more or less of rod **45** into the reactor. Safety rods, not shown, are also arranged to be pulled into the reactor by gravity upon an excessive rise in neutron density as indicated by the ionization chamber. Full details of the reactor control are given in our co-pending application cited above.

Having discussed several forms for the present invention, what will hereafter be called plate slab geometry as used in these reactors will next be discussed.

In reactors operating continuously at high powers (high neutron densities), radioactive elements of extremely high neutron capture cross section are formed inside the reactor within a few hours, notably the gaseous fission product Xe^{135}. The production of Xe^{135} and its destruction by neutron absorption and by decay creates an equilibrium poisoning effect in the reactor determined as to value by the power at which the reactor is to be continuously operated. It can thus be seen that a neutronic reactor can have a satisfactory operating size when operated intermittently and at low power which is entirely too small to support a chain reaction when operated at some elevated power where a noticeable equilibrium amount of Xe^{135} is formed. It is, therefore, necessary, to obtain a continuously operating reactor when a high power is required, to determine the operating sizes required to maintain the desired high power output. Thus, it will be seen there is actually a minimum critical (and operating) size for a neutronic reactor based on operating power, that is larger than the critical size for the reactor if zero power level is not to be substantially exceeded.

However, the reduction in the reproduction ratio due to the Xe^{135} equilibrium amount present when the neutron density is theoretically infinity in the reactor, has been found to be about 0.03, which means that to obtain a rise in neutron density to any desired density up to infinity, the control rod would have to be eventually removed by an amount corresponding to an increase in the reproduction ratio of about 0.03, and somewhat less than 0.03 when finite densities are to be obtained. In accordance with the density desired, the size of the reactor has to be initially big enough to provide the increase. For example, in a continuously operated water cooled uranium graphite reactor, the poisoning effect due to Xe^{135} at equilibrium in terms of the reproduction ratio is about 0.0012 at 10,000 kilowatts, 0.009 at 100,000 kilowatts, and about 0.013 and 0.020 at 200,000 kilowatts and 500,000 kilowatts, respectively. As before stated, operating sizes ordinarily are not sufficiently large to provide maximum reproduction ratios of over 1.01 with all controls removed. However, if power outputs over 100,000 kilowatts are desired, the reactor must have its critical and operating sizes defined as set forth herein, using final constants decreased by the Xe^{135} factor for the power desired, even though the amount of reproduction is over 0.01. In other words a significant impurity is added during operation at high powers, and must be compensated for by enlargement of the reactor.

This may lead to the requirement for a reactor of such size that, if it did not acquire Xe^{135} during operation, could attain a maximum reproduction ratio of over 1.01 with all control rods removed, and which, before a substantial amount of Xe^{135} is formed, could in consequence attain a dangerous neutron density if controls were entirely removed.

Such a reactor can be adequately safeguarded by the use of "shim" or limiting rods of neutron absorbing material inserted in the reactor preferably to depths that will not permit a reproduction ratio of about 1.01 to be attained at any time during the operation of the reactor, even when the control rod is completely removed. Then, if a reproduction ratio of more than unity cannot be attained by outward movement of the control rod alone, due to the build-up of the Xe^{135} effect, the shim rod can be withdrawn to compensate for the Xe^{135} effect, but still be left in a position where the reproduction ratio cannot exceed 1.01, when the control rod is completely removed.

It can thus be seen that compensation for the Xe^{135} effect is obtained, first by considering the Xe^{135} impurity factor for the power desired, as a reduction in reproduction ratio to determine a proper operating size for a desired power; and second, by initially providing in the reactors, impurities that can be removed by amounts compensating for the Xe^{135} equilibrium amount acquired at a given power output. Clearly then, there are two operating sizes; one, that is able to sustain a chain reaction at low power in absence of any substantial amount of Xe^{135}; and two, a larger size able to sustain chain reaction with Xe^{135} present. Both must be known prior to building the reactor.

As D_2O is the best moderator so far known, the range of plate thicknesses and volume ratios for uranium in heavy water will first be discussed with reference to Fig. 9.6. In this figure, a family of K factor curves is shown as graphs on coordinates where the ordinates represent half thicknesses of the uranium slabs, and the abscissae the ratio of volume of the moderator to uranium volume.

Curves showing the K factors from unity to the optimum are shown, and it will be seen that the maximum K factor of somewhat more than 1.14 is found with a high volume ratio of about $25 \div 1$ and a half thickness of about 0.6 cm and also at a lower volume ratio of about $7\frac{1}{2} \div 1$ with a half thickness of about 3.0 cm.

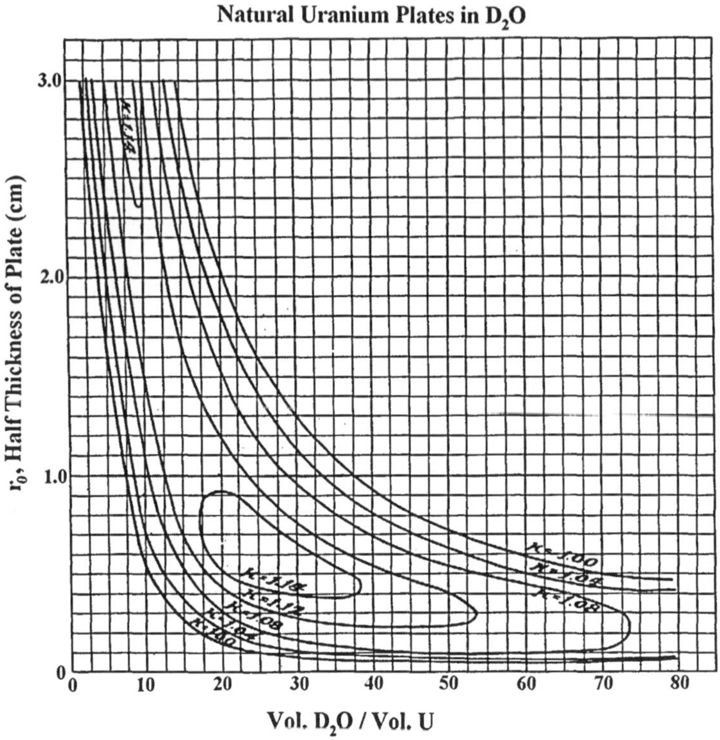

Fig. 9.6

As an example of the use of the curves of Fig. 9.6, when uranium slabs or plates are to be used as a reactive composition spaced 17 cm in a D_2O moderator, the volume ratio will be about 43 to 1 with the uranium plates 0.4 cm thick. The K factor will be about 1.1 from the curves in in Fig. 9.6. If spaced 33 cm, the plates will be about 0.76 cm thick and the K factor will be about 1.13.

The reduction of resonance absorption due to the use of slabs is sufficient to permit a chain reaction to be sustained in a reactor of practical size when beryllium metal is used as a moderator, with K factors obtainable for the reactive composition up to slightly over 1.08, as indicated in Fig. 9.7. In Fig. 9.7, K curves are plotted as graphs in the same manner as Fig. 9.6. It will be noted that in this case the $K = 1$ curve is closed at the high volume ratio end, thus clearly indicating uranium units smaller than

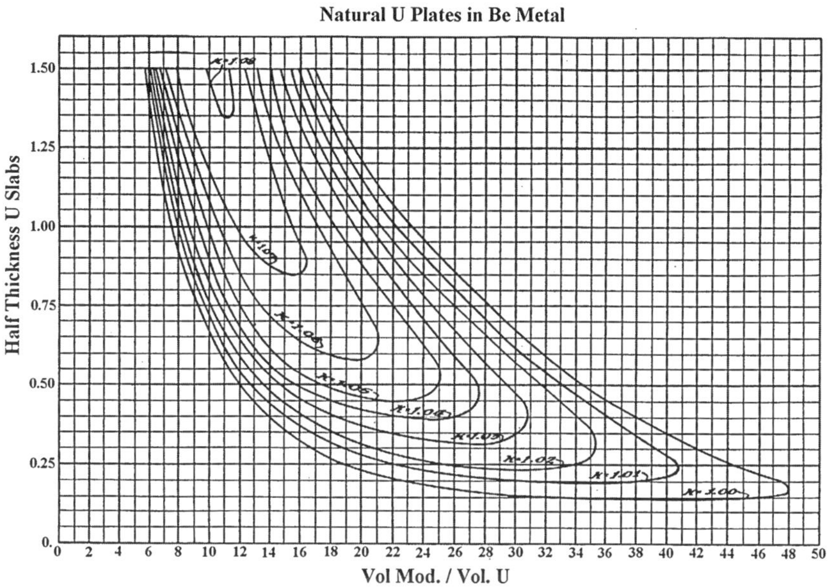

Fig. 9.7

0.15 cm half thickness will not sustain a chain reaction with any volume ratio or in any size. Thus, a mixture of fine particles of natural uranium and beryllium will not be a reactive composition able to sustain a chain reaction. The reduction in resonance absorption by the use of alternate layers of uranium and beryllium does, however, permit the creation of the desired chain reaction.

Uranium plates can also be used with a graphite moderator, and K curves for uranium plates in graphite similar to those shown in Figs. 9.6 and 9.7, are shown in Fig. 9.8. Here it will be seen that optimum geometry occurs with a volume ratio of about $55 \div 1$ and a half thickness of the uranium plates of about 0.4 cm.

When the K factor is known, operating sizes for various reactive compositions can be found by the use of critical size formulae, modified to include size increase due to absorption in a cooling system, absorption caused by Xe^{135}, and excess reproduction ratio desired while operating as follows:

$$K - C - X - r = \frac{M}{R^2}$$

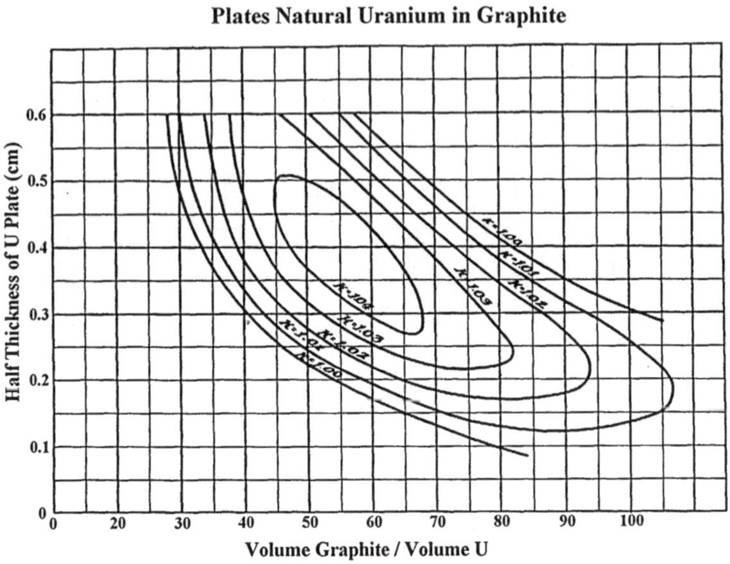

Fig. 9.8

for spherical reactors where R is the radius in cm,

$$K - C - X - r = M\left(\frac{1}{a^2} + \frac{1}{b^2} + \frac{1}{c^2}\right)$$

for a parallelepiped where a, b, and c are the sides thereof in cm and

$$K - C - X - r = M\left(\frac{1}{H^2} + \frac{0.59}{R^2}\right)$$

for cylinders where H is the height and R is the radius in cm.

In the above formulae:

K = Reproduction factor for a reactive composition of infinite size;

C = Reduction in K due to absorption in coolant tubes, coolant and in jackets to protect uranium from corrosion by the coolant, when any of these items are used inside the reactor;

X = Reduction in reproduction ratio due to Xe^{135} poisoning at desired operating power;

r = Excess of reproduction ratio desired while reactor is operating with Xe^{135} equilibrium established. This amount is usually not greater than 0.005 and may be 0.0005;

M = a constant embodying the effect of the moderator and slab geometry, as follows:

M for $D_2O \simeq 2500$
M for Be metal $\simeq 3000$
M for graphite $\simeq 6500$

However, reactive compositions by themselves are seldom used in a neutronic reactor because of the cost of materials and because of the fact that operating sizes can be very considerably reduced in size by surrounding the reactive mass with a layer of neutron scattering material to form what is known as a neutron reflector, as shown in Figs. 9.1-9.3. All good moderators are good reflectors.

The reduction in size due to the use of a reflecting layer surrounding the reactive mass is given for reflecting layers, which are thin with respect to the extent of the reactive mass, by the formula

$$\Delta r' \approx L \tanh \frac{\Delta r}{L},$$

where $\Delta r'$ is the reduction in radius of the reactor due to surrounding the reactive mass by a reflecting layer Δr thick and where L is the diffusion length for thermal neutrons in the reflecting material.

Using data given above, the reactor shown in Fig. 9.1, when surrounded on five sides with a 1 ft reflector of graphite, and when using uranium plates having a thickness of 1.2 cm, a volume ratio of 15 total volume to 1 uranium and 1 mm aluminum jackets will have an operating size for continuous operation up to 5000 kilowatts, for example, of 12 cubic meters for the active portion and will use 12.4 metric tons of D_2O and 15.2 metric tons of uranium. Similarly, a uranium slab-Be metal reactor such as shown and described herein having a 1 ft Be reflector on all sides, and using uranium slabs 3 cm thick with a volume ratio of about 10 to 1 will have a volume of 33.7 cubic meters in the active portion and will contain 59.0 metric tons of uranium and 56.6 metric tons of beryllium exclusive of 42.2 metric tons of beryllium in the reflector. This latter reactor, being uncooled, cannot be operated continuously at high power outputs, but reactors of this type can dissipate a few watts continuously, and up to 10,000 kilowatts intermittently for short periods.

When graphite is used as a moderator, in the form of bricks, for example, a neutronic reactor similar to that shown in Fig. 9.5 can be built. In this case, using optimum geometry as shown in Fig. 9.7, the K factor will be about 1.05, when uranium layers 0.8 cm thick are used with a volume ratio of $55 \div 1$. The reactor, with a 1 foot graphite reflector, will have a volume of about 258 cubic meters, in the active portion and will contain 87.4 metric tons of uranium and 390 metric tons of graphite inclusive of 115 metric tons of graphite in the reflector.

What is claimed is:

1. A neutronic reactor consisting essentially of plates of natural uranium separating a moderator mass into distinct and individual layers, the width and the length of each plate being substantially equal to the width and length of adjoining moderator layers, and each plate having an uninterrupted surface area substantially equal to the cross-sectional area of the active portion of the reactor, said moderator mass selected from the group consisting of heavy water, beryllium and graphite, the thickness of the uranium plate and the volume ratio of moderator to uranium being within the area encompassed by the branches of the $K = 1.00$ curves of Figs. 9.6-9.8, and the purity of the moderator and the uranium and the total mass thereof being sufficient to sustain a chain reaction.

2. A neutronic reactor in the shape of a rectangular parallelepiped comprising a plurality of plates of natural uranium uniformly disposed in parallel relationship to each other in a moderator of heavy water, said plates separating the moderator into distinct and individual layers, and each plate having an uninterrupted surface area substantially equal to the cross-sectional area of the active portion of the reactor, the width and the length of each plate being substantially equal to the width and length of adjoining moderator layers, tubes communicating with the moderator to provide an inlet and an outlet for movement of the heavy water through the reactor, jackets on said uranium plates to isolate the uranium from the moderator, the volume ratio of moderator to uranium and the thickness of the uranium plate being within the area encompassed by the branches of the $K = 1.00$ curve of Fig. 9.6, the excess reproduction ratio, r, being between 0.0005 and 0.005, the neutron absorption of the jackets, and the tubes diminishing the K factor by an amount C, the Xe^{135} concentration at the desired operating power diminishing the K factor by an amount X, and the dimensions a, b and c of the reactor active portion being given by

the equation

$$K - C - X - r = M\left(\frac{1}{a^2} + \frac{1}{b^2} + \frac{1}{c^2}\right),$$

where M is a constant embodying the effect of the moderator and slab geometry, and having a value of 2500.

REFERENCES CITED in the file of this Patent

UNITED STATES PATENTS

2,708,656 Fermi et al. .... May 17, 1955

FOREIGN PATENTS

114,150 Australia .... May 2, 1940
861,390 France .... Oct. 28, 1940
233,011 Switzerland .... Oct. 2, 1944

OTHER REFERENCES

Physical Review, Aug. 1, 1939, vol. 56 (1939), pp. 284-286. An article by Anderson, Fermi & Szilard.

Journal de Physique Series 7, 10, pp. 428-429 (1939), by Von Halban, Joliot, Kowarski, Perrin.

H.D. Smith: "A General Account of the Development of Methods of Using Atomic Energy for Military Purposes", 1940-1945. Supt. of Doc., Washington, D.C., August 1945, pp. 22, 177.

Swain: "Power", July 1940, article titled "Uranium 235 – Power Fuel of the Future?", pp. 56-59.

Kelly et al.: Phy. Rev., vol. 73, pp. 1135-9 (1948).

Chapter 10

Neutronic Reactor III

Enrico Fermi, Santa Fe, N. Mex., Walter H. Zinn, Chicago, Ill., and Herbert L. Anderson, Hartford, Conn., assignors to the United States of America as represented by the United States Atomic Energy Commission

Application October 11, 1945, Serial No. 621,837

U.S. Patent No. 2,852,461 - Patented Sept. 16, 1958

An improvement of the reactors described in the previous Patents, aimed at increasing the reproduction factor, is reported here, such improvement being obtained by diminishing the neutron loss due to impurities within the reactor. This is achieved by encasing the reactor in a rubberized balloon cloth housing (or something like this) in order to eliminate the atmospheric air therefrom, thus eliminating both the effect of the danger coefficient of nitrogen (70% of the atmospheric air) and that of the argon present in the air, which can become radioactive. Since the removal of the air from the reactor may result in structural problems, caused by the forces brought into play by that evacuation, the reactor is then filled with a non-reactive (from a chemical and nuclear standpoint) gas such as helium or carbon dioxide.

It is interesting to point out that the authors consider also the possibility to control (a little) the reproduction ratio of the reactor by varying the air content of it.

Just a rapid mention of the main idea of the present Patent (i.e. the encasing of the pile in a balloon cloth) appeared in [Fermi (1942f)], but no detailed description of the system considered here is reported in any other published paper.

This invention relates to the general subject of nuclear fission, and more particularly to a novel means for improving the establishment of self-sustaining nuclear fission chain reactions.

In order to attain a self-sustaining chain reaction in a system of practical size, the ratio of the number of neutrons produced in one generation by the fissions, to the original number of neutrons initiating the fissions, must be known to be greater than unity after all neutron losses are deducted, and this ratio is, of course, dependent upon the values of the pertinent constants.

In a self-sustaining chain reaction of uranium with slow neutrons, as presently understood, 92^{238} is converted by neutron capture to the isotope 92^{239}. The latter is converted by beta decay to 93^{239} and this 93^{239} in turn is converted by beta decay to 94^{239}. Other isotopes of 93 and 94 may be formed in small quantities. By slow or thermal neutron capture, 92^{235} on the other hand, can undergo nuclear fission to release energy appearing as heat, gamma and beta radiation, together with the formation of fission fragments appearing as radioactive isotopes of elements of lower mass numbers, and with the release of secondary neutrons.

The secondary neutrons thus produced by the fissioning of the 92^{235} nuclei have a high average energy, and must be slowed down to thermal energies in order to be in condition to cause slow neutron fission in other 92^{235} nuclei. This slowing down, or moderation of the neutron energy, is accomplished by passing the neutrons through a material where the neutrons are slowed by collision. Such a material is known as a moderator. While some of the secondary neutrons are absorbed by the uranium isotope 92^{238} leading to the production of element 94, and by other materials such as the moderator, enough neutrons can remain to sustain the chain reaction when proper conditions are maintained.

Under these proper conditions, the chain reaction will supply not only the neutrons necessary for maintaining the neutronic reaction, but also will supply the neutrons for capture by the isotope 92^{238} leading to the production of 94, and excess neutrons for use as desired.

As 94 is a transuranic element, it can be separated from the unconverted uranium by chemical methods, and as it is fissionable by slow neutrons in a manner similar to the isotope 92^{235}, it is valuable, for example, for enriching natural uranium for use in other chain reacting systems of smaller overall size. The fission fragments are also valuable as sources of radioactivity.

The ratio of the fast neutrons produced in one generation by the fissions to the original number of fast neutrons in a theoretical system of infinite size where there can be no external loss of neutrons is called the reproduction or multiplication factor or constant of the system, and is denoted by the symbol K. For any finite system, some neutrons will escape from the periphery of the system. Consequently a system of finite size may be said to have a K constant, even though the value thereof would only exist if the system as built were extended to infinity without change of geometry or materials.

Thus when K is referred to herein as a constant of a system of practical size, it always refers to what would exist in the same type of system of infinite size. If K can be made sufficiently greater than unity to indicate a net gain in neutrons in the theoretical system of infinite size, and then an actual system is built to be sufficiently large so that this gain is not entirely lost by leakage from the exterior surface of the system, then a self-sustaining chain reacting system of finite and practical size can be built to produce power and related by-products by nuclear fission of natural uranium. The neutron reproduction ratio in a system of finite size therefore, differs from K by the external leakage factor, and by a factor due to the neutron absorption by localized neutron absorbers; and the reproduction ratio must still be greater than unity to permit the neutron density to rise exponentially with time in the system as built.

More specific details of the theory and of the essential characteristics of neutronic reactors and neutronic reactor systems are set forth in the co-pending application of Enrico Fermi and Leo Szilard, Serial No. 568,904, filed December 19, 1944, now Patent No. 2,708,656, dated May 17, 1955.[1]

During the interchange of neutrons in a system of finite size, comprising bodies of any size disposed in a neutron moderator, neutrons may be lost to the chain reaction in four ways:

(1) By absorption or capture in the uranium content of the bodies without producing fission;

(2) By absorption or capture in the moderator material itself;

(3) By absorption or capture by the impurities present in both the uranium bodies and the moderator;

(4) By leakage out of the system through the periphery thereof.

The present invention is directed to a novel means and method for increasing the reproduction ratio of a finite system by diminishing the neutron loss due to the third factor above listed, namely, impurities within the reactor.

The effect of impurities on the optimum reproduction factor K may be conveniently evaluated by means of certain constants known as "danger

[1]@ See page 217.

coefficients" which are assigned to the various elements. These danger coefficients for the impurities are each multiplied by the percent by weight of the corresponding impurity, with respect to the weight of uranium in the system, and the total sum of these coefficients gives a value known as the total danger sum. This total danger sum is subtracted from the reproduction constant K as calculated for theoretically pure materials and for the specific geometry under consideration.

The danger coefficients are defined in terms of the ratio of the weight of impurity per unit mass of uranium and are based on the cross section for absorption of neutrons of the various elements. These values may be obtained from published literature on the subject, and by direct measurement, and the danger coefficient computed by the formula[2]

$$\frac{\sigma_i}{\sigma_u}\frac{A_u}{A_i},$$

wherein σ_i and σ_u represent the cross sections for the impurity and the uranium respectively, A_i the atomic weight of the impurity and A_u the atomic weight for uranium. In general, whether the impurities are in the moderator or in the uranium, they may be computed as their fraction by weight, of the uranium in the system, and the result will be sufficiently accurate for design purposes.

Danger coefficients for some elements are given in the following table 10.1, wherein the elements are listed in order of their atomic number.

The sum of the danger coefficients of the impurities in any given composition entering into a reactor as multiplied by the fraction by weight of the uranium in the reactor, is known as total danger sum of the composition. This figure is a dimensionless constant like K and accordingly can be directly subtracted from K. The danger coefficients given are related to a neutron absorption value of unity for uranium.

As a specific example of the use of danger coefficients, if the materials of a system under consideration have 0.01 percent by weight of each of the elements Co, Ag and N with respect to the weight of the uranium in the system, then the total danger sum in K units for such an analysis would be:

$$0.0001 \times 17 + 0.0001 \times 18 + 0.0001 \times 4 = 0.0039.$$

This figure can then be subtracted from the K calculation for a particular geometry of theoretically pure materials to give the actual K constant for the materials used.

[2]@ In the original text the symbols T_i, T_u appear instead of σ_i, σ_u.

Table 10.1

Element	Danger Coefficient	Element	Danger Coefficient
Li	310	Ni	3
B	2150	Cu	1.8
C	0.012	Zn	0.61
N	4	Ga	~ 1
O	0.002	As	2
F	0.02	Se	6.3
Na	0.65	Br	2.5
Mg	0.48	Rh	50
Al	0.30	Ag	18
Si	0.26	Cd	870
P	0.3	In	54.2
S	0.46	Sn	0.18
Cl	31	Sb	1.6
K	2.1	I	1.6
Ca	0.37	Ba	0.30
Ti	3.8	Sm	~ 1430
V	4	Eu	435
Cr	2	Gd	~ 6320
Mn	7.5	Pb	0.03
Fe	1.5	Bi	0.0025
Co	17	Th	1.1

Therefore, a principal object of the present invention is to reduce the total danger sum of the impurities in a neutronic reactor by the novel method and means of eliminating atmospheric air therefrom, thereby eliminating a substantial quantity of nitrogen which comprises more than 70 percent of the atmospheric air. It may be noted in the above table 10.1 that nitrogen has a danger coefficient of 4.0, and it will, therefore, be readily understood that by eliminating atmospheric air from a reactor, the K thereof will be substantially increased.

Another object is to provide a novel method and means of varying the reproduction ratio of a neutronic reactor by varying the air content of the reactor.

Other objects and advantages will be apparent from the following description taken in conjunction with the accompanying drawing, in which:

Fig. 10.1 is a perspective view of a reactor incorporating a preferred embodiment of the present invention, parts being broken away and parts being shown in section for clarity;

Fig. 10.2 is an enlarged section partly in elevation, on the line **2-2** of Fig. 10.1, taken as indicated;

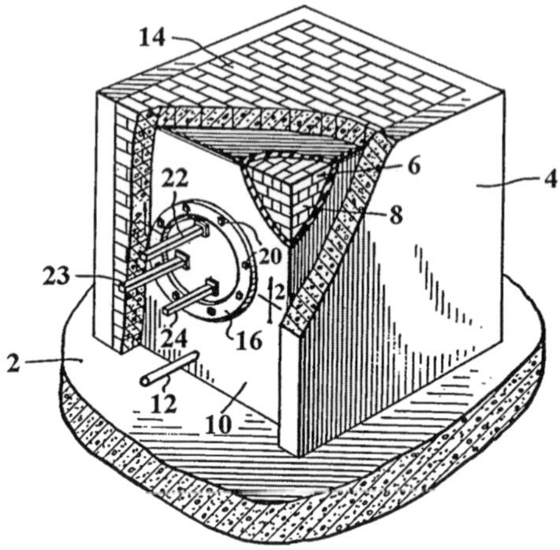

Fig. 10.1

Fig. 10.3 is an enlarged section partly in elevation, on the line **3-3** of Fig. 10.2, taken as indicated; and

Fig. 10.4 is a section on the line **4-4** of Fig. 10.3.

Describing the invention in detail, and referring first to Fig. 10.1, on a concrete foundation **2** is provided an upstanding, concrete vault or shield **4** receiving therewithin a neutronic reactor or pile, generally designated **6**, said reactor **6** being per se no part of the present invention, and being fully described and claimed in co-pending application, Serial No. 568,904, filed December 19, 1944, now Patent No. 2,708,656, dated May 17, 1955,[3] in the joint names of Enrico Fermi and Leo Szilard. The reactor **6** shown is constructed as a lattice of uranium-containing slugs or lumps in a moderating material, such as graphite, formed as blocks. Around the moderator is a reflector **8** formed of graphite blocks, or the like.

A fluid-tight casing or envelope **10** is disposed around the four sides, top, and bottom of the reflector **8** so that the entire reactor **6** is contained within the casing **10**, which is preferably formed of balloon cloth but may be constructed of any other suitable material, such as steel. In this connection, it may be noted that the casing **10** is preferably formed of material which is substantially neutron pervious or, in other words, material having a negligible neutron absorption, inasmuch as a neutron absorbent mate-

[3]@ See page 217.

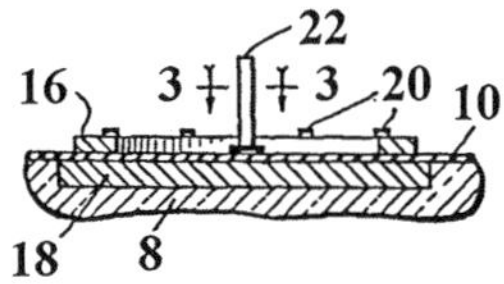

Fig. 10.2

rial, after operation of the pile, would become highly radioactive, thereby constituting a serious hazard to operating personnel. The casing **10** may enclose the vault **4** and be anchored to the foundation **2** if preferable. A suitable valve **12** is provided for evacuation of the casing **10**.

The casing **10** is clamped between a clamping ring **16** and a plate **18** (Fig. 10.2) by means of stud bolts **20**, which extend through the plate **18** and into the reflector **8**. The plate **18** is inset into the reflector **8** to provide an even surface for the casing **10**. Three rods **22**, **23** and **24** of neutron absorbent material, preferably cadmium, extend into slots within the reactor **6** through openings in the plate **18** and the casing **10**, each of said openings being sealed by a gland **26** (Figs. 10.3 and 10.4) clamped to the plate **18** by a rectangular frame **28**, which is secured to the plate **18** by stud bolts. A Wilson seal, which is well-known in the art, may be employed instead of the seal just described.

The rods **23** and **24** may function as safety rods and are connected to suitable actuating means (not shown) for urging them to their innermost positions within the reactor **6**, thereby reducing the reproduction ratio thereof to a value below unity, whenever the reactor attains a temperature of predetermined maximum value as the result of the neutron density within the reactor. The rod **22** may function as a control rod connected to a suitable actuating means (not shown) for moving said rod **22** inwardly and outwardly of the reactor **6**, thereby controlling the reproduction ratio

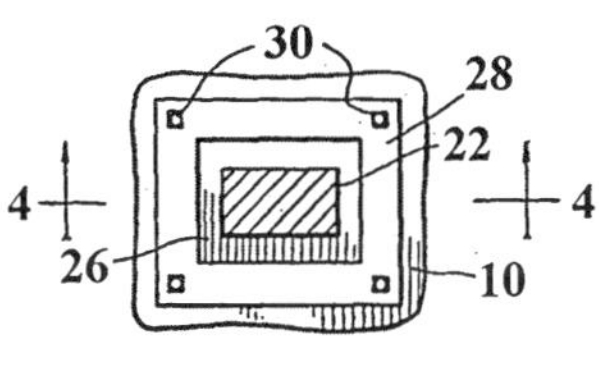

Fig. 10.3

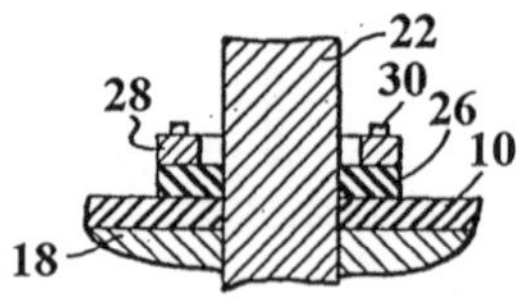

Fig. 10.4

thereof for the purpose of regulating the neutron density within the reactor **6**. Hence, the rods **22**, **23**, and **24** must be able to be moved into and out of the reactor **6**. Concrete blocks **14** are placed against the open wall and on top of the reactor **6**. The valve **12** includes a stem of sufficient length to extend through the blocks **14**. Those blocks **14** covering the ring **16** are recessed to receive the ring **16** and studs **20**.

In constructing the above-described system embodying the present invention, the blocks of the reactor **6** including those of the reflector **8** are piled or stacked within the vault **4** on the casing **10**. Certain of the blocks are, of course, slotted to provide horizontal passages through the reactor **6** for the reception of the rods **22**, **23**, and **24**. During construction of the reactor **6** within the casing **10**, the rods **22**, **23**, and **24** are inserted through provided slots in the casing **10** and into the respective slots in the reactor **6** as soon as that part of the reactor **6** containing the rod slots is completed. Each rod seal is completed so that the rods may be used as controls in the remaining construction of the reactor **6**.

After the reactor **6** is completed, the casing **10** is sewed or otherwise sealed. The concrete blocks **14** as placed as shown, completing the vault **4**. The casing **10** is evacuated through the valve **12**, thereby reducing the impurities in the reactor **6**, and improving the reproduction ratio thereof. Manifestly, the reproduction ratio of the reactor **6** may be varied by the present invention within the two extremes of an evacuated state and of a state of air saturation. Thus, the present invention is a delicate control within such limits.

It is to be noted that a simple form of reactor **6** is illustrated and described in this application. However, it will be readily understood by any one skilled in the art that the present invention can be utilized with other types of reactors.

It will be understood that, if desired, the reactor **6** may be built up to a point at which the reproduction ratio thereof is equivalent to unity or a value slightly lower than unity, with the control and safety rods withdrawn to their outermost positions, provided that the substantial removal of atmosphere from the reactor **6** is sufficient to raise said ratio to a value greater than unity. Then, by evacuating the casing **10** an operative system may be obtained. However, such a practice is generally impractical because of the close tolerances involved, and in most cases it is preferable to build the reactor **6** to a size at which its reproduction ratio is greater than unity, with the control rods removed, even though the chamber is not evacuated. By atmospheric air is meant the usual atmospheric gases containing oxygen, nitrogen, and commonly a small amount of argon.

The enormous forces brought into play by the evacuation of a large neutronic reactor system or pile present problems of design resulting in involved structural forms. Furthermore, it is difficult to make gas tight. It is better to displace the air by a non-reactive gas, that is non-reactive from both a chemical and nuclear standpoint. Suitable gases are helium and carbon dioxide. This also has the advantage of increasing K of the pile.

Other advantages of this invention include the removal from the reactor of argon which, if it remains, become radioactive, diffuses out of the reactor, and thus constitutes a health hazard to the operators. In addition, the removal of the oxygen in the air prevents oxidation of component parts of the reactor such as the uranium or container for the uranium.

While the theory of the nuclear chain fission reaction set forth herein is based on the best presently known experimental evidence, the present invention is not limited thereto, except by the scope of the claims, inasmuch as additional experimental data later discovered may modify the theory disclosed and other means and methods for practicing this invention within the scope of the claims may later be developed by those skilled in the art.

What is claimed is:

A neutronic chain reacting device comprising a rectangular shaped neutronic reactor containing uranium disposed within a graphite block moderator and surrounded by a graphite block reflector, said reactor having a plurality of parallel adjacent slots extending therein from one surface thereof, a fluid-tight casing disposed about the reactor having an aperture therein confronting each slot in the reactor, a flat recess in the reactor confronting the apertures in the casing, a plate disposed in the recess abutting the reactor and the casing, the plate having orifices therein aligned with the slots and apertures, a retaining ring disposed about the slots on the side of the casing opposite the plate and secured to the reactor to support the casing, a neutron absorbing control rod slidably disposed in each slot of the reactor and extending outwardly therefrom through the confronting orifice in the plate and aperture in the casing, a gland having an opening disposed about each control rod in contact with the surface of the casing opposite the plate, a frame disposed about each control rod in contact with the surface of the gland opposite the casing, a plurality of stud bolts extending through each frame and gland and anchored in the plate, and means for removing atmospheric air from the reactor.

REFERENCES CITED in the file of this Patent

UNITED STATES PATENTS

Re. 21,781	Toulmin	Apr. 22, 1941
1,306,568	Weintraub	June 10, 1919
2,127,193	Toulmin	Aug. 16, 1938
2,205,854	Kroll	June 25, 1940
2,708,656	Fermi et al.	May 17, 1955

FOREIGN PATENTS

114,150	Australia	May 2, 1940
114,151	Australia	May 3, 1940
233,011	Switzerland	Oct. 2, 1944
861,390	France	Oct. 28, 1940

OTHER REFERENCES

Smyth: "Atomic Energy for Military Purposes", pp. 177-181, 70, 82-87, August 1945. Copy may be purchased from Supt. of Documents, Washington 25, D.C..

The Merriam-Webster Pocket Dictionary, p. 123, Pocket Books, Inc., N. Y. (1947).

Chapter 11

Neutronic Reactor IV

Enrico Fermi, Santa Fe, N. Mex., and Walter H. Zinn, Chicago, Ill., assignors to the United States of America as represented by the United States Atomic Energy Commission

Application November 2, 1945, Serial No. 626,383

U.S. Patent No. 2,714,577 - Patented Aug. 2, 1955

This Patent describes a series of technical improvements of a chain reacting pile. Some attention is paid, for instance, to the shielding of the active part of the reactor, the design of the uranium-containing rods and the recombination of D_2 and O_2 in D_2O (heavy water is expensive).

Particularly interesting, from the scientific point of view, is the opportunity to have a well inside the active part of the reactor (which is the part most rich in neutrons and gamma rays), where objects to be bombarded with neutrons or γ rays may be placed, and from which collimated beams of such particles may be formed, to be used outside the reactor.

The description of the mentioned technical improvements is not reported in any other published paper (see, however, [Fermi (1943c)] for the radiation shield, [Anderson (1944)] for the dissociation of (light) water and [Fermi (1944c)] for the collimation of a neutron beam).

The present invention relates generally to neutronic reactors and, more particularly, to novel articles of manufacture used in and in combination with such reactors, and to the combination of such novel articles of manufacture with neutronic reactors.

In neutronic reactors a thermal neutron fissionable isotope such as U^{233},

U^{235}, or 94^{239} or mixtures thereof is subjected to fission by absorption of neutrons and a self-sustaining chain reaction is established by the neutrons evolved by the fission. In general, such reactors comprise bodies of compositions containing such fissionable material, for example, natural uranium, disposed in a neutron slowing material which slow the neutrons to thermal energies. Such a slowing material is termed a neutron moderator. Carbon, beryllium, and D_2O (heavy water) are typical moderator suitable for such use. Heat is evolved during the reaction which is removed by passage of a coolant through the reactor or in heat exchange relationship therewith. Specific details of the theory and essential characteristics of such reactors are set forth in copending application of Enrico Fermi and Leo Szilard, Serial No. 568,904, filed December 19, 1944.[1]

More specifically, an object of the present invention is to provide novel shielding means for the active portion of a neutronic reactor adapted to be used in combination therewith.

Another object is to provide in a neutronic reactor a novel cooled shield.

Another object is to provide a novel composite rod adapted to be used as part of the active portion of a neutronic reactor.

Another object is to provide a novel rod for use as part of the active portion of a neutronic reactor which is constructed with fissionable material in a portion thereof only.

Another object is to provide in a neutronic reactor novel means for introducing foreign subject matter into the active portion of the neutronic reactor for bombardment by neutrons.

Another object is to provide in a neutronic reactor a novel collimated beam for utilizing the active effects of the neutronic reactor upon objects exposed exteriorly of the reactor.

Other objects and advantages will become more apparent from a study of the following description taken in conjunction with the accompanying drawings, in which:

Fig. 11.1 is a central vertical cross-sectional view of a neutronic reactor embodying the present invention, being off-center as to the rod and well;

Fig. 11.2 is an enlarged cross-sectional view of a rod assembly designated in Fig. 11.1;

Fig. 11.3 is an enlarged cross-sectional view of a modification of the ball valve seal shown in Fig. 11.2; and

Fig. 11.4 is an enlarged sectional view of a supporting rod equipped

[1]@ The authors refer to E. Fermi and L. Szilard, *Method of operating a neutronic reactor*, U.S. Patent No. 2,708,656, May 17, 1955 (see page 217).

with an adapter for the drawing process.[2]

Referring to the drawing, numerical **1** denotes a neutronic reactor tank or container of cylindrical shape and of a material that is relatively non-corrodible at low temperatures and that is relatively non-absorbent with respect to neutrons, such as, for example, aluminum. A suitable size of such tank for a self-sustaining chain reaction is one that is 6 feet in diameter and 7 feet high, although other sizes may be used as well, particularly larger sizes. Suspended in tank **1** is a plurality of composite rods **2** of suitable configuration. Composite rods **2**, described in detail hereinafter, are immersed in a neutron moderator, such as, for example, a body of deuterium oxide **3** substantially filling the tank **1** except for a small space above the surface which space is filled with helium, described later. More specifically, sufficient deuterium oxide is introduced into tank **1** so as to obtain a volume of uranium and heavy water slightly over the critical size.

Immediately surrounding tank **1** is a neutron reflector **4** of graphite, having substantially a cup shape. Reflector **4** and tank **1**, together with its contents, is referred to as the neutronic reactor. The active portion of the reactor is the volume of moderator and wetted uranium. The thickness of the graphite reflector may be of the order of two or three feet or more, depending on the size of the active portion of the reactor and the degree of scattering required. The reflector **4** reflects into the active portion of the reactor some of the neutrons that might otherwise diffuse out of the system. Graphite is a good scattering material having a low neutron absorption characteristic. Such elements as lead or bismuth may also be used. By using this peripheral layer of scattering or reflecting material, the overall size of the active portion of the reactor may be made somewhat smaller than in a case where no scattering layer is employed, since neutron losses to the exterior are somewhat reduced. Or tank **1** and its contents may be built slightly below critical size so that the addition of reflector **4** with its neutron reflecting action will convert the reactor from one that is not self-sustaining to one that is self-sustaining.

A concrete shield **5** surrounds the graphite reflector **4** and serves to prevent gamma and other penetrating radiations from escaping to the outside of the structure. The carbon in reflector **4**, being a relatively light element, absorbs gamma rays only to a relatively small extent. The concrete shield

[2]@ In the original text, at this point the following sentence is present: "Fig. 5 is a sectional view taken along the line **5-5** of Fig. 3 in the direction indicated by arrows.". However, it seems quite evidently a typographical mistake, since no figure 5 (nor references to it in the subsequent text) is included in the present Patent.

may be of the order of 20 or 30 feet in thickness. The water of crystallization in the concrete absorbs escaping neutrons.

Generally speaking, the higher the atomic weight of an element, the better it serves as a shield to prevent escape of penetrating radiations, such as gamma radiations. Lead of a particular thickness is, therefore, an excellent shield for certain purposes while water of the same thickness is only fair, for example. However, thick water shields are convenient and satisfactory. Interposed between the concrete shield **5** and graphite reflector **4** is a cup-shaped cooled metal shield **7**, preferably of a lead-cadmium alloy, having cooling tubes passing therethrough, through which light water, that is, ordinary water, or other coolant may be circulated. Shield **7** is especially adapted not only to cool the neutronic reactor but to minimize the escape of gamma and other penetrating radiations. A suitable thickness may be 4 inches or more.

A cover plate **8**, for example, of stainless steel, is provided at the top of tank **1** and serves as a thermal shield as well as a support for control and safety rods pivotally supported thereunder, as is described later. Above the cover plate **8** there is a space **10**, the use of which is pointed out below. A shield **9** of lead blocks is disposed above the reflector **4** and the shield **7**. The space **10** is filled with lead, or the like. A biological shield **9a** having alternate layers of iron and Masonite for preventing escape of gamma rays and other penetrating radiations through the top of tank **1** to the outside caps the reactor. A well **9b** of any suitable diameter, for example 4 inches, extends through shield **9a**, space **10** and plate **8** into the tank **1**, preferably axially of the tank. Objects may be introduced from the exterior through the tube **9b** into the center of the tank **1** so that they may be bombarded by high intensity neutron radiations.

In order to extract some of the heat developed as the result of the chain reaction and to provide a thermal operating equilibrium, a shielded heat exchanger (not shown) is provided, and the heavy water moderator **3** is circulated through conduits to the heat exchanger to be cooled thereby. A minimum of about 5 tons of heavy water and about 2 tons of uranium are necessary in reactor tank **1** to make the neutronic reaction self-sustaining. About 6 tons of heavy water is provided in the entire system, including the heat exchanger and other external parts. Of course, if additional cooling tubes are introduced into tank **1** to more effectively cool the uranium rods and heavy water, the neutron multiplication ratio (reproduction ratio) will be diminished because such cooling tubes absorb an appreciable percentage of neutrons, thereby acting as impurities, thus necessitating the addition of more heavy water and uranium to provide a larger active portion so that the reactor may be self-sustaining.

A pan **15** of stainless steel or other suitable material is located at the bottom of reflector **4** for collecting any heavy water that may leak from tank **1**, inasmuch as heavy water, at present, is relatively scarce and expensive. The collected heavy water is drained as indicated by the arrow **A**.

Helium at substantially atmospheric pressure is introduced through pipes **16** into the top of reactor tank **1**, that is, above the level of the heavy water, and thence is circulated to the exterior of the tank **1** through pipes **17**. As the result of high neutron densities and heat developed during the operation of the chain reaction in tank **1** some of the deuterium oxide will decompose into D_2 and O_2. These uncombined gases will collect at the top of tank **1**. In order to remove such uncombined gases, recombine and recondense them without explosive effect, a gas circulating system of any well-known type (not shown) may be used for circulating the helium together with the uncombined gases into a recombiner of any well-known type (not shown) such as a hot grid or platinum-charcoal catalyst, or both, for effecting recombination of the D_2 and O_2 into heavy water, and for returning the condensed D_2O into tank **1**, or, in fact, to any other suitable place, for example, to a heavy water storage tank (not shown). Helium may be circulated, for example, at the rate of 3 cubic feet per minute. A suitable ratio of D_2 to helium may be 1 to 150 although other dilutions may be used instead.

Immediately below coverplate **8** there is provided a cadmium sheet **18** to act as a shield for minimizing the escape of slow neutrons. Immediately below cadmium sheet **18** there is provided an aluminum sheet **19** which is useful primarily to prevent electrolytic action between otherwise dissimilar metals (cadmium and aluminum) in the interior of tank **1**. Cadmium sheet **18** is sandwiched between coverplate **8** and aluminum sheet **19** by bolts and nuts (not shown) or other suitable fastening means. Shield **9a** is supported by two pairs of crossed I beams **20** and **21**. The lower I beams **21** have their extremities supported by the concrete shield **5** – in other words, the ends of these I beams extend beyond the periphery defined by the shield **7**, hence there are provided four supporting points for the lower pair of I beams **21**.

Referring to Fig. 11.2, there is shown one of the plurality of composite uranium containing rods **2** that are suspended in the deuterium oxide moderator **3** as shown in Fig. 11.1. Each composite rod **2** comprises a cylindrical rod of uranium or uranium containing material **22** that is screw-threaded

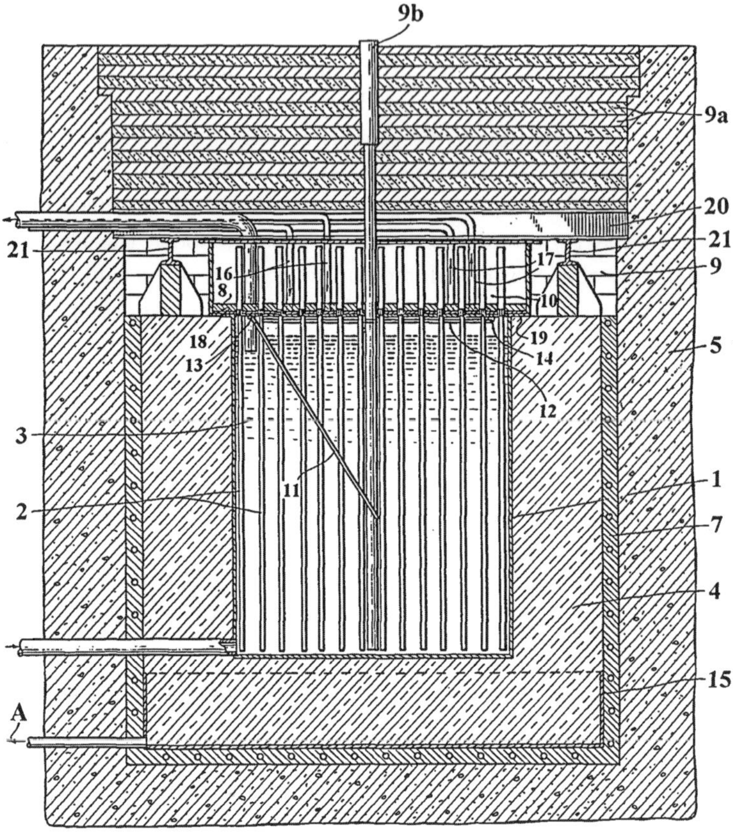

Fig. 11.1

into a supporting rod **23** of non-fissionable material such as, for example, aluminum. A thin tubing of aluminum **24** is drawn on to the outer surface of uranium rod **22** and supporting rod **23** by any well-known drawing process by screwing on an attachment **38** (see Fig. 11.4) that is subsequently removed. Thereafter, the joints formed at the top of supporting member **23** and at the bottom of rod **22** are welded so as to form an airtight seal in tubing **24**, thus protecting the uranium rod **22** from the effects of the surrounding medium. Narrow longitudinal connecting grooves **22a** and **23a** are provided in the connected elements **22** and **23**. An axial groove **25** and communicating radial groove **25a** are provided near the top of supporting

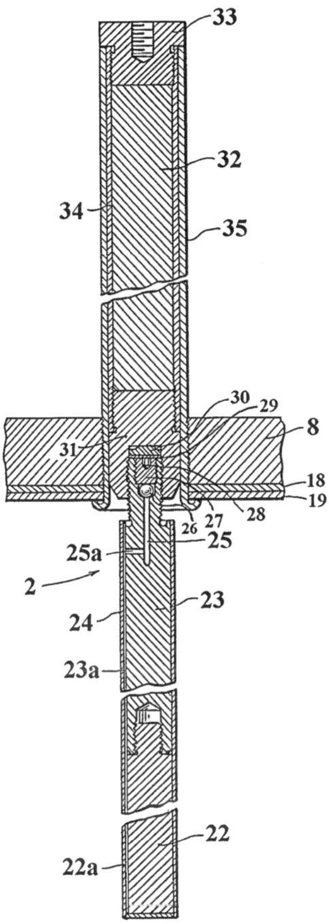

Fig. 11.2

rod **23** and communicate with the grooves **22a** and **23a** thereby making it possible to evacuate any air, and thereby to test for possible leaks in the welded joints in tubing **24** by noting the interior pressure after a predetermined evacuation. A ball valve **26** is provided to seal the groove **25** from the surrounding atmosphere. A set screw **27** having a hexagonal socket **28** is screwed downwardly of supporting rod **23** so as to firmly seat the ball valve **26**. A lead gasket **29** with a serrated surface is placed above set screw **27** and a steel disk **30** is placed on top of lead gasket **29** so as to complete an airtight seal between set screw **27** and supporting member **23** as shank

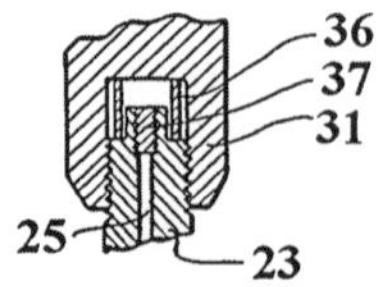

Fig. 11.3

31 and supporting member **23** are screw-threaded together. An aluminum sleeve **34** is screw-threadedly connected to shank **31**, the latter receiving screw-threadedly a member **33**. A lead rod **32** is disposed between the shank **31** and flanged member **33** as a supplemental biological shield. The flange of member **33** is supported on the upper end of a tube **35** which is threadedly supported by the plate **8**, the bottom edge of the tube **35** beng peened as shown. The aluminum sleeve **34** surrounds the lead rod **32** so as to prevent direct contact between the lead rod **32** and any medium surrounding the sleeve **34** when removed from tube **35**. As is clear from the above description, the entire composite rod **2** may be removed for replacement by sliding the rod **2** upwardly through the tube **35**. While sleeve **34** and tubing **24** have been described as being of aluminum, it will be apparent that other non-fissionable materials that are relatively non-corrosive with respect to heavy water vapor and that have a small absorption cross-section for neutrons may be used, such as, for example, beryllium, or even

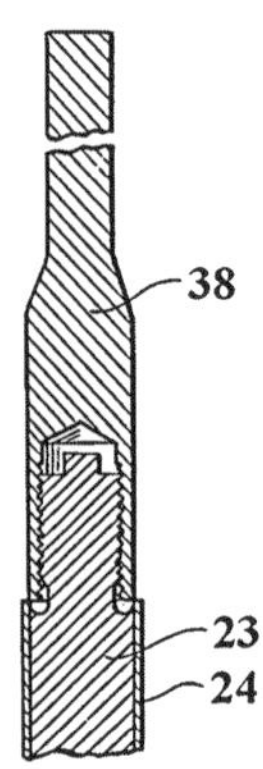

Fig. 11.4

stainless steel.

The heavy water level is located well above the top of uranium rods **22**, so that it wets a substantial portion of the aluminum covering of supporting rod **23**. The volume of heavy water above the top surface of uranium rods **22** will act as a neutron scattering or reflecting material rather than as an additional portion of the active uranium-heavy water volume of the neutronic reactor. By raising the level of the heavy water, greater neutron reflecting action may be obtained without adding to the active volume of the neutron reactor.

Fig. 11.3 shows an alternate form of seal which may be used in place of the ball valve **26** shown in Fig. 11.2. The seal comprises a stud **37** of aluminum, for example, that is screw-threaded into a projecting collar portion of supporting member **23**. A gasket or sealing ring **36** of lead, for example, is provided between elements **23** and **31** so that as the latter-mentioned elements are screw-threaded together, they compress gasket **36** thereby forming an airtight seal between supporting rod **23** and shank **31**.

The neutron chain reaction may be stopped merely by immersing into the heavy water one or more rods, such as rods **11** and **12** (Fig. 11.1) of neutron absorbing material, pivotally supported by the stainless steel plate **8**. A suitable neutron absorbing material is cadmium, for example. Rod **11**, for example, may be operated as a control or regulating rod, being intermittently immersed into and withdrawn out of the body of heavy water, as desired. Rod **12** (shown out of the heavy water) may be considered as a safety rod which normally is held out of contact with the body of heavy water in tank **1** and is immersed into the heavy water only for emergency purposes, that is, when control rod **11** per se, is insufficient for immediate stoppage of the chain reaction. Rod **11** may be pivotally mounted and rigidly secured to a shaft **13**, extending through the tank **1** through stuffing box types of bearings (not shown) and through reflector **4** and the concrete shield **5** to the exterior so that shaft **13** may be rotated from the exterior of the concrete shield either manually or automatically, as desired. Shaft **14** is rigidly connected to rod **12** and extends through the reactor and concrete wall **5** in a manner similar to shaft **13**. A method of automatic control is to make rotation of shaft **13** responsive to the neutron density at a peripheral portion of reflector **4**, for example, as indicated by ionization chambers having suitable amplifiers (not shown) so as to be effective to control the neutron density of the reactor and keep it substantially constant. Such means of automatic control, however, forms no part of the present invention.

In utilizing the output of the reactor, well **9b** plays an important role.

It extends through the center of the reactor where the highest neutron density exists, and intense neutron bombardment of material inserted into this well will take place, even at relatively low reactor powers. A sufficiently high reproduction ratio r is provided to supply the extra neutrons required. Furthermore, the well acts to collimate the fast neutrons released at the center of the reactor and a high density collimated beam of neutrons emerges through the open end of well **9b**, projected upwardly, this collimated beam having a far greater neutron density than any neutron produced at a given reactor power, can be utilized outside of the reactor for nuclear research in all of its aspects.

In addition, extremely high energy gamma rays are emitted during nuclear fission. These rays also escape through well **9b** to the exterior of the reactor and can there be used for taking radiographs through large castings, for example, with relatively short exposure, during operation of the neutronic reactor. The neutrons coming from the reactor can be screened out of the gamma ray beam by the use of relatively thin sheets of materials having high neutron absorption cross-sections without substantially reducing the gamma ray intensity. In addition, a bismuth filter has been found to effectively reduce the gamma rays, without substantial interference with the neutron beam. Thus well **9b** can be used either to produce a high intensity collimated neutron beam, or to produce a high energy beam of gamma rays, as desired, both for use either inside or outside the reactor.

While the theory of nuclear fission of uranium as set forth herein, is based on the best presently known experimental evidence, we do not wish to be bound thereby, since additional experimental evidence, later discovered, may modify the theory disclosed. Any such modification of theory however, will in no way affect the results to be obtained in the practice of the invention herein described and claimed.

It should be noted that the system described above is merely exemplary, and not limiting insofar as our invention is concerned, since it will be apparent that other similar systems will be suggested to those skilled in the art. Hence, our invention is limited only insofar as is set forth in the following claim.

What is claimed is:

In a neutronic reactor, a container, a cover therefor, a plurality of composite rods suspended from the cover so as to extend well into the tank, each rod being composed of an aluminum portion extending into the tank from the cover and a thermal-neutron-fissionable portion secured to a region of the aluminum portion spaced from the cover and extending from the aluminum portion in a direction away from the cover, and heavy water filling the container to a level above the juncture of the fissionable portion with the aluminum portion for each composite rod so as to cause the fissionable portion to be wholly immersed, variation in the level of the heavy water serving to vary the amount of heavy water above the fissionable portions and thus to vary the neutron-reflecting action of the heavy water.

REFERENCES CITED in the file of this Patent

UNITED STATES PATENTS

507,374	Lyte	Oct. 24, 1893
578,070	Woolf	Mar. 2, 1897
667,498	Chapman	Feb. 5, 1901
983,962	Werner	Feb. 14, 1911
1,970,804	Kerk	Aug. 21, 1934

FOREIGN PATENTS

114,150	Australia	May 2, 1940
861,390	France	Oct. 28, 1940
233,011	Switzerland	Oct. 2, 1944
233,278	Switzerland	Oct. 16, 1944

OTHER REFERENCES

Atomics, vol. 6, No. 6, November-December 1950, pages 10-15. Pub. bi-monthly by Technical Pub. Co., Chicago, Ill. (Copy in Patent Office Library).

Kelly et al., Phy. Rev. 73, 1135-9 (1948) (Copy in Patent Office Library).

Chapter 12

Method of Testing Thermal Neutron Fissionable Material for Purity

Enrico Fermi, Santa Fe, N. Mex., and Herbert L. Anderson, Hartford, Conn., assignors to the United States of America as represented by the United States Atomic Energy Commission

Filed Nov. 21, 1945, Ser. No. 630,123

U.S. Patent No. 2,969,307 - Patented Jan. 24, 1961

The main aim of this Patent is to outline a method for determining the "neutronic purity" (i.e., with respect to elements with an high cross section for neutron capture) of given materials to be used in a pile.

The "shotgun" test is conducted by placing an indium foil (as a neutron detector) near a neutron source, and measuring its induced radioactivity with a Geiger-Muller counter. The same measure is performed when a given quantity of boron (a standard neutron absorbing pellet) is placed near the detector foil and, subsequently, by replacing the boron with the material containing impurities. A direct comparison between the absorption caused by the unknown composition and the standard boron absorber gives the desired result for the sum of the danger coefficients of the impurities (in terms of boron equivalent).

Some theoretical developments show, as well, that the fractional absorption of the impurities with respect to uranium is approximately equal to the variation of the reproduction factor in the pile, induced by the presence of the impurities themselves.

Neutron absorption by impurities is considered in [Anderson (1947)] (published in 1947 but referring to work made in 1943-4), but the method adopted is completely different from that described in the present Patent, which is not reported in any other published paper.

This invention relates to a novel method of testing the neutronic purity of uranium or other material to be used in a neutronic reactor.

In neutronic reactors a neutron fissionable isotope such as U^{233}, U^{235}, or 94^{239} or mixtures thereof is subjected to fission by absorption of neutrons, and a self-sustaining chain reaction is established by the neutrons evolved by the fission. In general such reactors comprise bodies of compositions containing such fissionable material, for example, natural uranium, disposed in a material that slows the neutrons to thermal energies. Such a material is termed a neutron moderator. Carbon and D_2O (heavy water) are typical moderators suitable for such use. Heat is evolved during the reaction which is removed by passage of a coolant through the reactor in heat exchange relationship therewith. Specific details of the theory and essential characteristics of such reactors are set forth in copending application of Enrico Fermi and Leo Szilard, Serial No. 568,904, filed Dec. 19, 1944, now Patent 2,708,656, dated May 17, 1955.[1]

It has been found that certain impurities or neutron absorbers, which do not take part in the reaction involving fission of an isotope and formation of a fissionable isotope, exert a marked depressing action upon the neutron chain reaction. Unless the amount of such absorbers is controlled within proper limits, a self-sustaining reaction cannot be maintained. Where a material such as natural uranium is used to establish the reaction even very small amounts of certain impurities or absorbers will prevent maintenance of the reaction. Furthermore, they are specific to natural uranium and will vary with a change in the fissionable material used in the pile.

It is an object of the invention to provide a novel method of determining the neutronic purity of material, such as uranium, to be used in a neutronic reactor. Neutronic purity, as hereinafter more fully discussed, has no necessary connection with chemical purity, i.e., a composition having high neutronic purity is one which is substantially free from particular elements having relatively great neutron capture cross-sections.

An initial number of fast neutrons in the system, by going through the process of absorption and fission, produce in the next generation a number of neutrons generally different from the initial number. The ratio of the number produced after one generation to the initial number for a system of infinite size is called the reproduction or multiplication factor of the system and is denoted by the symbol K.

For any finite system, some neutrons will escape from the periphery of

[1]@ See page 217.

the system. Consequently a system of finite size may be said to have a K constant, even though the value thereof would only exist if the system as built were extended to infinity without change of geometry or materials. Thus when K is referred to herein as a constant of a system of practical size, it always refers to what would exist in the same type of system of infinite size. If K can be made sufficiently greater than unity to indicate a net gain in neutrons in the theoretical system of infinite size, and then an actual system is built to be sufficiently large so that this gain is not entirely lost by leakage from the exterior surface of the system, then a self-sustaining chain reacting system of finite and practical size can be built to produce power and related by-products by nuclear fission of natural uranium. The neutron reproduction ratio in a system of finite size, therefore, differs from K by the external leakage factor, and by a factor due to the neutron absorption by localized neutron absorber, and the reproduction ratio must still be sufficiently greater than unity to permit the neutron density to rise exponentially with time in the system as built.

During the interchange of neutrons in a system of finite size, comprising bodies of any size disposed in a neutron moderator, neutrons may be lost to the chain reaction in four ways:

(1) By absorption or capture in the uranium content of the bodies without producing fission,

(2) By absorption or capture in the moderator material itself,

(3) By leakage out of the system through the periphery thereof, and

(4) By absorption or capture by the impurities present in both the uranium bodies and the moderator.

However, even when the first three above-mentioned losses are reduced to a practical minimum, no self-sustaining chain reaction can be obtained in any system unless impurities in the material used for the reaction are reduced to such an extent that the loss by parasitic capture by such impurities will not prevent the reaction from becoming self-sustaining. Impurities present in both the uranium and the moderator consequently constitute a very important neutron loss factor in the chain. The effectiveness of various elements as neutron absorbers varies tremendously.

Certain elements such as boron, cadmium, samarium, gadolinium, and some others, if present even in a few parts per million, could very likely prevent a self-sustaining chain reaction from taking place. It is highly important, therefore, to remove as far as possible all impurities capturing neutrons to the detriment of the chain reaction from both the slowing material and the uranium. If these impurities are present in too great quantity, the self-sustaining chain reaction cannot be attained. The permissible amounts of impurities will vary for each specific geometry, depending upon such considerations as the form in which the uranium is used – that is, whether natural or enriched, whether as metal or oxide. The type of slowing-down material used, for example, whether deuterium, graphite, or beryllium, also influences the effect of impurities, as do the weight ratios between the uranium and the slowing-down material.

The effect of impurities on the optimum reproduction factor K may be conveniently evaluated by means of certain constants known as "danger coefficients" which are assigned to the various elements. There danger coefficients for the impurities are each multiplied by the fraction by weight of the corresponding impurity, with respect to the weight of uranium in the system; and the total sum of these coefficients gives a value known as the total danger sum. This total danger sum is subtracted from the reproduction factor K as calculated for pure materials under consideration.

The danger coefficients are defined in terms of the ratio of the weight of impurity per unit mass of uranium and are based on the cross-section for absorption of neutrons of the various elements. These values may be obtained from published literature on the subject such as Physikalische Zeitschrift, vol. 43, p. 440 (1942), or Zeitschrift für Physik, vol. 121, p. 201 (1943), and the danger coefficient computed by the formula[2]

$$\frac{\sigma_i}{\sigma_u}\frac{A_u}{A_i},$$

wherein σ_i represents the cross section for the impurity and σ_u the cross-section for the uranium, A_i the atomic weight of the impurity and A_u the atomic weight for uranium. In general, whether the impurities are in the moderator or in the uranium, they may be computed as their fraction by weight of the uranium in the system and the result will be sufficiently accurate for design purposes.

The approximate danger coefficients for some elements in a natural uranium system are given in table 12.1, wherein the elements are listed in order of their atomic number.

The sum of the danger coefficients of the impurities in any given composition entering into a reactor as multiplied by the fraction by weight of the uranium in the reactor is known as total danger sum of the composition. This figure is a dimensionless constant like K and accordingly can be directly subtracted from K. It will be noted that the danger coefficients given are related to the neutron danger coefficient of unity for uranium.

[2]@ In the original text the symbols T_i, T_u appear instead of σ_i, σ_u.

Table 12.1

Element	Danger Coefficient	Element	Danger Coefficient
Li	310	Ni	3
B	2150	Cu	1.8
C	0.012	Zn	0.61
N	4.0	Ga	$\sim$ 1
O	0.002	As	2
F	0.02	Se	6.3
Ne	3	Br	2.5
Na	0.65	Ag	18
Mg	0.48	Cd	870
Al	0.30	In	54.2
Si	0.26	Sn	0.18
P	0.3	Sb	1.6
S	0.46	I	1.6
Cl	31	Ba	0.3
A	0.8	Sm	$\sim$1430
K	2.1	Eu	$\sim$435
Ca	0.37	Gd	$\sim$6320
Ti	3.8	Pb	0.03
V	4	Bi	0.0025
Cr	2	Th	1.1
Mn	7.5	Te	0.5
Fe	1.5	Hg	82
Co	17	Au	16
		Px	$\sim$2.5
		Ir	$\sim$70
		Ta	2.7
		W	$\sim$18
		Dy	$\sim$200

As a specific example, if the materials of a system under consideration have 0.01 percent by weight of each of the elements N, Co, and Ag with respect to the weight of the uranium in the system, the total danger sum in K units for such an analysis would be:

$$0.0001 \times 4 + 0.0001 \times 17 + 0.0001 \times 18 = 0.0039.$$

This figure can then be subtracted from the K calculated for theoretically pure materials to give the actual K factor for the geometry and material used. This would be a rather unimportant reduction in the reproduction factor K unless the reproduction factor K for a given geometry and materials without considering impurities, is very nearly unity. If, on the other hand, the impurities in the uranium are Li, Co, and Ag in the same percentage, the total danger sum would be:

$$0.0310 + 0.0017 + 0.0018 = 0.0345 \text{ reduction in } K \text{ due to impurities.}$$

The maximum possible K factors for natural uranium neutronic reaction systems when optimum geometry is used and where the materials used are assumed to be theoretically pure, have been calculated as shown in the following table 12.2:

Table 12.2

Materials	K for Pure Materials
U metal–graphite moderator	1.1
U oxide–graphite moderator	1.07
U metal–beryllium moderator	1.18
U metal–beryllium oxide moderator	1.09
U metal–heavy water moderator	1.3

It can be readily seen from the above tabulation 12.2 that the total danger sum for impurities in both the uranium and moderator must be less than 0.3 in order that the K factor remain equal to or greater than unity with a moderator having a minimum absorption of neutrons, a deuterium moderator, less than 0.09 for a beryllium-oxide moderator, and less than 0.1 for a graphite moderator. Even when neutron losses due to moderator and leakage have been substantially eliminated, the danger sum of the impurities or absorbers in the natural uranium can not exceed about 0.3.

The present invention comprehends a very practical method of testing material, such as uranium, said method being known as the "shotgun" test. In this test a thin neutron detector, e.g., a piece of indium foil, placed near a neutron source inside a block of paraffin is made radioactive by an amount proportional to the density of thermal neutrons at the foil and the time of exposure. This radioactivity can be measured. A neutron absorbing pellet such as one of a standard amount of boron, when placed close to the detector foil, decreases the thermal neutron density in the neighborhood of the foil and lowers the induced radioactivity of the detector foil. By replacing the standard boron absorbing pellet with a pellet containing the impurities from a known amount of uranium composition to be tested, and again measuring the radioactivity of the neutron detector, a direct comparison is obtained between the absorption caused by the unknown composition and the standard boron absorber. From this comparison the danger sum of the impurities in the uranium composition can be calculated in terms of boron equivalent. From the danger coefficient of boron the K reduction can be calculated.

One suitable means for carrying out the "shotgun" test is described be-

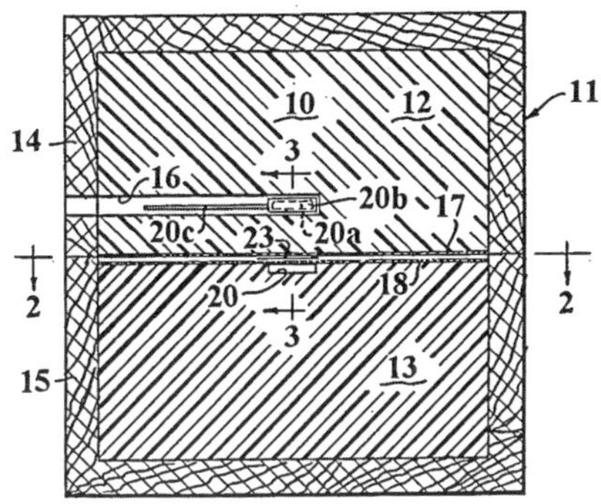

Fig. 12.1

low, although such means is not part of the present invention. It is also illustrated in the accompanying drawing, wherein Figure 12.1 is a cross-sectional view of the apparatus during measurement of the radioactivity on a indium foil without a pellet; Figure 12.2 is a top view of the lower box along line **2-2** of Figure 12.1; Figure 12.3 is an enlarged view of the apparatus along line **3-3** of Figure 12.1 after insertion of the pellet of uranium impurities; and Figure 12.4 is a cross-sectional view of the apparatus like that of Figure 12.3 after insertion of a standard boron pellet.

A block **10** of paraffin 30 cm on each side is molded in a wooden box **11**. The box and the paraffin box are cut in halves **12**, **13** and **14**, **15**, respectively, horizontally so that each half of the paraffin remains in its own portion of the box. A hole **16** is formed through one side of the upper box half and into the paraffin 3 cm above and parallel to the lower surface of the paraffin; this hole **16** extends into the center of the paraffin. The adjacent paraffin surfaces of the block halves are covered, the upper with onion skin paper **17**, the lower with cellophane **18** linked at the back with Scotch tape **19** (Figure 12.2) to permit lifting for access to the paraffin surface. The upper half of the box is supported so that it can be readily moved up and down, and the lower half of the box is supported so that it can slide in and out like a drawer when the upper half is up. A small square hole **20** is cut in the center of the upper surface of the lower block of paraffin a little more than $\frac{3}{16}$ of an inch deep.

It is known that 3 cm of paraffin will slow down the neutrons evolved from a radium-beryllium source to slow or thermal speeds. A radium-

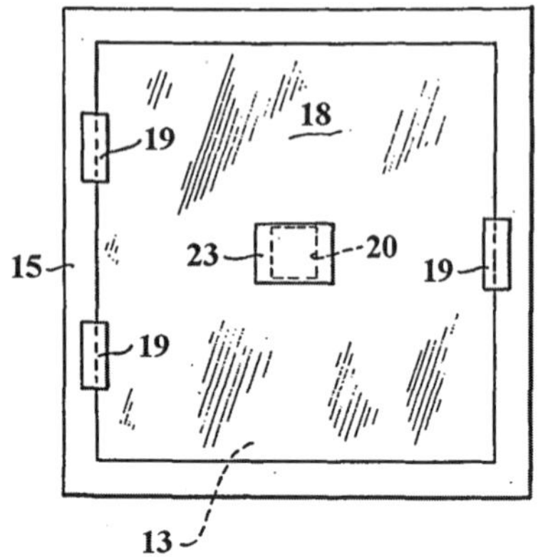

Fig. 12.2

beryllium source **20a** consisting of 200.5 mg of radium and 0.6 g of beryllium emitting about 2×10^6 neutrons per second may be used as the source. This source is contained in a brass capsule **20b** at the end of a long handle **20c**.

In order to make a measurement, the composition to be measured is formed into a pellet **21** of predetermined size and shape by hydraulic pressure. This pellet **21** is placed in an aluminum container **22** which just fits the square hole **20** in the upper surface of the lower half **15** of the paraffin block **13** and a piece of indium foil **23** is placed carefully over the aluminum container **22** on the surface of the lower paraffin block **13**. The lower block **13** is then put in place and the upper block **12** is lowered on top of it. The radium-beryllium source **20a** is then inserted all the way into the hole **16** in the upper block, and exact notations are made of the time. The irradiation is allowed to continue for a set time, usually 15 minutes. It is important that the time be measured accurately and preferably should be the same for all comparable tests.

Preferably, exactly 15 minutes after starting the irradiation, the source is removed. The upper paraffin block **12** is lifted, the lower paraffin block **13** is slid out, and the indium foil **23** is removed. The foil is then wrapped around the tube of a Geiger-Müller counter (not shown), and preferably exactly 3 minutes after the end of the irradiation counting is begun and continued for exactly 21 minutes.

The counting is preferably started exactly 3 minutes after the end of the irradiation in order to allow the short-lived radiation of the indium, which

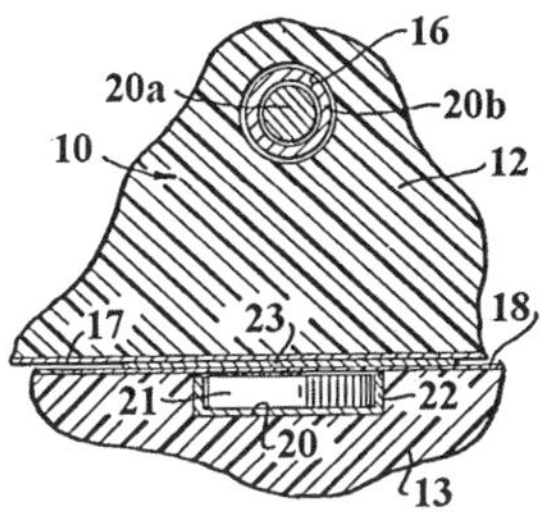

Fig. 12.3

has a half-life of 13 seconds, to substantially disappear, although mathematical correction can be made, if the time element of either irradiation or counting is charted.

It is important that every precaution be taken to avoid contamination of the foil **23** of Geiger-Müller counter with UX or other foreign substances. Every precaution must be taken to obtain strict cleanliness and to repeat each test in exactly the same way, so as to make all measurements comparable. The details of technique necessary to obtain these results will be obvious to those skilled in the art.

In order to use the "shotgun" test so as to determine the neutronic purity of uranium to be used in a neutronic reactor, a practical procedure is to take a representative sample of material having a 10 kg uranium content after large scale purification as by the methods disclosed in the parent case. The sample is submitted to an additional extremely exhaustive purification

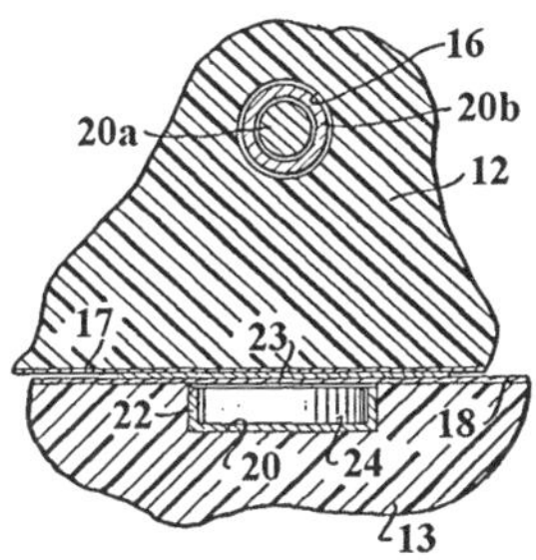

Fig. 12.4

on a laboratory scale. In this way, practically all of the impurities left in the uranium composition to be tested can be removed and the impurities are removed, recovered, and formed into pellet **21** (Figure 12.3). The sample pellet is then compared with the absorption of the standard boron pellet **24** (Figure 12.4) in the apparatus described.

The results of the "shotgun" test are usually reported as percent absorption which is equal to

$$\frac{\text{Absorption of impurities expressed in equivalent milligrams of boron}}{\text{Absorption of 10 kilograms of uranium expressed in equivalent milligrams of boron}}.$$

The absorption of 10 kg of uranium in terms of boron can be readily calculated from the danger coefficients given above as equal to 4560 mg of boron.

As shown in the derivation given below, the percent absorption from the "shotgun" test is approximately the same term as the K factor for a neutronic reactor. It has been found that this approximation is very close and is sufficiently accurate for a practical and periodic check on uranium compositions produced on a large scale for use in an operating neutronic reactor.

In a neutronic reactor:

$$K = f\,n\,E\,P,$$

where

f = thermal utilization, or the fraction of neutrons absorbed in the uranium;
P = probability of a fast neutron escaping resonance capture;
E = No. of fast neutrons produced by fast fission per fast neutron produced by thermal fission;
n = No. of fast neutrons produced by thermal fission per thermal neutron absorbed.

$$\frac{\Delta K}{K} = \frac{n\,E\,P\,\Delta f}{n\,E\,P\,f} = \frac{\Delta f}{f},$$

where Δ indicates a small increment in the associated term.

$$\frac{\Delta K}{K} = \frac{f_1 - f_2}{f_1},$$

where

$$f_1 = \frac{\text{neutrons absorbed in U}}{\text{neutrons absorbed in U and moderator}},$$

$$f_1 = \frac{N_0\,\sigma_{a\,0}}{N_0\,\sigma_{a\,0} + N_m\,\sigma_{a\,m}};$$

$$f_2 = \frac{\text{neutrons absorbed in U}}{\text{neutrons absorbed in U, moderator and impurities}},$$

$$f_2 = \frac{N_0\,\sigma_{a\,0}}{N_0\,\sigma_{a\,0} + N_m\,\sigma_{a\,m} + N_i\,\sigma_{a\,i}},$$

where

N_0, N_m, N_i = average number of atoms per cm^3 for the uranium, the moderator, and the impurities in the uranium, respectively, and

$\sigma_{a\,0}$, $\sigma_{a\,m}$, $\sigma_{a\,i}$ = absorption cross sections for the uranium, the moderator, and the impurities in the uranium (considered as one impurity).

Substituting

$$\frac{\Delta K}{K} = \frac{\dfrac{N_0\,\sigma_{a\,0}}{N_0\,\sigma_{a\,0} + N_m\,\sigma_{a\,m}} - \dfrac{N_0\,\sigma_{a\,0}}{N_0\,\sigma_{a\,0} + N_m\,\sigma_{a\,m} + N_i\,\sigma_{a\,i}}}{\dfrac{N_0\,\sigma_{a\,0}}{N_0\,\sigma_{a\,0} + N_m\,\sigma_{a\,m}}}$$

$$= \frac{N_i\,\sigma_{a\,i}}{N_0\,\sigma_{a\,0} + N_m\,\sigma_{a\,m} + N_i\,\sigma_{a\,i}}.$$

Fractional absorption (shotgun test)

$$= \frac{\text{absorption of impurities}}{\text{absorption of uranium}} = \frac{n\,v\,N_i\,\sigma_{a\,i}}{n\,v\,N_0\,\sigma_{a\,0}},$$

where

n = No. of neutrons per cm^3
v = velocity (cm/s).

$$\text{Fractional absorption} = \frac{N_i\,\sigma_{a\,i}}{N_0\,\sigma_{a\,0}}.$$

As a close approximation, the absorption in the moderator and in impurities may be ignored as compared to the absorption in the uranium, i.e., $N_m\,\sigma_{a\,m}$ and $N_i\,\sigma_{a\,i}$ may be considered negligible.

Then

$$\frac{\Delta K}{K} = \frac{N_i\,\sigma_{a\,i}}{N_0\,\sigma_{a\,0}}.$$

But

$$\text{Fractional absorption (shotgun test)} = \frac{N_i\,\sigma_{a\,i}}{N_0\,\sigma_{a\,0}}.$$

Therefore, as a close approximation, fractional absorption from shotgun test $= \Delta K$, provided K is approximately unity.

While the theory of nuclear reactions set forth herein is based on the best presently known experimental evidence, the invention is not limited thereto, as additional experimental data later discovered may modify the theory disclosed.

Obviously, many modifications may be made in the specific embodiments disclosed without departing from the intended scope of the invention.

What is claimed is:

A method of testing the neutronic purity of a thermal neutron fissionable material sample comprising the steps of producing a neutron flux field, inserting a standard neutron absorber of known characteristics and amount in said field, extracting substantially all of the chemical impurities from said sample, replacing said standard absorber with all of the impurities from the sample of the fissionable material, and producing successive indications of the neutron intensity by bombardment of a neutron detector foil placed at a region of said field with both the absorber and the impurity sample removed, with the absorber present and the impurity sample removed, and with the absorber removed and the impurity sample present, whereby the neutronic purity of the sample may be calculated from said indications.

REFERENCES CITED in the file of this Patent

UNITED STATES PATENTS

1,471,514 Elliot .... Oct. 23, 1923
2,288,717 Kallmann et al. .... July 7, 1942
2,288,718 Kallmann et al. .... July 7, 1942

OTHER REFERENCES

Burger et al.: "Determination of Hydrogen Content by Slow-Neutron Scattering", AECD-2319, U.S.A.E.C., 9 pages, date of manuscript, October 21, 1943; date declassified, September 22, 1948. (Copy in Patent Office Library).

Booth et al.: Phy. Rev. 57, 748 (1940). (Copy in Patent Office Library).

Phy. Rev. 55, 511-512 (1939). (Copy in Patent Office Library).

Kennedy et al.: "Component Analysis of Small Uranium Samples", MDDC-973, U.S.A.E.C., date of manuscript, March 26, 1943, date declassified, May 23, 1947.

Naturwissenschaften, vol. 27, 1939, pages 402-410. (Copy in Division 56 of the Patent Office).

Physical Review, March 15, 1940, page 546. (Copy in Patent Office Library).

Chapter 13

Method of Sustaining a Neutronic Chain Reacting System

Enrico Fermi, Santa Fe, N. Mex., and Miles C. Leverett, Oak Ridge, Tenn., assignors to the United States of America as represented by the United States Atomic Energy Commission

Application November 28, 1945, Serial No. 631,406

1 Claim. (Cl. 204 – 154.2)

U.S. Patent No. 2,813,070 - Patented Nov. 12, 1957

This Patent gives a general discussion of a reactor with variable critical dimensions. The pile considered is an uranium-graphite one, cooled by air and with control rods of cadmium or boron (the uranium rods are placed in aluminium jackets).

Of particular interest is the discussion of the variation of the reproduction factor K due to long and short term effects. Long term effects are, for instance, the increase of K due to the production of plutonium and its decrease due to the production of fission impurities. Instead, among the short term effects considered are the production of xenon, which absorbs neutrons, and the effect of retarded neutrons.

It is also of some relevance the pointing out that a moderator with a thickness of 1-2 feet around the uranium in the reactor acts as a reflecting screen for neutrons, with the same efficiency of an infinite thickness one. From this it follows that by using a moderator of 10 feet, for instance, the uranium content of the pile may be increased, with no relevant consequence on the efficacy of the screen.

A peculiar curiosity is the suggestion that the presence of nitrogen (as an impurity) in the reactor, which may change due to changes in the atmospheric pressure, could be used to obtain a sensitive barometer.

Some partial results may be found already in other Patents and/or in several papers of Volume II of [Fermi (1962)] (in particular, the realization of xenon poisoning is narrated on pages 428-429 of this reference). No published reference article behind the present Patent exists.

The present invention relates to devices of primary use for the production of neutrons by virtue of a self-sustaining chain reaction through fission of uranium or other fissionable isotopes with slow neutrons, known as neutronic reactors.

Natural uranium may be used in the reaction. Said natural uranium contains the isotopes 92^{238} and 92^{235} in the ratio of approximately 139 to 1. Hereinafter in the specification and the claim the term uranium is to be understood as referring to uranium and its chemical compositions of normal isotopic content unless otherwise indicated by the context.

In a self-sustaining chain reaction with slow neutrons, 92^{238} is converted by neutron capture to the isotope 92^{239}. The latter is converted by beta decay to 93^{239} and this 93^{239} in turn is converted by beta decay to the transuranic element 94^{239}. By slow neutron capture, 92^{235} on the other hand undergoes nuclear fission to release energy appearing as heat, gamma and beta radiation, together with the formation of fission fragments appearing as radioactive isotopes of elements of lower mass numbers and with the release of secondary neutrons.

The secondary neutrons thus produced by the fissioning of the 92^{235} nuclei have a high average energy, and must be slowed to thermal energies by passing said neutrons through a material in which the neutrons are slowed by collisions. Such a material is known as a moderator.

Among the basic parts of a neutronic reactor system may be listed the following.

(1) A fissionable material such as uranium is properly dispersed in a moderator such as carbon or deuterium oxide.

(2) In most cases, a neutron reflecting or scattering layer is provided around the active portion for returning at least a portion of escaping neutrons into the active portion. The reflecting layer can be an extension of the moderator beyond the active portion.

(3) In all but the lowest power output reactors, coolant channels or pipes are passed through, or in heat exchange relationship to the active portion, in order that a stable temperature can be maintained in the reactor.

(4) In most cases, loading and unloading mechanisms are provided to charge and remove uranium from the active portion.

(5) A shield is almost invariably provided around the reactor to minimize escape of radiations biologically harmful to operating personnel. Such shields may comprise layers of iron and hydrogenous material closely surrounding the reactor, these layers in turn being surrounded with from 5 to 20 feet of concrete. Loading and unloading devices are usually operated

from outside this massive shield. The size and weight of the shields make them permanent, and therefore it is very difficult to change the volume inside of the shields once they are constructed.

(6) A control rod of neutron absorbing material insertable into the reactor for the purpose of maintaining the neutron reproduction in an average state of balance is provided.

In the discussion to follow, the term neutronic reactor is used to mean the active portion and the reflector if used, and the term neutronic reactor system is used to mean the entire system including shields, charging and cooling systems, etc.

In considering the requirements for an operating neutronic reactor, the ratio of secondary neutrons produced by the fissions to the original number of primary neutrons initiating the fissions in a chain reacting system of infinite size using specific materials is called the reproduction or multiplication factor of the system and is denoted by the symbol K. If K can be made sufficiently greater than unity to create a net gain in neutrons, and the system made sufficiently large so that this gain is not entirely lost by leakage from the exterior surface of the system, then a self-sustaining chain reacting system can be built to produce power by nuclear fission of natural uranium. The neutron reproduction ratio (r) in a system of finite size differs from K by the leakage factor and operating poisoning factor, and it, too, must be greater than unity to permit the neutron density to rise exponentially. Such rise will continue if not stabilized at a desired density corresponding to a desired power output.

During the interchange of neutrons in a system comprising bodies of uranium of any size dispersed in a neutron moderator, neutrons may be lost in four ways: by absorption in the uranium metal or compound, by absorption in the moderator, by absorption in impurities present in the system or added as a result of operation, and by leakage from the system.

Natural uranium, particularly by reason of its U^{238} content, has an especially strong absorbing power for neutrons when they have been slowed down to moderate energies, these energies being termed resonance energies. The absorption of neutrons in uranium at these energies is termed the uranium resonance absorption or capture. It is caused by the isotope U^{238} and does not result in fission but creates the relatively stable nucleus 94^{239}. It is not to be confused with absorption or capture of neutrons by impurities, referred to later. Neutron resonance absorption in uranium may take place either on the surface of the uranium bodies, in which case the absorption is known as surface resonance absorption, or it may take place further in the interior of the uranium body in which case the absorption is known as volume resonance absorption. Volume resonance absorption is due to the fact that some neutrons make collisions inside the uranium body and may thus arrive at resonance energies therein. After successfully reaching thermal velocities, a large percent of the neutrons are also subject to capture by U^{238} without fission to produce 94^{239}.

It is possible by proper physical arrangement of the materials in the moderator to control the amount of uranium resonance absorption. By the use of light elements such as graphite and beryllium for example, fewer collisions are required to slow the neutrons to thermal energies, thus decreasing the probability of a neutron being at a resonance energy as it enters a uranium atom. During the slowing process, however, neutrons are diffusing through the slowing medium over random paths and distances so that the uranium is not only exposed to thermal neutrons but also to neutrons of energies varying between the energy of fission and thermal energy. Neutrons at uranium resonance energies will, if they enter uranium at these energies, be absorbed on the surface of a uranium body whatever its size, giving rise to surface absorption. Any substantial change of overall surface of the same amount of uranium will change surface absorption. Thus the volume ratio of moderator to uranium will control surface resonance absorption losses of neutrons in the uranium. The uranium is placed in the system in the form of spaced uranium masses or bodies of substantial size, either of metal, oxide, carbide, or combinations thereof. The uranium bodies can be in the form of layers, rods or cylinders, cubes or spheres, or approximate shapes, dispersed throughout the graphite, preferably in some geometric pattern.

The term geometry is used to mean any pattern or arrangement wherein the uranium bodies are distributed in the moderator with at least a roughly uniform spacing and are roughly uniform in size and shape, or are systematic in variation of size, shape or spacing to produce a volume pattern conforming to a generally symmetric system. If the pattern is a repeating or rather exactly regular one, the structure may be conveniently described as a lattice. The uranium bodies can be in the form of layers, rods, or cylinders, cubes or spheres, or approximate shapes, dispersed throughout the moderator. Optimum conditions are obtained with natural uranium by using metal spheres.

The resonance losses in uranium constitute one of the critical factors in coordinating the total losses permissible in a neutronic reactor.

The thermal neutrons are also subject to capture by the moderating material. While carbon and beryllium have very small capture cross-sections for thermal neutrons, an appreciable fractions of thermal neutrons (about 10 percent of the neutrons present in the system under best conditions with graphite) is lost by capture in the slowing material during diffusion therethrough. This means that when volume ratios are changed, the absorption in the moderator will also change.

In addition to the above-mentioned losses, that are inherently a part of the nuclear chain reaction process, impurities present in both the moderator and the uranium add a very important neutron loss factor in the chain. Such impurities may be originally present, or be formed during operation. The effectiveness of various elements as neutron absorbers varies tremendously. Certain elements such as boron, cadmium, samarium, gadolinium, and some others, if present even in a few parts per million, could very likely prevent a self-sustaining chain reaction from taking place. If impurities, solid, liquid or gaseous, and in elemental or combined form, are present in too great quantity, in the uranium bodies or the slowing material or in, or by absorption from, the free spaces of the system, the self-sustaining chain reaction cannot be attained. The amounts of impurities that may be permitted in a system, vary with a number of factors, such as the specific geometry of the system, and the form in which the uranium is used – that is, whether natural or containing more than the natural amount of a fissionable isotope, whether as metal or oxide – and also factors such as the weight ratios between the uranium and the slowing down material, and the type of moderating material used – for example, whether deuterium, graphite or beryllium. Although all of these considerations influence the actual permissible amount of each impurity material, it has fortunately been found that in general the effect of any given impurity or impurities can be correlated directly with the weight of the impurity present and with the K factor of the system, so that knowing the K factor for a given geometry and composition, the permissible amounts of particular impurities can be readily computed without taking individual account of the specific considerations named above. Different impurities are found to affect the reaction to widely different extents; for example, relatively considerable quantities of elements such as hydrogen may be present, and, as previously suggested, the uranium may be in the form of oxide, such as UO_2 or U_3O_8, or carbide, although the metal is preferred. Nitrogen may be present to some extent, and its effect on the chain reaction is such that the neutron reproduction ratio of the system may be changed by changes in atmospheric pressure. This effect may be eliminated by enclosing or evacuating the system if desired, or may be utilized by determining changes in a particular system in the reproduction ratio as changes occur in the atmospheric pressure. A sensitive barometer is thus obtained. In general, the inclusion of combined nitrogen is to be avoided.

The effect of impurities on the optimum reproduction factor K may be conveniently evaluated to a good approximation, simply by means of certain constants known as "danger coefficients" which are assigned to the various elements. These danger coefficients for the impurities are each multiplied by the percent by weight of the corresponding impurity, and the total sum of these products gives a value known as the total danger sum. This total danger sum is subtracted from the reproduction factor K as calculated for pure materials and for the specific geometry under consideration.

The danger coefficients are defined in terms of the ratio of the weight of impurity per unit mass of uranium and are based on the cross section for absorption of thermal neutrons of the various elements. These values may be obtained from physics textbooks on the subject and the danger coefficient computed by the formula[1]

$$\frac{\sigma_i}{\sigma_u}\frac{A_u}{A_i},$$

wherein σ_i represents the cross sections for the impurity and σ_u the cross-section for the uranium, A_i the atomic weight of the impurity and A_u the atomic weight for uranium. If the impurities are in the carbon, they are computed as their percent of the weight of the uranium in the system. Thus the effect of all impurities originally in the system as placed in operation can be evaluated in terms of the K factor. This fact is used to ensure control of the reactor, by deliberately inserting neutron absorbing materials to a greater or less extent in the reactor when built to an increase in neutron density, with subsequent stabilization of the neutron density at a desired value, as will be later explained.

The size of the system for proper operation will vary depending upon the K factor of the system, and upon other things such as, for example, the type of moderator and power of operation. If the reproduction factor K is greater than unity, the number of neutrons present will increase exponentially, provided the structure is made sufficiently large. If, on the contrary, the structure is small, with a large surface-to-volume ratio, there will be a rate of loss of neutrons from the structure by leakage through the outer surfaces, which may overbalance the rate of neutron production inside the structure so that a chain reaction will not be self-sustaining. For each value

[1]@ In the original text the symbols T_i, T_u appear instead of σ_i, σ_u.

of the reproduction factor K greater than unity, there is thus a minimum overall size of a given active structure known as the critical size, above which the rate of loss of neutrons by diffusion to the walls of the structure and leakage away from the structure is less than the rate of production of neutrons within the system, thus making the chain reaction self-sustaining. The rate of diffusion of neutrons away from a large structure in which they are being created through the exterior surface thereof may be treated by mathematical analysis when the value of K and certain other constants are known.

In the case of a spherical structure employing uranium bodies embedded in graphite in the geometries disclosed herein and without an external reflector the following formula gives the critical overall radius when K is known (critical sphere of radius R in feet)

$$K - 1 = \frac{C}{R^2},$$

where C is a constant that varies slightly with geometry and for normal uranium-graphite lattices may have a value close to 7.4.

For a rectangular parallelepiped structure rather than spherical, the critical size can be computed from the formula

$$K - 1 = C\left(\frac{1}{a^2} + \frac{1}{b^2} + \frac{1}{c^2}\right),$$

where a, b, and c are the lengths of the sides in feet. The critical size for a cylindrical structure is given by the formula, irrespective of the shape of the uranium bodies (cylinder height h in feet, radius R in feet)

$$K - 1 = C\left(\frac{1}{h^2} + \frac{0.59}{R^2}\right).$$

However, when critical size is attained by definition, no rise in neutron density can be expected. It is therefore necessary to increase the size of the structure beyond the critical size but not to the extent that the period for doubling of the neutron density is too short, as will be explained later. A desirable reproduction ratio (r) for an operating structure at a stabilized power with all control absorbers removed is about 1.005. This will permit the neutron density to rise in a reasonable time and yet permit control by insertion of sufficient neutron absorbers to bring the effective reproduction ratio to unity for continued operation at some predetermined neutron density. The size at which the reproduction ratio of 1.005 can be obtained may be computed from modifications of the above formulae for critical size when corrected for power of operation. For example, for spherical active structure the formula $K - r = 7.4/R^2$ may be used to find R when K is known and r is somewhat over unity. The same formula will, of course, give r for given structures for which K and R are known.

It can be seen from the above discussion that if, after a neutronic reactor has been built to a fixed operating size, the K factor should change for any reason, critical size would change. The reproduction ratio r would then change in accordance with the difference between critical size and operating size. If K increases, critical size decreases, and r decreases. If K decreases critical size increases, r decreases, and if K decreases sufficiently, critical size may coincide with or become less than operating size and the reactor will become inoperative unless increased to a new operating size where r is again greater than unity.

During the lifetime of operation of a neutronic reactor, K may change for several reasons which may be grouped under long term effects and short term effects.

During the neutronic reaction fission products, including such extremely absorptive elements as samarium and gadolinium, for example, may form from the fission of the uranium isotopes U^{235} and 94^{239}. Samarium, as well as some other elements formed in this manner, has very high neutron absorption characteristics and may increase the total parasitic neutron absorption in the reactor, with consequent decrease in the reproduction factor. However, the absorption of neutrons by the uranium isotope U^{238} leads to the production of 94^{239}. Plutonium is fissionable to an ever greater degree than the isotope U^{235} so that formation of plutonium tends to increase the reproduction factor K, notwithstanding decrease in the U^{235} content of the uranium in the reactor due to destruction by fission. The increase in K over a long period of production of 94^{239} may run as high as between 1 and 2 percent.

Still another factor tends to increase the reproduction factor in an operating reactor. Certain neutron absorbing impurities are present in graphite moderators, for example, sufficient to reduce K by from 1 to $1\frac{1}{2}$ percent. A substantial portion of this reduction is due to boron inherently present in the graphite. This boron is gradually transmuted, probably to lithium, during the operation of the reactor, and as lithium has a very much less neutron capture probability than boron, K will increase.

The long term effects can be summarized by stating that K will probably initially increase due to production of 94 and then decrease as the fission products build up.

In addition to the long term changes in K to be expected, there are shorter and even more significant changes in K that take place in reactors operating continuously at high neutron densities, as for example, U-graphite reactors operating at from 5,000 to 500,000 kilowatts power output.

The short term effect of xenon 135 formed as a radioactive decay element from iodine produced in the reactor during operation, becomes significant after 4 to 5 hours of continuous operation and changes K by an amount equal to the equilibrium effect of the amount of xenon 135 produced and destroyed at the power of operation. Xenon 135 decays to barium, but is changed by neutron absorption to xenon 136. Xenon 135 has an extremely high neutron capture cross-section of about $2,500,000 \times 10^{-24}$ cm^2 whereas the capture cross-sections of iodine, barium and xenon 136 are relatively small. Thus while the xenon 135 is present, due to the operation of the reactor, K is reduced by an amount related to the continuous power output in a graphite moderated reactor as follows:

$$\begin{aligned} \infty \text{ kW} &= 0.03K \\ 500,000 \text{ kW} &= 0.02K \\ 100,000 \text{ kW} &= 0.013K \\ 10,000 \text{ kW} &= 0.0012K \end{aligned}$$

The end result of this xenon 135 effect is that a reactor having an actual size capable of supporting a chain reaction at low, intermittent powers, may not be large enough to support a chain reaction at higher continuous power outputs. In consequence, the active portion of the reactor must be made large enough in the first place, if continuous high power operation is to be attained.

While it is, of course, apparent that a decrease in K might cause a system fixed in size to become inoperative it is not so apparent that an increase in K would be undesirable. However an increase in K is undesirable for the reason that if K increases, r will also increase and the reactor may become dangerous to operate, if r reaches or exceeds 1.01.

In order that the significance of a reproduction ratio of 1.01 be more fully understood, the mechanism of fission will be discussed further. Not all of the fast neutrons originating in the uranium body, as the result of fission, leave the uranium immediately. With natural uranium about one percent are "delayed neutrons". These delayed fast neutrons may appear at any time up to several minutes after the fission has occurred. Half these neutrons are emitted within six seconds and 0.9 within 45 seconds. The mean time of delayed emission is about 5 seconds. The neutron reproduction cycle is completed by 99 percent of the neutrons in about 0.0015 seconds, but if the reactor is operating with a reproduction ratio near unity, the extra 1 percent may make all the difference between an increase or a decrease in the activity of the reactor. The fact that the last neutron in the cycle is held back, as it were, imparts a slowness of response to the pile that would not be present if the fission neutrons were all emitted instantaneously.

For cases in which the reproduction ratio (r) differs from unity by appreciably less than 1 percent, the rise of neutron density, or more specifically the value N to which the number of neutrons has risen from an original value N_0, after a lapse of time of t seconds during and before which the pile has operated at a fixed value of r (N_0 being the number of neutrons at the beginning of t, i.e., after disappearance of transient effects due to any preceding change in r) is given by

$$N = N_0 \, e^{w\,t},$$

where

$$w = \frac{r-1}{\alpha-(r-1)} \cdot \frac{1}{T}.$$

In this formula α is the fraction of the neutrons that are delayed, i.e., $\alpha = 0.0067$, and T is the mean time of delayed emission of the delayed neutrons = 5 seconds. The above formula is only approximate because it uses an average delay time.

As an example, suppose as a result of moving the control rod, r becomes 1.001, and assume that the system has settled down to a steady exponential rise in neutron density. Then

$$w = \frac{0.001}{0.0067-0.001} \cdot \frac{1}{5} = \frac{1}{28.5}, \tag{13.1}$$

that is, $N/N_0 = 2.72$ in 28.5 seconds. Hence doubling of the neutron density occurs about every 20 seconds and continues indefinitely. The above formula thus indicates the rate of rise for relatively low values of r and shows how the reduction of the rate of the delayed neutron effect is particularly significant in the stated lower range of r values. Strictly speaking, the given equation holds only for the steady state, i.e., where r has been held constant for some time; an additional transient term must be included to obtain an accurate representation of the neutron density during the first few seconds after a sudden change of r.

If r were made exactly 1.01, a more detailed theory shows that the neutron density would be more than tripled each second. However, if the reproduction ratio r is several percent greater than unity, so that the one percent delayed neutrons are unimportant compared with $r-1$, the density increases at a much more rapid rate as given approximately by $r^{t/l}$ where l is 0.0015 seconds, the normal time to complete a cycle. Thus if r were to be made 1.04, the neutron density would increase in 1.5 seconds by a factor of approximately 10^{17} over its original level. However, if r were 1.02 and 1.03, the factor by which the neutron density would be multiplied each second, would be 1100 and 700,000 respectively. It is thus apparent that too high a reproduction ratio in a practical system leads to the necessity of inserting what may be considered as an excessive amount of controlling absorbers to reduce the effective reproduction ratio to unity. An exceedingly dangerous condition could exist if by accident these absorbers were suddenly completely removed, as the time required for reinserting the absorbing material might be too long to prevent destruction of the system. As the same eventual density can be obtained with a reproduction ratio only slightly over unity, as with a higher ratio, only at a slower rate, the lower reproduction ratios which exceed unity by not substantially more than 0.01, an amount equal to the percentage of the neutrons formed which are "delayed neutrons" are preferred in practice in the interest of safety.

It is therefore desirable to provide a reactor in which compensation for variation in the reproduction factor may be made when necessary.

As massive iron and concrete shields, which may be of the order of 20 feet in thickness, customarily surround an operating reactor, any requirement that a reactor be increased in size beyond a size limited by the shielding structures could not be met because of the immobility of said shields. Furthermore, if a decrease in size is to be made, the entire system must be such as to be fully as efficiently operated after decrease as before.

It is the main object of the present invention to provide a novel neutronic reactor that may be initially constructed in such a manner as to allow for either an increase or decrease in the size of the active portion.

Broadly stated, compensation for changes in the K factor with a consequent requirement for change in operating size can be accomplished in accordance with the present invention by initially providing a moderator mass greater in size than would be required for a uranium-moderator structure of a given initial K factor, and then initially loading uranium into that mass only in an amount and with such distribution to provide a reactor of a predetermined operating reproduction ratio r. Then, if K should change, the amount of uranium in the moderator mass can be increased or decreased to provide continued operation with the predetermined r.

The amount of uranium loaded into a moderator mass can be changed because any portion of the moderator extending beyond the uranium containing mass acts as a reflector, returning by the scattering action of the moderator nuclei a large number of the escaping neutrons. The use of a reflecting layer of moderator 2 feet thick, for example, around a neutronic reactor can reduce the critical size of the reactor by several percent of the radius thereof.

It is not the reduction in size of the reactor that is important in this instance. The main consideration is that a reflecting layer one to two feet thick around a reactor is nearly as efficient a reflector as a reflecting layer of infinite thickness. As a close approximation, reflecting layers 2 to 10 feet thick, for example, may be considered equivalent. In consequence, when the moderator mass is larger than that portion of the moderator containing uranium, the active portion always has an efficient reflector, providing only that a moderator layer of sufficient thickness remains around the uranium bearing portion to give proper reflecting efficiency.

As a result, a mass of moderator can be charged or loaded with varying amounts of uranium to form a portion of operating size, as needed, an all cases being surrounded by a reflecting layer of substantially equal reflecting efficiency.

It is therefore another object of the present invention to provide a neutronic reactor system of fixed dimensions wherein the size of the active portion supporting the chain reaction can be changed to suit circumstances without loss of efficiency.

There are other advantages of a system wherein the active portion can be changed to suit conditions. As noted above, the active portions of neutronic reactors can be in the overall shape of a cube, sphere or cylinder. Further, it has been stated that the uranium can be placed in the reactor in the form of spheres, rods or layers. Thus, it is desirable that provision be made that a moderator mass be provided such that active portions can be loaded, not only to varying overall sizes, but to varying overall shapes, with varying types of uranium and with varying geometries of the uranium with respect to the moderator in the portion loaded.

It follows that another object of the present invention is to provide a neutronic reactor system in which uranium or other fissionable material can be loaded with varying shapes of the loaded portions, and with varying sizes and shapes of the uranium bodies themselves, as desired.

Other objects and advantages of the present invention may be more clearly understood by references to the following description and the attached drawings which illustrate, as an example, one form the invention may take. This example is not to be taken as limiting, as other forms within the scope of the appended claim will be readily apparent to those skilled in the art.

In the drawings:

Fig. 13.1 is a longitudinal view partly in section and partly in elevation of an air cooled neutronic reactor system illustrating the present invention;

Fig. 13.2 is a cross-sectional view, partly in elevation, taken as indicated by the line **2–2** in Fig. 13.1;

Fig. 13.3 is a plan view of the system shown in Figs. 13.1 and 13.2;

Fig. 13.4 is a cross-sectional view taken as indicated by the line **6–6** in Fig. 13.8;

Figs. 13.5 and 13.6 are diagrammatic cross-sections taken as indicated by lines **14–14** and **15–15** in Figs. 13.14 and 13.15, respectively, section lines being omitted for clarity;

Fig. 13.7 is a longitudinal sectional view partly in elevation of a jacketed slug;

Fig. 13.8 is a longitudinal sectional view, partly in elevation, of a horizontal channel during a loading and unloading operation;

Fig. 13.9 is a longitudinal sectional view, partly in elevation, of one form of loading device;

Fig. 13.10 is a view partly in section and partly in elevation taken as indicated by the line **8–8** in Fig. 13.9;

Fig. 13.11 is a top plan view of the loading device shown in Figs. 13.9 and 13.10;

Fig. 13.12 is an end view of the loading device shown in Figs. 13.9, 13.10 and 13.11;

Fig. 13.13 is a view partly in section and partly in elevation of a friction drive used in the loading device;

Figs. 13.14 and 13.15 are diagrammatic longitudinal sections showing loadings of different sizes, the section lines being left off for purposes of clarity; and

Fig. 13.16 is a diagrammatic cross-section of a cubically loaded neutronic reactor, section lines being omitted for clarity.

Referring to the drawings, the present invention is illustrated by reference to an air cooled graphite-uranium reactor, sometimes known as a pile. The operation and construction of such reactors are more fully described in Fermi et al. Patent 2,708,656, dated May 17, 1955.

Such a reactor broadly comprises a mass of graphite blocks closely piled or stacked into a cube **10** as shown in Figs. 13.1 and 13.2. This graphite

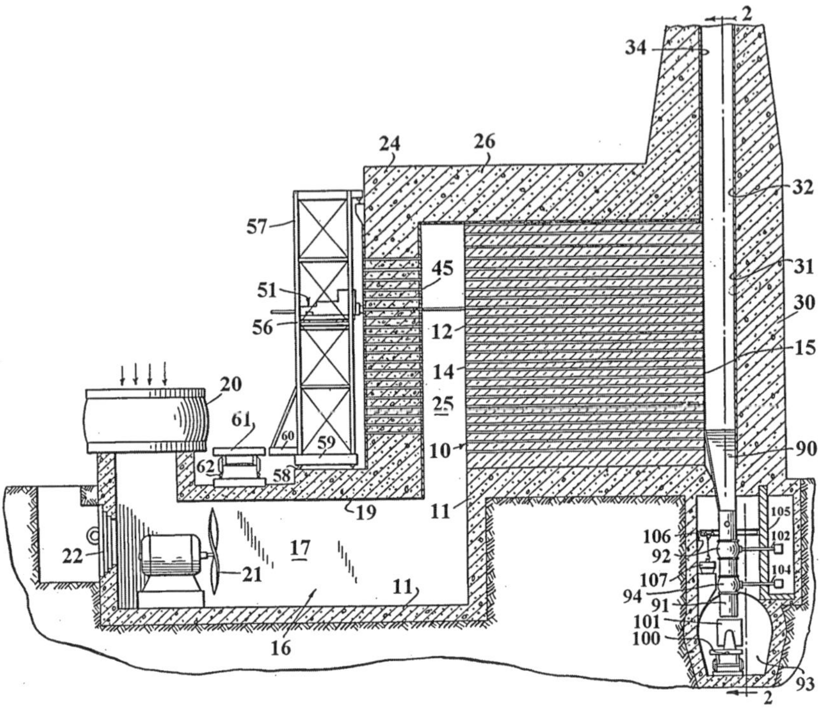

Fig. 13.1

cube may be, for example, 24-26 feet on a side and rest on a concrete foundation **11**. The graphite cube **10** is pierced with horizontal air channels **12**, of square cross-section, with one of the diagonals vertical. The channels may be readily made by grooving adjacent blocks (Fig. 13.4). The channels are 0.175 inches on a side and extend completely through the reactor, from an inlet face **14** to an outlet face **15**. Two thousand channels may be provided, for example, and as will later be brought out, any unused channels can be plugged. Only a few of the channels are shown in the drawings for the sake of clarity.

Adjacent the inlet face **14** of the cube, the foundation is continued downwardly to form the floor of an inlet air duct **16** extending outwardly. The inlet air duct **16** is completed by concrete side walls **17** and top **19**.

At some distance away from the graphite cube **10** the inlet duct is turned upwardly to terminate in an air filter **20**, relatively close to the surface of

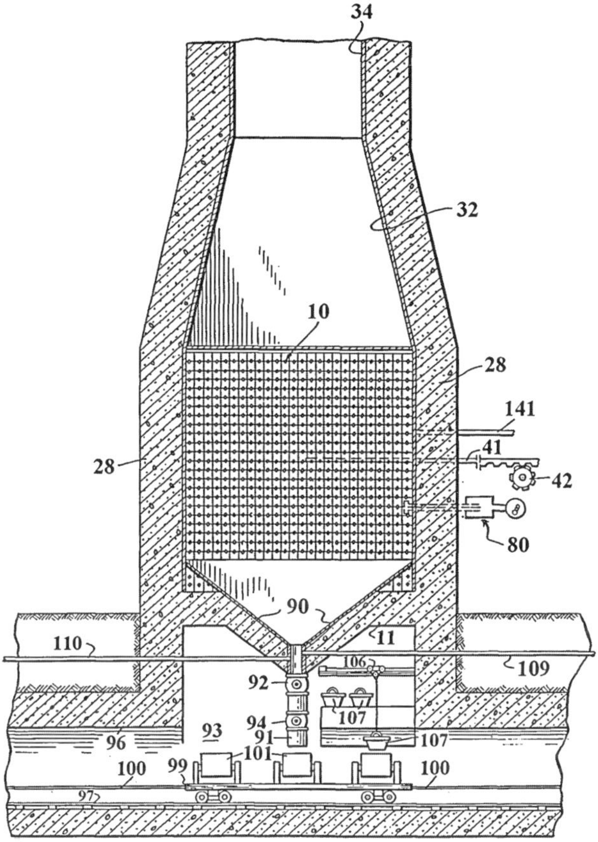

Fig. 13.2

the ground. A fan or blower **21**, here illustrated as electrically driven, is installed on the floor of the inlet duct just below the air filter, access to the fan being conveniently obtained through duct door **22**, behind the fan.

The concrete top **19** of the inlet air duct is continued upwardly as an inlet end shield **24**, positioned parallel to but spaced away from inlet face **14** of the cube **10** to form an inlet chamber **25** communicating with the air channels **12**.

Above the inlet chamber **25** and the cube **10** the concrete is continued horizontally to form a top shield **26**, and side shields **28** are built up from

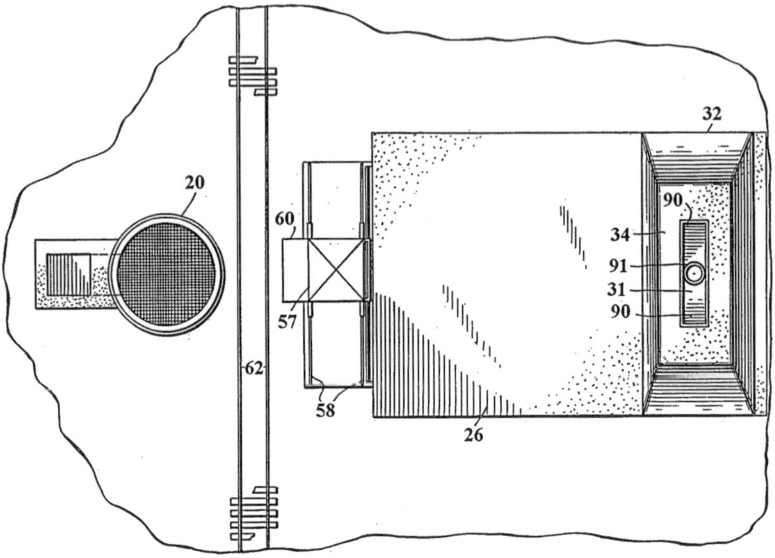

Fig. 13.3

the foundation **11** to enclose cube **10**. Shields **26** and **28** closely approach the top and side faces of the cube, to minimize air flow around the outside of the cube. A small amount of air circulation, however, may be desirable over the top and side faces to cool these faces.

At the outlet face **15**, an outlet end shield **30** of concrete is provided. End shield **30** is parallel to and spaced from the outlet face **15** of the graphite cube to form an outlet chamber **31** communicating above with the base **32** of a stack **34** projecting upwardly and formed as a continuation of the concrete top side and outlet end shields. Thus the cube **10** is completely enclosed by concrete shields, with a duct system operating by virtue of pressure provided by fan **21** to conduct air from close to ground level through channels **12** into the stack and then into the atmosphere well above ground level at the top of the stack. The concrete shields may be from ten to twenty feet thick in accordance with the maximum desired operating power of the reactor and serve as shields to reduce escape of neutrons and gamma radiation.

As a neutronic reaction will take place when uranium bodies are prop-

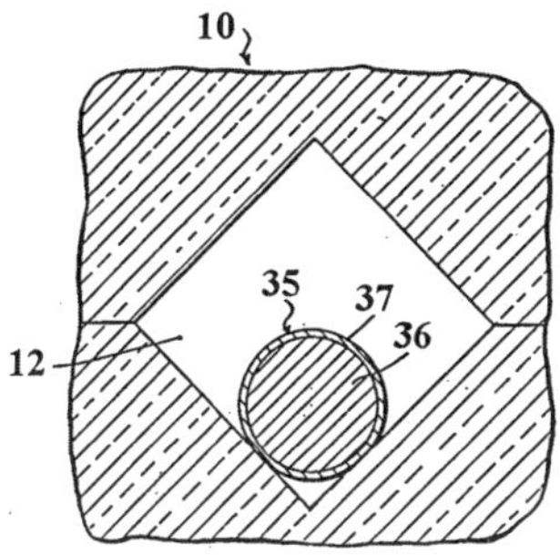

Fig. 13.4

erly spaced in a moderator mass of a certain finite size, the above described device can be made chain reacting by placing uranium bodies in the horizontal channels in such a manner and in such an amount that a neutron reproduction ratio of slightly over unity is obtained, exclusive of all neutron losses within the reactor and from the exterior of the reactor. This reproduction ratio may be defined as the ratio of the number of neutrons gained by fission to the total number of neutrons lost by absorption in the uranium, absorption in the moderator, absorption by impurities in the reactor or added as a result of operation, and by leakage from the reactor for a reactor of finite size.

Using the graphite mass as the moderator to slow fast neutrons to energies where they again are able to create fission in 92^{235}, the device will have a reproduction ratio of unity when approximately 700 of the channels **12** in the graphite cube are each loaded with 68 aluminum jacketed uranium slugs **35** lying end to end, with a channel spacing of 7 inches measured center to center, and with the loaded channels roughly forming a cylinder (Figs. 13.5 and 13.6). Both graphite and uranium should be of highest possible purity. The diameter of the cylinder will be about 18 feet.

However, more than a unity reproduction ratio is required, as when the reproduction ratio is exactly unity no rise in neutron density will occur. Under such conditions the device will not develop high neutron densities or power in the form of heat. By loading additional channels, i.e., making the active portion greater than critical size, however, the reproduction ratio within the reactor can be brought above unity in order that a rise in density can occur. Then this excess neutron reproduction can be absorbed by

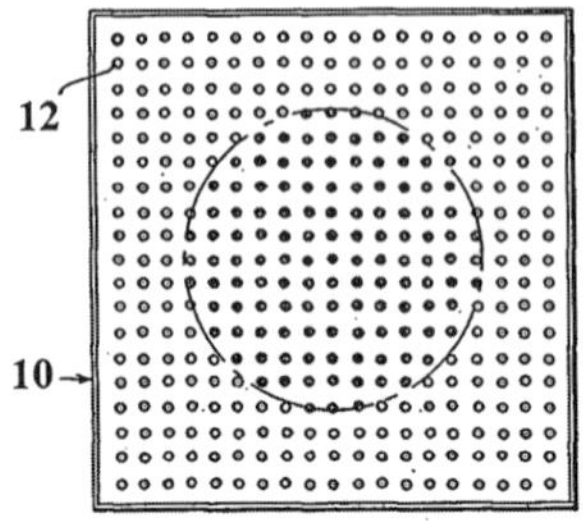

Fig. 13.5

neutron absorbing materials deliberately inserted into the reactor in order to hold the reproduction ratio at an average value of unity after a desired power output has been obtained, as a result of the initial rise in density.

Consequently, in accordance with the amount of excess reproduction ratio desired, about 1,000 channels may be loaded with uranium slugs. The diameter of the cylinder will now be about 21 feet, leaving a minimum of 3 to 4 feet of graphite around the cylinder except at the ends. Most of the channels not loaded with uranium may be closed by inserting plugs, preferably of graphite, in such channels in order to conserve air. Some of the channels, however, in the peripheral portions of the cube may be left open for cooling of the graphite in those portions.

One preferred form of slug construction is shown in Fig. 13.7. Each uranium metal slug is 1.1 inches in diameter and 4 inches long covered with an aluminum jacket approximately 20 mm thick in good heat conductive

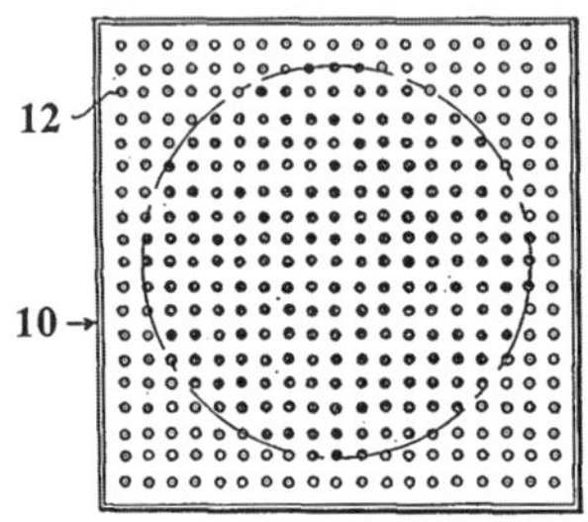

Fig. 13.6

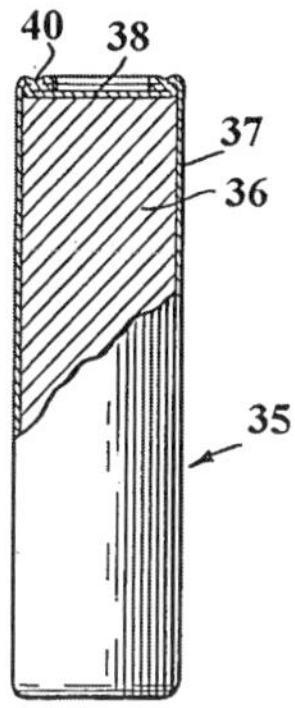

Fig. 13.7

relation to the uranium. The slugs weight about $2\frac{1}{2}$ pounds each.

In forming the slugs **35**, the uranium portion **36** is machined to size, cleaned in trisodium phosphate and then washed in water. Aluminum or other non-fissionable metal jacket cans **37** are provided having an inside diameter somewhat larger than the uranium portion. This can **37**, open at one end only, is slipped over the uranium after being cleaned in benzine and hot water. The can **37** with the uranium inside is then passed through a sizing die of 1.134 inches diameter. This die, being of smaller diameter than the 1.1 inch uranium portion plus the two 20 mm walls, draws the can in tight thermal contact with the uranium.

A cup-shaped cap **38** is then placed base down inside the projecting portion of the can **37** and is seam welded to the can. The projecting portion is then cut off above the seam weld **40** and the remaining projecting portion including the weld, spun over the adjacent end of the slug. Thus each jacket completely encloses and seals the uranium, preventing air from corroding the uranium and, as will be pointed out later, also preventing fission fragments created by nuclear fission at the surface of the uranium from entering the air stream.

The channels are loaded with uranium until the reproduction ratio, with neutron absorbers removed, is about 1.005 to 1.006. This means that for every two hundred neutrons starting in each neutron generation about two hundred and one neutrons are produced in the reactor over and above all losses. Under these conditions and taking into account the fact that about one percent of the neutrons of fission are delayed in their emission for a mean time of about 5 seconds, the neutron density of the reactor will double every 8 to 15 seconds. With some part of the neutron absorbers inserted but with the insertion of less than the amount of neutron absorbers required to make the reproduction ratio unity, the rise is slower. When the neutron absorbers are almost but not entirely inserted the doubling of the neutron density may take several hours. Then when a desired density has been reached, the reproduction ratio can be reduced to unity so that the desired density is continuously maintained by the introduction of neutron absorbing material into the reactor.

The neutron absorbing material is introduced into the reactor by means of a control rod **41** as shown in Fig. 13.2. This control rod extends into the graphite cube, sliding in a channel therein and is operated from outside of side shield **28** as by rack and pinion **42**. The rod is made from, or incorporates therein, an efficient neutron absorber, such as for example, cadmium or boron. A sheet of cadmium riveted to a steel strip forms a satisfactory control rod. As the depth of insertion of the rod determines the amount of neutron absorbing material inside the reactor, the critical position of the rod is where the rate of neutron absorption by the rod balances the reproduction ratio at unity. Thus, by moving the rod outwardly from the critical position the neutron density in the reactor will rise. Moving the rod inwardly from the critical position causes the reproduction ratio to fall below unity, and the reaction stops. Thus the reaction is always under control, and as the rise in neutron density is exceptionally slow as the rod approaches the critical position, manual control is possible. Other and similar rods **141**, shown schematically in Fig. 13.2, may be provided, if desired, for rapid progression into the reactor to stop the reaction in case of failure of the control rod to stop the rise in neutron density for any reason. Such rods are termed safety rods.

During operation heat is released in the reactor in accordance with the neutron density therein. Most of the heat arises from the kinetic energy of the fission fragments and about 92 percent of the energy is released in the uranium. About 6 percent is released in the graphite due to neutron absorption therein and about 2 percent escapes from the reactor in the form of neutrons and gamma radiation. Consequently, the reactor can only be operated at a power dependent upon heat removal to the point where a stable temperature obtains. Otherwise, the reactor will accumulate heat to the point that the device may be damaged. Since aluminum melts at 658°C, stable temperature below this value should be used although with jackets of other non-fissionable metals, such as beryllium, the stable temperature may

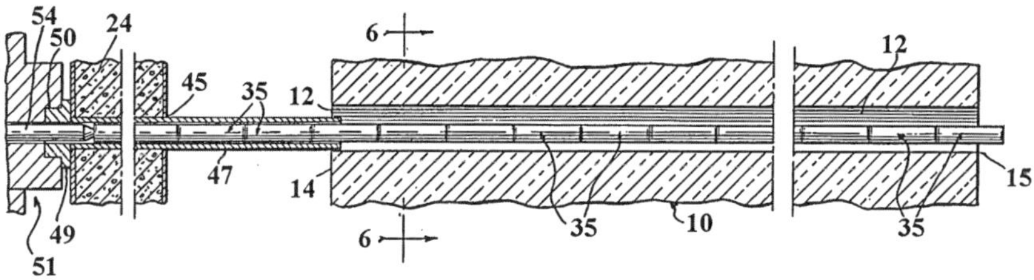

Fig. 13.8

be increased, although if the temperature should rise too high the uranium bodies might be damaged even when using beryllium jackets as uranium of the type used in neutronic reactors melts at about 1100°C.

A stable temperature is obtained in the device of the present invention by passing atmospheric air through the reactor, and in the specific example shown and described, the air is passed through the graphite channels and directly in contact with the aluminum jackets of the slugs. Under these circumstances the reactor can be operated continuously at 250 kilowatts electrical equivalent of heat by passing 32,000 cubic feet per minute through the reactor with a maximum temperature rise of the slugs to about 100°C, and at 500 kilowatts continuously with about 50,000 cubic feet per minute of air with a maximum metal temperature of 200°C. The output of the reactor can be stabilized at still higher powers by the use of larger fans if desired.

Having discussed generally the operation of the reactor, and the temperature stabilization thereof by air cooling at elevated powers and neutron densities, there is now described one means and method by which the reactor can be loaded and unloaded, in order that the neutron irradiated uranium can be removed for further processing such as the recovery of the 94^{239} formed in the uranium, and fresh uranium inserted for subsequent operation of the reactor, and in any amount desired.

To accomplish loading of the slugs **35** into the various air channels **12**, the concrete of the inlet end shield **24** is pierced with a plurality of loading apertures **45**, as shown in Figs. 13.1 and 13.8, each aperture being aligned with the axis of slug positions in the air channels **12**. Normally, during operation of the reactor, each aperture **45** is closed by a removable lead plug extending through the shield **24** only.

When it is desired to load a channel with new slugs, the lead plug in

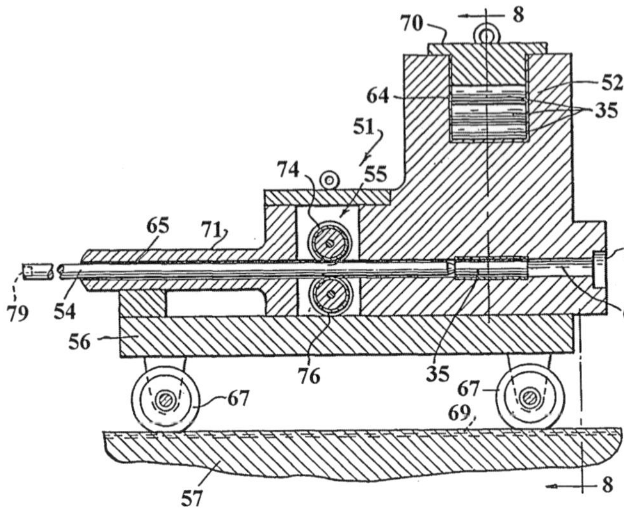

Fig. 13.9

shield **24** for that channel alone is removed, and a charging tube **47** inserted, extending through the inlet end shield **24**, across the inlet chamber **25** and entering the corresponding air channel **12** as shown in Fig. 13.8. The outer end of charging tube **47** is provided with a flanged nipple **49** shaped to engage a nipple recess **50** of a loading mechanism indicated generally by numeral **51**. It will be noted that the charging tube is smaller than the air channel **12** and that air can pass through the channel being unloaded. The air should circulate during unloading, although it may be at reduced velocity.

Loading mechanism **51** comprises a loading magazine **52**, a loading plunger **54**, and a plunger drive **55**, as shown in Figs. 13.9 to 13.12 inclusive.

The loading mechanism **51** is mounted on an elevator platform **56** mounted to be raised and lowered in an elevator frame **57** capable of moving along the outside of inlet end shield **24** on elevator tracks **58**. Base **59** of the elevator frame is provided with a platform **60** projecting outwardly on the same level as the top of a supply car **61** travelling on supply car tracks **62**. Supply car **61** is used to bring a supply of slugs to the elevator for use in the loading mechanism **51**.

The slugs **35** when received at the elevator, are loaded into an inclined loading channel **64** in the loading magazine, in side by side relationship and

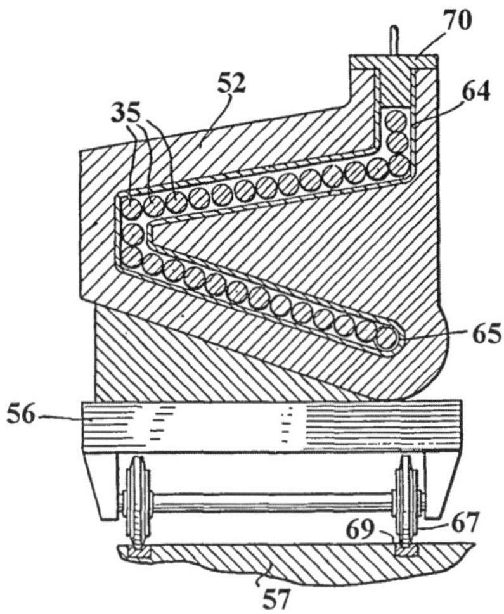

Fig. 13.10

feed by gravity to the bottom thereof. The bottom of loading channel **64** is a part of a plunger bore **65** extending through the loading magazine ending in the nipple recess **50** cooperating with flanged nipple **49** on charging tube **47** so that the plunger bore **65** and the loading bore in guide tube **47** are in concentric alignment. To provide engagement and disengagement of nipple recess **50** and nipple **49**, the entire loading mechanism is movable with

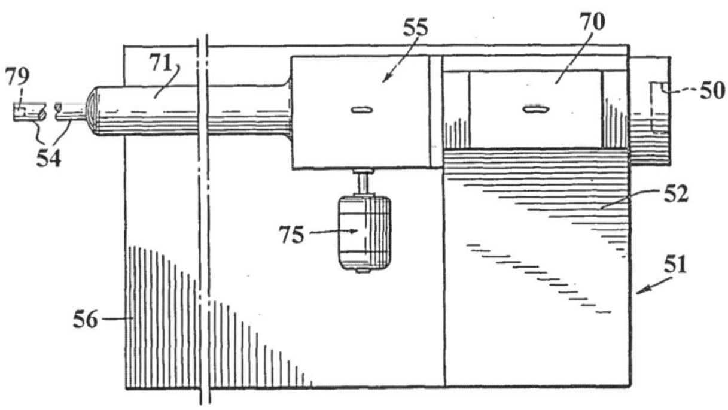

Fig. 13.11

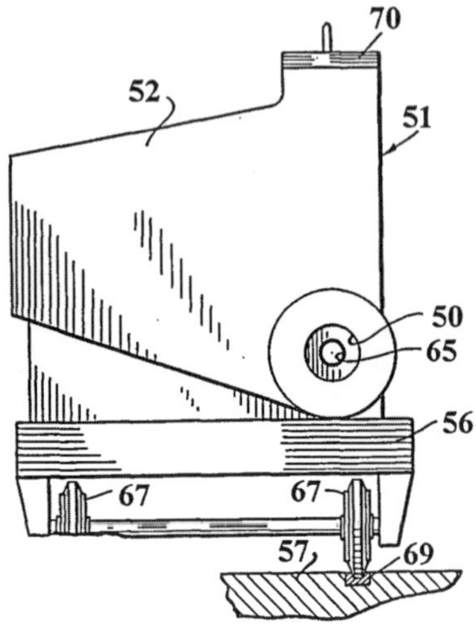

Fig. 13.12

respect to elevator platform **56** on wheels **67** running in guides **69** on the elevator platform.

It will be noted that loading magazine **52** is massive. In some instances it may be desirable to load slugs already partially irradiated and in consequence radioactive. The thick walls of the magazine then act as a shield for the radioactive slugs, and in this case a heavy cap **70** may close the upper opening of the loading channel **64**. Iron or lead may be used for the body of the magazine. In addition, the use of thick metal in the magazine, particularly around the plunger bore **65**, reduces radiation that might pass through the interconnected loading aperture **45** and guide tube **47** either from radioactive slugs therein or from the irradiated slugs in the reactor when charging tube **47** is empty.

The slugs are fed from magazine **52** by a reciprocating motion of plunger **54** operating in plunger bore **65**. Plunger **54** may be of iron to act as a shield when inserted into charging tube **47** and is supported outwardly by plunger bearing **71** on the opposite side of a plunger drive **55**.

Plunger drive **55** in simplified form, may be a friction wheel **74** driven by motor **75** as shown in Figs. 13.9, 13.11 and 13.13 opposed by an idler wheel **76** pressed against plunger **54** by spring **77**. Motor **75** is reversible and under control of the loading operator. Plunger **54** is sectional, having a

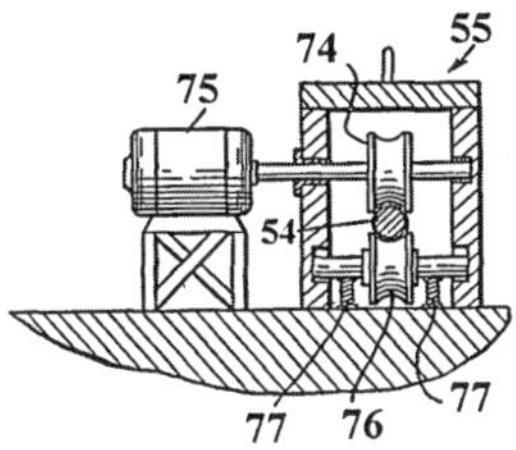

Fig. 13.13

threaded end **79** capable of making connection with additional plunger sections. Sufficient sections are provided to insert the plunger entirely through an air channel **12** when required.

In the initial loading of the graphite cube **10**, loading is started with the more central air channels, and 68 slugs are placed in the guide tube **47** and pushed into each connected channel. Plunger **54** is operated to push the slugs into the channel until the outer end of the first slug is at the outlet face **15**. The plunger is then withdrawn, leaving the outer end of the last slug about 16 inches from the inlet face **14**, for purposes explained later.

Proceedings outwardly and preferably concentrically, additional channels are loaded, meanwhile checking the neutronic activity of the reactor. As the activity increases as the loading approaches a critical size, that is, the size where the reproduction ratio will be exactly unity, the approach to critical size can be predicted by extrapolation of observed neutron density values with respect to the volume of the cube loaded with uranium. The neutron density values can be obtained from an ionization chamber **80** (Fig. 13.2), for example, using any well known indicating circuit, or by measuring the radioactivity of indium foils, for example, induced by neutron irradiation when inserted into the reactor.

As the critical size is approached, the control rod **41** is inserted deeply into the reactor to prevent a self-sustaining chain reaction; and loading is continued until the desired maximum reproduction ratio of, for example, from 1.005 to 1.006 is attained. This ratio can be checked by removal of the control rod and measuring the time taken by the reactor to double its neutron density. From this period, the reproduction ratio can be mathematically computed.

When the desired number of channels are loaded the active core of the reactor may contain from 34 to 50 tons of uranium, and will be ready for operation. Graphite plugs for the unused air channels may be loaded in a manner similar to that described for the uranium slugs.

It will be noted that on 4 sides of the graphite cube excess graphite will be present. On the fifth side, i.e., at the inlet face, graphite will also extend 16 inches beyond the uranium. On the remaining side, i.e., the outlet face, no graphite extends beyond the uranium. Thus, 5 sides of the active portion (the uranium bearing portion) are surrounded by graphite. This graphite constitutes a reflector and reduces the amount of uranium required to reach critical size.

Graphite and other neutron moderators can be used as reflectors around the active portions of a neutronic reactor, as such materials scatter neutrons passing through them and thereby change the direction of the neutrons. Statistically, due to the scattering action, some of the neutrons that otherwise would be lost to the active portion are returned to it, thus reducing exterior loss and thereby reducing critical size. In graphite-uranium reactors having roughly a spherical shape for the active portion there can be a reduction of 3 to $3\frac{1}{2}$ feet in diameter of the active portion when it is surrounded by a layer of graphite from 3 to 5 feet thick. In the present instance the reduction in size is slightly less as only 5 sides of the active portion are surrounded with the reflecting layer.

It will be noted that more air channels are originally provided in the graphite cube than are required for the disposition of the uranium containing slugs. The manner in which the uranium is disposed in the graphite is known as the geometry of the system, and this geometry may be varied to suit conditions, for example, the system as described may be said to have rod geometry with cylindrical loading as the end to end relationship of the slugs forms, in effect, a long rod of uranium. Further, the use of full length rods in all channels concentrically arranged around the center of the active portion, gives that portion a generally cylindrical shape.

Lump geometry may also be used in the reactor, and is accomplished by separating the individual slugs by graphite rods, for example 4 inches long. In this additional channels will be loaded to bring the amount of uranium almost to the same value as that used in the rod geometry, although the lump geometry is slightly more efficient than rod geometry.

Furthermore, a spherical shape is somewhat more efficient for the active portion than a cylindrical shape, and can be approached with either rod or lump geometry by shortening the extent of uranium loading in the channels as the peripheral concentric layers are added, until an approximate sphere is formed. A suitable compromise is the use of one or more concentric rings of outer channels around a cylindrically loaded central portion, with only half the number of slugs in the outer channels and disposed with the ends of the slug row equally distant from the ends of the fully loaded rows. Other arrangements will be apparent to those skilled in the art, and can be attained in the structure described because of the provision of more channels than are to be used.

After the reactor is loaded the fan is started and the control rod is withdrawn until a rise in neutron density to a desired power output is attained. The control rod is then moved forward into the reactor until a neutron balance is obtained with the reproduction ratio at unity, thus maintaining the chain reaction at the desired operating power. Small variations from the unity reproduction ratio will occur during operation, due to temperature variations of the cooling air, and to change in barometric pressure and to minor variations in air pressure delivered by the fan. However, such variations are compensated by slight inward or outward corrective movements of the control rod, either by hand in response to indicated variations in neutron density, or automatically by direct linkage of the control rod to the output of the ionization chamber **80**. However, such automatic control is no part of the present invention.

During operation of the reactor, the air passing through the reactor becomes radioactive due to the fact that it is subjected to intense neutron irradiation. Investigation has proved that the only significant radioactivity present in the air after having passed through the operating reactor is that of Ar^{41}, having a 110 minute half-life. At high power output, however, this activity may be present in the exhaust air to the point that it would be biologically dangerous to operating personnel unless highly diluted during its radioactive decay. For that reason the air passing through the operating reactor is not delivered to the atmosphere at ground level but is exhausted at a substantial distance above ground, such as for example from the top of a 200 feet stack, with the result that when and if any of the radioactive Ar^{41} reaches ground level it is so dispersed in and diluted by fresh atmospheric air that less than 0.1 Roentgen per day will be received by any persons on the ground, either close to or away from the stack. Thus, the air is passed only once through the reactor and does not acquire excessive radioactivity.

The sole presence of the above noted type of radioactivity, however, is predicated on the use of the jackets sealed around the uranium bodies. Aluminum is preferred for the jackets, as aluminum has a relatively low neutron capture capability and, consequently, can be used in substantial amounts in the reactor without absorbing or capturing sufficient neutrons to prevent a self-sustaining chain reaction from occurring. Aluminum also corrodes very slowly in hot air.

The jackets have two functions, both of which reduce radioactivity of the cooling air. The first is to prevent oxidation of the uranium. While considerable oxide could be tolerated in the reactor itself if the uranium were to be used in unprotected condition, some of the oxide particles would be picked up by and exhausted in the cooling air. As these particles would be highly radioactive and relatively heavy, the proper dispersal thereof would be a difficult problem.

In addition, if operation should be accomplished in the reactor with bare uranium, fission fragments from nuclear fissions occurring on the surface of the uranium would also be projected into the air stream and would be carried out by the air stream. These fragments are exceptionally radioactive and could not safely be dispersed into the atmosphere. When jackets are used, these fragments are stopped by the jackets and cannot enter the air stream.

Thus, the jackets prevent corrosion of the uranium and prevent fission fragments and corrosion products of uranium from entering the air stream. As fission fragments will pass through an extremely small hole, one method of monitoring the reactor for jacket failure, such as for example, a weld crack is to monitor the radioactivity of the stack gas. If the stack gas shows any substantially radioactivity other than that of Ar^{41} then it is clear that a jacket failure has occurred. Such monitoring of the stack gas is normally a routine procedure of an operating air cooled reactor, but forms no part of the present invention.

After operation of the reactor for a sufficient length of time for an amount of 94^{239} to be created sufficient for chemical separation, such as for example 100 days at 500 kilowatts, the reactor is shut down by inserting the control rod fully into the reactor. After about one half hour's wait, during which all delayed neutron omission will have ceased and the more violent radioactivity subsided, the reactor may be unloaded.

The unloading may be accomplished in two ways, either by using the plunger to push the slugs out of the channels so that they fall by gravity out of the outlet face **15**, or by using the plunger to insert new slugs in the channels, each slug so inserted pushing an irradiated slug out of the outlet face **15**. In the first instance the graphite cube will be left empty after unloading. In the second instance the insertion of new slugs is continued until all or a predetermined part of the irradiated slugs are out of the reactor, having been replaced by fresh material. Thus, the reactor is left ready for the next run. Under ordinary circumstances the latter manner of unloading is preferred.

In either case the slugs drop by gravity from the outlet face into outlet chamber **31**, falling onto two angularly disposed pad plates **90** positioned to intersect the falling slugs, in the bottom half of outlet chamber **31** as shown in Figs. 13.1 and 13.2. The two plates slant to a centrally disposed outlet pipe **91** extending downwardly through foundation **11** and provided with spaced valves **92** and **94**. The slugs fall by gravity into pipe **91** above valve **92**.

Outlet pipe **91** opens into a lower coffin chamber **93** that in turn connects with a tunnel **96** carrying car tracks **97** on which a coffin car **99** may be moved by means of cable **100**. Coffin car **99** supports a plurality of slug coffins **101** in position to be successively positioned beneath the lower opening of pipe **91**.

Valves **92** and **94** are operated by means of rods **102** and **104**, respectively, from behind a heavy led shield **105**, as shown in Fig. 13.1. A crane **106** is used for placing coffin caps **107** on each coffin after it has been filled with irradiated slugs.

Before unloading is started, both valves **92** and **94** are closed, and the upper portion of pipe **91** is filled with water from water inlet pipe **109**. A proper water level is maintained above valve **92** by water outlet pipe **110**. The air circulation is maintained, although it may be reduced to about 25 percent of the operating value. Slugs are then pushed out of the reactor to fall on to pad plates **90** and then roll by gravity into the water in the upper part of outlet pipe **91**.

In order that there be no material damage to the jacketing of the slugs, plates **90** are preferably padded with a soft material that does not deteriorate under neutron irradiation, and that will be able to withstand the slug impacts. A satisfactory pad has been found to be $\frac{1}{4}$ inch cotton duck on felt laid on wood backed by steel. Combinations of various synthetic elastic materials have also been found satisfactory.

After a number of slugs have been collected above valve **92** sufficient to fill a coffin **101**, unloading is stopped and valve **92** is opened, permitting the slugs and the water to drop through the valve and remain in the space between valves **92** and **94**. Valve **92** is then closed, the water level reestablished and unloading continued. In the meantime, valve **94** is opened permitting the slugs and water to fall into one of the coffins **101**. The car is then moved to register the opening of the next coffin with the end of the outlet pipe and the first coffin is capped. The procedure is continued until all of the irradiated slugs are in coffins. These coffins may then be taken to a soaking pit (not shown) to remain until the radioactivity has decayed to a point where the slugs can be submitted to chemical removal of the products formed therein by irradiation. After 100 days' operation the aging period may be about 30 days.

Removal of the irradiated slugs under the conditions specified is performed for two reasons. Firstly, the slugs are so highly radioactive that they cannot be safely approached by personnel without adequate shielding being interposed, and, secondly, for some time after removal from the reactor this radioactivity is so intense that self absorption of the emitted radiations causes self heating of the slugs possibly sufficient to melt the slugs if not cooled in some manner. By unloading during maintenance of the air stream, by dropping the slugs at once into water, and by keeping the slugs in water until the more violent radioactivity has subsided, melting is prevented, as the slugs are cooled as they boil the water in which they are immersed. The slugs are then stored or aged under water until ready for chemical treatment, as for example, for thirty days.

It is to be noted that reactor operation at 250 to 500 kilowatts heat equivalent has been mentioned. These powers, however, are in no way maxima as the operating powers are dependent solely on the air supply available and the permissible maximum slug temperatures. Ordinarily slug temperatures of from 100°C to slightly over 400°C are permissible without special treatment of the uranium bodies before jacketing. In case, however, it is desired to operate the reactor with a maximum slug temperature above about 430°C, then precautions should be taken to remove occluded hydrogen from the uranium to less than 0.0002 percent in order to prevent swelling of the jackets by release of this hydrogen. At temperatures below 430°C no swelling occurs from the release of occluded hydrogen, as uranium hydride is formed, preventing rise of internal pressure. Above 430°C, however, internal pressure from released hydrogen may, with unprocessed uranium, swell the jackets. The occluded hydrogen, however, is easily removed from the uranium bodies by heating them to a temperature of from 580°C to 600°C under continuous evacuation by a vacuum pump until equilibrium is reached. The uranium bodies are then cooled 10 to 20 hours, preferably in an argon atmosphere, and then jacketed. When processed in this manner no swelling will occur at any temperature. Thus, if desired, the reactor can be operated at powers of from 1000 kilowatts to 5000 kilowatts, when the proper amount of air is supplied, and the hottest slugs permitted to rise in temperature to from 400°C to 500°C.

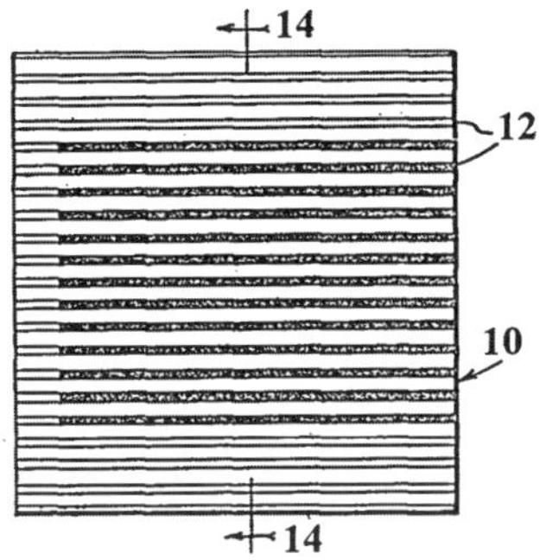

Fig. 13.14

Having described the reactor with one type of loading, changes in loading required by operating conditions can be made, as illustrated by the diagrams shown in Figs. 13.14 to 13.16.

In Fig. 13.14 and 13.5, a loading is shown, for example, where only minor amounts of intermittent power are required. For example, 800 tubes can be loaded providing an excess r of just enough to reach the required power, the power being lowered or the reaction stopped before the appearance of the Xe^{135} effect. The solid black areas indicate uranium loading.

If, however, even higher powers are desired, the moderator blocks can, for example, be almost completely filled up, as shown in Figs. 13.15 and 13.6, as for example 1600 channels can be loaded to provide an excess r of about 0.01. This will permit, other conditions being consistent, a large

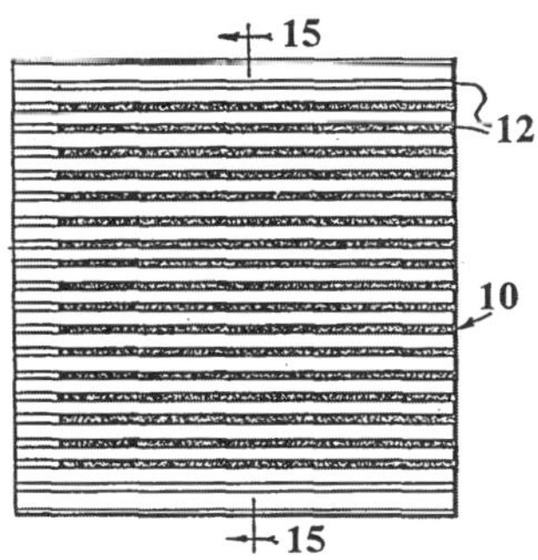

Fig. 13.15

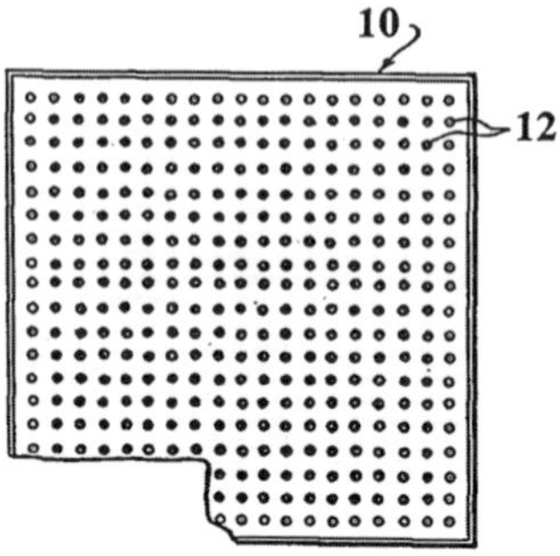

Fig. 13.16

continuous power output with some excess r available after Xe^{135} poisoning is taken care of, for control of minor variations by the control rod.

Thus by making the moderator blocks large, so that at least sufficient channels are provided to allow enough channels to be loaded for the maximum power contemplated and for maximum expected poisoning both short and long term, the reactor structure except for the loading, need not be changed.

Furthermore, the moderator blocks pierced with more channels than may be initially required, permits loading to be made in various patterns.

The loading in Figs. 13.14 and 13.5, and 13.15 and 13.6 is what is known as cylindrical loading, in that the active portion defines a cylinder. However, cubical or parallelepiped loading may be desirable, as shown by the diagram of Fig. 13.16, in cross section. The longitudinal section will be the same as shown in Fig. 13.15. It will be seen that the cubical loading can also be contracted or expanded in size to meet operating conditions. By progressively shortening peripheral rods an approximation of a spherical overall shape of the active portion can be obtained.

By so providing space for increasing or decreasing the size of the active portion, all presently known operating conditions requiring change in critical or operating size can be obtained. Thus, when the neutron reproduction ratio decreases due to impurities formed during operation of the reactor some plugs of graphite are removed and additional bodies of thermal neutron fissionable material are added to the passages thus provided.

While the theory of the nuclear chain fission mechanism in uranium set forth herein is based on the best presently known experimental evidence, the present invention is not to be bound thereby, as additional experimental data later discovered may modify the theory disclosed. Any such modification of theory, however, will in no way affect the results to be obtained in the practice of the invention herein described and claimed.

What is claimed is:

The method of sustaining a neutronic chain reacting system comprising operating a reactor containing a solid neutron moderator of graphite having a plurality of passages adapted to receive bodies of thermal neutron fissionable material, and containing sufficient bodies of thermal neutron fissionable material in certain of said passages and solid moderator plugs of graphite in other of the passages to achieve a chain reaction having a neutron reproduction ratio greater than unity, removing some plugs of solid moderator from some of the other said passages, and adding more bodies of thermal neutron fissionable material, said additional bodies being added to the passages thus provided when the neutron reproduction ratio decreases due to impurities formed during operation of said reactor and said additional fissionable material being sufficient to maintain the neutron reproduction ratio at greater than unity.

REFERENCES CITED in the file of this Patent

UNITED STATES PATENTS

2,708,656	Fermi et al.	May 17, 1955

FOREIGN PATENTS

114,150	Australia	May 2, 1940
114,151	Australia	May 3, 1940
861,390	France	Oct. 28, 1940
233,011	Switzerland	Oct. 2, 1944

OTHER REFERENCES

Power, July 1940, pp. 56-59.

Smyth: "Atomic Energy for Military Purposes", Aug. 1945.

Goodman: "The Science and Engineering of Nuclear Power", vol. 1, page 275, Addison-Wesley (1947).

Kelly et al.: Phy. Rev. 73 1135-9 (1948).

Chapter 14

Neutronic Reactor Shield

Enrico Fermi, Santa Fe, N. Mex., and Walter H. Zinn, Chicago, Ill., assignors to the United States of America as represented by the United States Atomic Energy Commission

Application January 16, 1946, Serial No. 641,625

2 Claims. (Cl. 250 – 108)

U.S. Patent No. 2,807,727 - Patented Sept. 24, 1957

The argument of the present Patent is a radiation shield suitable for protection of personnel from both gamma rays and neutrons. Such a shield from dangerous radiations is achieved to the best by the combined action of a neutron slowing material (a moderator) and a neutron absorbing material. Hydrogen is particularly effective for this shield since it is a good absorber of slow neutrons and a good moderator of fast neutrons. The neutrons slowed down by hydrogen may, then, be absorbed by other materials such as boron, cadmium, gadolinium, samarium or steel. Steel is particularly convenient for the purpose, given its effectiveness in absorbing also the gamma rays from the reactor (both primary gamma rays and secondary ones produced by the moderation of neutrons).

In particular, in the present Patent a shield is described, made of alternate layers of steel and Masonite (an hydrolized ligno-cellulose material).

The object of the present Patent is not discussed in any other published paper.

This invention relates to radiation shielding devices and more particularly to a radiation shield that is suitable for protection of personnel from both gamma rays and neutrons.

In the operation of certain devices known as neutronic reactors a neutron fissionable isotope such as U^{233}, U^{235}, or Pu^{239} or mixtures thereof is subjected to fission by absorption of neutrons and a self-sustaining chain reaction is established by the neutrons evolved by the fission. In general such reactors comprise bodies of compositions containing such fissionable material, for example, natural uranium, disposed in a neutron slowing material which slows the neutrons to thermal energies. Such a slowing material is termed a neutron moderator. Carbon, beryllium, and D_2O (heavy water) are typical moderators suitable for such use. Specific details of the theory and essential characteristics of such reactors are set forth in copending application of Enrico Fermi and Leo Szilard, Serial No. 568,904, filed December 19, 1944, now Patent No. 2,708,656.[1]

In the operation of a neutronic reactor, a large quantity of neutrons and gamma rays as well as alpha and beta rays are released. It is essential that personnel working around said reactors be protected from these types of radiation as they have detrimental biological effects.

Therefore, it is an object of this invention to provide a device that will protect personnel from radiations, especially all types of radiation which emanate from a neutronic reactor.

Another object of the invention is to provide a type of shielding that will be easy to construct, simple to assemble, low in cost, and occupy a relatively small space.

These and other objects are obtained by the novel construction, combination of materials, and arrangement of said materials hereinafter described and shown in the accompanying drawings in which:

Fig. 14.1 is an enlarged plan view of one section of the shield shown in Figs. 14.4 and 14.5;

Fig. 14.2 is an elevational view of the shield section shown in Fig. 14.1;

Fig. 14.3 is an enlarged fragmentary elevational view of the shield section shown in Fig. 14.2;

Fig. 14.4 is a vertical sectional view through the center of a neutronic reactor equipped with a shield constructed in accordance with the teachings of the present invention, the shield being shown diagrammatically;

Fig. 14.5 is a plan view of the neutronic reactor and shield shown in Fig. 14.4, part being broken away to show the top construction of the shield; and

Fig. 14.6 is a perspective view of one sheet of the hydrogenous material of the shield shown in Figs. 14.1 and 14.2.

---

[1]@ See page 217.

In the past, gamma rays have been absorbed according to well established principles by shields of metal, usually lead, but these shields are inadequate for the absorption of fast and slow neutrons. Neutrons may be absorbed by very thick shields of concrete, but the mass of such a shield makes it unwieldy, if it must be moved in operation of the reactor. Some of the neutrons emanating from a neutronic reactor known as "fast" neutrons may have energies over one million electron volts. These fast neutrons are not readily absorbed by any single material; but it has been discovered that by using a composite shield of a neutron slowing material, also called a moderator, and a neutron absorbing material, it is possible to reduce the amount of radiation passing through said shield to a point at which it is safe for personnel to work close to the outside of said shield even when a neutronic reactor is operating at a high power output of 100,000 kilowatts or more inside of said shield. To be equally effective in stopping radiations, a concrete shield would have to be 3 to 4 times as thick as the described composite shield.

Light elements are suitable for the moderating material in such a shield because they tend to slow the fast neutrons to thermal energies by means of elastic collisions. A fast neutron colliding with the nuclei of light elements in a series of collisions gradually loses its speed until it reaches thermal energy. Light elements such as carbon, beryllium, or deuterium, are suitable for use as a slowing material, but they do not absorb either fast or slow neutrons to a high degree and they are all comparatively expensive; therefore, it is preferred to use a hydrogen-containing material, hereinafter called a hydrogenous material, for the slowing material in the shield. Hydrogen is a relatively good absorber of slow neutrons as well as a moderator of fast neutrons and is readily available in the form of hydrocarbons, cellulose or water. It will be understood that hydrogen has the same moderating and absorbing qualities in the form of an element or combined in a compound.

When the fast neutrons have been slowed by the slowing material to the thermal energies, they are readily absorbed by materials such as boron, cadmium, gadolinium, samarium and iron. Any one of several materials could be used with the slowing material to form a shield but it is preferred to use an iron-containing material, because it is cheap and a good absorber of gamma rays. Iron in the form of steel is very advantageous because it has high structural strength; and it has a relatively high capture cross-section for thermal neutrons and therefore is a good absorber of thermal neutrons. Iron is also more effective than most metals in slowing fast neutrons by means of inelastic collisions. In these collisions the neutron's energy is used

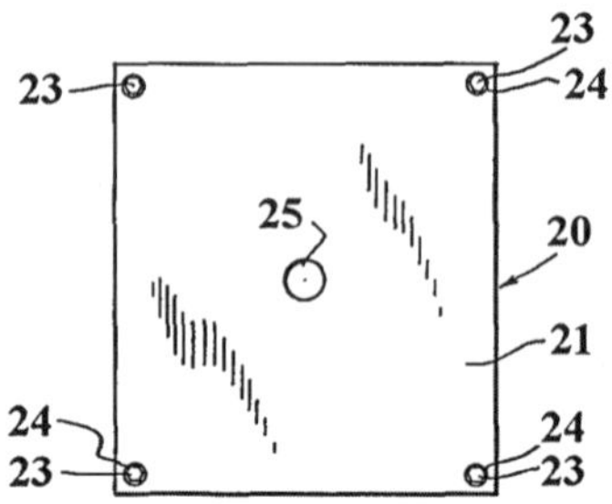

Fig. 14.1

up in producing an excited state of the iron nucleus, which later emits the energy as a gamma ray. For this reason the iron will act as a moderator as well as an absorber.

As has been mentioned, iron is also a good absorber of gamma rays. This is an essential feature because the absorption of a neutron by an iron nucleus causes the emission of a gamma ray of approximately two million electron volts energy. As has been explained above, said gamma rays are dangerous to personnel and therefore said rays must be absorbed by the shield. A proper thickness of iron will absorb said gamma rays.

In a preferred embodiment of the present invention, the iron is contained in the form of steel sheets and the hydrogenous material is in the form of a wood-fibre board such as Masonite. (Masonite is a hydrolized ligno-cellulose material compressed under heat.) A Masonite board which contains about 6 percent hydrogen and has a density of about 1.3 grams per cubic centimeter has been found satisfactory. The steel sheets and the Masonite board are assembled in several alternate layers to form sections of the shield. Thus,

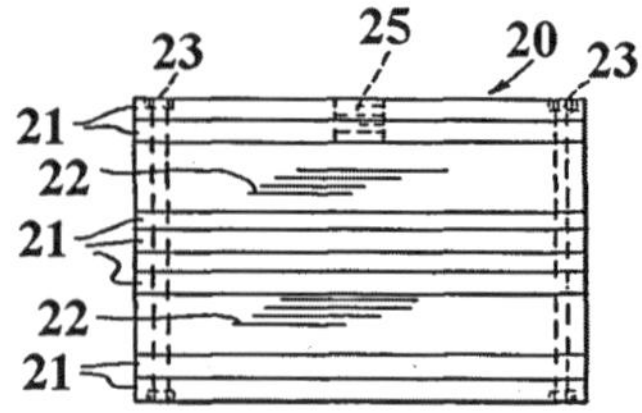

Fig. 14.2

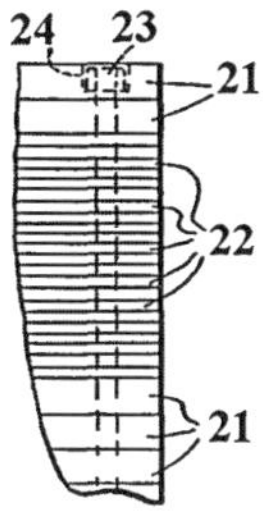

Fig. 14.3

a fast neutron entering the shield is slowed to thermal energy either by inelastic collisions with the iron nuclei in which the neutron energy is used up in producing an excited state of the iron nucleus or it may be slowed by elastic collisions with the light nuclei such as hydrogen and carbon contained in the wood fibre. Upon reaching thermal energies the neutron is readily absorbed by an iron nucleus or it may possibly be absorbed by a hydrogen nucleus.

Since both steel and Masonite have excellent structural qualities, a shield composed of layers of these materials is very strong mechanically and may be self-supporting.

Turning now to the drawings, one exemplification of a layered shield of alternate sheets of steel and Masonite is shown. In Figs. 14.1, 14.2, and 14.3, one section **20** of such a shield is disclosed. Reading from top to bottom of Fig. 14.2 the section **20** is composed of two sheets **21** of steel, each $\frac{3}{4}$ of an inch thick, then twenty-two sheets **22** of Masonite each $\frac{1}{8}$ of an inch thick, then four sheets **21** of steel each $\frac{3}{4}$ of an inch thick, then another twenty-two sheets **22** of Masonite each $\frac{1}{8}$ of an inch thick, and finally two sheets **21** of steel each $\frac{3}{4}$ of an inch thick. Thus the total assembly is $11\frac{1}{2}$ inches in thickness. The dimensions given above are to be taken as illustrative. The assembly is secured together by four bolts **23** positioned at the corners of each sheet and passing through each sheet. The heads of the bolts **23** are countersunk in depressions **24** in the upper and lower sheets so that the heads are flush with the surface of upper and lower sheets **21**. A threaded hole **25** is provided in the center of the upper steel plate **21** in which an eye bolt may be screwed for use in positioning the section in place in the shield. For clarity, the individual sheets **22** of Masonite are not indicated in Fig. 14.2, but are in Fig. 14.3.

Figs. 14.4 and 14.5 disclose diagrammatically three layers of shield

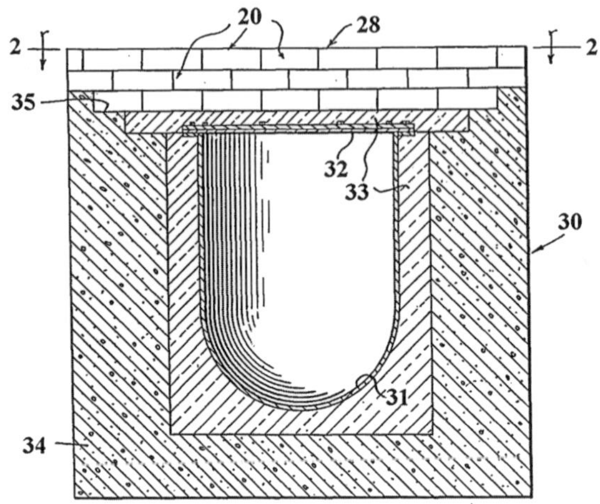

Fig. 14.4

sections **20** laid in place as the upper shield **28** of a neutronic reactor **30**. The reactor **30** shown is a deuterium oxide moderated type in which uranium (not shown) is suspended in an aluminum tank **31** partially filled with deuterium oxide (not shown). Such a reactor is fully described in the Fermi-Szilard Patent mentioned above. The reactor tank **31** is provided with a heavy metal top **32** from which the uranium in the form of long rods is suspended. At intervals in the operation of the reactor **30**, it is necessary to remove this top **32** and with it the uranium rods in order that said rods may be replaced. Therefore, it is essential that any shielding placed over the reactor top **32** must be easily removable.

A neutron reflector **33** of graphite surrounds the tank **31** on all sides including bottom and top. A heavy concrete shield **34** may surround the sides and bottom of the tank **31** of the present improved shield may be used. Since weight is of little moment on the sides and accessibility is not necessary, the concrete side shield may be made several times as thick as the top layered shield. The concrete shield should be at least three times as thick to give equivalent protection. The upper wall of the side shield **34** is formed into steps **35** for supporting the exterior sections **20** of the top shield **28**. As shown in the drawings, the first layer of shield **28** is laid above the reactor **30** with the edges of the sections **20** aligned. The second layer is then laid so that each section **20** is centered over the intersection

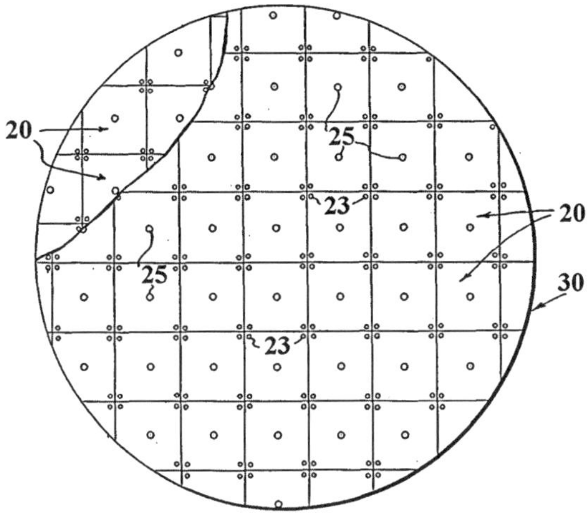

Fig. 14.5

of four sections **20** of the lower layer (Fig. 14.5). Thus, the spaces between sections **20** are effectually blocked by alternate layers so that there is no clear path for neutrons or other radiations to escape through the shield **28**. Each neutron or ray which passes through the spaces of one layer of the shield is absorbed by the second layer which covers said spaces. It will be understood that the number of layers of sections **20** used will depend upon the intensity of the radiation being emitted by the source to be shielded. This adaptability to various conditions is one of the principal advantages of the invention, because the shield **28** may be easily designed to fit a specific condition.

It will be noted that a substantial thickness of steel is provided in the exterior layers of the section **20**. Thus when the section **20** is positioned on a reactor **30**, the interior sheets next to the reactor are of steel and the exterior sheets are also of steel. A fast neutron escaping form the reactor **30** first passes through iron before reaching the hydrogenous material. This is important because under a high concentration of neutron bombardment the hydrogenous material tend to dissociate releasing hydrogen and other gases. By placing the steel adjacent the source of neutrons enough neutrons are absorbed by the iron nuclei so that the number of neutrons striking the hydrogenous material is considerably reduced and the possibility of

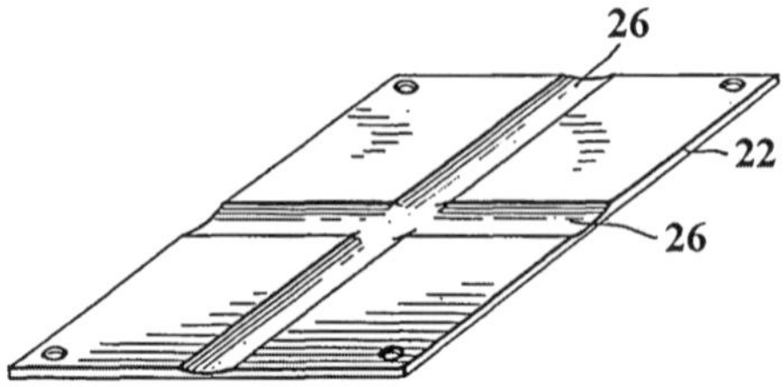

Fig. 14.6

dissociation is lessened. As shown in Fig. 14.6 transverse grooves **26** are provided in each Masonite board **22** so that any gas evolved in said board may escape to the exterior without causing warping of said board.

The exterior thickness of iron is required as explained above because gamma rays originate in the shield when a neutron is absorbed. The exterior layer of iron reduces the quantity of gamma rays escaping to a safe amount.

The structure and configuration of the shield sections **20** and of the shield **28** may be changed, and other arrangements of the hydrogenous material and the dense material may be used without departing from the spirit of the present invention. The arrangements and dimensions of the hydrogenous and dense materials have been given by way of examples and will vary depending on types of materials used and the shielding job to be accomplished. The above disclosure is to be regarded as descriptive and illustrative only, and not as restrictive of the invention which is limited only by the appended claims.

It is claimed:

1. In combination with a neutronic reactor, a radiation shield adjacent said reactor, said radiation shield comprising a plurality of layers of shielding blocks, each of said blocks comprising a plurality of layers of an iron-containing material and a compressed cellulosic material secured together in the following order, first a layer of said iron-containing material approximately one and one-half inches in thickness, then a layer of said cellulosic material approximately two and three-quarter inches in thickness, then a second layer of said iron-containing material approximately three inches in thickness, then a second layer of cellulosic material approximately two and three-quarter inches in thickness and finally a third layer of said iron-containing material approximately one and one-half inches in thick-

ness.

2. The radiation shield as described in claim 1 wherein each of said layers of cellulosic material comprises flat sheets, each of said sheets being provided with a transverse groove in its surface, said groove extending to at least one edge of the sheet.

REFERENCES CITED in the file of this Patent

UNITED STATES PATENTS

710,201	Lowder	Sep. 30, 1902
1,580,857	Richards	Apr. 13, 1926
1,844,512	Mains	Feb. 9, 1932
2,183,790	Dillehay	Dec. 19, 1939
2,708,656	Fermi et al.	May 17, 1955

Appendix A: Other Documents not Reported in this Volume

The recent careful work performed by one of us (S.E.), devoted to bring to light previously unknown material (earlier classified) regarding Fermi's activity, has lead to the retrieval of a consistent amount of papers, reports or other documents written by Fermi himself and/or his collaborators, which is not limited to the lecture notes and to the patents reported in the present volume. About 600 original pages in all have been discovered, all of them directly pointing out the peculiar activity of the Italian scientist about different aspects of Nuclear Physics, but which are *not* limited to the scientific point of view (in a strict sense).

Apart from the Papers of Sir James Chadwick at the Churchill Archives Centre in Cambridge (U.K.) and the United States Patent and Trademark Office, the other major source of documents has been the Albert Wattenberg Papers at the University Library of the University of Illinois at Urbana-Champaign. In this archive, a number of papers are deposited donated by Albert Wattenberg (1917-2007), a former collaborator of Fermi, who collected documents pertaining to the Manhattan Project. Part of them were retrieved by Wattenberg as joint editor of the Fermi's *Collected Papers*, and then published in [Fermi (1962)]. However, among these documents we found many unpublished notes and reports, all dealing with the activity by Fermi and others on nuclear fission topics, ranging from 1942 to 1944. In particular 23 notes have been found on meetings of different Councils, where explicit interventions by Fermi were annotated, 2 scientific/technical reports written by Fermi and collaborators, 5 periodic reports edited by Fermi and others, 1 anonymous scientific/technical report classified by Wattenberg among the Fermi papers.

Other "minor" documents have been recovered as well, which will be discussed in some detail in the following.

The notes on meeting, however, are of particular relevance for the history of the achievement of the knowledge on chain reactions (with particular reference to the construction of the first chain reacting pile in Chicago at the end of 1942) and its application in the Manhattan Project. In some documents, explicit references to weapons, their use during the Second World War, and related matters appear. Quite a persistent "obsession", even as early as in 1942, for the production of fissile material (mainly plutonium) for military uses emerges from many documents, a feature which was not at all considered in previous historical reconstructions. The attitude of Fermi on this point comes out very clear: he is not "obsessed" at all by military applications (like, instead, several other colleagues), but rather by civil use of nuclear energy (for "the heating of towns") and, quite unexpectedly, by the physiological effects of radiations. Quite important (and, again, unexpected) are, as well, the discussions at several meeting of long term physics research and post-war research policy, and those regarding the relationship, about nuclear power for pacific and/or military use, between U.S. and Britain just after the end of the war.

For coding and specific symbols used in the following, the reader is referred to Appendix B.

Scientific reports

We start with an account of three scientific reports from the Wattenberg archive, not comprised in the Fermi's *Collected Papers* (the first of these reports not showing the list of authors).

WAT1043v * *The Fourth Intermediate Pile* (15 pages),
Metallurgical Project,
Report C-102.

"The experiments reported here belong to a set of experiments designed to establish the arrangement of 3″ cubes of alloy oxide in a graphite moderating medium which will produce the highest multiplication factor. Previous experiments were made on a simple cubic lattice of these alloy oxide cubes of such dimensions that the ratio between the volume of the graphite and the volume of the oxide was roughly 20/1. This gave a multiplication factor of 0.94. Half of the oxide cubes were then removed, leaving a face-centered lattice in which the volume ratio of graphite to oxide was 40/1. In this arrangement the multiplication factor fell to 0.86. In the present report, a body centered structure was assembled in which the volume ratio of graphite to oxide was about 10/1. It is found that the multiplication factor is again close to 0.86. From the results of these structures it is concluded that the simple cubic lattice in which the volume ratio was 20/1 represents closely the optimum conditions. Theoretical calculations support this experimental result."

This paper did not report explicitly the list of the authors, but it was classified by Wattenberg among the Fermi papers. A careful analysis has shown that, apart from indirect information on Fermi's activity, it was to some extent effectively written (or, at least, "inspired") by Fermi. The date of writing was, as well, not given but, according to the material presented in the paper, it was likely written in 1942.

The results presented here, very well summarized in the abstract above, were not reported in any other published paper.

WAT1043t *Report of the Committee for the Examination of the Moore-Leverett Design of a He Cooled Plant* (18 pages),
by E. Fermi, S.K. Allison, C. Cooper and E.P. Wigner,
Report CE-324 (1942).[2]

This report was likely written by Fermi (the chairman of the Committee) with memoranda by Allison, Cooper, Wigner, and a letter from Szilard.

In it a pile of dimensions considerably larger than that originally planned in the Moore-Leverett design is considered, this urging for a re-design of the lattice, for a reduction of the amount of uranium metal, and the consideration of the possibility to use a non cubic cell (as stated in previous conferences).

[2]This report was likely written around October 29, 1942.

The employment of centrifugal (turbo) compressors (for the coolant) is considered, instead of reciprocating compressors, with high purity helium to avoid corrosion of uranium.

A number of technical problems, such as that of an adequate radiation shielding, the production of radioactive materials in the reactor which can be collected by helium during the shut-down of operations, or precaution on helium released in atmosphere are discussed. Problems of emergency measures for serious loss of helium and to prevent the activated uranium from melting (if the cooling system with helium is switched off) are as well considered.

It is here pointed out that the operation of the control rods takes place by looking at the neutron density, rather than at the temperature of the reactor. Attention is also paid to possible displacements in the arrangement of the graphite due to the thermal expansion, that can cause damages to the structure and interfere with the operation of the control rods. Wigner, in particular, proposes a cylindrical arrangement instead of the spherical one.

As a conclusion, the Moore-Leverett design of a He-cooled power plant can work satisfactorily, although several details have still to be worked out.

The object of the present report is not discussed in any other published paper.

WAT149 *Measurement of the Cross Section of Boron for Thermal Neutrons* (4 pages),
by E. Bragdon, E. Fermi, J. Marshall and L. Marshall,
Report CP-1098 (January 11, 1944).

"Measurements of the boron cross section have been made for slow neutrons from different sources. The cross section of boron for neutrons of velocity = 2kT/m = 2200 meters/second at 293° K is found to be 705×10^{-24} cm^2/atom. The cross section varies widely with different moderators, due to the fact that the temperature of the thermal neutrons depends on the nature of the moderator."

As stated in the abstract, this papers deals with the measurement of the cross section of thermal neutrons on boron for different velocities of the neutrons. Velocities ranging from 1700 to 5000 m/s were obtained with a velocity selector, not described in this paper (see, however, USP8). The relevant measurements are done by varying also the pressure.

The results of the present paper converged later in the published article in [Fermi (1947)] (see also the comment to this paper in [Fermi (1962)], noting the different number of authors), where the velocity selector was described as well.

Notes on meetings

A substantial part of the documents testifying for Fermi's activity and retrieved in the Wattenberg archive consists of many notes on meetings about nuclear piles and related matters, attended by Fermi mostly in 1942. Some of these notes were already published in [Fermi (1962)], but many of them were not included in the Fermi's *Collected Papers*, probably because the corresponding material does not present itself as reports, but largely as minutes of discussions. However, these documents are of great importance both from a purely scientific point of view and for historical reasons, since all the notes but the last one (accounting for a meeting of April 1944) directly reported on the activity that lead to the achievement of the first chain reaction, ranging from May to November, 1942. In fact, although the final scientific results obtained in Chicago were later collected and discussed in subsequent papers (patents or, in few cases, published articles), the present notes testify on *how* those results were obtained and, in some cases, also give detailed information on further, practically unknown, achievements, not reported in published papers.

Just to quote few examples, we mention an interesting trick suggested by Fermi for lowering the temperature in the pile, inspired by what happens in wind tunnels; or the control of the multiplication factor by means of the pressure of nitrogen in a liquid cooled pile. Much attention was, indeed, paid to the problem of heat transfer in the planned power and production plants, and to that of an effective and easy control of the chain reaction. Some discussions were also carried out on chain reacting piles working with fast (instead of slow) neutrons, and on different schemes for the uranium-graphite pile.

From an history of science viewpoint, these notes also present very interesting information, not available from other sources, on the Metallurgical Project, its formation and development, social and political implications (interventions of General Groves to one of the Meetings considered are registered in the corresponding notes), and so on. Several interesting and annoying discussions reveal, in fact, the urgency of the production of plutonium or other fissile material for military rather than civil applications already in 1942, the position of the problem of the moral effect of the operation and that of the relations with Army, including the issue of security and the distribution of information. However, different matters related to the physiological effect of the radiations developed in a pile (a problem raised more than once by Fermi), were considered as well, along with discussions about power utilization and long term research after the conclusion of the war.

A number of other interesting topics treated in the Meetings may be found in the detailed account of any document which follows, including (at the end) the notes on the Meeting of April 1944.

WAT1043a ◇ *Meeting of Engineering Council* (4 pages),
Present: Moore, Allison, Fermi, Leverett, Wheeler, Compton, Hilberry and Doan,
Report CE-106 (May 28, 1942).

The main discussion is on the cooling of the uranium-graphite pile by water, helium or both; some discussion is present on problems related to the pumping of the coolant. An interesting remark by Fermi is the following: since the temperature in wind tunnels is controlled by changing the cross section of the tubes, this trick can be used as well in piles for obtaining lower temperatures.
Minor discussions are on the design of a pilot power plant and a pilot extraction plant, with a remark by Fermi on the possibility of long lived activity induced in iron. Fermi also suggested a way for avoiding non uniform production of power in piles, just by blocking part of channels by graphite.
Other minor discussion is on leakage, where Fermi suggested an external graphite layer of 1 feet.

WAT1043b ◇ *Meeting of the Planning Board* (3 pages),
Present: Hilberry, Spedding, Allison, Wigner, Doan, Szilard, Wheeler, Fermi, Moore and Compton,
Reports CS-112, CS-185 (June 6, 1942).

Discussion on the status and organization of the activities: first self-sustaining pile, open pile working at a low rate of operation, helium cooling, other best cooling agents.
Re-organization of the work at Chicago (concentration of physics under Fermi), with discussion of problems on finding location, which should be chosen according to facilities and personnel at disposal (the decision, however, lies with Washington).
Discussion on the steps to undertake to protect Government's position about patents and on the understanding with British (a common patent pool).
Other scientific discussions are on the purity of the graphite supplied from various factories, with a remark by Fermi on the properties of different samples. Wigner discussed the results by Creutz on resonance absorption.

WAT1043c ◇ *Meeting of the Engineering Council* (9 pages),
Present: Moore, Wilson, Seaborg, Wheeler, Leverett, Fermi, Hilberry, Compton, Spedding and Allison,
Report CS-131 (June 11, 1942).

The choice of the site location of the first pile is discussed in detail, with reference to: water power for the cooling system; alternative schemes for supplying the power needs of the plant (about 65000 kW obtained either directly from the electric lines or from boilers by using engines producing mechanical power); schedule of plants (100 W, 100 kW, 100-100000 kW, 10^6 kW); housing and list of the personnel; advantages and disadvantages of the possible association of the separation project to the atomic power project. Alternatives on the site location are Chicago or Tennessee Valley: the majority of the presents is for the first one.

WAT1043d ◇ *Meeting of the Engineering Council* (7 pages),
Present: Moore, Leverett, Fermi, Wheeler, Seaborg, Doan, Wilson and Spedding,
Report CS-135 (June 18, 1942).

Discussions on: optimum lattice constants, neutrons losses due to ducts and channels, batch versus continuous operation of the pile and relation with

possible "chemical" experiments to plan (especially on the transformations undergone by U and Pu).
Fermi made some estimates on optimum lattice constants and gave relations among the multiplication factor K, the critical size and the neutron density. He favored the possibility to have more small rectangular ducts instead of less large squared ones.
Moreover, Fermi also suggests to use nitrogen as well for controlling a pile with liquid coolant, since K depends on pressure and would control up to 4% in K.

WAT1043e ◇ *Meeting of the Engineering Council* (4 pages),
Present: Moore, Leverett, Doan, Szilard, Allison, Teller, Seaborg, Wheeler, Fermi and Wilson,
Report CS-147 (June 25, 1942).

Fermi describes the advantages of the possibility to work with a pile *not* operating at optimum K.
Discussions on cooling by gas or liquid (but people later focused on the former): problems with hydrogen that reacts with uranium metal, with graphite (producing methane), etc.; problems with helium about leakage, large sound velocity (in relation to the use of blowers or compressors). The general agreement is to use helium for the 100000 kW pile.
Further discussions are on the diffusion of the fission products (in the circulating helium, the walls of the pile, etc.), with a suggestion by Fermi to not use water spray, and about blowers and compressors.

WAT1043f ◇ *Meeting of the Engineering Council* (6 pages),
Present: Fermi, Allison, Seaborg, Whitaker, Doan, Wilson, Moore, Leverett, Wheeler, Szilard, Compton, Spedding, Hilberry and Wollan,
Report CS-163 (July 2, 1942).

Discussion on the construction, installation (at Chicago) and operation of pile I, pile II and pilot plant. During the discussion on the necessary facilities, Fermi recommends to avoid limitations on water and electric supply in order not to "cut the wings".

Several studies and possible experiments for the piles I and II are considered. Fermi proposes to: study the thermal stability in pile I or intermediate pile with heat supplied from external sources; test different cooling mechanisms on the various piles; have experiments with the piles with the precaution that the produced radioactivity does not influence them. Fermi also suggests to: measure K by using the theory of anisotropic pile or, otherwise, preferably depend altogether on theoretical calculations; design pile I for evacuation (that is: for extracting uranium oxide from the pile); etc. (gas tight, sheet metal, balloon cloth).

WAT1043g ◇ *Meeting of the Engineering Council* (4 pages),
Present: Moore, Leverett, Fermi, Wigner, Allison, Wollan, Wheeler, Seaborg, Spedding, Szilard, Steams and Wilson,
Report CS-174 (July 9, 1942).

Fermi reports on a test of chemical stability of uranium, its reaction with graphite, etc. He also discusses a number of other topics as follows.
About the problem of heat transfer in an energy producing pile, Fermi proposes to study the behavior of the pile both when its temperature is large and when it is small, noting that the Reynolds number depends on temperature.
About the control mechanisms, he suggests an alternative scheme of control by putting in an absorbing gas, as well as to regulate the pile by raising or lowering the water level in it.
Finally, about cooling, Fermi observes that calculations are sufficient for helium but not for water, for which the intermediate experiment is required. Also, he points out that the exponential experiment is not suitable for measuring the effect of the helium diffusion through ducts.

WAT1043h ◇ *Meeting of the Technical Council* (7 pages),
Present: Fermi, Compton, Allison, Moore, Szilard, Wigner and Wheeler,
Report CS-184 (July 14, 1942).

Fermi discusses the use of beryllium (giving also some data) as moderator and neutron reflector, pointing out that it is not convenient to have an

all-beryllium structure, but rather a pile with 2 cm of beryllium around the uranium metal, since the thermal absorption is not compensated by the $n \to 2n$ reaction.
A discussion on the shielding of experimental plant and the handling of materials (with respect to protection problems) follows, with remarks by Fermi.
About the problem of cooling, a general discussion is made on the temperature dependence of conductivity of uranium oxide and graphite. In particular, Fermi considers the possible behavior of U_3O_8 in a pile working at 100 W (corresponding to a temperature of about 1000°C), and alternative choices of cooling. For the last point, Szilard proposes the use of bismuth as coolant.
About the control system, Fermi also observes that, for proper consideration of the problem of removing oxide, it is important to decide if control columns have to take out from the pile horizontally or vertically, such a choice coming out from practical attempts (build different structures and try out).
A minor discussion on the necessity of more people involved in operations is also present.

WAT1043i ◇ *Meeting of the Engineering Council* (3 pages),
Present: Moore, Leverett, Lewis, Fermi, Wollan, Hilberry, Whitaker, Wilson, Wheeler, Allison, Wigner, Seaborg and Doan,
Report CE-194 (July 21, 1942).

General discussions on power plant, extraction plant (by using fluorination, precipitation with a carrier), and pile problems (radiation, recharging, breakdown, etc.) are present.
Fermi observes that it would be useful to make some of the discussed problems clear to a radiologist.

WAT1043j ◇ *Meeting of the Technical Council* (5 pages),
Present: Fermi, Szilard, Wigner, Compton, Whitaker, Allison, Moore, Wheeler and Doan,
Report CS-202 (July 25, 1942).

Fermi discusses the value for the multiplication factor K (equal to 1.06) obtained by using several technical solutions and with UO_2 compressed in pseudo-spheres.
A general discussion on the providers of uranium metal and the product provided then follows.
Fermi raises also the question of the physiological effects of radiations, by claiming the need for a physician. On this point, Compton displays data from several medical institutions (National Cancer Institute and Chicago Tumor Institute).

WAT1043k ◇ *Meeting of the Planning Board* (4 pages),
Present: Spedding, Fermi, Szilard, Wigner, Doan, Moore, Wheeler, Compton and Hilberry,
Report CS-213 (August 1, 1942).

About the problem of heat transfer and cooling, Fermi reports some values for the conductivity of the materials involved in different conditions.
A proposal comes out to study the chemistry of the pile under radiation, with Fermi observing that two tubes leading into the pile were planned for insertion of samples. Other technical topics under discussion regards the use of carbide in pilot plant, but not in experimental plant, the recommendation to use graphite around the pile (for reflecting escaping neutrons), and the problem of carbide production.
Also interesting are the discussions about the responsibility for clearance and security (leaved to Army), and the authorization for giving information about the pile to "anyone", Conant being recognized as the final authority on the distribution of information. A mention is present about the use of the pile for power but not for production of explosives.

WAT1043l ◇ *Meeting of the Engineering Council* (2 pages),
Present: Moore, Leverett, Steinbach, Fermi, Spedding, Wheeler, Wigner, Seaborg and Wilson,
Report CE-229 (August 8, 1942).

Minor discussions are present about safety and control rods, the use of a covering layer of graphite, possible external (rather than inside the pile)

cooling with oil, etc.
Fermi estimates a 90% probability for achieving thermal stability in the pile.

WAT1043m ♢ *Meeting of the Technical Council* (2 pages),
Present: Nichols, Hilberry, Spedding, Doan, Fermi, Steinbach, Grafton, Boyd, Moore and Wheeler,
Report CS-251 (September 4, 1942).

Fermi reports on the status of the values obtained in the multiplication factor K. Other interesting discussions regard the possible combination of water cooling and bismuth cooling (with bismuth circulating in the uranium-containing channels, between the aluminium jacket and the uranium rod), the status of production of uranium (and, in particular, about material provided by Alexander), and problems with radiation protection.
Fermi proposes to consider the possibility to test uranium metal with an exponential pile, with a remark that "the metal won't be better than the oxide".

WAT1006 ♢ *Discussion of Helium Cooled Power Plant* (3 pages),
Present: Leverett, Cooper, Moore, Wigner, Steinbach, Fermi, Szilard and Wheeler,
Report CS-267 (September 16, 1942).

Several problems related to He cooled power plant (steel and compressors, control and safety rods, oil unaffected by radiation, etc.) are discussed in this meeting. A list of topics to be studied (and tasks to assign) is reported. It is also interesting to point out a remark by Szilard, which argued that 10^5 kW of power were necessary to win the war.

WAT1043n ♢ *Meeting of the Technical Council* (11 pages),
Present: Compton, Wheeler, Moore, Allison, Szilard, Fermi and Spedding,
Report CS-274 (September 18, 1942).

General discussions on technical details, including a water cooled plant, are made.
Quite interesting is the discussion of several points regarding the policy for the site X. These points included: triple extraction plant; water supply; association with the production unit.
It is pointed out that, while the work by Lawrence and Urey was carried out under the only control of OSRD, that performed by the Metallurgical Laboratory was under the control of OSRD through the Army. Possible alternatives were considered to propose to work not under the direct supervision of the Army, Szilard suggesting to take no action until the pile was effectively operating. Discussions about the "future" of the people working in the project are made, with particular reference to the research policy of the next 20 years.
After other discussions about alternative locations for production piles (Argonne, Tennessee, Palos Park), the attention is turned on the committee proposed by Bush for the decision about the destination of the fissile material: this had to be employed for the design and production of bombs.
Finally, the problem of further man power is discussed. Szilard recommends to associate Auger, Rasetti, Goldhaber and Rossi, taking care of an advice by Conant that the inclusion of such people in the project was possible only if the work is made in a restricted area. Another name proposed by Szilard is that of Lewis, although it is recognized that he had already too many duties, among which the development of methods for explosives in China.

WAT10430 ◊ *Meeting of the Technical Council* (8 pages),
Present: Allison, Fermi, Moore, Wigner, Compton, Wheeler and Oppenheimer,
Report CS-281 (September 29, 1942).

The main discussions deals mainly with the work related to the pile operating with fast neutrons, and the question of the move of the work to Site X (logistic problems with the Site X, etc.). To this regard, it has to be pointed out that Fermi dissented about the shift of the work on fast neutrons to Site X.
The choice of the Site X as the new main basis of the Metallurgical Project, had by now been definitive (a preliminary, rough map was also included), this having been favored to a certain extent by Army, who preferred to have

all the relevant work in only one enclosure.
Other non physics arguments, touched in the meeting, are about the patent rights assignment to the American government and the collaboration of British engineers. Instead, minor topics of scientific interest are about helium and bismuth cooled plants and water cooling.

WAT1043p ◇ *Meeting of the Technical Council* (10 pages),
Present: Allison, Fermi, Wigner, Compton, Whitaker, Moore, Cooper, Szilard, Manley, McMillan, Wheeler and Doan,
Report CS-284 (October 1, 1942).

As in the meeting of two days before, the arguments of the discussion deals with work on fast neutrons and, especially, with the move to Site X (with evaluation of alternative, possible options). The majority of the presents prefer Chicago as a suitable site in order not to waste time and, moreover, they express their preference to work not under the supervision of Army and independently of industry.
Other discussions concern the work by Fermi on slow neutrons. The development work for piles 1 and 2 was assumed to be out of way by March 1 (1943); after this was completed, the schedule comprised to work on other piles, including fast neutron pile and heavy water pile.
Mentions are made to bismuth cooling and to other technical issues.

WAT1043q ◇ *Meeting of the Technical Council* (9 pages),
Present: Allison, Fermi, Szilard, Moore, Wigner, Whitaker, Wheeler, Steinbach, Compton and Groves,
Report CS-286 (October 5, 1942).

Detailed discussions regard results about different schemes for the pile: 1) (Wigner) uranium rods, water cooling with pipes, graphite as moderator; 2) (Fermi) uranium lumps imbedded in graphite, cooling by occasional water pipes; 3) (Cooper) metal pipes, shot, and graphite as moderator, with removal of uranium metal and recharging; 4) external cooling (which gives about 300 kW power only). No definitive decision is adopted on this argument.
At a certain point of the meeting, Compton and General Groves came in

reporting, quite interestingly, that War Department considered the Metallurgical Project important. The discussion then changed a little, with Allison's claims that one couldn't win the war with an externally cooled plant and Fermi's remark that the program will be delayed by several months due to change plan at the Argonne site.

WAT1043r ◇ *Meeting of the Technical Council* (9 pages),
Present: Allison, Wigner, Compton, Whitaker, Szilard, Wheeler, Fermi, Moore, Cooper, Steinbach and Kirkpatrick,
Report CS-290 (October 7, 1942).

Various arguments are treated, all related (directly or indirectly) to the move to the Site X.
Groves is said to be eager to have explosives, by June 15, 1943, several plants being assumed to be operating after June 15 (dates by Groves have not been reported in the minutes). Some discouragement in Army is reported, that the project had not achieved more so far. Moral effect of operations is discussed as well.
Technical council recommends the construction at site X of a 300 kW pile (Pile 2) by March 15 and, to this end, Fermi notes that there was no need that such a pile be made with uranium only in the form of metal.
The decision by Groves about production and extraction of fissile material at Site X is discussed, as well as concentration of the work on fast neutrons at the same site (Groves' decision urged by Oppenheimer). Topics related to power utilization and long term research (not) at the same Site X, with possible "countermeasures" to undertake regarding this last point, are also considered.
The time schedule for Pile 1 is committed to Fermi, that for Pile 2 to Whitaker, etc. In connection to possible changes in the plans due to move of the work, Fermi notes that the original plan (for Pile 1) was to prove that chain reaction goes and to flash pile for a limited time.
A minor discussion on the question of plating uranium (and possible reactions of it) is also made.

WAT1043s ◇ *Meeting of the Technical Council* (6 pages),
Present: Allison, Szilard, Wigner, Moore, Wheeler and Fermi,
Report CS-294 (October 12, 1942)

The questions of machining graphite and sintered uranium metal are discussed in some detail.
Several options for an externally cooled pile are considered by using: 1) copper pipes; 2a) copper shell cooled by air; 2b) copper shell cooled by water spray. Fermi's preference is for choice 2a), assuming that copper didn't leak; however he himself points out that it would be difficult to find leak (then, the rubber cap method was considered). A general agreement is expressed to leave this problem to Fermi to look into.
A peculiar remark by Fermi is about his feeling that time estimates for pile to work were not certain, since the amounts of uranium metal required (to prevent loss in K) were probably underestimated. Nevertheless, he considered a mistake to wait any time at all for producing the chain reaction in this way, while he favored to put the uranium oxide in form of spheres.
Other interesting suggestions by Fermi are about the change in picture of K, and the use of a "sandwich" experiment with 4 layers of uranium metal (about 1 ton); U_3O_8 pile was suggested to be used as a standard, and a gain of 0.8% in K was estimated just by removing nitrogen.
More discussions are again about the move to Site X: people felt not to transfer until the working situation was not clear. Fermi observes that all matter about Site X appeared to arise from a mistaken impression that experimental work was practically finished.

WAT1008 ◇ *Conference on Lattice Spacing* (2 pages),
Present: Steinbach, Leverett, Fermi, Wigner and Wheeler,
Memo #15 (October 21, 1942).

Estimates about optimum lattice spacing in the pile and C/U ratio for the helium cooled plant are presented, by taking into account the request of a minimum total amount of uranium metal. Fermi suggests to save the amount of metal by diminishing the proportion of U to C toward the outside of the pile.

WAT1043u ◇ *Meeting of the Technical Council* (4 pages),
Present: Allison, Compton, Fermi, Moore, Spedding, Szilard and Wigner,
Report CS-356 (November 19, 1942).

Preliminary agreements are discussed in this meeting about the preparation of a report for a committee (which included industrial specialists on production problems) formed for examining the Metallurgical project in Chicago (alternative to the project based at Berkeley), that will come on November 26. The topics discussed regard the purity of the final product (plutonium), its radioactivity, spontaneous heating, etc., by pointing out that very little was known about the metallurgy of plutonium, and the processes proposed for producing it seemed very far from industrial possibilities.
Remarks are present about the availability in U.S.A. of 1000 tons of uranium, and that no slowing down of neutrons would be required for Piles 1 and 2, although Wigner pointed out that studies on fast neutron reactions were still preliminary.
About non strictly scientific issues, it is quite interesting to note the invitation from Washington authorities to go ahead with the production of plutonium, and the remark on General Groves who was interested in all possible military uses of what studied in Chicago, rather than applications to power production.
In these notes it is pointed out that a report on the chain reaction of about 10 pages should have been prepared by Fermi (along with the contributions of other scientists on other subjects), this report being unknown.

WAT ◇ *Notes on Meeting of April 26-28, 1944* (5+8 pages),
Present: Fermi, Allison, Wigner, Smyth, Szilard, Morrison, Watson, Feld, Hogness, Young, Weinberg, Creutz, Cooper, Vernon and Ohlinger,
Report N-1729, Eck-209.

This report is composed of two papers accounting for a two-day meeting (April 26-28, 1944).
A large part of the first paper, with much of the Fermi's intervention at the two-day meeting, was already published in the *Collected Papers* [Fermi (1944a)]. There, the discussion focused on chain reaction for the produc-

tion of a power of about 10^6 kW. A large mother plant was conceived for producing plutonium to be used as fissile material in smaller plants; Fermi noted that this arrangement could be useful for the heating of towns. Then, after a brief theoretical discussion, with numerical estimates and data, for the full metal utilization, Fermi focused on four different types of piles, both operating with slow neutrons and fast neutrons, and depending on the percentage of the enrichment of the fissile material and the moderator employed. The remaining part of the first paper, not published in [Fermi (1944a)], dealt with alternatives to what discussed by Fermi proposed by Szilard, coming out by "assuming more optimistic values of the constants so as to indicate other potentialities".
The second paper dealt mainly with a discussion by Morrison on several scientific, economic, and social issues related to pile producing power, resulting quite interesting from an historical point of view. Among the remarks to the Morrison's relation, we mention that "Fermi questioned the estimated value of [the number of neutrons produced/number of fissionable atoms used up] = 2.5 on the ground that it might be too optimistic and pointed out that there is a long range future in developing the full utilization of [^{238}U] and thorium".

Edited reports

In the Wattenberg archive several reports are present, describing some of the work performed in 1943 and 1944 by the division of the Metallurgical Laboratory headed by Fermi. As it is already known (see, for example, [Segrè (1970)] or the introduction to the papers of Volume II of [Fermi (1962)]), Fermi played a very active role in the work officially assigned to others, so that although none of these reports describes (theoretical or experimental) activities directly performed by him, they nevertheless reveal precious information about part of the work done under the supervision of Fermi. Of course, these few reports account only for activities not directly related to military applications, although they were finalized primarily to the production of fissile material for explosives.

The detailed description of the four reports available follows below.

WAT1023 * *Report for Month Ending September 25, 1943* (55 pages),

edited by A.H. Compton, S.E. Allison and E. Fermi,
Report CP-964.

This report does not contain the description of the work performed directly by Fermi, but rather it describes some of the activities performed in the month of September 1943 by different people under Fermi's supervision.[3] The most interesting ones are summarized below.

Much work, both experimental and theoretical, was devoted to the study of a so called P-9 pile, that is a chain reacting system with heavy water as a coolant. The Zinn group was involved in making plans for an experimental P-9 plant at Argonne, while the Young group worked on the design of a P-9 pile, both for a heterogeneous and a homogeneous pile. In a P-9 pile more fissile material had to be used for several technical reasons (related to pumps and heat exchangers employed), but this was compensated by a higher value of the effective multiplication factor K. Other problems to be solved were that of the separation of the uranium oxide from the circulating heavy water and the method to choose for separating the heavy water from the cooling liquid (in order to use it again after a given cycle), the determination of the critical size for a P-9 pile, etc. A sketch showing one possible arrangement for a (near) homogeneous P-9 slurry pile was presented as well.

Another research conducted by some people of the Zinn group regarded the "cell saturation" effect, induced by increasing the absorption cross section of a single cell (to be used in the lattice of a pile) to such an extent that the change in reactivity of the pile was no longer proportional to the amount of impurity added, but rather to its square.

The group headed by Anderson studied, among the other things, the residual radioactivity of control rods made of different materials, the effect of fast fission on the multiplication factor, and remeasured the ratio of the absorption cross sections of boron and hydrogen.

Marshall, instead, studied and prepared a velocity selector consisting of a sandwich of aluminium and cadmium sheets for obtaining measurements on neutrons with definite energy (see previous section), while Morrison performed an experimental study on the boundary conditions for the neutron density between paraffin and graphite for a study on a neutron reflector, with the determination of the temperature effect on the diffusion length in

[3]In this report, the summary of these activities was written (on September 27, 1943) by Wigner "in the absence of Mr. Fermi".

graphite.
The Feld group got involved in the investigation on inelastic cross sections of several heavy elements (lead, bismuth, iron and uranium), which were relevant for fast neutron chain reaction, while started novel measurements on the $(n, 2n)$ and (γ, n) reactions on beryllium with the paraffin pile technique.
In this report, a large account is also given to some theoretical activity performed by several people, as emphasized by Wigner himself: "the Theoretical Physics Section's report for this month is in considerably more detail than was the custom in previous months". Then Wigner interestingly continues: "It is not expected to report in similar detail in the future, as a good part of the work done by us is principally for our own use. However, it was intended to give a more adequate picture of the work that we are doing".
As already mentioned, much of the theoretical work was devoted to the P-9 pile, but another interesting investigation (by some people of the Weinberg group) dealt instead with the fast chain reacting pile, having found that the measured cross section for fast fission was smaller than previously assumed (the ratio of the fast fission neutrons to the thermal fission ones was previously measured incorrectly).
Other researches regarded the study of possible danger situations in which control rods could not be governed due to a pressure damping in the cooling circuit, or the studies on the preferred geometrical form of a pile, its dimensions or other technical details (pumps, valves, heat exchangers, etc.)

WAT1027 * *Report for Month Ending December 25, 1943 - Part II* (6 pages), edited by A.H. Compton, S.E. Allison and E. Fermi, Report CN-1190.

This report does not contain the description of the work performed directly by Fermi, but rather it describes some of the activities performed in the month of December 1943 by different people under Fermi's supervision. The summary of the first part of this report is in [Fermi (1943g)], while the second part, dealing with the work performed by the Anderson group, is considered here.
The Anderson group determined the yield of plutonium per kWh for the Argonne pile, and predicted that the Hanford 250 kW operations should

produce about 230 grams of plutonium per day. They also studied neutron yields from polonium by irradiating different samples (lithium sulphur, chlorine and argon) with alpha particles.

WAT1037 * *Report for Month Ending August 26, 1944* (38 pages),
edited by A.H. Compton, E. Fermi and W.H. Zinn,
Report CP-2081.

This report does not contain the description of the work performed directly by Fermi, but rather it describes some of the activities performed in the month of August 1944 by different people under Fermi's supervision. The most interesting ones are summarized below. It was here noted that the Chicago Pile N.3 completed its second month of operation, while a silver tape recorder was completed and installed on CP-2.
The Seren group studied the properties of the thermal column (see the previous section), while Zinn performed measurements on the Bragg reflection of a highly collimated beam of thermal neutrons and, more in general, on neutron spectroscopy.
Licthenberger studied, instead, the scattering from strong absorbers, Morrison and Teller having identified the isotope ^{112}Cd as that responsible for the strong capture of thermal neutrons, while Wattenberg prepared photoneutron sources made by activation of several nuclides.
Finally, Anderson focused on the neutrons from the reaction ^{3}H + ^{2}D $\rightarrow ^{4}$He + n and, in general, studied possible transformations of thermal into fast neutrons, while the Nagle's group measured the yield of delayed neutrons.

WAT1039 * *Report for Month Ending October 28, 1944* (27 pages),
edited by A.H. Compton, E. Fermi and W.H. Zinn,
Report CP-2301.

This report does not contain the description of the work performed directly by Fermi, but rather it describes some of the activities performed in the month of October 1944 by different people under Fermi's supervision. The most interesting ones are summarized below.
The group guided by Zinn studied the poisoning of the chain reaction by ^{135}Xe in CP-3 (see the previous section) and related arguments, and May

and Anderson measured the nuclear constants of ^{233}U, whose behavior they found similar to that of ^{235}U but giving larger values for K, so that the use of ^{233}U was suggested to be more favorable than that of Pu.
Langsdorf studied, instead, the resonance scattering of neutrons, while the Seren group measured the activation cross section of columbium Cb, which resulted to be a useful information for producing stainless alloy with uranium.

WAT1040 * *Report for Month Ending November 25, 1944* (24 pages),
edited by A.H. Compton, E. Fermi and W.H. Zinn,
Report CP-2436.

This report does not contain the description of the work performed directly by Fermi, but rather it describes some of the activities performed by different people under Fermi's supervision. The most interesting ones are summarized below.
The Zinn group continued their studies on the Bragg reflection of thermal neutrons from a crystal (considered as a neutron spectrometer), and observed also the total reflection by Cu, Al, Be, glass and graphite mirrors.
The Lichtenberger group, instead, made boron absorption measurements in order to study the variation with energy of the resonance absorption of ^{238}U, while the Wattenberg group mainly focused on photo-neutrons from U+Be sources.
Finally, the Hill group studied the tuning of coincidences in α-chambers and made an analysis of a number of pictures from cloud chambers searching for ternary fissions (and possible appearance of three-particle fissions).

Other contributions

Few other minor documents, of different nature and relevance, have come to light during our research.

Although not properly a paper or a report, the first document we point out here is a letter written by Fermi to Lord Rutherford as early as in 1934, when he and his group in Rome started to study the radioactivity induced by neutron bombardment. As recalled above, these studies led, in October 1934, to the discovery of the important properties of slow neutrons (see the first patent). This document is presently conserved among the Rutherford Papers at the Cambridge University Library (U.K.). The text of the letter is as follows:

Rome, June 15th, 1934

> My dear Prof. Rutherford,
> I enclose a reprint of a paper on the present status of our researches on the activation of uranium. The same results shall appear shortly in Nature.
>
> We have been forced to publish these results of a research which is actually not yet finished by the fact that the newspapers have published so many phantastic [sic!] statements about our work that we found it necessary to state clearly our point of view.
>
> We are now engaged in trying to understand the influence of the neutron energy on the activation of elements. We try to do this using neutrons from a source of Em+B.
>
> We are interested in this problem not only as it can throw some light on the processes involved, but also because we plan to construct a neutron tube similar to that of the Cavendish laboratory.
>
> In this connection I would be much obliged if in case you have tested your tube for activating elements, you would let me know some data on the intensities of the activations.
>
> The construction of this tube would be much facilitated for us if it were possible for some assistants of our laboratory (Drs. Amaldi and Segré) to come this summer to the Cavendish Laboratory in order to see the apparatus and possibly be instructed about its use.
>
> I would be very grateful to you if you will give me an answer on this point.
>
> With kindest regards
>
> yours very truly
>
> Enrico Fermi

Here, the relevant information, mainly from an historical point of view, is the reference to the construction of "a neutron tube similar to that of the Cavendish laboratory" and, particularly, the request by Fermi of "some data on the intensities of the activations" ought to be obtained by the Rutherford group at the Cavendish Laboratory. Indeed, this testify for an attempt made by Fermi to set up some collaboration among the two groups even *before* Amaldi and Segrè came to Cambridge in the summer of 1934. Although four letters by Rutherford to Fermi were known (they are conserved at the Domus Galilaeana in Pisa, Italy), such collaboration at a distance had not been addressed previously. The reply to the letter by Fermi is, indeed, known[4], but this fact cannot be deduced from it. The answer by Rutherford to the specific request by Fermi was negative, denoting the advantage of the Rome group over that of the Cavendish Laboratory on this point (probably unexpected by Fermi): "I cannot at the moment give you definite statement as to the output of the neutrons from our tube but it should be of the same order as from an Em+Be tube containing 100 millicurie and may be pushed much higher."

Other papers we consider here in the following are not, strictly speaking, archive documents, since they were already published and may be found in the original literature but, for some reasons, do not appear among the Fermi's *Collected Papers*, so that they are practically unknown. For completeness, we think it is useful to briefly consider here even such papers.

The first paper we consider is the following:

> *Moments magnétiques des noyaux*,
> by E. Fermi,
> in *Le magnétisme*, Rapports et discussions du Sixieme Conseil de Physique, tenu a Bruxelles du 20 au 25 Octobre 1930, sous les auspices de l'Institut International de Physique Solvay, Gauthier-Villars, Paris, 1932.

The general argument of the talk delivered by Fermi at the mentioned conference (Fermi was a member of the Institut de Physique Solvay) is the same of that treated in other three papers [Fermi (1930a)], [Fermi (1930b)], [Fermi (1930c)], that is the hyperfine structures of atomic spectra, so that we refer to the introduction to these papers in [Fermi (1962)] (see page 328 of Volume I of this book) for further details. The exposition of the argument is somewhat different from that of the other papers, and interesting discussions (from an historical point of view) with Pauli, Stern, Kramers, Richardson, Zeeman, Dirac and Heisenberg were reported.

[4]It is reported by E. Amaldi in his not very known paper *Neutron Work in Rome in 1934-36 and the Discovery of Uranium Fission*, Riv. Stor. Sci. **1**, 1-24 (1984).

Particularly interesting is a proposal of a "theoretical experiment" made by Fermi at the discussion section of the mentioned conference after a talk by Pauli, regarding the measure of the magnetic moments of electrons oriented in a given magnetic field. The comment by Fermi is as follows (see page 276 of the book above; see also the comments by Bohr and others to the Fermi proposal):

> On peut aisément se rendre compte, par une expérience théorique, de la possibilité d'obtenir un faisceau d'électrons polarisé, et cela de la manière suivante. Au-dessus d'un pôle magnétique se trouve une chambre contenant des électrons. A l'intérieur agit le champ magnétique, que l'on peut considérer comme uniforme, produit par le pôle magnétique. Nous pouvons toujours supposer la température assez basse pour que dans l'état d'équilibre thermique les électrons se disposent presque tous avec leur moment propre parallèle à la direction du champ. Lorsque léquilibre thermique est complètement établi nous ouvrons l'enceinte vers l'exteriéur: les électrons se diffusent lentement dans l'espace environnant en conservant leurs spins orientés parallèlement aux lignes de force.

The second paper is, instead, the following:

Nuclei ed elettroni,
by E. Fermi,
Gerarchia **11**, 879-885, (1931).

This is the first popular article by Fermi devoted to nuclear physics, illustrating its status at the end of 1931. It was written just after the first international conference on this topic, held in Rome in October 1931. This was inaugurated by Prime Minister Benito Mussolini, and the paper mentioned is practically an account of the results presented at the conference, written for the political journal "Gerarchia" which was founded and edited by Mussolini himself. It has been recently reproduced in [Barone (2009)].

The third paper by Fermi we consider is the following:

Total Collision Cross Section of Negative Pions on Protons,
by D.E. Nagle, H.L. Anderson, E. Fermi, E.A. Long and R.L. Martin,

Phys. Rev. **86**, 603 (1952).

"The transmission of negative pions in liquid hydrogen has been measured using the pion beams of the Chicago synchrocyclotron. Pion beams with energies from 60 to 230 MeV were used. The transmissions were measured using scintillation counting techniques. The total collision cross section increases with energy starting from small values at 30 MeV and rising to the "geometrical" value of about 60×10^{-27} cm^2 at about 160 MeV. Thereafter up to 220 MeV, the cross section remains close to this value. The steep energy dependence at low energies is consistent with interpretation that the pion is pseudoscalar with a pseudovector interaction."

As the companion paper in [Lundby (1952)], it was presented at the 1952 Annual Meeting of the American Physical Society held at New York on January 31 - February 2, 1952; in the mentioned journal, only the abstract of both papers were reported, as custom for the proceedings of that meeting. It testifies for some of the work performed at Chicago by Fermi and his collaborator on pion physics; the results are summarized in the abstract reported entirely above. Strange enough, the paper considered does not appear among the *Collected Papers* [Fermi (1962)], contrary to what happen for the paper in [Lundby (1952)], although both abstracts were published in the journal on the same page.

The last paper is a popular article written by Fermi for a newspaper, in the occasion of the tenth anniversary of the operation of the first chain reacting pile at Chicago, on December 2, 1942:

Fermi's own story,
by E. Fermi,
Chicago Sun-Times, November 23, 1952.

Here Fermi gave a personal description of that event, preceded by a short story of the main stepping-stones that leaded to the realization of the first chain reaction, starting from the discovery of radioactivity by H.A. Becquerel. It is particularly interesting the conclusion of this article, where Fermi stated his view (and hope) about science and possible military applications of it:

> The further development of atomic energy during the next three years of the war was, of course, focused on the main objective of producing an effective weapon.
>
> At the same time we all hoped that with the end of the war emphasis would be shifted decidedly from the weapon to the peaceful aspects of atomic energy.
>
> We hoped that perhaps the building of power plants, production of radioactive elements for science and medicine would become the paramount objectives.
>
> Unfortunately, the end of the war did not bring brotherly love among nations. The fabrication of weapons still is and must be the primary concern of the Atomic Energy Commission.
>
> Secrecy that we thought was un unwelcome necessity of the war still appears to be an unwelcome necessity. The peaceful objectives must come second, although very considerable progress has been made also along those lines.
>
> The problems posed by the world situation are not for the scientist alone but for all people to resolve. Perhaps a time will come when all scientific and technical progress will be hailed for the advantages that it may bring to man, and never feared on account of its destructive possibilities.

Appendix B: The Complete List of Fermi's Papers

In the following we report the complete list of works by Enrico Fermi, as recognized on December 2009. This list is probably incomplete; a number of papers issued in U.S.A during the Second World War, for example, have been treated as restricted data for some time by the competent governmental authorities.

In this list we have included also papers not directly written by Fermi, but directly related to works performed by him, such as lecture notes, reports or notes on meetings, and so on. These papers have been pointed out by special symbols. In particular, we have denoted with a $\diamondsuit$ those where the contribution by Fermi is explicitly recognizable (typically, notes on meetings), and with a * those where such contribution can be deduced only indirectly (lecture notes or edited reports).

The following list has been compiled by adopting a full chronological criterion. We have ordered the papers according to: 1) the explicit date reported in it; 2) the date of reception of the paper by the publishing house; 3) the publication date. If none of this applies, we have made recourse to some internal analysis of the given paper and to the comparison with other papers, also taking into account (if possible) the ordering of the Fermi's *Collected Papers* [Fermi (1962)] prepared by E. Segrè et al. (which, however, is not strictly chronological).

The papers already present in the *Collected Papers* have been pointed out by the code FNM followed (if applicable) by the related list number. Those unpublished present only (for what we know) in the Wattenberg Archive have been denoted by the code WAT followed (if applicable) by the list number of that archive. Finally, the patents registered at the

U.S. Patent Office have been pointed out by the code USP followed by the chronological list number.

For books, we have reported only the first, original edition of them, omitting the subsequent translations.

Books

Book1 *Lezioni di elettrodinamica*, pubblicate a cura dello studente Adelino Morelli, pp. 95, Stabilimento tipo-litografico del Genio Civile, Roma, [s.d.].

Book2 *Lezioni di Fisica teorica*, dettate dal Prof. E. Fermi, raccolte dai Dott. Dei e Martinozzi, pp. 60, [s.n.], Roma, 1927.

Book3 *Introduzione alla fisica atomica*, pp. 330 Zanichelli, Bologna, 1928.

Book4 *Fisica ad uso dei Licei*, vol. I, pp. 239 and vol. II, pp. 243, Zanichelli, Bologna, 1929.

Book5 *Molecole e cristalli*, pp. 303 Zanichelli, Bologna, 1934.

Book6 *Thermodynamics*, pp. VII-160, Prentice-Hall, New York, 1937.

Book7 E. Fermi and E. Persico, *Fisica per le Scuole Medie Superiori*, pp. 314 Zanichelli, Bologna, 1938.

Book8 * *Nuclear Physics. A Course given at the University of Chicago*, Notes compiled by J. Orear, A.H. Rosenfeld, and R.H. Shulter, pp. VII+246, The University of Chicago Press, Chicago, 1949.

Book9 *Elementary Particles*, pp. XII+110, Yale University Press, New Haven, 1951.

Book10 *Notes on Quantum Mechanics*, pp. VII+171, The University of Chicago Press, Chicago, 1961.

Papers

1921

P1. [FNM1] *Sulla dinamica di un sistema rigido di cariche elettriche*, Nuovo Cimento **22**, 199-207 (1921).

P2. [FNM2] *Sull'elettrostatica di un campo gravitazionale uniforme e sul peso delle masse elettromagnetiche*,
Nuovo Cimento **22**, 176-188 (1921).

1922

P3. [FNM3] *Sopra i fenomeni che avvengono in vicinanza di una linea oraria*,
Rend. Lincei **31**(1), 21-23, 51-52, 101-103, (1922).

P4. [FNM4b] *Correzione di una grave discrepanza tra la teoria delle masse elettromagnetiche e la teoria della relatività. Inerzia e peso dell'elettricità*,
Rend. Lincei **31**(1), 184-187 (1922).

P5. [FNM4b] *Correzione di una grave discrepanza tra la teoria elettrodinamica e quella relativistica delle masse elettromagnetiche. Inerzia e peso dell'elettricità*,
Rend. Lincei **31**(1), 306-309 (1922).

P6. [FNM38b] *Un teorema di calcolo delle probabilità ed alcune sue applicazioni*,
Tesi di abilitazione della Scuola Normale Superiore. Pisa, 1922.

P7. [FNM4a] *Über einen Winderspruch zwischen der elektrodynamishen und der relativistishen Theorie der elektromagnetischen Masse*,
Phys. Zeits. **23**, 340-344 (1922).

P8. [FNM6] *I raggi Rontgen*,
Nuovo Cimento **24**, 133-163, (1922).

P9. [FNM7] *Formazione di immagini coi raggi Rontgen*,
Nuovo Cimento **25**, 63-68, (1923).

P10. [FNM14] *Sulla teoria statistica di Richardson dell'effetto fotoelettrico*,
Nuovo Cimento **26**, 97-104, (1923).

1923

P11. [FNM5] *Le masse nella teoria della relatività.* Da A. Kopff, *I fondamenti della relatività Einsteiniana* Edizione italiana a cura di R. Contu e T. Bembo. Hoepli, Milano 1923, 342-344

P12. [FNM8] *Sul peso dei corpi elastici*,
Mem. Lincei **14**, 114-124, (1923).

P13. [FNM9] *Sul trascinamento del piano di polarizzazione da parte di un mezzo rotante*,
Rend. Lincei **32**(1), 115-118, (1923).

P14. [FNM10] *Sulla massa della radiazione in uno spazio vuoto,*
E. Fermi and A. Pontremoli,
Rend. Lincei **32**(1), 162-164, (1923).

P15. [FNM4c] *Correzione di una contraddizione tra la teoria elettrodinamica e quella relativistica delle masse elettromagnetiche,*
Nuovo Cimento **25**, 159-170, (1923).

P16. [FNM12] *Il principio delle adiabatiche ed i sistemi che non ammettono coordinate angolari,*
Nuovo Cimento **25**, 171-175, (1923).

P17. [FNM11b] *Dimostrazione che in generale un sistema meccanico normale è quasi ergodico,*
Nuovo Cimento **25**, 267-269, (1923).

P18. [FNM13] *Alcuni teoremi di meccanica analitica importanti per la teoria dei quanti,*
Nuovo Cimento **25**, 271-285, (1923).

P19. [FNM11a] *Beweis dass ein mechanisches Normalsystem im allgemeinen quasi-ergodisch ist,*
Phys. Zeits **24**, 261-265, (1923).

P20. [FNM15] *Generalizzazione del teorema di Poincarè sopra la non esistenza di integrali uniformi di un sistema di equazioni canoniche normali,*
Nuovo Cimento **26**, 105-115, (1923).

P21. [FNM16] *Sopra la teoria di Stern della costante assoluta dell'entropia di un gas perfetto monoatomico,*
Rend. Lincei **32**(2), 395-398, (1923).

P22. [FNM17a] *Sulla probabilità degli stati quantici,*
Rend. Lincei **32**(2), 493-495, (1923).

1924

P23. [FNM18] *Sopra la riflessione e la diffusione di risonanza,*
Rend. Lincei **33**(1), 90-93, (1924).

P24. [FNM19] *Considerazioni sulla quantizzazione dei sistemi ch contengono degli elementi identici,*
Nuovo Cimento **1**, 145-152, (1924).

P25. [FNM20] *Sull'equilibrio termico di ionizzazione,*
Nuovo Cimento **1**, 153-158, (1924).

P26. [FNM11a] *Über die existenz quasi-ergodisher Systeme*,
Phys. Zeits. **25**, 166-167, (1924)

P27. [FNM17b] *Über die Wahrscheinlichkeit der Quantenzustande*,
Z. Physik **26**, 54-56, (1924).

P28. [FNM21a] *Berekeningen over de intensiteiten van spektraallijnen*,
Physica **4**, 340-343, (1924).

P29. [FNM23b] *Über die Theorie des Stosses zwischen Atomen und elektrisch geladenen Teilchen*,
Z. Physik **29**, 315-327, (1924).

1925

P30. [FNM22] *Sui principi della teoria dei quanti*,
Rend. Seminario matematico Università di Roma **8**, 7-12, (1925).

P31. [FNM24] *Sopra l'urto tra atomi e nuclei di idrogeno*,
Rend. Lincei **1**, 77-80, (1925).

P32. [FNM21b] *Sopra l'intensità delle righe multiple*,
Rend. Lincei **1**, 102-124, (1925).

P33. [FNM25] *Una relazione tra le costanti delle bande infrarosse delle molecole triatomiche*,
Rend. Lincei **1**, 386-387, (1925).

P34. [FNM23a] *Sulla teoria dell'urto tra atomi e corpuscoli elettrici*,
Nuovo Cimento **2**, 143-158, (1925).

P35. [FNM26] *Effect of an Alternating Magnetic Field on the Polarization of the Resonance Radiation of Mercury Vapour*,
E. Fermi and F. Rasetti,
Nature (London), **115**, 764, (1925).

P36. [FNM28] *Effetto di un campo magnetico alternato sopra la polarizzazione della luce di risonanza*,
E. Fermi and F. Rasetti,
Rend. Lincei **1**, 716-722, (1925).

P37. [FNM28] *Ancora dell'effetto di un campo magnetico alternato sopra la polarizzazione della luce di risonanza*,
E. Fermi and F. Rasetti,
Rend. Lincei **2**, 117-120, (1925).

P38. [FNM29] *Sopra la teoria dei corpi solidi,*
Periodico di Matematiche **5**, 264-274, (1925).

P39. [FNM34] *Problemi di chimica, nella fisica dell'atomo,*
Periodico di Matematiche **6**, 19-26, (1926).

P40. [FNM27] *Über den Einfluss eines wechselnden magnetischen Feldes auf die Polarization der Resonanzstrahlung,*
Z. Physik **33**, 246-250, (1925).

1926

P41. [FNM30] *Sulla quantizzazione del gas perfetto monoatomico,*
Rend. Lincei **3**, 145-149, (1926).

P42. [FNM35] *Sopra l'elettrone rotante,*
F. Rasetti and E. Fermi,
Nuovo Cimento **3**, 226-235, (1926).

P43. [FNM32] *Sopra l'intensità delle righe proibite nei campi magnetici intensi,*
Rend. Lincei **3**, 478-483, (1926).

P44. [FNM38a] *Sopra una formula di calcolo delle probabilità,*
Nuovo Cimento **3**, 313-318, (1926).

P45. [FNM37] *Il principio delle adiabatiche e la nozione di forza viva nella nuova meccanica ondulatoria,*
E. Fermi and E. Persico,
Rend. Lincei **4**(II), 452-457, (1926).

P46. [FNM31] *Zur Quantelung des idealen einatomigen Gases,*
Z. Physik **36**, 902-912, (1926).

P47. [FNM39] *Quantum Mechanics and the Magnetic Moment of Atoms,*
Nature (London) **118**, 876, (1926).

P48. [FNM33] *Argomenti pro e contro la ipotesi dei quanti di luce,*
Nuovo Cimento **3**, XLVII-LIV, (1926).

1927

P49. [FNM40b] *Una misura del rapporto h/k per mezzo della dispersione anomala del tallio,*
E. Fermi and F. Rasetti,
Rend. Lincei **5**, 566-570, (1927).

P50. [FNM42] *Sul meccanismo dell'emissione nella meccanica ondulatoria*, Rend. Lincei **5**, 795-800, (1927).

P51. [FNM36] *Zur Wellenmechanik des Stossvorganges*, Z. Physik **40**, 399-402, (1927).

P52. [FNM40a] *Eine Messung des Verhaltnisses h/k durch die anomale Dispersion des Thalliumdampfes*, E. Fermi and F. Rasetti, Z. Physik **43**, 379-383, (1927).

P53. [FNM41] *Gli effetti elettro e magnetoottici e le loro interpretazioni*, Fascicolo speciale dell'Energia Elettrica nel 1° centenario della morte di A. Volta, Uniel, Roma (1927).

P54. [FNM43] *Un metodo statistico per la determinazione di alcune proprietà dell'atomo*, Rend. Lincei **6**, 602-607, (1927).

1928

P55. [FNM44] *Sulla deduzione statistica di alcune proprietà dell'atomo. Applicazione alla teoria del sistema periodico degli elementi*, Rend. Lincei **6**, 602-607, (1927).

P56. [FNM47] *Eine statistische Methode zur Bestimmung Einiger Eigenschaften des Atoms und ihre Anwendung auf die Theorie des periodischen System der Elemente*, Z. Physik **48**, 73-79, (1928).

P57. [FNM46] *Anomalous Group in the Periodic System of Elements*, Nature (London) **121**, 502, (1928).

P58. [FNM45] *Sulla deduzione statistica di alcune proprietà dell'atomo. Calcolo della correzione di Rydberg per i termini s*, Rend. Lincei **7**, 726-730, (1928).

P59. [FNM48] *Statistische Berechnung der Rydbergkorrektionen der s-Terme*, Z. Physik **49**, 550-554, (1928).

P60. [FNM49] *Über die Anwendung der statistischen Methode auf die Probleme des Atombaues*, Falkenhagen, Quantentheorie und Chemie. Leipziger Vortrage, 95-111, (1928). Hirzel, Leipzig, 1928

1929

P61. [FNM50] *Sopra l'elettrodinamica quantistica*,
Rend. Lincei **9**, 881-887, (1929).

P62. [FNM51] *Sul moto di un corpo di massa variabile*,
Rend. Lincei **9**, 984-986, (1929).

P63. [FNM52] *Sulla teoria quantistica delle frange di interferenza*,
Rend. Lincei **10**, 72-77, (1929);
Nuovo Cimento **7**, 153-158, (1930).

P64. [FNM53] *Sul complesso 4d della molecola di elio*,
Rend. Lincei **10**, 515-517, (1929);
Nuovo Cimento **7**, 159-161, (1930).

P65. [FNM55] *Magnetic Moments of Atomic Nuclei*,
Nature (London) **125**, 16, (1930).

P66. [FNM56] *I fondamenti sperimentali delle nuove teorie fisiche*,
Atti Soc. It. Progr. Sci., 18ª Riunione, vol. 1, 365-371, (1929).

P67. [FNM58] *Problemi attuali della fisica*,
Annali dell'istruzione media **5**, 424-428, (1929).

1930

P68. [FNM57a] *Sui momenti magnetici deli nuclei atomici*,
Mem. Accad. d'Italia **1** (Fis.), 139-148, (1930).

P69. [FNM63] *Sul calcolo degli spettri degli ioni*,
Mem. Accad. d'Italia **1** (Fis.), 149-156 (1930);
Nuovo Cimento, **8**, 7-14, (1931).

P70. [FNM54b] *Sul rapporto delle intensità nei doppietti dei metalli alcalini*,
Nuovo Cimento **7**, 201-207, (1930).

P71. [FNM66] *La thèorie du rayonnement*,
Annales de l'Inst. H. Poincarè **1**, 53-74, (1930).

P72. [FNM57b] *Über die magnetischen Momente der Atomkerme*,
Z. Physik **60**, 320-333, (1930).

P73. [FNM59] *L'interpretazione del principio di causalità nella meccanica quantistica*,
Rend. Lincei **11**, 980-985, (1930).

P74. [FNM54a] *Über das Intensitatsverhaltnis der Dublettkomponenten der Alkalien*, Z. Physik **59**, 680-686, (1930).

P75. [FNM64] *Sopra l'elettrodinamica quantistica*, Rend. Lincei **12**, 431-435, (1930).

P76. [FNM60] *Atomi e stelle*, Atti. Soc. It. Progr. Sci., 19ª Riunione, vol. 1, 228-235, (1930).

P77. [FNM61] *I fondamenti sperimentali della nuova meccanica atomica*, Periodico di matematiche **10**, 71-84, (1930).

P78. [FNM62] *La fisica moderna*, Nuova Antologia **65**, 137-145, (1930).

1931

P79. [FNM68] *Über den Ramaneffekt des Kohlendioxyds*, Z. Physik **71**, 250-259, (1931).

P80. [FNM65] *Le masse elettromagnetiche nella elettrodinamica quantistica*, Nuovo Cimento **8**, 121-132, (1931).

P81. [FNM69] *Über den Ramaneffekt des Steinsalzes*, E. Fermi and F. Rasetti, Z. Physik **71**, 689-695, (1931).

P82. *Nuclei ed elettroni*, Gerarchia **11**, 879-885, (1931).

1932

P83. *Moments magnétiques des noyaux*, in *Le magnétisme*, Rapports et discussions du Sixieme Conseil de Physique, tenu a Bruxelles du 20 au 25 Octobre 1930, sous les auspices de l'Institut International de Physique Solvay, Gauthier-Villars, Paris, 1932.

P84. [FNM67] *Quantum Theory of Radiation*, Rev. Mod. Phys. **4**, 87-132, (1932).

P85. [FNM71] *L'effetto Raman nelle molecole e nei cristalli.*, Mem. Accad. d'Italia **3** (Fis.), 239-256, (1932).

P86. [FNM70] *Über die Wechselwirkung: von zwei Elektronen*, H. Bethe and E. Fermi, Z. Physik **77**, 296-306, (1932).

P87. [FNM72a] *La physique du noyau atomique.*,
Congrès International d'électricité. Paris (1932), C.R., 1 Sect., Rep. 22, 789-807.

P88. [FNM72b] *Lo stato attuale della fisica del nucleo atomico*,
Ricerca Scientifica **3**(2), 101-113, (1932).

P89. [FNM73] *Sulle bande di oscillazione e rotazione dell'ammoniaca.*,
Rend. Lincei **16**, 179-185, (1932);
Nuovo Cimento **9**, 277-283, (1932).

1933

P90. [FNM74] *Azione del campo magnetico terrestre sulla radiazione penetrante.*,
E. Fermi and B. Rossi,
Rend. Lincei **17**, 346-350, (1933).

P91. [FNM75b] *Sulla teoria delle strutture iperfini*,
E. Fermi and E. Segrè,
Mem. Accad. d'Italia **4**(Fis), 131-158, (1933).

P92. [FNM77b] *On the Recombination of Electrons and Positrons*,
E. Fermi and G. Uhlenbeck,
Phys. Rev. **44**, 510-511, (1933).

P93. [FNM75a] *Zur Theorie der Hyperfeinstrukturen*,
E. Fermi and E. Segrè,
Z. Physik **82**, 11-12, 729-749 (1933).

P94. [FNM77a] *Sulla ricombinazione di elettroni e positroni*,
E. Fermi and G. Uhlenbeck,
Ricerca Scientifica **4**(2), 157-160, (1933).

P95. [FNM78] *Uno spettrografo per raggi ≪gamma≫ a cristallo di bismuto*,
E. Fermi and F. Rasetti,
Ricerca Scientifica **4**(2), 299-302, (1933).

P96. [FNM76] *Tentativo di una teoria di emissione dei raggi ≪beta≫*,
Ricerca Scientifica **4**(2), 491-495, (1933).

P97. [FNM79] *Le ultime particelle costitutive della materia*,
Atti Soc. It. Progr. Sci., 22ª Riunione, vol. 2, 7-14, (1933);
Scientia **55**, 21-28 (1934).

1934

P98. [FNM80a] *Tentativo di una teoria dei raggi* β,
Nuovo Cimento **11**, 1-19, (1934).

P99. [FNM80b] *Versuch einer Theorie der* β*-Strahlen. I*,
Z. Physik **88**, 161-171, (1934).

P100. [FNM95] *Sopra lo spostamento per pressione delle righe elevate delle serie spettrali*,
Nuovo Cimento **11**, 157-166 (1934).

P101. [FNM84] *Radioattività provocata da bombardamento di neutroni. I*,
Ricerca Scientifica, **5**(1), 283, (1934).

P102. [FNM85] *Radioattività provocata da bombardamento di neutroni. II*,
Ricerca Scientifica, **5**(1), 330-331, (1934).

P103. [FNM93] *Radioactivity Induced by Neutron Bombardment*,
Nature (London) **133**, 757 (1934).

P104. [FNM86] *Radioattività provocata da bombardamento di neutroni. III*,
E. Amaldi, O. D'Agostino, E. Fermi, F. Rasetti and E. Segrè,
Ricerca Scientifica, **5**(1), 452-453, (1934).

P105. [FNM82] *Le orbite* ∞ *s degli elementi*,
Mem. Accad. d'Italia **6**(1) (Fis.), 119-149, (1934).

P106. [FNM94] *Sulla possibilità di produrre elementi di numero atomico maggiore di 92*,
E. Fermi, F. Rasetti, O. D'Agostino,
Ricerca Scientifica **5**(1), 536-537 (1934).

P107. [FNM99] *Possible Production of Elements of Atomic Number Higher than 92*,
Nature (London) **133**, 898-899 (1934).

P108. [FNM87] *Radioattività provocata da bombardamento di neutroni. IV*,
E. Amaldi, O. D'Agostino, E. Fermi, F. Rasetti and E. Segrè,
Ricerca Scientifica, **5**(1), 652-653, (1934).

P109. [FNM96] *Radioattività prodotta da bombardamento di neutroni*,
Nuovo Cimento **11**, 429-441 (1934).

P110. [FNM97] *Nuovi radioelementi prodotti con bombardamenti di neutroni*,
E. Amaldi, E. Fermi, F. Rasetti and E. Segrè,
Nuovo Cimento **11**, 442-447 (1934).

P111. [FNM88] *Radioattività provocata da bombardamento di neutroni. V*,
E. Amaldi, O. D'Agostino, E. Fermi, F. Rasetti and E. Segrè,
Ricerca Scientifica, **5**(1), 21-22, (1934).

P112. [FNM98] *Artificial Radioactivity Produced by Neutron Bombardment*,
E. Fermi, E. Amaldi, O. D'Agostino, F. Rasetti and E. Segrè,
Proc. Roy. Soc. (London) Series A *146*, 483-500 (1934).

P113. [FNM81] *Zur Bemerkung von G. Beck und K. Sitte*,
Z. Physik **89**, 522, (1934).

P114. [FNM101] *Conferencias*, Faculdad de Ciencias Exactas Fisicas y Naturales,
Publicacion **15**, Buenos Aires (1934).

P115. [FNM102] *Natural Beta Decay*,
International Conference on Physics, London 1934. Vol. 1. Nuclear Physics, 66-71,
Physical Society (London).

P116. [FNM103] *Artificial Radioactivity Produced by Neutron Bombardment*,
International Conference on Physics, London 1934. Vol. 1. Nuclear Physics, 75-77,
Physical Society (London).

P117. [FNM104] *La radioattività artificiale*,
Atti Soc. It. Progr. Sci., 23ª Riunione, vol. 1, 34-39.

P118. [FNM105] *Azione di sostanze idrogenate sulla radioattività provocata da neutroni. I*,
E. Fermi, E. Amaldi, B. Pontecorvo, F. Rasetti and E. Segrè,
Ricerca Scientifica **5**(2), 282-283 (1934).

P119. [FNM100] *Artificial Radioactivity Produced by Neutron Bombardment*,
Nature (London) **134**, 668 (1934).

P120. [FNM106] *Effetto di sostanze idrogenate sulla radioattività provocata da neutroni. II*,
E. Fermi, E. Amaldi, B. Pontecorvo and F. Rasetti,
Ricerca Scientifica **5**(2), 380-381 (1934).

P121. [FNM89] *Radioattività provocata da bombardamento di neutroni. VII*,
E. Amaldi, O. D'Agostino, E. Fermi, B. Pontecorvo, F. Rasetti and E. Segrè,
Ricerca Scientifica, **5**(1), 467-470, (1934).

1935

P122. [FNM90] *Radioattività provocata da bombardamento di neutroni. VIII*,
E. Amaldi, O. D'Agostino, E. Fermi, B. Pontecorvo, F. Rasetti and E. Segrè,
Ricerca Scientifica, **6**(1), 123-125, (1935).

P123. [FNM109] *On the Velocity Distribution Law for the Slow Neutrons*,
Zeeman Verhandelingen, p. 128-130, Martinus Nijhoff, the Hague, 1935.

P124. [FNM107] *Artificial Radioactivity Produced by Neutron Bombardment. Part II*,
E. Amaldi, O. D'Agostino, E. Fermi, B. Pontecorvo, F. Rasetti and E. Segrè,
Proc. Roy. Soc. (London) Series A **149**, 522-558 (1935).

P125. [FNM108] *Ricerche sui neutroni lenti*,
E. Fermi and F. Rasetti,
Nuovo Cimento **12**, 201-210 (1935).

P126. [FNM91] *Radioattività provocata da bombardamento di neutroni. IX*,
E. Amaldi, O. D'Agostino, E. Fermi, B. Pontecorvo and E. Segrè,
Ricerca Scientifica, **6**(1), 435-437, (1935).

P127. [FNM92] *Radioattività provocata da bombardamento di neutroni. X*,
E. Amaldi, O. D'Agostino, E. Fermi, B. Pontecorvo and E. Segrè,
Ricerca Scientifica, **6**(1), 581-584, (1935).

P128. [FNM110] *On the Recombination of Neutrons and Protons*,
Phys. Rev. **48**, 570 (1935).

P129. [USP1] *Process for the production of radioactive substances*,
E. Fermi, E. Amaldi, B. Pontecorvo, F. Rasetti and E. Segrè,
Patent filed October 3, 1935 (Patent No. 2,206,634; July 2, 1940).

P130. [FNM112] *Sull'assorbimento dei neutroni lenti. I*,
E. Amaldi and E. Fermi,
Ricerca Scientifica **6**(2), 344-347 (1935).

P131. [FNM113] *Sull'assorbimento dei neutroni lenti. II*,
E. Fermi and E. Amaldi,
Ricerca Scientifica **6**(2), 443-447 (1935).

P132. [FNM111] *Recenti risultati dalla radioattività artificiale*,
Ricerca Scientifica **6**(2), 399-402 (1935);
Atti. Soc. It. Progr. Sci., 24ª Riunione, vol. 3, 116-120.

1936

P133. [FNM114] *Sull'assorbimento dei neutroni lenti. III*,
E. Amaldi and E. Fermi,
Ricerca Scientifica **7**(1), 56-59 (1936).

P134. [FNM115] *Sul cammino libero medio dei neutroni nella paraffina*,
E. Amaldi and E. Fermi,
Ricerca Scientifica **7**(1), 223-225 (1936).

P135. [FNM116] *Sui gruppi di neutroni lenti*,
E. Amaldi and E. Fermi,
Ricerca Scientifica **7**(1), 310-315 (1936).

P136. [FNM117] *Sulle proprietà di diffusione dei neutroni lenti*,
E. Amaldi and E. Fermi,
Ricerca Scientifica **7**(1), 393-395 (1936).

P137. [FNM118a] *Sopra l'assorbimento e la diffusione dei neutroni lenti*,
E. Amaldi and E. Fermi,
Ricerca Scientifica **7**(1), 454-503 (1936).

P138. [FNM119] *Sul moto dei neutroni nelle sostanze idrogenate*,
Ricerca Scientifica **7**(2), 13-52 (1936).

P139. [FNM118b] *On the Absorption and the Diffusion of Slow Neutrons*,
E. Amaldi and E. Fermi,
Phys. Rev. **50**, 899-928 (1936).

P140. [FNM83] *Statistica, meccanica.*,
Enciclopedia Italiana di Scienze, Lettere ed Arti, Istituto G. Treccani, Roma, vol. 32, 518-523, (1936).

1937

P141. [FNM120] *Un Maestro: Orso Mario Corbino*,
Nuova Antologia **72**, 313-316 (1937).

P142. [FNM121] *Un generatore artificiale di neutroni*,
E. Amaldi, E. Fermi and F. Rasetti,
Ricerca Scientifica **8**(2), 40-43 (1937).

P143. [FNM122] *Neutroni lenti e livelli energetici nucleari*,
Nuovo Cimento **15**, 41-42 (1938).

P144. [FNM123] *(Tribute to Lord Rutherford)*,
Nature (London) **140**, 1052 (1937).

1938

P145. [FNM125] *On the Albedo of Slow Neutrons*,
E. Fermi, E. Amaldi and G.C. Wick,
Phys. Rev. **53**, 493 (1938).

P146. [FNM124] *Azione del boro sui neutroni caratteristici dello iodio*,
E. Fermi and F. Rasetti,
Ricerca Scientifica **9**(2), 472-473 (1938).

P147. [FNM126] *Prospettive di applicazioni della radioattività artificiale*,
Rendiconti dell'Istituto di Sanità Pubblica, vol. 1, 421-432 (1938).

P148. [FNM127] *Guglielmo Marconi e la propagazione delle onde elettromagnetiche nell'alta atmosfera*,
Soc. It. Progr. Sci., Collectanea Marconiana, Roma, 1-5 (1938).

P149. [FNM] ◇ *Intervista con Enrico Fermi, 11 Novembre 1938*,
Ricerca Scientifica, **9**(2), 638-639 (1938).

P150. [FNM128] *Artificial Radioactivity Produced by Neutron Bombardment*,
Les Prix Nobel en 1938. Les Conférences Nobel, Stockholm p. 1-8 (1939).

1939

P151. [FNM129] *The Fission of Uranium*,
H.L. Anderson, E.T. Booth, J.R. Dunning, E. Fermi, G.N. Glasoe and F.G. Slack,
Phys. Rev. **55**, 511-512 (1939).

P152. [FNM130] *Production of Neutrons in Uranium Bombarded by Neutrons*,
H.L. Anderson, E. Fermi and H.B. Hanstein,
Phys. Rev. **55**, 797-798 (1939).

P153. [FNM131] *Simple Capture of Neutrons by Uranium*,
H.L. Anderson and E. Fermi,
Phys. Rev. **55**, 1106-1107 (1939).

P154. [FNM132] *Neutron Production and Absorption in Uranium*,
H.L. Anderson, E. Fermi and L. Szilard,
Phys. Rev. **56**, 284-286 (1939).

P155. [FNM133] *The Absorption of Mesotrons in Air and in Condensed Materials*,
Phys. Rev. **56**, 1242 (1939).

1940

P156. [FNM134] *The Ionization Loss of Energy in Gases and in Condensed Materials*,
Phys. Rev. **57**, 485-493 (1940).

P157. [FNM137a] *Reactions Produced by Neutrons in Heavy Elements*,
Nature **146**, 640-642 (1940); Science **92**, 269-271 (1940).

P158. [FNM136] *Production and Absorption of Slow Neutrons by Carbon*,
H.L. Anderson and E. Fermi, Report A-21 (September 25, 1940).

P159. [FNM137] *Branching Ratios in the Fission of Uranium (235)*,
H.L. Anderson, E. Fermi and A.V. Grosse,
Phys. Rev. **59**, 52-56 (1941).

1941

P160. [FNM138] *Production of Neutrons by Uranium*,
H.L. Anderson and E. Fermi,
Report A-6 (January 17, 1941).

P161. [FNM135] *Fission of Uranium by Alpha-Particles*,
E. Fermi and E. Segrè, Phys. Rev. **59**, 680-681 (1941).

P162. [FNM139] *Capture of Resonance Neutrons by a Uranium Sphere Imbedded in Graphite*,
E. Fermi, H.L. Anderson, R.R. Wilson and E.C. Creutz,
Appendix A of Report A-21 to The National Defense Research Committee
by H. D. Smith, Princeton University, (June 1, 1941).

P163. [FNM140] *Standards in Slow Neutron Measurements*,
H.L. Anderson and E. Fermi,
Report A-2 (June 5, 1941).

P164. [FNM141] *Some Remarks on the Production of Energy by a Chain Reaction in Uranium*,
Report A-14 (June 30, 1941).

P165. [FNM142] *The Absorption of Thermal Neutrons by a Uranium Sphere Imbedded in Graphite*,
E. Fermi and G.L. Weil,
Report A-1 (July 3, 1941).

P166. [FNM143] *Remarks on Fast Neutron Reactions*,
Report A-46 (October 6, 1941).

P167. [FNM146] *Absorption Cross Section for Rn + Be Fast Neutrons*,
H.L. Anderson, E. Fermi and G.L. Weil, Report C-72.

P168. [FNM145] *Fission Cross Section of Unseparated Uranium for Fast Rn + Be Neutrons*,
H.L. Anderson and E. Fermi, Report C-83.

P169. [FNM144] *The Effect of Chemical Binding in the Scattering and Moderation of Neutrons by Graphite*,
Report C-87.

1942

P170. [FNM148] *The Absorption Cross Section of Boron for Thermal Neutrons*,
H.L. Anderson and E. Fermi,
Report CP-74.

P171. [FNM153] *A table for Calculating the Percentage of Loss Due to the Presence of Impurities in Alloy*,
Report C-5 (February 10, 1942).

P172. [FNM154] *The Temperature Effect on a Chain Reacting Unit. Effect of the Change of Leakage*,
Report C-8 (February 25, 1942).

P173. [FNM155] *The Use of Reflectors and Seeds in a Power Plant*,
G. Breit and E. Fermi,
Report C-11 (March 9, 1942)

P174. [FNM156] *Slowing Down and Diffusion of Neutrons*,
Report C 29 (Notes on Lecture of March 10, 1942).

P175. [FNM149] *Neutron Production in a Lattice of Uranium and Graphite*,
Theoretical Part
Report CP-12 (March 17, 1942).

P176. [FNM157] *Determination of the Albedo and the Measurement of Slow Neutron Density*,
Report C-31 (Notes on Lecture of March 17, 1942).

P177. [FNM158] *The Number of Neutrons Emitted by a Ra + Be Source (Source I)*,
H.L. Anderson, E. Fermi, J.H. Roberts and M.D. Whitaker,

Report C-21 (March 21, 1942).

P178. [FNM150] *Neutron Production in a Lattice of Uranium Oxide and Graphite (Exponential Experiment)*,
H.L. Anderson, B.T. Feld, E. Fermi, G.L. Weil and W.H. Zinn,
Report CP-20 (March 26, 1942).

P179. [FNM151] *Preliminary Report on the Exponential Experiment at Columbia University*,
Report CP-26 (March, April 1942).

P180. [FNM159] *The Determination of the Ratio Between the Absorption Cross Sections of Uranium and Carbon for Thermal Neutrons*,
Report C-84 (May 15, 1942).

P181. [FNM152] *Effect of Atmospheric Nitrogen and of Changes of Temperature on the Reproduction Factor*,
Report CP-85 (May 19, 1942).

P182. [FNM147] *Neutrons Emitted by a Ra + Be Photosource*,
B.T. Feld and E. Fermi,
Report CP-89 (later issued on November 5, 1948).

P183. [WAT1043] ◇ *Meeting of Engineering Council*,
Present: Moore, Allison, Fermi, Leverett, Wheeler, Compton, Hilberry and Doan,
Report CE-106 (May 28, 1942).

P184. [WAT1043] * *The Fourth Intermediate Pile*,
Metallurgical Project (no explicit list of authors; classified by Wattenberg among the Fermi papers),
Report C-102.

P185. [WAT1043] ◇ *Meeting of the Planning Board*,
Present: Hilberry, Spedding, Allison, Wigner, Doan, Szilard, Wheeler, Fermi, Moore and Compton,
Report CS-112, CS-185 (June 6, 1942).

P186. [WAT1043] ◇ *Meeting of the Engineering Council*,
Present: Moore, Wilson, Seaborg, Wheeler, Leverett, Fermi, Hilberry, Compton, Spedding and Allison,
Report CS-131 (June 11, 1942).

P187. [WAT1043] ◇ *Meeting of the Engineering Council*,
Present: Moore, Leverett, Fermi, Wheeler, Seaborg, Doan, Wilson and Spedding,
Report CS-135 (June 18, 1942)

P188. [FNM164] *Status of Research Problems in Experimental Nuclear Physics*,
Report C-133 for Week Ending (June 20, 1942).

P189. [WAT1043] ◇ *Meeting of the Engineering Council*,
Present: Moore, Leverett, Doan, Szilard, Allison, Teller, Seaborg, Wheeler, Fermi and Wilson,
Report CS-147 (June 25, 1942)

P190. [FNM160] *The Absorption of Graphite for Thermal Neutrons*,
Report C-154 (Notes on Lecture of June 30, 1942).

P191. [WAT1043] ◇ *Meeting of the Engineering Council*,
Present: Fermi, Allison, Seaborg, Whitaker, Doan, Wilson, Moore, Leverett, Wheeler, Szilard, Compton, Spedding, Hilberry and Wollan,
Report CS-163 (July 2, 1942).

P192. [FNM161] *Longitudinal Diffusion in Cylindrical Channels*,
E. Fermi and A. M. Weinberg,
Report C-170 (July 7, 1942).

P193. [WAT1043] ◇ *Meeting of the Engineering Council*,
Present: Moore, Leverett, Fermi, Wigner, Allison, Wollan, Wheeler, Seaborg, Spedding, Szilard, Steams and Wilson,
Report CS-174 (July 9, 1942).

P194. [WAT1043] ◇ *Meeting of the Technical Council*,
Present: Fermi, Compton, Allison, Moore, Szilard, Wigner and Wheeler,
Report CS-184 (July 14, 1942).

P195. [FNM162] *The Number of Neutrons Emitted by Uranium per Thermal Neutron Absorbed*,
Report C-190 (July 16, 1942).

P196. [WAT1043] ◇ *Meeting of the Engineering Council*,
Present: Moore, Leverett, Lewis, Fermi, Wollan, Hilberry, Whitaker, Wilson, Wheeler, Allison, Wigner, Seaborg and Doan,
Report CE-194 (July 21, 1942).

P197. [WAT1043] ◇ *Meeting of the Technical Council*,
Present: Fermi, Szilard, Wigner, Compton, Whitaker, Allison, Moore, Wheeler and Doan,
Report CS-202 (July 25, 1942).

P198. [WAT1043] ◇ *Meeting of the Planning Board*,
Present: Spedding, Fermi, Szilard, Wigner, Doan, Moore, Wheeler, Compton and Hilberry,
Report CS-213 (August 1, 1942).

P199. [WAT1043] ◊ *Meeting of the Engineering Council*,
Present: Moore, Leverett, Steinbach, Fermi, Spedding, Wheeler, Wigner, Seaborg and Wilson,
Report CE-229 (August 8, 1942).

P200. [FNM166] *Status of Research Problems of the Physics Division*,
Report CP-235 for Month Ending August 15, 1942.

P201. [FNM167] *Exponential Pile No. II*,
Report CA-247 for Week Ending August 29, 1942.

P202. [WAT1043] ◊ *Meeting of the Technical Council*,
Present: Nichols, Hilberry, Spedding, Doan, Fermi, Steinbach, Grafton, Boyd, Moore and Wheeler,
Report CS-251 (September 4, 1942).

P203. [FNM163] *Effect of Temperature Changes on the Reproduction Factor*,
R.F. Christy, E. Fermi and M. D. Whitaker,
Report CP-254 (September 14, 1942).

P204. [FNM168] *Status of Research Problems of the Physics Division*,
Report CP-257 for Month Ending September 15, 1942.

P205. [WAT1006] ◊ *Discussion of Helium Cooled Power Plant*,
Present: Leverett, Cooper, Moore, Wigner, Steinbach, Fermi, Szilard and Wheeler,
Report CS-267 (September 16, 1942).

P206. [WAT1043] ◊ *Meeting of the Technical Council*,
Present: Compton, Wheeler, Moore, Allison, Szilard, Fermi and Spedding,
Report CS-274 (September 18, 1942).

P207. [FNM169] *Purpose of the Experiment at the Argonne Forest. Meaning of the Reproduction Factor "k"*,
Report CP-283 (Notes on Lecture of September 23, 1942).

P208. [WAT1043] ◊ *Meeting of the Technical Council*,
Present: Allison, Fermi, Moore, Wigner, Compton, Wheeler and Oppenheimer,
Report CS-281 (September 29, 1942).

P209. [FNM170] *The Critical Size-Measurement of "k" in the Exponential Pile*,
Report CP-289 (Notes on Lecture of September 30, 1942).

P210. [WAT1043] ◊ *Meeting of the Technical Council*,
Present: Allison, Fermi, Wigner, Compton, Whitaker, Moore, Cooper, Szilard, Manley, McMillan, Wheeler and Doan,

Report CS-284 (October 1, 1942).

P211. [WAT1043] ◇ *Meeting of the Technical Council*,
Present: Allison, Fermi, Szilard, Moore, Wigner, Whitakcr, Wheeler, Steinbach, Compton and Groves,
Report CS-286 (October 5, 1942).

P212. [FNM176] *Methods of Cooling Chain Reacting Piles*,
Memo #10 (October 5, 1942).

P213. [FNM171] *Problem of Time Dependance of the Reaction Rate: Effect of Delayed Neutrons Emission*,
Report CP-291 (Notes on Lecture of October 7, 1942).

P214. [WAT1043] ◇ *Meeting of the Technical Council*,
Present: Allison, Wigner, Compton, Whitaker, Szilard, Wheeler, Fermi, Moore, Cooper, Steinbach and Kirkpatrick,
Report CS-290 (October 7, 1942).

P215. [WAT1043] ◇ *Meeting of the Technical Council*,
Present: Allison, Szilard, Wigner, Moore, Wheeler and Fermi,
Report CS-294 (October 12, 1942).

P216. [FNM175] *The projected Experiment at Argonne Forest and the Reproduction Factor in Metal Piles*,
Report CP-297 for Month Ending October 15, 1942.

P217. [FNM172] *A simplified Control. Optimum Distribution of Materials in the Pile*,
Report CP-314 (Notes on Lecture of October 20, 1942).

P218. [WAT1008] ◇ *Conference on Lattice Spacing*,
Present: Steinbach, Leverett, Fermi, Wigner and Wheeler,
Memo #15 (October 21, 1942).

P219. [FNM177] *The Effect of Bismuth on the Reproduction Factor*,
Report CA-320, Bulletin for Week Ending October 31, 1942

P220. [FNM173] *Design of the Graphite-Uranium Lattice: Experimental Determination of ft from the Cd Ratio*,
Report CP-337 (Notes on Lecture of October 27 and November 3, 1942).

P221. [WAT1043] *Report of the Committee for the Examination of the Moore-Leverett Design of a He-Cooled Plant*,
E. Fermi, S.K. Allison, C. Cooper and E.P. Wigner,
Report CE-324 (1942).

P222. [FNM174] *Calculation of the Reproduction Factor*,
Report CP-358 (Notes on Lecture of November 10, 1942).

P223. [FNM178] *The Experimental Chain Reacting Pile and Reproduction Factor in Some Exponential Piles*,
Report CP-341 for Month Ending November 15, 1942.

P224. [WAT1043] ◇ *Meeting of the Technical Council*,
Present: Allison, Compton, Fermi, Moore, Spedding, Szilard and Wigner,
Report CS-356 (November 19, 1942).

P225. [FNM179] *Feasibility of a Chain Reaction*,
Report CP-383 (November 26, 1942).

P226. [FNM180] *Work Carried out by the Physics Division*,
Report CP-387 for Month Ending December 15, 1942.

1943

P227. [FNM182] *Summary of Experimental Research Activities*,
Report CP-416 for Month Ending January 15, 1943.

P228. [FNM183] *Summary of Experimental Research Activities*,
Report CP-455 for Month Ending February 6, 1943.

P229. [FNM184] *The Utilization of Heavy Hydrogen in Nuclear Chain Reactions*,
Memorandum of Conference between Prof. E. Fermi and Prof. H. C. Urey on March 6, 7, and 8, 1943, Report A-544.

P230. [FNM185] *The Slowing Down of Neutrons in Heavy Water*,
Report CP-530 (March 19, 1943).

P231. [FNM186] *Summary of Experimental Research Activities*,
Report CP-570 for Month Ending April 17, 1943.

P232. [FNM187] *Summary of Experimental Research Activities*,
Report CP-641 for Month Ending May 10, 1943.

P233. [FNM188] *Standardization of the Argonne Pile*,
H.L. Anderson, E. Fermi, J. Marshall and L. Woods,
Report CP-641 for Month Ending May 10, 1943.

P234. [FNM189] *Tests on a Shield for the Pile at Site W*,
E. Fermi and W. H. Zinn,
Report CP-684 for Month Ending May 25, 1943.

P235. [FNM190] *Summary of Experimental Research Activities*,
Report CP-718 for Month Ending June 12, 1943.

P236. [FNM192] *Summary of Experimental Research Activities*,
Report CP-781 for Month Ending July 10, 1943.

P237. [FNM193] *Range of Indium Resonance Neutrons from a Source of Fission Neutrons*,
E. Fermi and G. L. Weil,
Report CP-871 for Month Ending August 14, 1943.

P238. [WAT1023] * *Report for Month Ending September 25, 1943*,
edited by A.H. Compton, S.K. Allison and E. Fermi,
Report CP-964.

P239. [FNM194] *Summary of Experimental Research Activities*,
Report CP-1016 for Month Ending October 23, 1943.

P240. [FNM195] *Summary of Experimental Research Activities*,
Report CP-1088 for Month Ending November 23, 1943.

P241. [FNM196] *The Range of Delayed Neutrons*,
E. Fermi and G. Thomas,
Report CP-1088 for Month Ending November 23, 1943.

P242. [FNM197] *Slowing Down of Fission Neutrons in Graphite*,
E. Fermi, J. Marshall and L. Marshall,
Report CP-1084 (November 25, 1943).

P243. [FNM199] *Summary of Experimental Research Activities*,
Report CP-1175 for Month Ending December 25, 1943.

P244. [WAT1027] * *Report for Month Ending December 25, 1943 - Part II*,
edited by A.H. Compton, S.K. Allison and E. Fermi,
Report CN-1190

P245. [FNM198] *Fission Cross-Section and ν-Value for 25*,
E. Fermi, J. Marshall and L. Marshall,
Report CP-1186 (December 31, 1943).

1944

P246. [WAT149] *Measurement of the Cross Section of Boron for Thermal Neutrons*,
E. Bragdon, E. Fermi, J. Marshall and L. Marshall,
Report CP-1098 (January 11, 1944).

P247. [FNM201] *Summary of Experimental Research Activities*,
Report CP-1255 for Month Ending January 24, 1944.

P248. [FNM202] *Summary of Experimental Research Activities*,
Report CP-1389 for Month Ending February 24, 1944.

P249. [FNM203] *Report of Fermi's Activities with The Marshall Group*,
Report CP-1389 for Month Ending February 24, 1944.

P250. [FNM204] *Summary of Experimental Activities*,
Report CP-1531 for Month Ending March 25, 1944.

P251. [FNM205] *Range of Fission Neutrons in Water*,
H.L. Anderson, E. Fermi and D.D. Nagle,
Report CP-1531 for Month Ending March 25, 1944.

P252. [FNM206] *Evidence for the Formation of 26*,
E. Fermi and L. Marshall,
Report CP-1531 for Month Ending March 25, 1944.

P253. [FNM207] *Summary of Experimental Research Activities*,
Report CP-1592 for Month Ending April 24, 1944.

P254. [FNM208] *Absorption of 49*,
Report CP-1592 for Month Ending April 24, 1944.

P255. [FNM209] *Comparison of the Ranges in Graphite of Fission Neutrons from 49 and 25*,
E. Fermi, A. Heskett and D.E. Nagle,
Report CP-1592 for Month Ending April 24, 1944.

P256. [FNM211] ◇ *Notes on Meeting of April 26, 1944*,
Present: Fermi, Allison, Szilard, Wigner, Weinberg, Seitz, Morrison, Cooper, Vernon, Tolman, Watson and Ohlinger,
Report N-1729, Eck-209.

P257. [WAT] ◇ *Notes on Meeting of April 28, 1944*,
Present: Fermi, Allison, Wigner, Smyth, Szilard, Morrison, Watson, Feld, Hogness, Young, Weinberg, Creutz, Cooper, Vernon and Ohlinger,
Report N-1729, Eck-209.

P258. [USP2] *Test exponential pile*,
Patent filed May 4, 1944 (Patent No. 2,780,595; February 5, 1957).

P259. [FNM212] *Report on Recent Values of Constants of 25 and 49*,
Report CK-1788 (May 19, 1944).

P260. [FNM213] *Summary of Experimental Research Activities*,
Report CP-1729 for Month Ending May 25, 1944.

P261. [FNM214] *Dissociation Pressure of Water due to Fission*,
H.L. Anderson and E. Fermi,
Report CP-1729 for Month Ending May 25, 1944.

P262. [FNM215] *Summary of Experimental Research Activities*,
Report CK-1761 for Month Ending May 25, 1944.

P263. [FNM216] *Summary of Experimental Research Activities*,
Report CK-1827 for Month Ending June 25, 1944.

P264. [FNM217] *Collimation of Neutron Beam from Thermal Column of CP-3 and the Index of Refraction for Thermal Neutrons*,
E. Fermi and W.H. Zinn,
Report CK-1965 for Month Ending July 29, 1944.

P265. [FNM218] *Methods for Analysis of Helium Circulating in the 105 Unit*,
Document HW 3-492 (August 7, 1944).

P266. [WAT1037] * *Report for Month Ending August 26, 1944*,
edited by A.H. Compton, E. Fermi and W.H. Zinn,
Report CP-2081.

P267. [FNM219] *Boron Absorption of Fission Activation*,
H.L. Anderson and E. Fermi,
Report CF-2161 for Month Ending September 23, 1944.

P268. [WAT1039] * *Report for Month Ending October 28, 1944*,
edited by A.H. Compton, E. Fermi and W.H. Zinn,
Report CP-2301

P269. [WAT1040] * *Report for Month Ending November 25, 1944*,
edited by A.H. Compton, E. Fermi and W.H. Zinn,
Report CP-2436

P270. [USP3] *Neutronic reactor*,
E. Fermi and L. Szilard,
Patent filed December 19, 1944 (Patent No. 2,708,656; May, 17 1955).

1945

P271. [USP4] *Chain reacting system*,
E. Fermi M.C. Leverett,
Patent filed February 16, 1945 (Patent No. 2,837,447; June 3, 1958).

P272. [USP5] *Neutronic reactor*,
Patent filed May 12, 1945 (Patent No. 2,931,762; April 5, 1960).

P273. [USP6] *Air cooled neutronic reactor*,
E. Fermi and L. Szilard,
Patent filed May 29, 1945 (Patent No. 2,836,554; May 27, 1958).

P274. [FNM221] ◇ *Relation of Breeding to Nuclear Properties*,
Present: Fermi *et al.*,
Report CF-3199 (June 19-20, 1945).

P275. [USP7] *Testing material in a neutronic reactor*,
E. Fermi and H.L. Anderson,
Patent filed August 28, 1945 (Patent No. 2,768,134; October. 23, 1956).

P276. [USP8] *Neutron velocity selector*,
Patent filed September 18, 1945 (Patent No. 2,524,379; October 3, 1950).

P277. [USP9] *Neutronic reactor*,
E. Fermi and L. Szilard,
Patent filed October 11, 1945 (Patent No. 2,807,581; September 25, 1957).

P278. [USP10] *Neutronic reactor*,
E. Fermi, W.H. Zinn and H.L. Anderson,
Patent filed October 11, 1945 (Patent No. 2,852,461; September 16, 1958).

P279. [USP11] *Neutronic reactor*,
E. Fermi and W.H. Zinn,
Patent filed November 2, 1945 (Patent No. 2,714,577; August 2, 1955).

P280. [USP12] *Method of testing thermal neutron fissionable material for purity*,
E. Fermi and H.L. Anderson,
Patent filed November 21, 1945 (Patent No. 2,969,307; January 24, 1961).

P281. [USP13] *Method of sustaining a neutronic chain reacting system*,
E. Fermi and M.C. Leverett,
Patent filed November 28, 1945 (Patent No. 2,813,070; November 12, 1957).

1946

P282. [USP14] *Neutronic reactor shield*,
E. Fermi and W.H. Zinn,
Patent filed January 16, 1946 (Patent No. 2,807,727; September 24, 1957).

P283. [FNM222] * *Neutron Physics. A Course of Lectures by E. Fermi*,
Notes by I. Halpern. Part I, Document LADC-225 (February 5, 1946); Part II (Declassification in 1962).

P284. [CHAD] * *Neutron Physics. A Course of Lectures by E. Fermi*,
Notes by A.P. French (June 23, 1947).

P285. [FNM223] *The Development of the First Chain Reacting Pile*,
Proc. Amer. Philosophical Soc. **90**, 20-24 (1946).

P286. [FNM224] *Atomic Energy for Power*,
The George Westinghouse Centennial Forum. *Science and Civilization - The Future of Atomic Energy*; also MDDC-1 (May 27, 1946).

P287. [FNM225] *Elementary Theory of the Chain-Reacting Pile*,
Proceedings of the American Physical Society, Meeting at Chicago, June 20-22, 1946 (see Document MDDC-74 and the announcement in Phys. Rev. **70**, 99 (1946));
Science **105**, 27-32 (1947).

P288. [FNM220] *Reflection of Neutrons on Mirrors*,
E. Fermi and W.H. Zinn,
International Conference on Fundamental Particles and Low Temperature Physics, Cambridge, July 1946. Physical Society (Cambridge), p.92 (1947).

P289. [FNM227] *Phase of Neutron Scattering*,
E. Fermi and L. Marshall,
International Conference on Fundamental Particles and Low Temperature Physics, Cambridge, July 1946. Physical Society (Cambridge), pp.94-97 (1947).

P290. [FNM191] *Production of Low Energy Neutrons by Filtering Through Graphite*,
H.L. Anderson, E. Fermi and L. Marshall,
Phys. Rev. **270** 815-817 (1946).

1947

P291. [FNM226] *The Transmission of Slow Neutrons through Microcrystalline Materials*,
E. Fermi, W.J. Sturm and R.G. Sachs,
Phys. Rev. **71**, 589-594 (1947).

P292. [FNM232] *The Decay of Negative Mesotrons in Matter*,
E. Fermi, E. Teller and V. Weisskopf,
Phys. Rev. **71**, 314-315 (1947).

P293. [FNM228] *Inference Phenomena of Slow Neutrons*,
E. Fermi and L. Marshall, Phys. Rev. **71**, 666-667 (1947).

P294. [FNM210] *Method for Measuring Neutron-Absorption Cross Section by Effect on the Reactivity of a Chain-Reacting Pile*,
H.L. Anderson, E. Fermi, A. Wattenberg, G.L. Weil and W.H. Zinn,
Phys. Rev. **72**, 16-23 (1947).

P295. [FNM231] *Further Experiments with Slow Neutrons*,
E. Fermi and L. Marshall,
Reports CF-3574 (July 26, 1946), CP-3750 (January 17, 1947) and CP-3801 (April 14, 1947).

P296. [FNM200] *A thermal Neutron Velocity Selector and its Application to the Measurement of the Cross-Section of Boron*,
E. Fermi, J. Marshall and L. Marshall,
Phys. Rev. **72** 193-196 (1947).

P297. [FNM229] *Phase of Scattering of Thermal Neutrons by Aluminium and Strontium*,
E. Fermi and L. Marshall,
Phys. Rev. **71**, 915 (1947).

P298. [FNM230] *Spin Dependence of Scattering of Slow Neutrons by* Be, Al *and* Bi,
E. Fermi and L. Marshall,
Phys. Rev. **72**, 408-410 (1947).

P299. [FNM233] *The Capture of Negative Mesotrons in Matter*,
E. Fermi and E. Teller,
Phys. Rev. **72**, 399-408 (1947).

P300. [FNM234] *On the Interaction Between Neutrons and Electrons*,
E. Fermi and L. Marshall,
Phys. Rev. **75**, 1139-1146 (1947).

1948

P301. [FNM236] *Note on Census-Taking in Monte-Carlo Calculations*,
E. Fermi and R.D. Richtmyer,
Document LAMS-805 (July 11, 1948).

P302. [FNM235] *Spin Dependence of Slow Neutrons Scattering by Deuterons*,
E. Fermi and L. Marshall,
Phys. Rev. **75**, 578-580 (1949).

1949

P303. [FNM237] *On the Origin of the Cosmic Radiation*,
Phys. Rev. **75**, 1169-1174 (1949).

P304. [FNM238] *An Hypothesis on the Origin of the Cosmic Radiation*,
Nuovo Cimento **6**, Suppl. 317-323 (1949).

P305. [FNM239] *Are Mesons Elementary Particles?*,
E. Fermi and C.N. Yang,

Phys. Rev. **76**, 1739-1743 (1949).

P306. [FNM240a] *La visita di* Enrico Fermi *al Consiglio Nazionale delle Ricerche*, Ricerca Scientifica **19**, 1113-1118 (1949).

P307. [FNM240] *Conferenze di Fisica Atomica*, (Fondazione Donegani). Accademia Nazionale dei Lincei (1950).

1950

P308. [FNM241] *High Energy Nuclear Events*, Prog. Theor. Phys. **5**, 570-583 (1950).

P309. [FNM242] *Angular Distribution of the Pions Produced in High Energy Nuclear Collision*, Phys. Rev. **81**, 683-687 (1951).

1951

P310. [FNM243] *Lecture on Taylor Instability*, given during the Fall of 1951 at Los Alamos Scientific Laboratory, issued on November 1955.

P311. [FNM244] *Taylor Instability of an Incompressible Liquid*, Part 1 of Document AECU-2979 (September 4, 1951)

P312. [FNM246] *Fundamental Particles*, International Conference on Nuclear Physics and the Physics of Fundamental Particles, The University of Chicago (September 17 to 22, 1951).

1952

P313. [FNM251] *Letter to Feynman* (January 18, 1952).

P314. [FNM248] *Total Cross Sections of Negative Pions in Hydrogen*, H.L. Anderson, E. Fermi, E.A. Long, R. Martin and D.E. Nagle, Phys. Rev. **85**, 934-935 (1952).

P315. [FNM249] *Ordinary and Exchange Scattering of Negative Pions in Hydrogen*, E. Fermi, H.L. Anderson, A. Lundby, D.E. Nagle and G.B. Yodh, Phys. Rev. **85**, 935-936 (1952).

P316. [FNM250] *Total Cross Sections of Positive Pions in Hydrogen*, H.L. Anderson, E. Fermi, E.A. Long, and D.E. Nagle, Phys. Rev. **85**, 936 (1952).

P317. [FNM249a] *Scattering of Negative Pions by Hydrogen,*
A. Lundby, E. Fermi, H.L. Anderson, D.E. Nagle and G. Yodh,
Phys. Rev. **86**, 603 (1952).

P318. [] *Total Collision Cross Section of Negative Pions on Protons,*
D.E. Nagle, H.L. Anderson, E. Fermi, E.A. Long and R.L. Martin,
Phys. Rev. **86**, 603 (1952).

P319. [FNM252] *Deuterium Total Cross Sections for Positive and Negative Pions,*
H.L. Anderson, E. Fermi, D.E. Nagle, and G.B. Yodh,
Phys. Rev. **86**, 413 (1952).

P320. [FNM247] *The Nucleus,*
Physics Today **5**, 6-9 (March 1952).

P321. [FNM253] *Angular Distribution of Pions Scattered by Hydrogen,*
H.L. Anderson, E. Fermi, D.E. Nagle and G.B.Yodh,
Phys. Rev. **86**, 793-794 (1952).

P322. [FNM254] *Scattering and Capture of Pions by Hydrogen,*
H.L. Anderson and E. Fermi,
Phys. Rev. **86**, 794 (1952).

P323. [FNM181] *Experimental Production of a Divergent Chain Reaction,*
Am. J. Phys. **20**, 536-558 (1952).

P324. [FNM256] *Numerical Solution of a Minimum Problem,*
E. Fermi and N. Metropolis,
Document LA-1492, (November 19, 1952).

P325. *Fermi's own story,*
Chicago Sun-Times, November 23, 1952.

P326. [USP15] *Method of operating a neutronic reactor,*
E. Fermi and L. Szilard,
Patent filed December 1, 1952 (Patent No. 2,798,847; July 9, 1957).

P327. [FNM255] *Report on Pion Scattering,*
Proceeding of the Third Annual Rochester Conference (December 18-20, 1952).

1953

P328. [FNM257] *Angular Distribution of Pions Scattered by Hydrogen,*
H.L. Anderson, E. Fermi, R. Martin and D.E. Nagle,
Phys. Rev. **91**, 155-168 (1953).

P329. [FNM261] *Magnetic Fields in Spiral Arms*,
S. Chandrasekhar and E. Fermi,
Astrophysical Journal **118**, 113-115 (1953).

P330. [FNM262] *Problems of Gravitational Stability in the Presence of a Magnetic Field*,
S. Chandrasekhar and E. Fermi,
Astrophysical Journal **118**, 116-141 (1953).

P331. [FNM258a] *Scattering of 169 and 192 MeV Pions by Hydrogen*,
R.L. Martin, E. Fermi and D.E. Nagle,
Phys. Rev. **91**, 467 (1953).

P332. [FNM258] *Nucleon Polarization in Pion-Proton Scattering*,
Phys. Rev. **91**, 947-948 (1953).

P333. [FNM259] *Scattering of Negative Pions by Hydrogen*,
E. Fermi, M. Glicksman, R. Martin and D.E. Nagle,
Phys. Rev. **92**, 161-163 (1953).

P334. [FNM264] *Multiple Production of Pions in Nucleon-Nucleon Collision at Cosmotron Energies*,
Phys. Rev. **92**, 452-453 (1953);
Errata Corrige Phys. Rev. **93**, 1954, pp. 1434-1435.

P335. [FNM245] *Taylor Instability at the Boundary of Two Incompressible Liquids*,
E. Fermi and J. Von Neumann,
Part 2 of Document AECU-2979 (August 19, 1953).

1954

P336. [FNM267] *Polarization of High Energy Protons Scattered by Nuclei*,
Nuovo Cimento **11**, 407-411 (1954).

P337. [FNM268] *Polarization in the elastic Scattering of High Energy Protons by Nuclei*,
Private Communication (March 24, 1954).

P338. [FNM260] *Phase Shift Analysis of the Scattering of Negative Pions by Hydrogen*,
E. Fermi, N. Metropolis and E.F. Alei,
Phys. Rev. **95**, 1581-1585 (1954).

P339. [FNM263] *Multiple Production of Pions in Pion-Nucleon Collision*,
Academia Brasileira de Ciencias **26**, 61-63 (1954).

P340. [FNM265] *Galactic Magnetic Fields and the Origin of Cosmic Radiation*,
Astrophysical Journal **119**, 1-6 (1954).

1955 (posthumous)

P341. [FNM270] *Lectures on Pions and Nucleons*,
Nuovo Cimento **2**, Suppl., 17-95 (1955).

P342. [FNM266] *Studies of Nonlinear Problems. I*,
E. Fermi, J. Pasta and S. Ulam,
Document LA-1940 (May 1955).

P343. [FNM269] *Physics at Columbia University - The Genesis of the Nuclear Energy Project*,
Physics Today **8**, 12-16 (November 1955).

Bibliography

Amaldi E., D'Agostino O., Fermi E., Pontecorvo B., Rasetti F. and Segrè E. (1935), Proc. Roy. Soc. London **A149** (1935) 522-538. Reproduced in Volume II of [Fermi (1962)] on page 765.

Anderson H.L. and Fermi E. (1939a), Phys. Rev. **55**, 1106-1107. Reproduced in Volume II of [Fermi (1962)] on page 8.

Anderson H.L., Fermi E. and Szilard L. (1939b), Phys. Rev. **56**, 284-286. Reproduced in Volume II of [Fermi (1962)] on page 11.

Anderson H.L. and Fermi E. (1940), Report A-21, in Volume II of [Fermi (1962)] on page 32.

Anderson H.L., Fermi E. and Grosse A.V. (1941a), Phys. Rev. **59**, 52-56. Reproduced in Volume II of [Fermi (1962)] on page 141.

Anderson H.L. and Fermi E. (1941b), Report A-6, in Volume II of [Fermi (1962)] on page 50.

Anderson H.L. and Fermi E. (1941c), Report A-2, in Volume II of [Fermi (1962)] on page 76.

Anderson H.L. and Fermi E. (1941d), Report C-83, in Volume II of [Fermi (1962)] on page 107.

Anderson H.L. and Fermi E. (1942a), Report CP-74, in Volume II of [Fermi (1962)] on page 116.

Anderson H.L., Feld B.T., Fermi E., Weil G.L. and Zinn W.H. (1942b), Report C-20, in Volume II of [Fermi (1962)] on page 128.

Anderson H.L. and Fermi E. (1944), Report CP-1729, in Volume II of [Fermi (1962)] on page 421.

Anderson H.L., Fermi E. and Marshall L., Phys. Rev. **270**, 815-817. Reproduced in Volume II of [Fermi (1962)] on page 337.

Anderson H.L., Fermi E., Wattenberg A., Weil G.L. and Zinn W.H. (1947), Phys. Rev. **72**, 16-23. Reproduced in Volume II of [Fermi (1962)] on page 397.

Anderson H.L. (1984) in [Sachs (1984)].

Barone, V. (2009), *Enrico Fermi – Atomi, nuclei, particelle*, Bollati Boringhieri.

Breit G. and Fermi E. (1942), Report C-11, in Volume II of [Fermi (1962)] on page 152.

Christy R.F., Fermi E. and Whitaker M.D., Report CP-254, in Volume II of

[Fermi (1962)] on page 195.

De Gregorio A. (2006), Stud. Hist. Philos. Mod. Phys. **37**, 330-346.

Fermi E. (1930), Nature (London) **125**, 16. Reproduced in Volume II of [Fermi (1962)] on page 328.

Fermi E. (1930), Mem. Accad. d'Italia **1** (Fis.), 139-148. Reproduced in Volume II of [Fermi (1962)] on page 336.

Fermi E. (1930), Z. Physik **60**, 320-333.

Fermi E., Anderson H.L., Wilson R.R. and Creutz E.C. (1941), Report A-12, in Volume II of [Fermi (1962)] on page 70.

Fermi E. (1942a), Report C-5, in Volume II of [Fermi (1962)] on page 147.

Fermi E. (1942b), Report C-8, in Volume II of [Fermi (1962)] on page 149.

Fermi E. (1942c), Report CP-26, in Volume II of [Fermi (1962)] on page 137.

Fermi E. (1942d), Report CP-85, in Volume II of [Fermi (1962)] on page 144.

Fermi E. (1942e), Report CP-235, in Volume II of [Fermi (1962)] on page 206.

Fermi E. (1942f), Report CP-257, in Volume II of [Fermi (1962)] on page 212.

Fermi E. (1942g), (CP) Memo-10, in Volume II of [Fermi (1962)] on page 248.

Fermi E. (1942h), Report CP-387, in Volume II of [Fermi (1962)] on page 268.

Fermi E. (1943a), Report CP-416, in Volume II of [Fermi (1962)] on page 308.

Fermi E. (1943b), Report CP-455, in Volume II of [Fermi (1962)] on page 311.

Fermi E. (1943c), Report CP-570, in Volume II of [Fermi (1962)] on page 324.

Fermi E. (1943d), Report CP-641, in Volume II of [Fermi (1962)] on page 326.

Fermi E. and Zinn W.H. (1943e), Report CP-684, in Volume II of [Fermi (1962)] on page 333.

Fermi E. (1943), Report CP-1016, in Volume II of [Fermi (1962)] on page 347.

Fermi E. (1943), Report CP-1175, in Volume II of [Fermi (1962)] on page 364.

Fermi E. (1944a), Report N-1729, Eck-209, in Volume II of [Fermi (1962)] on page 411.

Fermi E. (1944b), Report CK-1788, in Volume II of [Fermi (1962)] on page 415.

Fermi E. and Zinn W.H. (1944c), Report CP-1965, in Volume II of [Fermi (1962)] on page 425.

Fermi E. (1944d), Document HW 3-492, in Volume II of [Fermi (1962)] on page 428.

Fermi E. (1946a), *Neutron Physics. A Course of Lectures by E. Fermi*, notes by I. Halpern. Reported in Volume II of [Fermi (1962)] on page 440.

Fermi E. (1946b), Proc. Amer. Philosophical Soc. **90**, 20-24. Reproduced in Volume II of [Fermi (1962)] on page 542.

Fermi E., Marshall J., and Marshall L. (1947), Phys. Rev. **72**, 193-196. Reproduced in Volume II of [Fermi (1962)] on page 366.

Fermi E. (1952), Am. J. Phys. **20**, 536-558. Reproduced in Volume II of [Fermi (1962)] on page 272.

Fermi, E. (1962), *Collected Papers*, edited by E. Amaldi *et al.*, The University of Chicago Press - Accademia Nazionale dei Lincei, Chicago - Rome, 1962-1965, 2 volumes.

Lundby A., Fermi E., Anderson H.L., Nagle D.E. and Yodh G. (1952), Phys. Rev. **86**, 603.

Sachs R.G. (1984), *The Nuclear Chain Reaction - Forty Years Later*, The Uni-

versity of Chicago Press.
Segrè, E. (1970), *Enrico Fermi, physicist*, The University of Chicago Press.
Szasz F.M. (1992), *British Scientists and the Manhattan Project: The Los Alamos Years*, Palgrave Macmillan.
Turchetti S. (2006), Isis **97**, 127.
Wattenberg A. (1984) in [Sachs (1984)].
Wellerstein A. (2008), Isis **99**, 5787.

Index

K, *see* multiplication factor
M, *see* migration length
Δ, 191, 198–206, 209–214, 233, 236–239, 307, 308
α-particles, 162, 163, 168, 250, 365, 461, 584
α-particles
  Gamow theory, 87
β-decay
  Mg^{27}, 11
β-particles, 162, 170, 172, 174, 175, 178, 179, 182, 197, 220, 224, 252, 343, 363, 369, 380, 389, 410, 448, 494, 514, 550, 584
  counting rate, 56
γ-rays, 162 164, 167, 168, 170–173, 175, 182, 220, 224, 244, 254, 256, 264, 267, 290, 293, 299, 323, 324, 343, 354, 369, 370, 380, 389, 400, 404, 405, 410, 425, 427, 448, 462, 494, 514, 523, 525, 526, 532, 550, 564, 568, 583–585, 590
r, *see* reproduction ratio

absorption mean free path, 71, 120
absorption volume, 110
age equation, 65, 71, 72, 77, 120, 124, 141
albedo, 81, 85
  measurement, 85
aluminium, 168–170, 342, 379, 390, 397, 401, 406, 426, 428, 437, 475, 481, 499, 525, 527, 528, 532, 568, 575
aluminium chloride, 168, 170
ammonium nitrate, 168
ammonium sulphide, 171
antimony, 173, 174
antimony sulphide, 174
argon, 409, 420, 437, 438, 440, 513, 521, 575, 576, 578
Argonne pile, 7, 119
arsenic, 172
arsenic sulphide, 172
arsenious anhydride, 172

barium, 174, 319, 376, 444, 557
barium fluoride, 168
barium hydroxide, 174
barium sulphate, 174
beryllium, 163–165, 167, 177–179, 183, 184, 188, 192, 199, 218, 228, 232, 240, 269, 276, 278, 293, 331, 334, 335, 373, 374, 380, 384, 385, 388, 411, 414, 415, 444, 448, 452, 475, 481, 493, 497, 503, 506, 509, 510, 524, 530, 538, 552, 553, 568, 584, 585
  moderator, 108
beryllium oxide, 540
beta decay, 11
BF_3 chamber, 41
binding energy, 12, 14, 22, 31, 36, 87
bismuth, 175, 400, 525

boron, 167, 168, 184, 218, 229, 238, 239, 250, 255, 282, 283, 285, 293, 316, 329, 335, 354, 364, 365, 385, 399, 409, 414, 427, 444, 452, 461, 475, 481, 535, 537, 540, 544, 549, 553, 556, 568, 583, 585
boron absorber, 41
boron fluoride, 250, 365, 399
boron trifluoride, 168, 461
Bragg relation, 41, 46
Bragg scattering, 42, 44
  isotopic effects, 43
Breit-Wigner formula, 36
  slow neutrons, 38
bromine, 172

cadmium, 76, 164, 173, 184, 218, 229, 232, 256, 264, 265, 268, 293, 312, 329, 335, 364, 385, 393, 409, 414, 427, 445, 446, 452, 453, 455, 458, 462, 475, 481, 500, 501, 519, 526, 527, 531, 537, 549, 553, 568, 583, 585
caesium, 319
calcium, 170, 281
calcium fluoride, 170
calcium oxide, 281
carbon, 164, 168, 177–179, 183, 184, 189, 213, 214, 228, 254, 264, 271, 273, 275, 276, 278, 323, 330, 331, 334, 336, 363, 374, 380, 384–386, 388, 389, 395, 401, 404, 405, 411, 414, 416, 434–436, 442, 444, 448, 451, 452, 454–458, 460–462, 470, 473, 497, 524, 525, 536, 538, 550, 552, 554, 584, 585, 587
  neutron scattering, 39
carbon dioxide, 513, 521
caustic soda, 169, 175
cerium, 174
chain
  convergent, 183, 251, 317
  divergent, 181, 183, 317
chain reaction, 105
  induced by fast neutrons, 106
  induced by thermal neutrons, 107
chlorine, 170
chromium, 170, 171
cinch, 465
Clinton pile, 119
cobalt, 171
cobalt sulphide, 171
collision parameter, 35
coolant, 214, 243, 264, 285, 287, 298, 300–303, 325, 330, 340, 353, 373–376, 380, 389, 390, 394, 397, 402, 404, 409, 419, 420, 427, 441, 442, 444, 447, 493, 498, 508, 524, 526, 536, 550
copper, 171, 172, 176, 379, 390, 397
copper sulphide, 171
counter, 481, 482, 484, 535, 542
critical size, 119
cross section
  scattering
    experimental determination, 18
    theoretical determination, 34
  transport, 141

de Broglie wavelength, 17, 32, 34
deuterium, 183, 184, 199, 228, 232, 331, 334, 335, 380, 384, 385, 388, 411, 414, 415, 448, 452, 497, 538, 540, 553, 585
deuterium oxide, 167, 242, 262, 265, 400, 406, 441, 525, 527, 550, 588
deuteron, 167
  triplet and singles states, 36
di-neutron, 14
diffusion
  integral equation method, 82, 144, 149
diffusion coefficient, 65, 71, 82
diffusion length, 71, 118, 144
  measurement, 76
  temperature variation, 76
dimethylglioxime, 173
diplon, *see* deuteron

electrons
  extra-nuclear, 28

eta constant, 339
excess reactivity, 128
experiment
  exponential, *see* exponential pile

fast reactors, 139
  critical size, 144
  equilibrium conditions, 144
filter, 344, 370–372
  air, 288, 423, 562
  bismuth, 324, 405, 532
fluorine, 168, 278, 281, 441
foil
  aluminium, 477, 479, 489, 527
  cadmium, 168, 477, 479, 489, 527, 568
  copper, 468
  indium, 181, 196, 197, 203, 208, 210, 211, 214, 215, 218, 237, 251, 252, 257, 258, 282, 363, 432, 535, 540–542, 573
  Masonite, 587
  steel, 586, 587

gadolinium, 174, 184, 229, 319, 335, 385, 414, 452, 475, 481, 537, 553, 556, 583, 585
gadolinium oxide, 174
gallium, 171, 172
gallium nitrate, 172
Gamow factor, 21
Gamow theory for α-particle decay, 87
gas recombination, 523, 527
germanium, 172
gold, 175
  (n, γ) resonance capture, 38
graphite
  dead, 241, 248–251, 253, 254, 346, 351, 354, 356, 357, 359, 363, 391, 395, 397, 399, 400, 459–461
  diffusion length of neutrons, 76
  live, 247–250, 261, 391, 395, 397, 399, 459–462, 471
  moderator, 108
  neutron diffusion, 77
  slow neutrons production, 46
  slowing down of neutrons, 53

hafnium, 174, 175
heavy water
  moderator, 108
helium, 177–179, 218, 265, 298, 325, 330, 342–344, 346, 347, 349, 352, 354, 356, 357, 360, 362, 368–370, 372, 373, 376, 377, 441, 500, 513, 521, 525, 527
  moderator, 107
  neutron scattering, 39
hydrochloric acid, 168, 171–176, 278
hydrofluoric tantalum, 176
hydrogen, 62, 370, 411, 415, 440, 496, 553, 578, 583
  neutron scattering, 39
  ortho- and para-states, 36
  slowing down of neutrons, 53
hydrogenous matter, 18, 49, 76, 77, 85, 161, 167, 169, 172, 177–179, 183, 188, 192, 199, 212, 215, 218, 222, 228, 232, 240, 254, 262, 264, 265, 268, 269, 272, 279, 280, 282, 285, 287, 297–299, 301–303, 311, 312, 318, 320, 321, 325, 361, 362, 370, 373, 388, 414, 441, 442, 444, 493, 498–500, 502, 505, 510, 523–526, 530, 531, 533, 536, 550, 584–586, 589, 590

ih., *see* inhour
impurities, 184, 185, 188, 189, 206, 212, 213, 217, 218, 226, 335, 336, 340, 342, 364, 374–376, 385, 386, 388, 392, 394, 397, 403, 412, 414–417, 425, 436, 441–444, 452, 453, 467, 505, 515, 516, 520, 526, 535–538, 540, 544–546, 549, 553, 554, 556, 580, 581
indium, 54, 56, 173, 174, 542
  (n, γ) resonance capture, 37
indium nitrate, 173
indium sulphide, 173

inhour, 447, 465–468, 470
  zero point, 467
iodine, 54, 174, 319–321, 376, 444, 557
ionization chamber, 168–170, 172, 173, 175, 246, 250, 269, 296, 301, 317, 363, 365, 367, 368, 370, 393, 399, 432, 437, 458, 461, 481, 482, 484, 487, 488, 502, 503, 531, 573, 575
iridium, 175
iron, 164, 171, 264, 281, 354, 409, 431, 526, 550, 559, 572, 585, 589, 590
isotopic chart, 9

lanthanum, 174
lead, 162, 164, 175–179, 245, 253, 264, 267, 268, 294, 297, 300, 379, 390, 397, 400, 405, 409, 431, 437, 500, 525, 526, 572, 585
liquid drop model, 87
  fission, 97
lithium, 167, 168, 285, 556
lithium hydroxide, 167

magnesium, 169, 281, 475, 481
magnesium chloride, 168, 169
magnesium hydroxide, 169
magnesium oxide, 169
manganese, 170, 171
manganese carbonate, 171
manganese dioxide, 171
manganese permanganate, 171
Masonite, 526, 583, 586, 590
mass defect, 91
Maxwell distribution, 41
mercury, 175, 176, 322
mercury sulphocyanate, 172
migration area, 313
migration length, 236, 238, 305, 306, 313
multiplication factor, 181, 182, 185, 189–192, 198–201, 205, 206, 212–214, 217–221, 226, 228, 230, 231, 233–235, 237–240, 242, 253, 256, 258, 269, 271–274, 276, 277, 279, 282, 283, 285, 291, 293, 295, 298, 302, 303, 305, 306, 308, 310, 312, 313, 315, 316, 318–321, 326, 330, 331, 336, 338, 356, 363, 364, 374, 376, 381, 383–386, 388, 396, 403, 404, 411, 414–419, 422, 434–436, 441–444, 447, 449, 451–455, 460, 463, 465, 467–471, 495, 497, 499, 505, 507, 510, 513–516, 521, 536, 538, 540, 544, 546, 549, 551, 553–557, 559

neon, 168
neutrino, 11
neutron
  mean free path, 61
  absorption cross section, *see* capture cross section
  absorption mean free path, 71
  albedo, 81, 85
    measurement, 85
  binding in nuclei, 91
  capture cross section, 497, 504, 516, 530, 532, 538, 545, 552, 554, 557, 585
  delayed, 293, 296, 309, 310, 317, 319, 364, 376, 427, 438, 442, 443, 455, 557–559, 567, 576
  diffusion, 65
    integral equation method, 82
  diffusion coefficient, 65, 71, 82, 198
  diffusion equation, 71, 79, 81, 123
    for a pile, 120
  diffusion length, 71, 118
  energy drift, 66
  fast, 18, 72, 161, 163, 164, 181–184, 186, 188, 195, 199, 200, 209, 218, 220, 221, 224–226, 232, 233, 235, 238, 271, 291, 309, 314, 324, 330, 331, 339, 340, 380, 410, 411, 425, 434, 448, 449, 494, 495, 514, 532, 536, 544, 557, 565, 583, 585, 587, 589
  interaction with a nucleus, 31
  mean free path, 56, 65, 144, 166,

199, 236
moderator, 107, 524–527, 536–538, 540, 545, 549–554, 556, 559, 560, 565, 574, 579, 581, 583–586
range, 78
reflection, 41, 45, 81, 165, 525, 533
refraction, 47
resonance absorption, 184, 187, 188, 217, 222, 225–229, 231, 237, 240, 271, 272, 278, 283, 332, 333, 381, 382, 412, 434, 435, 449, 450, 496–498, 506, 544, 551, 552
scattering, 17, 31, 58, 165, 218, 228, 233, 234, 263, 276, 312, 400, 404, 433, 525, 531, 550, 560, 574
  Bragg, 42, 44
  classical hard spheres, 17
  cross section, 34
  incoherent, 43, 46
  interference effects, 41
  isotopic effects, 43
scattering cross section, 497
slow, 17, 38, 41, 46, 72, 161, 163, 164, 167, 168, 170–173, 175, 218–221, 225, 226, 228, 233, 254, 265, 291, 293, 314, 323, 331–333, 340, 380–382, 404, 410–412, 425, 443, 448–450, 494–497, 514, 527, 550–552, 565, 583–585, 587
  spatial distribution, 54, 58, 71
slowing down, 49
  graphite, 53
  hydrogen, 53, 62
source, 5
  (γ, n), 7
  Oliphant tube, 167
  Po-Be, 6, 168
  Po-BF_3, 7
  pointlike, 68, 72
  Ra-Be, 6, 54, 64, 77, 207, 234, 541
  Rn-Be, 6, 165, 167
    strength, 69
  thermal
    spatial distribution, 71
neutron lens, 47
nickel, 171, 237
nickel sulphide, 171
niobium, 172
nitric acid, 171, 175, 176, 279, 280
nitric rhenium solution, 175
nitrogen, 168, 229, 322, 323, 335, 385, 403, 405, 415, 420, 463, 466, 513, 517, 521, 549, 553
nuclear explosion, 139
nuclear fission, 93
  cross section, 102
  delayed neutrons, 101
  induced by photons, 99
  limitations on the occurrence, 95
  liquid drop model, 97
  particle emission, 99
  prompt neutrons, 100
  spontaneous, 99
  stability condition, 99
nuclear potential, 31
nuclear reaction, 9
  $Al^{27}(n, \alpha)Na^{24}$, 9
  $Al^{27}(n, \gamma)Al^{28}$, 10
  $Al^{27}(n, p)Mg^{27}$, 10, 11, 169
  $B^{10}(n, He^3)Li^7$, 168
  $B^{10}(n, \alpha)Li^7$, 18
  $B^{11}(\alpha, n)N^{14}$, 7
  $Be^9(\alpha, n)3He^4$, 5
  $Be^9(\alpha, n)O^{12}$, 5
  $Be^9(\gamma, n)Bc^8$, 7
  $Be^9(\gamma, n)Be^8$, 14
  $Be^9(d, n)B^{10}$, 8
  $D^2(\gamma, n)H^1$, 7
  $D^2(\gamma, n)H^1$, 14
  $D^2(d, H^3)H^1$, 7
  $D^2(d, He^3)n$, 7
  energy balance, 10
  $F^{19}(\alpha, n)Na^{22}$, 7
  $Li^7(\alpha, n)B^{10}$, 5
  $Li^6(n, \alpha)H^3$, 18, 167
  resonance, 25, 36
nucleus

theoretical models, 27
compound, 25, 36
energy, 29
entropy, 29
lifetime, 26, 37
liquid drop model, 28, 87
mass defect, 91
odd-even effect, 89
potential well, 28
radius, 17
stability, 87

ortho-states
hydrogen, 36
oxygen, 168, 229, 278, 281, 322, 331, 370, 373, 385, 411, 441, 448, 466, 521
neutron scattering, 39

packing fraction, 12, 91, 93
palladium, 173
palladium nitrate, 173
para-states
hydrogen, 36
paraffin, 39
paraffin screen, 161, 164, 165, 167, 168, 178, 282, 540, 542
Pauli exclusion principle, 28
phase shift, 34
phosphoric acid, 170, 176
phosphorus, 169, 170
pile, 76, 77, 107
absorptivity, 133
Argonne, 7, 119, 136
Clinton, 119, 134–136
control rods, 132
costants, 116
critical, 181
critical radius, 182, 206, 311, 315, 338, 384, 418, 454, 508, 555
critical size, 119, 182, 185, 189–191, 193, 201, 205, 206, 213, 215, 222, 233, 240, 243, 251, 253, 254, 259, 262, 270, 272, 277, 281, 292, 295, 303, 306–308, 311, 312, 316–318, 320, 338, 339, 384, 388, 400, 403, 410, 418, 419, 425, 432, 433, 435, 436, 454, 455, 463, 496, 504, 507, 525, 549, 555, 556, 560, 565, 573, 574, 580
danger coefficients, 218, 230, 278, 282–284, 322, 330, 336, 342, 374, 385, 386, 415, 416, 420, 441, 453, 460, 462, 466, 513, 516, 535, 538, 540, 544, 554
danger sum, 230–232, 235, 259, 272, 278, 279, 282, 302, 336, 374, 375, 386, 388, 394, 416, 441, 443, 453, 516, 538–540, 554
diffusion equation for neutrons, 120
divergent, *see* divergent chain
e-folding, 139
enriched uranium, 137
exponential, 132, 181, 217, 218, 234, 235, 237–240, 259, 282, 307, 469
geometry
cylinder, 201, 206, 270, 302, 305, 307, 308, 312, 314, 338, 385, 418, 454, 508, 555, 560, 565, 566, 574, 580
parallelepiped, 182, 192, 201, 206, 215, 242, 258, 288, 305, 312, 338, 384, 418, 454, 508, 555, 560, 580
rectangular, 182
sphere, 201, 204, 206, 242, 258, 259, 305, 312, 314, 338, 339, 384, 418, 436, 454, 455, 457, 460, 508, 555, 556, 560, 574, 580
graphite, 108
C/U ratio, 108
Hanford, 134, 135
homogeneous, 112
intermediate, 185, 201
boudary corrections, 204, 211
leakage, 108, 122
Los Alamos, 137
lumps, 110, 115

neutron absorption, 181, 183, 188, 189, 198, 213, 218, 221, 226, 228, 230, 234, 238, 250, 254, 272, 276, 283, 293, 298, 303, 316, 319, 320, 324, 330–332, 335, 336, 339, 342, 356, 360, 373, 380, 381, 386, 399, 400, 404, 410, 411, 415, 420, 425–427, 441, 443, 444, 495, 498, 504, 507, 508, 510, 515, 518, 524, 525, 536, 537, 540, 551, 553, 556, 557, 568, 584, 586
neutron density, 181, 190–192, 195–198, 200–206, 208–211, 214, 215, 218, 219, 222, 223, 232, 234, 235, 237, 238, 242, 251–253, 257, 260, 262, 269, 285, 292, 293, 295, 296, 305, 307–310, 312–322, 324–326, 329, 332, 339, 340, 356, 359, 363–367, 369, 381, 384, 388, 392, 399, 401–403, 405, 410, 411, 417–420, 425, 427, 429, 432, 436, 440, 444, 449, 454, 455, 461, 463–467, 494, 495, 502–505, 515, 519, 531, 537, 540, 551, 554, 555, 558, 565, 568, 573, 575
neutron diffusion, 181, 184, 185, 198, 200, 226, 228, 313, 338, 363, 384, 414, 452, 454, 497, 553, 555
neutron diffusion length, 191, 199, 240, 509
neutron losses, 181–185, 187–189, 192, 205, 212, 217, 220, 222, 226, 229, 234, 246, 285, 286, 291, 298, 301, 307, 315, 316, 331, 332, 335, 380, 381, 385, 394, 400, 403, 410, 419, 427, 433, 448, 449, 451, 458, 462, 463, 471, 493, 494, 496, 514, 565, 567
  by Absorption in the Moderator, 228
  by absorption in the moderator, 184, 186, 187, 222, 224, 226, 332, 335, 375, 381, 412, 449, 497, 515, 537, 540, 545, 551, 552
  by Absorption in Uranium, 226
  by absorption in uranium, 183, 187, 222, 271, 332, 375, 381, 412, 449, 515, 537, 551, 552
  by capture by impurities, 184, 186, 187, 222, 224, 226, 229, 278, 332, 363, 375, 381, 385, 412, 414, 449, 452, 513, 515, 535, 537, 551, 553
  by leakage, 181, 184–187, 190, 191, 204, 222, 224–226, 233, 316, 332, 338, 375, 381, 384, 397, 411, 412, 417, 418, 436, 449, 454, 469, 495, 515, 525, 537, 540, 551, 554, 574
neutron production, 181, 182, 185, 190, 198, 205, 217, 222, 238, 324, 332, 338, 364, 375, 381, 417, 443, 454, 496, 550, 554
neutron range, 191, 199
neutron self-absorption, 111
neutron surface absorption, 184, 186, 189, 224, 227, 272, 332, 333, 382, 383, 412, 413, 435, 450, 496, 551, 552
neutron volume absorption, 184, 186, 224, 332, 382, 412, 450, 552
period, 125
relative excess reactivity, 128
shielding, 134
size, 132
supercritical, 151
tamper, 141
testing of the materials, 132
time dependence, 123
u-D_2O, 136
water boiler, 137, 139
platinum, 175, 265

plutonium, 99
polonium, 163
potassium, 170
potential barrier, 163
power production, 182, 189, 213, 218, 220, 221, 257, 276, 292, 293, 298, 317, 320, 324, 325, 330, 332, 340, 341, 343, 344, 350, 363–365, 369, 380, 381, 391, 399, 401, 405, 410, 411, 419, 425, 428, 429, 436, 437, 440, 444, 448, 449, 494, 495, 504, 505, 509, 510, 515, 537, 550, 551, 557, 565, 566, 568, 569, 575, 578, 580, 585
praseodymium, 174
probability
  neutron capture, 23
protoactinium, 176, 323, 404

radiation shield, 241, 244, 250, 253, 264, 265, 288, 289, 293, 294, 297, 299, 307, 344, 354, 360, 362, 365, 369, 370, 400, 404, 424, 427, 429, 430, 432, 439, 445, 456, 462, 500, 502, 503, 518, 523–525, 530, 531, 550, 559, 563, 568–570, 577, 578, 583–587, 590
radium
  decay series, 6
radon, 163, 165
range, 78
reflector, 234, 243, 248, 253, 258, 262, 264, 265, 269, 270, 284, 295, 299, 308, 312–315, 338, 388, 392, 397, 400, 418, 433, 436, 454, 457, 460, 462, 503, 509, 518, 520, 521, 525–527, 531, 551, 555, 560, 574, 588
refractive index, 47
reproduction factor, 107, 119, 128, 131, *see* multiplication factor
  for a pile with spherical lumps, 115
reproduction ratio, 309, 310, 312, 322, 332, 335, 339, 364, 374, 376, 381, 384, 385, 388, 396, 400, 402, 403, 411, 415, 417, 418, 425–427, 432, 436, 442, 444, 447, 449, 453, 455, 463–469, 471, 494, 495, 497, 502, 504, 505, 507, 509, 510, 515, 517, 519, 520, 526, 537, 551, 553, 555–559, 565, 567, 568, 573, 575, 580, 581
resonance
  curve, 25
  partial widths, 37
rhenium, 175
rhenium sulphide, 175
rhodium, 54, 172, 173
rhodium chloride, 173
rod
  control, 218, 226, 242, 245, 251, 253, 255, 256, 264, 268, 292, 293, 295, 296, 301, 316–322, 325, 329, 340, 344, 356, 360, 362–369, 376, 393, 399, 400, 402, 403, 409, 427, 433, 436, 438, 443, 444, 447, 453, 458, 461–466, 468–471, 473, 502–505, 519, 520, 522, 526, 531, 549, 551, 558, 568, 573, 575, 576, 580
  safety, 218, 242, 245, 251, 253, 256, 264, 268, 293, 301, 318, 321, 329, 356, 360, 362, 364–369, 391, 393, 399, 400, 402, 403, 409, 427, 458, 461–464, 502, 503, 519, 520, 526, 531, 568

S-wave
  cross section, 34, 35
  neutron-nucleus interaction, 31
samarium, 229, 335, 385, 414, 452, 537, 553, 556, 583, 585
saturation activity, 56, 197, 251, 252, 257
scandium, 170
scattering
  (n, γ), 22
  $(n, \text{heavy charged particle})$, 21
  (n, n) elastic, 20, 38, 106
  (n, n) inelastic, 21, 106
  (n, p), 35, 62, 64

- n from p-H_2/o-H_2, 36
- n from carbon, 39
- n from helium, 39
- n from oxygen, 39
- chemical binding, 38
- general equation, 20
- neutron-nucleus relative velocity, 22
- phase shift, 34
- thermal agitations of nuclei, 23

Schrödinger's equation, 31
selenious anhydride, 172
selenium, 172
sheet, *see* foil
shotgun test, 535, 540, 543–546
silicon, 164, 169, 177–179
silver, 173, 176
- (n, γ) resonance capture, 38

silver chloride, 173
silver nitrate, 173
slowing down density, 67
sodium, 168–170, 174
sodium carbonate, 168, 170
sodium chlorate, 171
sodium chloride, 174
sodium fluoride, 168
strontium, 172
sulphur, 170
sulphuretted hydrogen, 173
sulphurous anhydride, 172
supercritical reactors, 151

tantalum, 174, 175
tantalum pentoxide, 175
tellurium, 174, 319
thallium, 175
thorium, 162, 176, 323, 404, 443
tin, 173, 174
tin sulphide, 174
transition probability, 19
transport
- cross section, 141
- mean free path, 141

transuranic elements, 161, 176, 221, 356, 410, 495, 514, 550
trisodium phosphate, 292, 426, 567
tungsten, 175
tungstic anhydride, 175

uncertainty principle, 19, 25
uranium, 91, 99, 162, 176, 189
- enriched, 137, 188, 229, 277, 325, 335, 338, 374, 385, 415, 417, 442, 448, 452, 538
- natural, 105, 108, 136
- oxide, 176, 183, 188, 212, 213, 218, 222, 228, 240–242, 244, 248, 272, 273, 275, 276, 278–281, 283, 333, 335, 336, 383, 385, 413, 415, 416, 437, 451, 452, 456–460, 469, 470, 496, 538, 552, 553, 576

uranyl nitrate, 279, 280

vanadium, 170, 171, 218, 285
velocity selector, 475–477, 485, 486, 488, 489
virtual states, 36
- phase difference, 36

water
- diffusion length of neutrons, 76
- moderator, 108

Wilson seal, 519

xenon, 217, 319–321, 323, 376, 444, 504, 505, 507, 510, 557
- poisoning effect, 504, 505, 508, 549, 557, 579

yttrium, 164, 172
yttrium oxide, 172

zinc, 171, 172, 342, 426, 437
zinc mercury sulphocyanate, 172
zinc sulphide, 171
zirconium, 172, 176
zirconium hydroxide, 175
zirconium nitrate, 172, 176
zirconium phosphate, 176